PEPTIDES

POLYPEPTIDES

AND

PROTEINS

PEPTIDES

POLYPEPTIDES

AND

PROTEINS

Proceedings of the Rehovot Symposium on Poly(Amino Acids),
Polypeptides, and Proteins and Their Biological Implications
Israel, May 1974

EDITED BY

E. R. BLOUT
Harvard University Medical School
Boston, Massachusetts

F. A. BOVEY
Bell Laboratories
Murray Hill, New Jersey

M. GOODMAN
University of California, San Diego
La Jolla, California

N. LOTAN
The Weizmann Institute of Science
Rehovot, Israel

A WILEY-INTERSCIENCE PUBLICATION

JOHN WILEY & SONS, New York • London • Sydney • Toronto

Library of Congress Cataloging in Publication Data:

Rehovot Symposium on Poly(amino Acids), Polypeptides, and Proteins, and Their Biological Implications, 2d, 1974.
    Peptides, polypeptides, and proteins.

 "A Wiley-Interscience publication."
 Includes index.
    1. Amino acids—Congresses. 2. Polypeptides—Congresses. 3. Proteins—Congresses. I. Blout, Elkan Rogers, 1919-   ed. II. Title.
 [DNLM: 1. Amino acids—Congresses. 2. Peptides—Congresses. 3. Proteins—Congresses. W3 RE441k 1974p /QU60 R345 1974p]

QP561.R43 1974   574.1'9245   74-22202
ISBN 0-471-08387-9

Printed in the United States of America

10 9 8 7 6 5 4 3 2 1

PREFACE

The first international symposium devoted to poly(amino acids), polypeptides, and proteins was held in June 1961 in Madison, Wisconsin. Mark A. Stahmann, Elkan Blout, Paul Doty, Sidney Fox, and Ephraim (Katchalski) Katzir organized an exciting program. They brought together scientists interested in the organic, physical, and biological chemistry of poly(amino acids) and related biopolymers. The papers presented at that conference included definitive comments on the synthesis, structure, and properties of peptides and poly(amino acids).

The Wisconsin meeting had a strong international flavor. Ephraim (Katchalski) Katzir and his associates at the Weizmann Institute were among the major contributors to the development of the field. It was reasonable, therefore, to plan a second symposium to be held in Israel.

In 1972, Ephraim (Katchalski) Katzir, the late Arieh Berger,* and Noah Lotan began to form an international organizing committee for what was to become the Rehovot Symposium, the proceedings of which are contained in this volume. The organizing committee included C.B. Anfinsen, A. Berger, E.R. Blout, F.A. Bovey, E.M. Bradbury, J. Engel, M. Goodman, B.S. Hartley, K. Imahori, E. (Katchalski) Katzir, S.J. Leach, N. Lotan, H.A. Scheraga, M. Sela, and A.G. Walton.

The problems encountered in organizing an international symposium are substantial. Even after the October War in 1973, however, the Israeli scientists never waivered in their commitment to the symposium. Our meeting was held in an atmosphere of great anticipation and satisfaction because of the increased difficulties encountered in bringing this symposium to fruition. The opening session was held at the Weizmann Institute as a joint effort with the concluding session of the Thirteenth European Peptide Symposium. The remaining sessions were held at Kibbutz Ayelet Hashachar.

As our opening session Josef Rudinger presented the scientific highlights of the European peptide meeting, which was held at Kibbutz Kiryat Anavim near Jerusalem. Ephraim (Katchalski) Katzir then delivered an address that served to place our deliberations and reports in proper perspective. This lecture appears as an introduction to the volume. He traced the development of our field over the past twenty-five years, citing numerous achievements of scientists throughout the world. From this personal overview we can glean some of the joy and excitement of discovery in biopolymer research. Ephraim (Katchalski) Katzir stressed his commitment to understanding the biological properties of these

---

*We here take note of the untimely death of Arieh Berger, which removed from our midst a man of great insight and imagination. His participation and advice were sorely missed.

synthetic polymers and concluded with a challenge to establish novel applications for poly(amino acids).

Following (Katchalski) Katzir's lecture, Frank Bovey and Michael Sela made scientific presentations that set the tone for major sections of our symposium. The sections represent a broad coverage of the theoretical and experimental work with peptides, poly(amino acids) (polypeptides), and protein systems. Many papers within each part compare and contrast specific approaches to given topics. Thus, for example, complementary papers on calculations and predictions of most favored conformations are included in the program. Numerous spectroscopic techniques were described. The reader should be able to evaluate structural predictions based upon various techniques including x ray, optical activity, infrared, and nuclear magnetic resonance measurements. Recent results on the conformational characterizations of cyclic and linear peptides are presented. These investigations clearly establish that high-resolution nuclear magnetic resonance is a particularly valuable tool in the study of the conformations of these compounds.

A major focus of our symposium was the immunochemical properties of poly(amino acids) and peptides. Of particular interest were the papers dealing with the use of polypeptides in the study of genetic control of immune response.

The symposium concluded with a series of short communications about ongoing research. These reports bode well for the future of our field. Although much of the work was incomplete, it was possible to extrapolate to a third poly(amino acid) symposium in which many of these scientists will deliver major papers. The researchers involved in this session included H. Auer, C.M. Deber, A. Englert, C. Gilon, G. Jung, S. Koenig, S. Krimm, G. Lorenzi, B.R. Malcolm, W. Miller, D. Patel, E. Peggion, E. Ralston, M. Rigbi, W.P. Rippon, R. Roche, M. Sheinblatt, G. Spach, and J.T. Yang.

The Kibbutz Ayelet Hashachar is located in the upper Galilee. During our sessions, it was not uncommon to hear the firing of artillery. During recesses and lunch periods, we often saw large shells exploding on the horizon and military airplanes. Many of us reflected on the contrast between the military struggle around us and the scientific content of our symposium. The presence and participation of Ephraim (Katchalski) Katzir and the other Israeli scientists was a testimony to the importance of our work. Rather than being diminished by the difficult and tragic military struggles around us, our studies and research took on an added significance.

No international meeting of this scope can take place without sponsorship. We gratefully acknowledge the financial assistance and support given by the International Union of Pure and Applied Biophysics through its Commissions on Macromolecular Biophysics and Subcellular Biophysics, the Israel Academy of Sciences and Humanities, the Israel National Council for Research and Development, the Weizmann Institute of Science and the National Academy of Sciences of the United States. In addition, we received financial help from several chemical companies including Allied Chemical, Ethicon, General Foods, Hoffman-LaRoche, and Merck Sharp and Dohme.

The dedication of many individuals also was essential to the success of the symposium. We are particularly thankful to Harold Scheraga who worked tirelessly on many aspects of the organization and scientific content of the meeting. We are indebted also to Yitzchak Berman of the Public Relations Department at the Weizmann Institute who made many of the arrangements and anticipated the myriad requirements for an outstanding symposium.

August 1974   The Editors

CONTENTS

INTRODUCTION

POLY(AMINO ACIDS): ACHIEVEMENTS AND PROSPECTS

EPHRAIM (KATCHALSKI) KATZIR, Department of Biophysics, The Weizmann Institute of Science, Rehovot, Israel

Discussing the achievements and prospects in the field of poly(amino acids) is too difficult a task for a single speaker. I have decided, therefore, to limit myself to a description of my own experience in the field and some of my personal views concerning future developments. I have been engaged in the study of poly(amino acids) for more than 25 years. Therefore I am particularly glad to open this symposium on the recent theoretical, physicochemical, and biological discoveries in research on poly(amino acids).

When I started my work on amino acid polymers at the Hebrew University in Jerusalem (1), few scientists were interested in poly(amino acids) and almost no one shared my enthusiasm and my belief in the potential usefulness of this fascinating new group of high-molecular-weight model compounds. The many outstanding scientists participating in this meeting and the many excellent papers submitted show that the interest in poly(amino acids) has increased markedly and that theoreticians and experimentalists striving to understand the structure and function of proteins are using poly(amino acids) rather extensively in their studies. As a matter of fact, at present, it is difficult to keep up with the volume of information being published on the synthesis and the physicochemical and biological properties of poly(amino acids) and on their use as model compounds in the theoretical interpretation of the fundamental properties of proteins.

I recall the story told at the Weizmann Institute about Niels Bohr, who came to Rehovot to inaugurate our physics department. A symposium was arranged in his honor, in which many of our young physicists participated. During one of the sessions Mr. Weisgal, who was our chief administrator at the time, found Bohr walking in the beautiful surrounding garden. "Professor Bohr," said Weisgal, "I have come out to enjoy the Israeli sunshine, our trees and flowers, because I am a layman and I don't understand what our physicists are talking about. But what are you doing here?" The famous physicist replied, "To tell you the truth Mr. Weisgal, I also do not understand some of the topics being discussed by your youngsters, so I too prefer walking in the sun." I hope that during our meetings none of us will experience the perplexity of Niels Bohr and that we will seek out the beautiful gardens of Ayelet Hashachar only after the lectures.

As a biologist, I became interested in poly(amino acids) not only because I was convinced that they would shed light on the conformation of the polypeptide backbone and on the intra- and intermolecular forces operating among the various amino acid residues in proteins, but also because I hoped that the study of these high-molecular-weight model compounds would help us to understand some of the biological properties of proteins, and in particular, protein biosynthesis.

The work carried out by our group at the Weizmann Institute has been summarized on several occasions. In 1951, I published my first review (2) of the synthesis and chemical properties of poly(amino acids). It was my intention to arouse the interest of protein and polymer chemists in this new class of high-molecular-weight compounds. At the time it was difficult to find relevant papers, but as the study of protein model compounds developed, a number of papers dealing with the biological properties of poly(amino acids) appeared. Progress in the field was summarized by Sela and myself in 1958 (3) and 1959 (4). Gradually, poly(amino acids) acquired respectability in the eyes of protein chemists and biochemists. In 1964, together with my collaborators Sela, Silman, and Berger, I was gratified to review the new field thoroughly (5). Since then, a number of reviews on poly(amino acids) have appeared in the *Annual Review of Biochemistry* (6). The most recent among them, by Lotan, Berger, and myself (7) deals with the conformation and conformational transitions of poly(amino acids) in solution.

In addition to our own review articles, a considerable number of other surveys on poly(amino acids) have been published. Some deal with conformation in the solid state, others with physicochemical properties and conformational transitions in solution, and still others with biological properties. Special mention should be made of available books on poly(amino acids). The volume by C. H. Bamford, A. Elliott, and W. E. Hanby (8) deals mainly with solid state conformations and x-ray studies of fibers and films of poly(amino acids). The book edited by G. D. Fasman (9) discusses the spectroscopic, hydrodynamic, and biological properties of poly(amino acids), as well as experimental data and theoretical considerations concerning the conformation and conformational stability of these compounds. Detailed statistical-mechanical analyses of the order-disorder transitions in macromolecules, in general, and in polypeptides, in particular, are to be found in the books by P. J. Flory (10) and by D. Poland and H. A. Scheraga (11), which appeared in 1969 and 1970, respectively. An account of the results obtained in the study of high-molecular-weight polypeptides can be found in recently published books by A. G. Walton and J. Blackwell (12), A. J. Hopfinger (13), and F. A. Bovey (14).

As far back as the early 1960s, Murray Goodman, Elkan Blout, myself, and others recognized the need for a new journal on native and synthetic polymers of biological interest. To fill this need, *Biopolymers* was established and started to appear in 1963 (15). This journal has published many excellent articles dealing with high-molecular-weight polypeptides and related compounds.

Thirteen years ago a symposium on poly(amino acids), polypeptides, and proteins was organized by Mark Stahmann and held at the University of Wisconsin in Madison. This was the first international meeting devoted to high-molecular-weight protein model compounds, and it was there that the name "poly(amino acids)" was officially recognized as denoting synthetic polymers composed of amino acid residues linked by peptide bonds that have been prepared by a polymerization process. Like other synthetic polymers, these consist of mixtures of homologous macromolecules of varying chain length. The term "polypeptide" is used interchangeably with poly(amino acids). In addition, it refers to native or synthetic large peptides obtained by a stepwise synthesis, and generally consisting of molecules of the same size and chemical structure.

Let me now tell you how my own research on amino acids and amino acid polymers began and how it progressed with the aid of my students, friends, and collaborators. I started to work on protein building blocks and protein models at the Hebrew University on Mount Scopus in Jerusalem. The department was headed by the late Max Frankel and was devoted to synthetic polymers and theoretical organic chemistry. Frankel's only laboratory assistant was my late brother Aharon, who at that time was working on the interaction of amino acids and peptides with aldehydes and sugars. This was his first research project, and after some time he succeeded in persuading me that the most important task in life was the preparation of salt-free trifunctional amino acids from red blood cells. For about a year thereafter, I collected blood from the slaughter house, separated and hydrolyzed the red blood cells, and made every attempt to isolate from the hydrolysate the basic amino acids lysine, arginine, and histidine. This was no easy task, and when I found out that amino acids could be purchased, I asked Frankel to buy me an adequate supply. Amino acids were rather expensive and Frankel did not allow us to use large amounts at a time. Every morning he would check the amino acid bottles to ensure that the level of powder within them had not dropped markedly. To overcome this unforeseen obstacle in our research, we emptied the bottles of amino acids for our use and replaced them with talcum. Years later when I was working at the Weizmann Institute, I heard of a great scandal in Frankel's department at the Hebrew University: all of the amino acids that he had painfully acquired were no longer dissolving in water. However, our technique produced a suitable supply of amino acids during the early stages of our research.

After playing about for some time with the electrochemical properties of basic amino acids I became interested in the synthesis of water-soluble poly(amino acids). I had little success with the polycondensation of amino acid and peptide esters and therefore turned to the polymerization of N-carboxyamino acid anhydrides, which by this time are well known. I found that such anhydrides had been prepared by Leuchs in 1906 (16). Leuchs observed that the N-carboxyanhydrides that he synthesized from the simplest amino acids split off carbon dioxide upon heating and yielded insoluble residues, which I suspected represented the corresponding polymers. I turned to this technique to synthesize my first water-soluble poly(amino acid), poly(L-lysine) (17).

In 1947, I succeeded in synthesizing N^ϵ-benzyloxycarbonyl-N^α-carboxy-L-lysine anhydride, which yielded upon bulk polymerization poly(N^ϵ-benzyloxycarbonyl-L-lysine). I failed, however, in all my attempts to remove the carbobenzoxy-protecting groups by means of the known reducing procedures. Finally, we found that the desired polylysine can be derived from poly(carbobenzoxy-lysine) by treatment with phosphonium iodide in glacial acetic acid. During the reaction carbon dioxide evolved and benzyl iodide was formed. This observation led us to suspect that it is not the reducing phosphine but the hydrogen iodide derived from phosphonium iodide that attacks the carbobenzoxy-protecting groups. A few years later, Arieh Berger and Dov Ben-Ishai showed that this reaction does indeed take place and that anhydrous hydrogen bromide is a most suitable reagent for the removal of the benzyloxycarbonyl group from N-protected amino acids and peptides (18). The method developed by Berger and Ben-Ishai has become a standard procedure in classical peptide synthesis.

At this stage of research we had to determine whether the polymerization of N-carboxyanhydrides of optically active amino acids is accompanied by detectable racemization. In the case of poly(L-lysine) synthesis, we detected no racemization. Furthermore, no racemization could be detected in the polymerization of other optically active N-carboxyamino acid anhydrides. The way was thus paved for the preparation of optically pure D and L and poly(amino acids).

Since 1950 I have continued my work on poly(amino acids) at the Weizmann Institute with able scientists such as Arieh Berger, Michael Sela, Abraham Patchornik, and Joseph Kurtz. Within a period of approximately ten years we devised adequate protecting groups for the trifunctional amino acids and, consequently, synthesized the homopolymers of most of the natural amino acids, including poly(aspartic acid), polyhistidine, polyornithine, polyarginine, polycysteine, polymethionine, polytyrosine, polytryptophan, polyproline, and polyhydroxyproline. Random amino acid copolymers also can be synthesized from the corresponding N-carboxyamino acid anhydrides. By using the basic polylysine or polyornithine as multifunctional polymerization initiators, we were able to synthesize a great variety of multichain polyamino acids in addition to the above linear polymers.

In the meantime, Bamford, Waley, and Watson at Maidenhead in England, and Blout and Doty at Harvard studied the mechanism of the polymerization of N-carboxyamino acid anhydrides in solution. Of particular importance was the discovery by Blout and his collaborators that sodium methoxide in inert organic solvents initiates an ionic type of polymerization leading to poly(amino acid) preparations of high average molecular weight. These polymers can be drawn into fibers and films that readily undergo orientation under the appropriate conditions.

The remarkable mechanical and electric properties of oriented poly(γ-methyl-L-glutamate) and poly(γ-benzyl-L-glutamate) fibers, and their resemblance to native silk prompted the Courtaulds Co. to build a pilot plant to produce these synthetic fibers. While visiting the Courtaulds group, headed by Bamford in Maidenhead, I had the pleasure of receiving as a gift a piece of cloth made of poly(γ-methyl-L-glutamate) fibers. The plant has since been dismantled, most probably for economic reasons, but I do believe that when amino acids are available at a considerably lower price and the polymerization process has been improved, amino acid polymers will acquire practical importance as fibers, films and coating materials, because of their special characteristics and their expected degradability by microorganisms.

The availability of high-molecular-weight poly(amino acids) opened the way for a thorough investigation of their three-dimensional structure. Thus in 1951, by analyzing the x-ray-diffraction pattern of poly(γ-benzyl-L-glutamate) fibers, Perutz (19) observed an 1.5 Å reflection, which had been predicted by Pauling and Corey (20) for peptides and proteins whose backbone assumed an α-helical conformation characterized by 3.6 amino acid residues per turn and a residue translation of 1.5 Å along the helix axis. Thus poly(γ-benzyl-L-glutamate) was the first high-molecular-weight compound in which the existence of an α-helical polypeptide backbone was detected experimentally. The α helix has been subsequently detected in many other synthetic poly(amino acids), as well as in a great number of fibrous and globular proteins. In 1959, Elliott and Malcolm (21) carried out a most detailed analysis of the x-ray-diffraction pattern of oriented fibers of poly(L-alanine). All features of the observed pattern could be explained by assuming that each polypeptide chain has a right handed α-helical conformation and that the helices are packed within the fiber in a hexagonal arrangement. The above analysis was of considerable help to Kendrew and Perutz (22) in the deciphering of the x-ray patterns of myoglobin and hemoglobin.

Poly(amino acid) fibers and films were useful also in detecting the transformation from an α-helical conformation to the β structure in the solid state. Such transformations have been achieved by Bamford (23) and others by stretching, heating or treating with suitable solvents. The stretching in steam of poly(L-alanine) in the α-helical conformation, for example, yields poly(L-alanine) in the β conformation having an x-ray pattern similar to that of silk, in which a repeat pattern of 7 Å has been detected. The α-β transformation of poly(L-alanine) recalls the conformational change that occurs in hair keratin upon stretching in steam, as detected by Astbury in 1931 (24).

Poly(amino acids) were most useful as model compounds in the elucidation of the three-dimensional structure of the fibrous protein collagen. In 1954, Ramachandran and Kartha (25), and somewhat later Rich and Crick (26), suggested that collagen fibers have a triple-stranded helical conformation in which each of the separate peptide strands possesses a left-handed polyglycine-II–type helix with a residue translation of 3.1 Å along the helical axis. In our laboratory we were able to contribute to the understanding of the structure of collagen by synthesizing poly(L-proline), poly(hydroxy-L-proline), the sequential poly(Pro-Gly-Pro), and other sequential poly(amino acids) containing proline and glycine; we then studied their conformations in the solid state and in solution.

Poly(L-proline) was synthesized by the late Joseph Kurtz (27) through the polymerization of N-carboxyproline anhydride in pyridine. Upon dissolution of the polymer obtained in acetic acid or formic acid, it showed a specific optical rotation of $[\alpha]_D = +40°$. To our great surprise, however, the optical rotation was found to change rapidly with time and it reached a final value of approximately $-600°$ within two to three hours. The original positive optical rotation could be restored by addition of alcohols. Subsequently, by the use of x-ray techniques, Cowan and McGavin (28), and later, Traub (29) were able to show the existence of two stable conformations. Mutarotation thus results from the conversion of one form to the other. The polymer form with $[\alpha]_D = +40°$ was denoted as poly(L-proline) I, and its conformation was shown to be that of a right-handed helix in which all imide bonds are *cis* and the translation along the helical axis is 1.9 Å per residue. The other form, with $[\alpha]_D \cong -600°$, was denoted poly(L-proline) II and was attributed to a left-handed helix in which all imide bonds are in the *trans* configuration and the translation along the helical axis is 3.12 Å per residue. It is poly(L-proline) II that has a conformation similar to the single strands of collagen.

Poly(Pro-Gly-Pro) was the first polymer that in solution and in the solid phase was shown to form the triple-stranded helical conformation attributed to collagen (30). Therefore it has been used extensively by chemists and biologists as a model compound for collagen. In our laboratory, Segal and Kurtz (31) synthesized sequential poly(amino acids) containing proline, glycine, and other amino acids and analyzed their conformations, thereby making possible the determination of some of the essential factors contributing to the stability of the triple-stranded helical structure of collagen.

A considerable amount of work, both experimental and theoretical, has been done on the conformations and conformational transitions of poly(amino acids) not only in the solid state but also in solution. Studies with solvent systems were initiated experimentally by Doty (32), Blout (33), and their collaborators at Harvard, and theoretically by Moffitt and Kirkwood (34) and by Schellman (35). The existence, in appropriate solvent systems, of regular macromolecular conformations, such as right- and left-handed α helices, polyproline helices, and the β form, as well as the existence of the random-coil conformation, has been established experimentally by hydrodynamic, optical, electrical, and nuclear magnetic resonance methods. Moreover, conditions have been established for conformational transitions; these include changes in solvent composition, pH, temperature and pressure.

Theoretical analyses of the conformational transitions in poly(amino acids) have been carried out by Gibbs and DiMarzio (36), Zimm and Bragg (37), Lifson and Roig (38), Peller (39), and others, all of whom have used a statistical-mechanical approach. These studies are based on the assumption that the transformation of the polypeptide backbone from the disordered form to the α-helical one is a cooperative phenomenon resembling crystallization. Thus, the transition process requires an initiation step characterized by the arrangement of four consecutive amino acid residues into a helical loop, followed by a propagation step in which neighboring amino acid residues assume the helical conformation elongating the initially formed loop. The cooperative nature of the transition phenomenon is the result of the fact that the initiation step is a considerably less probable event than the propagation step. These theoretical considerations have contributed to a better understanding of the process of "denaturation" of synthetic biopolymers as well as proteins.

Concomitantly, the foundations have been laid for the computation of the stable conformations of macromolecules; for this purpose, use is made of basic information concerning chemical bonds and atomic interactions. Ramachandran (40) and his collaborators have played an important role in this area. They assumed a model in which each atom is represented by a hard sphere and were thereby able to divide the conformational space into allowed and forbidden domains. Subsequently, a more refined analysis of polypeptide conformation was developed by Liquori (41), Lifson (42), Scheraga (43), Flory (44), and others, based on the assumption that the most probable conformations are those that possess the lowest free energy. The latter approach takes into consideration torsional potentials, as well as nonbonded, electrostatic, and hydrogen-bond interactions. For the disordered state (the random coil), the calculations include a statistical-mechanical averaging over the available conformational space. Recently a more rigorous conformational analysis, based on molecular orbital theory, has been applied to low-molecular-weight model compounds, and in a few cases to poly(amino acids).

Theoretical developments briefly discussed above facilitated the successful correlation of the macromolecular conformations of poly(amino acids) in solution with their hydrodynamic properties, optical properties (UV, IR, CD spectra, and ORD), dipole moments, and nuclear magnetic properties (NMR). The fruitful collaboration of experimentalists and theoreticians in the study of the various properties of poly(amino acids) in solution undoubtedly should be continued. With suitable model compounds additional information undoubtedly will be acquired concerning the polypeptide backbone, as well as the various interactions among amino acid side chains. Such information should be of great value and may lead from basic principles to greater understanding of the chemical and physicochemical properties of proteins.

Finally, I shall turn to an old love of mine: the biological properties of poly(amino acids). My original aim in preparing poly(amino acids) was to obtain simple, water-soluble high-molecular-weight compounds usable in various biological studies. The first questions that I asked myself, once these compounds had been prepared, were whether they could be hydrolyzed by proteolytic enzymes and whether they could be used to detect new proteolytic enzymes. I still remember the excitement with which I followed the rapid hydrolysis of poly(L-lysine) by trypsin. As a matter of fact, poly(L-lysine) turned out to be one of the best substrates for trypsin, yielding mainly di- and trilysine upon exhaustive digestion. A careful analysis of the products formed as a result of the enzymic hydrolysis revealed that hydrolysis is accompanied by transpeptidation leading to the formation of new lysine oligopeptides. Poly(L-lysine) was found to be hydrolysed also by papain, whereas poly(L-glutamic acid) and glutamic acid copolymers were shown by us to be hydrolysed by pepsin and papain. Therefore it may be predicted that other water-soluble poly(amino acids) will be found to serve as substrates for many of the known proteolytic enzymes.

In an attempt to detect proteolytic enzymes capable of cleaving prolyl-peptide bonds, proline-rich polymers, such as poly(L-proline) and poly(Pro-Gly-Pro) were used as possible substrates. Poly(L-proline) was found to be digested to proline by $E.$ $coli$ extracts. It then became obvious that unknown hydrolytic enzymes were present within these extracts. Further studies by Yaron and Berger (45) led to the isolation in pure form of a Mn^{2+}-requiring amino peptidase, aminopeptidase P, which is capable of splitting any N-terminal amino acid (including proline) if followed by a proline residue. Thus, aminopeptidase P shows exclusive specificity toward amide bonds in which the proline nitrogen participates.

When the action of an $E.$ $coli$ extract on the sequential poly(amino acid) poly(Pro-Gly-Pro) was investigated, it was found that the extract contained another novel proteolytic enzyme, named *dipeptido-carboxypeptidase*, which is capable of splitting the C-terminal glycyl proline dipeptide of the polymer. The specificity of $E.$ $coli$ dipeptido-carboxypeptidase was found to resemble the so-called "converting enzyme" isolated from mammalian tissues, which is known to produce angiotensin II from angiotensin I.

The antigenicity of poly(amino acids) was of interest to us even in the earliest stages of our research work. As you know, poly(amino acids) were the first synthetic high-molecular-weight antigens, and their use in various immunological studies will be discussed in this symposium by Michael Sela. Allow me, however, to recall our first attempts in this intriguing field. Sela (46) described the synthesis of poly(L-tyrosine) and polymers of other aromatic amino acids. Their insolubility in water, however, prevented him from testing their antigenicity. Thus, he decided to prepare a soluble polytyrosyl derivative by the attaching of tyrosyl-containing side chains to gelatin, an immunologically inert protein. I still remember the first immunological experiments carried out at the Hebrew University in Jerusalem by Olitzki, who repeatedly injected guinea pigs with the polytyrosyl-gelatin preparations obtained in our laboratory. After the second and third injections the guinea pigs went into anaphylactic shock and some of them died. The first indication of the increase in the antigenicity of gelatin by attaching polypeptide side chains containing aromatic amino acids was thus obtained. The next step was obvious: preparation of a completely synthetic amino acid polymer antigen. This goal was achieved by the synthesis of a branched polymer in which polylysine served as backbone and the side chains contained tyrosine and glutamic acid. The availability of the synthetic antigen enabled Sela and his collaborators to study the effect on antigenicity of amino acid composition, molecular weight, steric configuration, conformation, and other parameters of their multichain compounds. Similar studies were carried out by Maurer (47), Gill (48), and others who have systematically investigated the antigenicity of various linear and branched amino acid polymers and copolymers.

An interesting discovery about poly(amino acids) is the fact that poly(Pro-Gly-Pro), which has been shown to mimic the conformation and other physico-chemical properties of collagen, also resembles the native protein immunological-ly. Thus, when injected into rabbits, poly(Pro-Gly-Pro) elicits antibodies that interact not only with the synthetic polymer, but also with natural collagens derived from several animal species. Furthermore, collagen, like poly(Pro-Gly-Pro), was shown to be representative of thymus-independent immunogens.

Poly(amino acids) have been useful also in the study of the genetic control of the immune response. The first such study was reported by Benacerraf (49), who observed in 1966 that guinea pigs of the inbred strain 2 react to poly(L-lysine) by producing a relatively large amount of antibody, whereas guinea pigs of strain 13 and some random-bred Hartley animals were nonresponders. The differences observed in the immune responsiveness were attributed to the presence or absence of a "polylysine gene," the expression of which is characterized by cellular immunity, as well as by the production of antibodies. Needless to say, I was delighted to learn that one of the genes controlling immune responsiveness was named for the first water-soluble poly(amino acid) that I synthesized.

Sela and his collaborators (50) obtained somewhat similar results with mice. When a branched polymer containing histidine and glutamic acid in its side chains was used as immunological probe, it elicited antibodies in CBA mice, whereas in C57 mice the response was poor. Both strains, however, responded well to a branched polymer containing phenylalanine and glutamic acid in its side chains. It is of interest to note that the synthetic branched polypeptide antigens were found to be closely linked to the major histocompatibility (H-2) locus of the mice.

The relative simplicity of the synthetic antigens and scientists' ability to design them to specification facilitate the study of the mechanisms involved in the genetically controlled immune response. It is no wonder, therefore, that a considerable amount of work going on with poly(amino acids) to determine whether the genes act via B cells, T cells, or by some other biological mechanism.

Many poly(amino acids) have been shown to interact with enzymes, nucleic acids, viruses, bacteria, fungi, red blood cells, and other eucaryotic cells. Some homo- and copolymers containing basic amino acids exhibit antiviral and anti-bacterial activity at considerably low concentrations. Of particular interest are the observations that, in the appropriate systems, attachment of certain poly(amino acids) to cell membranes leads to cell stimulation or to an increase in the permeability of the cell membrane to high-molecular-weight compounds, such as nucleic acids and proteins.

I cannot conclude this survey without emphasizing the role of poly(amino acids) in deciphering the genetic code. It was Nirenberg and Matthaei (51) who showed in 1961 that poly(uridylic acid) directs the synthesis of polyphenyl-alanine *in vitro* by a cell-free system from *E. coli* containing ribosomes, transfer-RNA, labeled amino acids, and an amino acid-activating system. Within two years, poly(adenylic acid) was shown to direct the synthesis of polylysine, and poly(cytidylic acid) the synthesis of polyproline. Under similar conditions, synthetic random ribopolynucleotides were shown to lead to a large number of corresponding random amino acid copolymers. From these, Nirenberg (52) and Ochoa (53) derived a considerable amount of information pertaining to the three nucleotides coding for each amino acid residue in protein biosynthesis. However, the most detailed information concerning the nucleotide triplets, which code for the various amino acids, has been obtained by Khorana (54) and his collabo-rators, who used sequential ribo-polynucleotides to direct the synthesis of the corresponding sequential polyamino acids in the Nirenberg-Matthaei cell-free system.

Having discussed the achievements of poly(amino acid) research, I would now like to suggest some of the research directions worthy of future pursuit. One of the most important aims of the modern biophysicist is to understand the correlation between the amino acid sequence of a given protein and its secondary, tertiary, and quaternary structure. Moreover, once the structure of the protein has been established, it is the hope of the theoretician to derive its chemical and physical properties. In such studies, amino acid polymers and oligomers more complex than those synthesized so far should be valuable. Among these are sequential poly(amino acids), cyclic peptides, and tailor-made polypeptides of predeter-mined amino acid sequence. Such model compounds will make possible the evaluation of the interactions occurring between the various amino acid side chains.

Recent studies show that the biological specificity and activity of proteins are determined not only by their characteristically static three-dimensional struc-ture, but also by their ability to undergo minor conformational changes during reaction. Knowledge of the detailed action mechanisms of biopolymers, in general, and of proteins, in particular, therefore, might be obtained via the study of the kinetics of the conformational changes involved. Again, thorough investi-gation of the kinetics and mechanisms involved in the conformational transitions occurring in judiciously chosen high-molecular-weight model compounds should be of considerable value.

A great variety of modern physicochemical techniques are being applied in the study of proteins. Polymers of amino acids have served in the past and will undoubtedly continue to serve in the future as test compounds for the determination of the usefulness and reliability of the methods under consideration.

Amino acid polymers have been used successfully to clarify the mode of action of known proteolytic enzymes, as well as to detect new peptidases and proteases. Because the active sites of the above enzymes have been shown to involve a number of subsites, it is clear that synthetic oligomers and polymers of well-defined amino acid sequence should play an even more important role in future investigations than in the past.

The original observation made in our laboratory that multichain poly(amino acids) containing aromatic amino acid residues stimulate antibody formation, has encouraged many researchers to use linear and branched poly(amino acids) as a tool in the study of the chemical and physical parameters determining antigenicity. A clearer picture of the nature of antigenic determinants of proteins will certainly continue to emerge from such investigations. The availability of a great variety of synthetic antigens, the composition and structure of which can be modified at will, is also expected to be of use in the elucidation of the cellular mechanisms involved in various immunological reactions.

Despite the advances in the field of poly(amino acid) research, it remains a challenge for peptide chemists to produce poly(amino acid) fibers and films by facile reactions that are easily scaled up to industrial proportions. The desirable mechanical and electrical properties of such preparations, which so far have been obtained only on laboratory or pilot-plant scale, as well as the great variety of structures and properties attained, appear to justify the continuation of the attempts by engineers and technologists to synthesize cheaply amino acid monomers, to develop new polymerization techniques and suitable work-up methods for the preparation of the desired fibers and films. In this respect, it is pertinent to note that poly(amino acids) belong to the class of biodegradable materials that undoubtedly will become increasingly important in the future. Because many poly(amino acids) represent biocompatible compounds, their possible use in medicine as structural or coating materials should be investigated.

To attain the goals outlined above, close cooperation between classical peptide chemists, modern polymer chemists, theoreticians and bioengineers is required. The time is ripe to achieve such cooperation.

REFERENCES

1. M. Frankel and E. Katchalski, *Nature* (London), **144**, 330 (1939).
2. E. Katchalski, *Adv. Protein Chem.*, **6**, 123 (1951).
3. E. Katchalski and M. Sela, *Adv. Protein Chem.*, **8**, 243 (1958).
4. M. Sela and E. Katchalski, *Adv. Protein Chem.*, **14**, 391 (1959).
5. E. Katchalski, M. Sela, H.I. Silman, and A. Berger, *The Proteins*, Vol. 2, H. Neurath, ed., Academic Press, New York, 1964, p. 406.
6. For example: C.M. Venkatachalam and G.N. Ramachandran, *Ann. Rev. Biochem.*, **38**, 45 (1969).
7. N. Lotan, A. Berger, and E. Katchalski, *Ann. Rev. Biochem.*, **41**, 869 (1972).

8. C.H. Bamford, A. Elliott, and W.E. Hanby, *Synthetic Polypeptides,* Academic Press, New York, 1956.
9. G.D. Fasman, ed., *Poly-α-Amino Acids,* Marcel Dekker, New York, 1967.
10. P.J. Flory, *Statistical Mechanics of Chain Molecules,* Wiley(Interscience), New York, 1969.
11. D. Poland and H.A. Scheraga, *Theory of Helix-Coil Transition in Biopolymers,* Academic Press, New York, 1970.
12. A.G. Walton and J. Blackwell, *Biopolymers,* Academic Press, New York, 1973, Chapter 10.
13. A.J. Hopfinger, *Conformational Properties of Macromolecules,* Academic Press, New York, 1973.
14. F.A. Bovey, *High Resolution NMR of Macromolecules,* Academic Press, New York, 1972.
15. M. Goodman, *Biopolymers,* **1,** 1 (1963).
16. H. Leuchs, *Ber.,* **39,** 857 (1906).
17. M. Frankel and E. Katchalski, *Scientific Papers Presented to C. Weizmann,* Y. Hirschberg, ed., p. 24 (1944); E. Katchalski, I. Grossfeld, and M. Frankel, *J. Am. Chem. Soc.,* **70,** 2094 (1948).
18. D. Ben-Ishai and A. Berger, *J. Biol. Chem.,* **17,** 1564 (1952).
19. M.F. Perutz, *Nature* (London), **167,** 1053 (1951).
20. L. Pauling and R.B. Corey, *Proc. Nat. Acad. Sci. U.S.,* **37,** 729 (1951).
21. A. Elliott and B.R. Malcolm, *Proc. Roy. Soc.* (London), **A249,** 30 (1959).
22. J.C. Kendrew, *Brookhaven Symp. Biol.,* **15,** 216 (1962). M.F. Perutz, *Sci. American,* 64 (1964).
23. C.H. Bamford, L. Brown, A. Elliott, W.E. Hanby and I.E. Trotter, *Nature* (London), **171,** 1149 (1953).
24. W.T. Astbury and A. Street, *Phil. Trans. Roy. Soc.* (London), **A230,** 75 (1931).
25. G.N. Ramachandran and G. Kartha, *Nature* (London), **176,** 593 (1955).
26. F.H.C. Crick and A. Rich, *Nature* (London), **176,** 780 (1955).
27. J. Kurtz, A. Berger and E. Katchalski, *Nature* (London), **178,** 1066 (1956).
28. P.M. Cowan and S. McGavin, *Nature* (London), **176,** 501 (1955).
29. W. Traub and U. Shmueli, *Nature* (London), **198,** 1165 (1963).
30. A. Yonath and W. Traub, *J. Mol. Biol.,* **43,** 461 (1969).
31. J. Engel, J. Kurtz, E. Katchalski, and A. Berger, *J. Mol. Biol.,* **17,** 255 (1966); D.M. Segal, *Ibid.,* **43,** 497 (1969).
32. For example: P. Doty, *Rev. Mod. Phys.,* **31,** 107 (1959); P.J. Urnes and P. Doty, *Adv. Prot. Chem.,* **16,** 401 (1961).
33. For example: E.R. Blout, in *Optical Rotatory Dispersion,* C. Djerassi, ed., McGraw-Hill, New York, 1960, p. 238; E.R. Blout, S. Farber, G.D. Fasman, E. Klein, and M. Narrod, in *Polyamino Acids, Polypeptides and Proteins,* M.A. Stahmann, ed., University of Wisconsin Press, Madison, Wis., 1962, p. 379.
34. For example: W. Moffitt, *Proc. Nat. Acad. Sci. U.S.,* **42,** 736 (1956); W. Moffitt and A. Moscowitz, *J. Chem. Phys.,* **30,** 648 (1959); J.G. Kirkwood and J. Riseman, *J. Chem. Phys.,* **16,** 565 (1948); J.G. Kirkwood, *J. Polymer Sci.,* **12,** 1 (1954).
35. J.A. Schellman and C. Schellman in *The Proteins,* Vol. 2, H. Neurath, ed., Academic Press, New York, 1964, p. 1.

36.   J.H. Gibbs and E.A. DiMarzio, *J. Chem. Phys.*, **28**, 1247 (1958) and
      J. Chem. Phys., **30**, 271 (1959).
37.   B.H. Zimm and J.K. Bragg, *J. Chem. Phys.*, **31**, 526 (1959).
38.   S. Lifson and A. Roig, *J. Chem. Phys.*, **34**, 1963 (1961).
39.   L. Peller, *J. Phys. Chem.*, **63**, 1194 (1959).
40.   G.N. Ramachandran and V. Sasisekharan, *Adv. Protein Chem.*, **23**, 283
      (1968).
41.   A.M. Liquori, in *Symmetry and Function of Biological Systems at the
      Macromolecular Level,* A. Engstrom and B. Strandberg, eds., Nobel Sym-
      posium II, Almquist and Wiksell, Stockholm, 1969, p. 101.
42.   A.T. Hagler, S. Lifson, and E. Huler, in *Peptides, Polypeptides, and Pro-
      teins,* Proceedings of the Rehovot Symposium 1974, Wiley(Interscience),
      New York, 1974.
43.   H.A. Scheraga, in *Peptides, Polypeptides, and Proteins,* Proceedings of the
      Rehovot Symposium 1974, Wiley(Interscience), New York, 1974.
44.   For example: D.A. Brant and P.J. Flory, *J. Am. Chem. Soc.*, **87**, 2791
      (1965); W.K. Olson and P.J. Flory, *Biopolymers*, **11**, 1 (1972); Ref. 10,
      this paper.
45.   A. Yaron and A. Berger, in *Methods in Enzymology,* Vol. 19, G.E. Perl-
      mann and L. Lorand, eds., Academic Press, New York, 1970, p. 521.
46.   M. Sela, Ph.D. thesis, the Hebrew University of Jerusalem, (1953); also,
      E. Katchalski and M. Sela, *J. Am. Chem. Soc.*, **75**, 5284 (1953).
47.   M. Sela, in *Peptides, Polypeptides, and Proteins,* Proceedings of the
      Rehovot Symposium 1974, Wiley(Interscience), New York, 1974.
48.   P.H. Maurer, *Progress in Allergy*, **8**, 1 (1964).
49.   T.J. Gill III, in *Immunogenicity,* F. Borek, ed., North-Holland, Amsterdam,
      1972.
50.   I. Green, W.E. Paul, and B. Benacerraf, *J. Exp. Med.*, **123**, 859 (1966).
51.   S. Fuchs and M. Sela, *Biochem. J.*, **93**, 566 (1964).
52.   M.W. Nirenberg and J.H. Matthaei, *Proc. Nat. Acad. Sci. U.S.*, **47**, 1588
      (1961).
53.   M.W. Nirenberg, *Sci. American,* **208**, 30 (1963).
54.   S. Ochoa, in *Informational Macromolecules,* H.J. Vogel, V. Bryson, and
      J.O. Lampen, eds., Academic Press, New York, 1963, p. 437.
55.   H.G. Khorana, *Fed. Proc.*, **24**, 1473 (1965).

PART I.   CONFORMATIONAL CALCULATIONS

ASPECTS OF PEPTIDE CONFORMATION*

G.N. RAMACHANDRAN, Molecular Biophysics Unit, Indian Institute of Science, Bangalore 560012, India, and Department of Biophysics and Theoretical Biology, University of Chicago, Chicago, Ill. 60637

SYNOPSIS: The paper is a critical review of some of the potential functions proposed for incorporation in the theory of peptide conformation. Although these functions all agree broadly in predicting the same low-energy regions of the (ϕ, ψ) map for an alanyl dipeptide, all of them disagree with the experimentally determined population distribution of such peptides in the conformational map of proteins. To resolve this discrepancy we suggest a new form for $V(\psi)$, namely, $V(\psi) = 2.0(1 - \cos 2\psi)$, based on quantum mechanical calculations and on the ψ values observed in crystals of amino acids and amides. Further, the need to have softer potentials for $H \cdots H$ and $H \cdots X$ interactions is pointed out. Such modifications in the potential functions result in much better agreement between the theoretically calculated and the experimentally observed conformational characteristics of the LL bend and of the randomly coiled polypeptide chains. A similar situation was encountered in the analysis of the crystal structures of benzene and N-methylacetamide. Consequently, the author believes that the potential functions used at present should be carefully revised.

The nonplanarity of the peptide unit was reinvestigated in about thirty examples of accurate crystal structures in light of results of the most recent quantum mechanical calculations (by CNDO/2 method). The findings indicate that deviations from planarity of up to $10°$ in ω and of up to $25°$ in θ_N (describing the nonplanarity of the three bonds meeting at the nitrogen atom) are quite feasible and lead only to destabilizing energies less than RT. Similar results are obtained for formamide, where INDO calculations correspond with microwave data on nonplanarity.

INTRODUCTION

In a paper submitted to the Conference on Conformation of Biological Molecules and Polymers (1) held in Jerusalem in 1973, the author discussed the accuracy of the empirical potential functions used in his laboratory to calculate the conformational energy of polypeptides. In fact, the form of other potential functions suggested (2-6), particularly for the nonbonded interactions, are essentially similar to one another, but the values of the constants are different (Fig. 1). We feel that a relatively stable minimum-energy conformation can be detected reasonably accurately by using potential functions having widely different parameters. The example discussed below is the orientation of the benzene molecule in its orthorhombic crystal structure.

A more complex situation is encountered with torsional potential functions, as applied to the conformation of a pair of peptide units. In this case also, the

---

*Communication No. 44 from the Molecular Biophysics Unit, Bangalore.

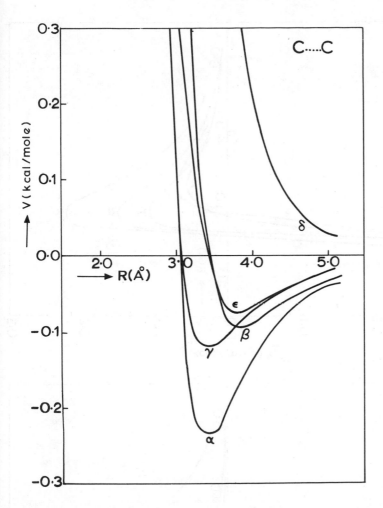

Figure 1. Variations among the potential functions for the nonbonded interactions H$\cdots$H and C$\cdots$H (see following page) and C$\cdots$C adopted in different
laboratories: α, ours (2); β, Williams (3); γ, Scheraga (4); δ, Scheraga (packing)
(5); and ϵ, Kitaigorodsky (6).

energy contours in the (ϕ,ψ) plane are very similar for the different types of
potential functions and constants used (2, 4, 7, 8), but none of them agrees with
the distribution of conformations observed in various globular proteins (9).
Thus, the validity of the potential function for $V(\psi)$ is questioned. In a recent
study of the barrier to rotation around the glycosidic C–N linkages in nucleosides and nucleotides (10), the experimental data seemed best fitted by using a
van der Waals radius of 1.0 Å for the hydrogen atom, rather than the normally
accepted value of 1.2 Å (2, 11).

The data indicate that the potential functions currently employed in the biopolymer field have to be revised. In this paper we will suggest a modified form

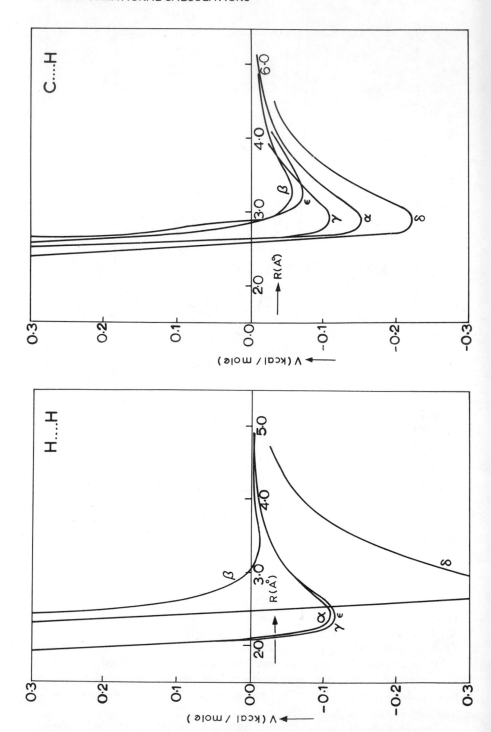

for $V(\psi)$, with special reference to the observed conformations of peptides and the molecular packing of N-methylacetamide in its crystal structure.

In addition, we shall consider the energy variation involved in the distortions of the planarity of the peptide unit itself. Quantum mechanical calculations were used originally to justify the proposal of a nonplanar peptide unit (12), and we shall present here an outline of experimental energy curves obtained recently from microwave and x-ray-crystallographic data that support the theoretical conclusions (13, 14).

MODIFIED ψ POTENTIAL

In studies involving the backbone atoms of two linked peptide units $C_1^\alpha-C_1-N_1-C_2^\alpha-C_2-N_2-C_3^\alpha$, the energy associated with torsion of the angles $\phi(C_1-N_1-C_2^\alpha-C_2)$ and $\psi(N_1-C_2^\alpha-C_2-N_2)$ has been usually described by the threefold potentials

$$V(\phi) = \tfrac{1}{2}V_\phi\,(1 \pm \cos 3\phi) \tag{1}$$

where V_ϕ is between 0.6 and 1.5 kcal/mole, and

$$V(\psi) = \tfrac{1}{2}V_\psi\,(1 - \cos 3\psi) \tag{2}$$

where V_ψ is between 0.25 and 1.0 kcal/mole. The shape of the energy contours in the (ϕ,ψ) plane are not appreciably modified by changing the values of V_ϕ and V_ψ within these ranges. Conclusions based on Eqs. (1) and (2) seem to be supported by some experimental results (1); however, infrared studies of amides (15) and data on the conformations of proteins (9), amino acids, and peptides (16, 17) all suggest for $V(\psi)$ a twofold potential, with minima at $\psi = 0°$ and $\psi = 180°$. Quantum mechanical calculations on model compounds performed in our laboratory (18), using the IEHT and an *ab-initio* method, indicate that, indeed, $V(\psi)$ has minima for two values of ψ, namely, $\psi = 0°$ and $180°$, with a barrier of approximately 4.5 kcal/mole between them. [Pople and Radom (19) have had similar results that point to a possibly greater barrier height.] These factors suggest a function of the form

$$V(\psi) = 2.0(1 - \cos 2\psi) \tag{3}$$

which we employed to calculate the conformational energy of the two peptide units mentioned above. The results, shown in Fig. 2(b), are markedly different from those obtained with the conventional form of the potential functions [Fig. 2(a)]. The experimental distribution of conformations obtained from data on globular proteins (9) is shown in Fig. 3. Obviously, Fig. 2(b) is in much better agreement with Fig. 3 than Fig. 2(a). Still better agreement with the experimental data of Fig. 3 is obtained by using softened potential functions, that is, by assuming for the hydrogen atom a shorter van der Waals radius, namely, 1.0 Å,

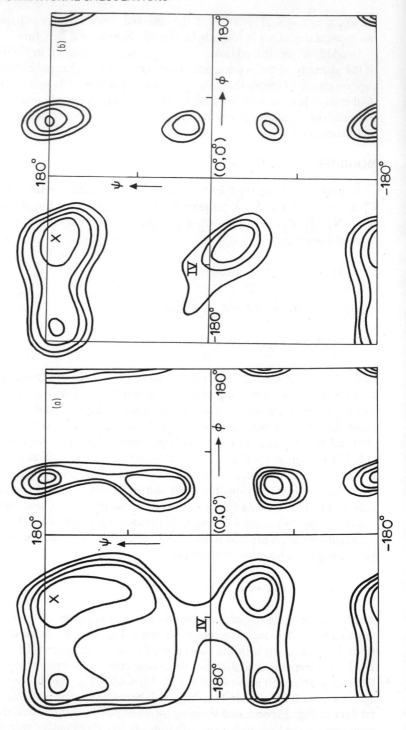

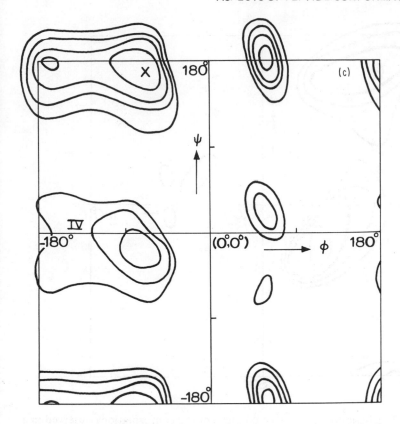

Figure 2. (a) Energy distribution in the (ϕ,ψ) plane obtained with conventional potential functions (2). (b) Energy map obtained with the potential function $V(\psi) = 2.0(1 - \cos 2\psi)$ [Eq. (3)], the other functions being the same as in (2). (c) Energy map obtained from $V(\psi) = 2.0(1 - \cos 2\psi)$, all interactions involving softened hydrogen atoms.

which is suggested also by other studies (10, 20); the results of these calculations (22) are presented in Fig. 2(c). In making this comparison, however, it should be remembered that the experimental distribution obtained from protein structures is modified by the occurrence of interactions not expected in an isolated pair of peptide units.

The improvement of theoretical predictions by the inclusion of the features mentioned above may be seen in another example, namely, the internally hydrogen-bonded hairpin bends (also known as beta bends). There are two types of hairpin bends, the so-called LL bend and the LD bend (Fig. 4), and in analyzing the experimental data a Gly residue was considered as either an L or a D unit. It is clear from the data given in Table 1 that the revised set of potential functions (18, 22) gives better agreement of the calculated conformation with experimentally obtained data, particularly for the angle ψ_3.

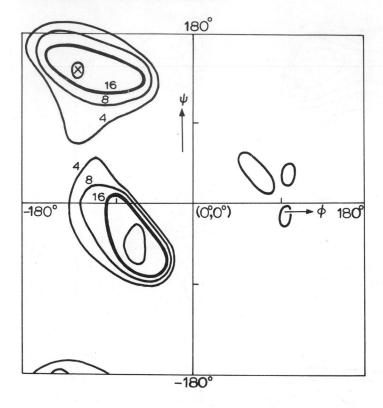

Figure 3. Distributions in the (ϕ, ψ) plane of the conformations observed in a number of crystal structures of globular proteins [redrawn from (9)].

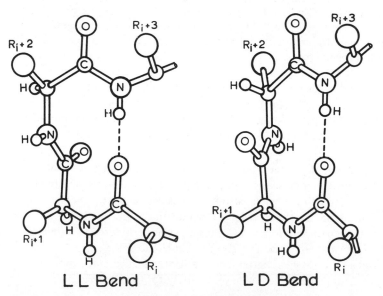

Figure 4. Typical conformations of LL and LD bends.

Table 1. Minimum-energy conformation of the LL bend of a peptide fragment
$C_1^\alpha - C_1 O_1 - N_2 H_2 - C_2^\alpha - C_2 O_2 - N_3 H_3 - C_3^\alpha - C_3 O_3 - N_4 H_4 - C_4^\alpha$ having a hydrogen bond of
the type $N_4 H_4 \cdots O_1$, as compared with experimentally observed data

Dihedral Angle	ϕ_2	ψ_2	ϕ_3	ψ_3
THEORY (Only for LL)				
Standard potential functions (2, 21)	–50°	–50°	–110°	+40°
Standard functions with new $V(\psi)$ (18)	–60°	–30°	–90°	20°
Standard functions with new $V(\psi)$ and softened hydrogen interactions (23)	–60°	–20°	–110°	10°
CRYSTAL STRUCTURE DATA (LL, LG, GL, or GG)				
Mean of observed data in lysozyme and chymotrypsin [data from (21)]	–49°	–40°	–96°	14°
Mean of eight molecules in crystal structures of small peptides [data from (21)]	–68°	–28°	–101°	9°

TEST OF POTENTIAL FUNCTIONS USING CRYSTAL PACKING

Because of appreciable differences in the variation of different nonbonded
potentials with interatomic distance, we assessed the validity of the functions by
analyzing the crystal structures of very simple compounds containing the atoms
normally found in biopolymers, such as C, H, O, N, and so on (24, 25). The test
cases were benzene and N-methylacetamide. In the case of benzene, whose space
group is *Pbca* (26, 27), the four molecules in the unit cell are related by symmetry in such a way that the centers of the molecules occur only at definite locations, namely, the inversion centers at 0,0,0; 0,1/2,1/2; 1/2,0,1/2; and 1/2,1/2,0.
There is, therefore, no translational freedom for the benzene molecules. The
only freedom available to each molecule is a possible rotation about three perpendicular axes, with the center of the molecule fixed at a point. The most convenient way of representing the rotations is by means of Eulerian angles, as
shown in Fig. 5. The values observed in the crystal for these three angles are
$\psi = 2.3°$, $\theta = 46.6°$, and $\phi = 104.4°$. The interaction energy of one molecule of
benzene with its surroundings in the crystal was calculated for the whole range
of Eulerian angles, at intervals of 10°. No appreciable electrostatic charges are
present in the benzene molecule, so the electrostatic interaction was neglected.
The total energy was calculated therefore only as the sum of the nonbonded
interactions $V(H,H)$, $V(C,H)$, and $V(C,C)$. The potential functions first used had
the form and values of parameters normally adopted in our laboratory (2), and
it was found that the lowest minimum occurred at $\psi = 0°$, $\theta = 50°$, and $\phi =$
$110°$ (70°), which corresponded closely to experimentally observed values. The
two possible values of ϕ correspond to identical structures, and therefore only

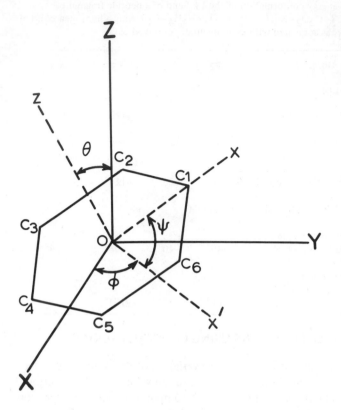

Figure 5. The Eulerian angles of rotation adopted for the benzene molecule.

the one with $\phi = 110°$ was examined further. The calculated energy values for the plane $\psi = 0°$ are shown in Fig. 6, where a second minimum is found at $\theta = 90°$ and $\phi = 50°$ (130°). It can also be seen that the lowest minimum is not appreciably lower than the next higher minimum; that is, the two differ only by about 0.9 kcal/mole and are connected together by a valley of low energy values. A more detailed investigation of the low-energy region (using Eulerian angles at intervals of 1°) revealed that the absolute minimum occurred at $\psi = 2°$, $\theta = 47°$, and $\phi = 106°$, angles that are extremely close to those observed in the crystal (see above).

These calculations appear to verify the correctness of the potential functions used. When the calculations were repeated (28) with other sets of potential functions (3–6), the results were not far different. In every case, the calculations at intervals of 10° gave the two minima at exactly the same place our functions did, that is, at $\psi = 0°$, $\theta = 50°$, $\phi = 110°$ (70°) and $\psi = 0°$, $\theta = 90°$, $\phi = 50°$ (130°). Interestingly, however, the difference between the absolute minimum and the second lowest minimum of energy varies appreciably—0.5 to 10 kcal/mole—from one set of functions to another (see Table 2). It should be mentioned that in these calculations we neglected the stabilizing energy from quadrupole-quadrupole interactions of the benzene molecules. This contribution is

$\psi = 0$

ϕ \ θ	0	10	20	30	40	50	60	70	80	90
0	–	–9.75	–27.74	–25.38	–	–	–	–	–	–9.14
10	–	–14.58	–28.24	–26.00	–2.44	–	–	–	–	–27.07
20	–13.92	–23.45	–29.00	–26.32	–8.32	–	–	–	–	–29.93
30	–24.01	–26.59	–28.66	–26.36	–12.26	–	–	–	–	–29.38
40	–13.92	–18.56	–25.41	–26.85	–20.48	–10.49	–9.50	–17.90	–24.99	–26.01
50	–	–2.59	–19.63	–27.14	–27.48	–26.23	–27.03	–28.79	–29.70	–20.39
60	–	–	–17.06	–27.55	–30.09	–30.41	–30.33	–29.95	–29.57	–17.33
70	–	–3.03	–20.53	–28.54	–30.61	–30.86	–30.36	–29.14	–27.24	–20.39
80	–13.92	–18.29	–26.21	–29.69	–30.60	–30.71	–30.06	–27.79	–23.29	–26.01
90	–24.01	–25.78	–28.68	–30.14	–30.50	–30.51	–29.90	–27.07	–21.21	–29.38
100	–13.92	–18.79	–26.21	–29.69	–30.60	–30.71	–30.06	–27.79	–23.29	–26.01
110	–	–3.03	–20.53	–28.54	–30.61	–30.86	–30.36	–29.14	–27.24	–20.39
120	–	–	–17.06	–27.55	–30.09	–30.41	–30.33	–29.95	–29.57	–17.33
130	–	–2.59	–19.63	–27.14	–27.48	–26.23	–27.03	–28.79	–29.70	–20.39
140	–13.92	–18.56	–25.41	–26.85	–20.48	–10.49	–9.50	–17.90	–25.00	–26.01
150	–24.01	–26.59	–28.66	–26.36	–12.26	–	–	–	–9.14	–29.38
160	–13.92	–23.45	–29.00	–26.32	–8.32	–	–	–	–	–29.93
170	–	–14.58	–28.24	–26.00	–2.44	–	–	–	–	–27.07
180	–	–9.75	–27.74	–25.38	–	–	–	–	–	–9.14

$\theta \longrightarrow$

$\phi \longrightarrow$

Figure 6. Crystal-packing energy of benzene (in kcal/mole). The calculations were performed assuming $\psi = 0°$. The lowest and the second lowest minima are underlined. The contact map corresponds to normal limits (——) and extreme limits (– – –).

Table 2. Conformational energy values of benzene in its orthorhombic crystal structure, calculated using the potential functions shown in Fig. 1

	Energy (kcal/mole)		
Potential Functions Used	Lowest Minimum	Next Lowest Minimum	Difference
Our laboratory (2)	−30.86	−29.93	0.93
Kitaigorodsky (6)	−13.46	−13.03	0.43
Scheraga (4)	−20.84	−20.28	0.56
Williams (3)	−17.98	−15.60	2.38
Scheraga (packing) (5)	−18.28	−7.75	10.53
Our usual potential functions (2), adding quadrupole-quadrupole interactions (27)	−31.91	−29.27	2.64

appreciably larger (negative) for the lowest minimum than for the second lowest one. When the quadrupole-quadrupole interaction is included in the calculations, the energy difference between the two minima is more than 2 kcal/mole for all the sets of potential functions. The consequence of all these factors, in relation to the stability of the observed crystal structure, is very significant. In fact, neutron diffraction studies conducted at different temperatures (27) indicate that the pattern of the crystal packing given by its space group, as well as the parameters defining the exact location of the molecules, do not change appreciably with temperature from −135°C to −3°C. At a higher pressure and temperature (25 kilobars and 21°C), however, benzene undergoes a phase transition (29) leading to a crystal having a space group $P2_1/c$. We have not yet examined either the stability of this structure or how well this structure is explained by theory.

The validity of our potential functions was tested in another example of crystal packing, namely, N-methylacetamide (25). This crystal has the symmetry of the space group $Pnma$ and it is a suitable case study because it has only three degrees of freedom (see Fig. 7): The coordinates x and z ($y = 0.25$ for this molecule) and the angle θ define the orientation of the C−N bond. The details of the calculations are described elsewhere (25). They will not be repeated here except to state that the calculations included the energies of nonbonded and electrostatic interactions (2) as well as the energies of the hydrogen bond (30) and that in the unit cell of lowest conformational energy none of the atoms was more than 0.5 Å away from its experimentally determined location. Similar results have been previously reported by other laboratories (4, 31, 32).

Although our calculated structure (25) differs very little from the one observed in the crystal, we believe that this agreement is still not good enough for such a simple structure. At this stage, it is not obvious what additional

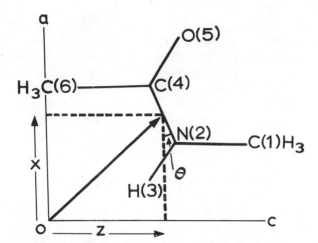

Figure 7. Definitions of the variable parameters x, z, and θ for the location and orientation of the molecule N-methylacetamide in its crystal structure.

refinements should be made without knowing more about the precise nature of the interatomic forces.

As in the case of the benzene crystal, different potential functions were found to yield similar conformational maps for the dipeptide. Moreover, the regions of low energy, in fact, could have been predicted merely by examining the inter-atomic contact distances, as is done in the so-called "contact maps." It seems, therefore, that for the particular structures considered, the precision of the potential functions used in the calculations is not too significant. This situation has both advantages and disadvantages. It means, for instance, that to a first approximation we could predict allowed protein or peptide structures reasonably well without being terribly concerned about the exact values of the parameters used in the potential functions. This approach, however, gives only what might be called "allowed conformations" and cannot be expected to yield the minimum-energy conformation with great accuracy. As a matter of fact, in the case of the hairpin bends, the use of nonappropriate potentials led to a lowest-energy conformation in which some of the dihedral angles were some 30° different from the values observed experimentally. We now believe that in view of the good agreement between the experimental data on the one hand and the values obtained using the softened potential functions and the high-barrier ψ-potential function on the other hand, there is good justification for the use of these new functions in calculations of protein and peptide energies. The precise way in which the softening is to be accomplished, however, requires further investigation.

TEST OF POTENTIAL FUNCTIONS ON DIMENSIONS OF POLYMERIC CHAINS

To determine the validity of the different calculation approaches, we felt it necessary to use the information provided by a variety of data, especially those derived from experiments of different types. Thus, the validity of our potential functions was tested by determining not only crystal structures, but also the dimensions of polymeric chains (33–38). The end-to-end distance of random-coil polypeptides was calculated using various refinements of our original potential functions (2). The results, expressed in terms of characteristic ratios, are summarized in Table 3. It can be seen that, unlike the crystal structure of benzene and the conformations of dipeptides, the characteristic ratio is significantly affected by the choice of potential functions. This type of analysis, therefore, is most useful for future developments of appropriate calculation procedures.

POSSIBLE NONPLANAR DISTORTIONS OF THE PEPTIDE UNIT

So far, we have considered only interactions between atoms that are not covalently bonded to one another, but our studies also included the peptide unit, in which the C—N bond has a partial double bond character. It is generally believed that the most stable conformation of the peptide unit is the planar one, in which all six atoms shown in Fig. 8(a) lie on a plane, and that the deviations of the atoms from this plane are relatively small. The possibility of the peptide unit assuming a nonplanar structure was recognized (2) as early as 1968, and the very occurrence of nonplanar peptide bonds in model amides (39–42) and proteins (43) has been reported. More recently, Winkler and Dunitz (44) have analyzed such distortions and indicated their importance. The torsion about the peptide C—N bond (defined as the dihedral angle ω) has also been investigated in other laboratories (including the author's previous laboratory in Madras) both theoretically, using a quantum mechanical approach (45, 46), and experimentally, by

Table 3. Calculated values of the characteristic ratio, using different types of potential functions

Nature of Potential Function	Characteristic Ratio
CALCULATED VALUES	
1. Usual potential functions (2, 33)	10
2. Twofold ψ potential ($\tau = 110°$) (34)	18
3. Twofold ψ potential and softening ($\tau = 110°$) (34)	14
4. Same as row 3, but with $\tau = 112°$ (34)	11
5. Same as row 4, but with barrier for $V(\psi) = 2$ kcal/mole (34)	8
EXPERIMENTAL DATA	
6. Range from several reported studies (35–38)	7.5–9.5

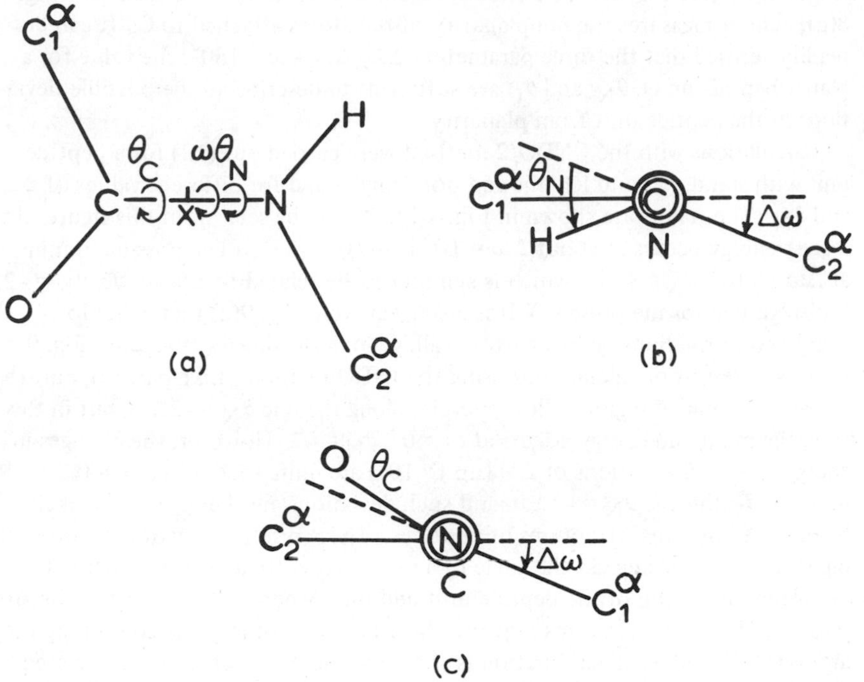

Figure 8. (a) Definition of the dihedral angles ω, θ_N, and θ_C in a peptide unit. (b) Definition of the angles $\Delta\omega$ and θ_N in the Newman projection down the bond direction NC. (c) Definition of the angles $\Delta\omega$ and θ_C in the Newman projection down the bond CN.

means of NMR (47) and IR spectroscopy (48). Only recently, however, have the precise nature of the distortions and the relative ease of their occurrence received the appropriate attention. The possibility that the three bonds meeting at the nitrogen atom might not all be coplanar was forcibly pointed out to the author by R.S. Mulliken in 1971. Mulliken suggested that the typical pyramidal character of the three bonds meeting at the nitrogen (as in the ammonia molecule) could not be completely obliterated by the N—C bond having a partial double bond character. During the last two years we have studied this possibility (12-14), and our results are summarized here.

In order to satisfactorily describe the phenomenon of nonplanarity, a more general definition was devised. Thus, the peptide unit is described (see Fig. 8) as consisting of two segments joined, say, at point X, which is the midpoint of the C—N bond [Fig. 8(a)]. The atoms C_1^α and O and the CX bond would be coplanar with C, but the atoms C_2^α and H and the NX bond would be coplanar with N, the two planes being twisted relative to one another by the angle ω about the C—N bond. Moreover, if the three bonds meeting at nitrogen were pyramidal, the resulting conformation could be described by the dihedral angle θ_N between the two planes defined by the atoms C, N, C_2^α and C, N, H, respectively [see Fig. 8(b)]. Similarly, a dihedral angle θ_C is defined [see Fig. 8(c)], and it measures the nonplanarity of the atoms attached to C. It can be readily verified that the three parameters $\Delta\omega$ ($\Delta\omega = \omega - 180°$, the value for a planar peptide unit), θ_N, and θ_C are sufficient to describe all the possible deviations of the peptide unit from planarity.

Calculations with the CNDO/2 method were carried out (12) for a peptide unit with standard bond lengths and bond angles and for different values of $\Delta\omega$ and θ_N. The results are shown in Fig. 9(a). As can be seen from this figure, the lowest energy occurs at about $\Delta\omega = 10°$ and $\theta_N = -25°$. (The possible minimum at $\Delta\omega = -10°$, $\theta_N = +25°$, which is symmetrically related to $\Delta\omega = 10°$, $\theta_N = -25°$, is disregarded for the present.) It is also clear from Fig. 9(a) that other low-energy conformations lie in a narrow valley along the line $\theta_N = -2\Delta\omega$. Fig. 9(b) shows the results of calculations using the INDO method. Like part (a), part (b) shows an extended region of low energies along the line $\theta_N = -2\Delta\omega$, but in this case, the minimum energy occurs at $\omega = 0°$, $\theta_N = 0°$. However, the changes in energy for small variations of $\Delta\omega$ (up to $10°$) are quite small (less than 0.5 kcal/mole) along the line $\theta_N = -2\Delta\omega$, and such deviations are therefore very likely to be observed in actual structures of peptides. These results bring out the interesting fact that the change in the angle ω alone is not sufficient to represent the nonplanar distortion of the peptide unit and that, when such a distortion occurs, both the H and the C_2^α atoms move to the same side of the plane containing the atoms C_1^α, C, and N. This situation is shown in Fig. 8(b), where it can be seen that the magnitude of θ_N is approximately double the value of $\Delta\omega$ and its sense is opposite to that of $\Delta\omega$. In other words, *the greater the value of $\Delta\omega$, the greater the pyramidal character of the bonds meeting at the nitrogen atom.* We consider this to be the most significant result of these quantum mechanical calculations.

Similar calculations made for the nonplanar distortions occurring at the carbonyl carbon and defined by the angles $\Delta\omega$ and θ_C revealed very large positive changes in energy even for small values of θ_C. Thus, very little nonplanarity of the three bonds meeting at C can actually occur. The conclusion is that an alteration in the planarity of the peptide bond is associated essentially with changes in the dihedral angles θ_N and $\Delta\omega$. These changes are related to one another ($\theta_N = -2\Delta\omega$) and are as great as $15°$ for the former and $25°$ for the latter. The correlated variation between θ_N and $\Delta\omega$, predicted from theory, has been confirmed by examining the crystal structures of small peptides (14, 49). The results of observations made in about 30 cases are plotted in Fig. 10(a). Note that the data all lie in the low-energy regions of the energy maps in Fig. 9. A similar analysis was carried out for the correlation of θ_C and ω. The data, which are shown in Fig. 10(b), indicate that, for variations of ω of up to $15°$, the variations in θ_C are small ($\pm 5°$) and randomly distributed, as predicted by the theory (see above).

From the population distribution of the experimental data [Fig. 10(a)] it is possible to deduce the variation of energy with $\Delta\omega$ (by using the standard Boltzmann distribution law). This variation is shown in Fig. 11, along with the variation of energy along the line $\theta_N = -2\Delta\omega$, as deduced from the INDO calculations. The very close correspondence between the two is extremely striking and

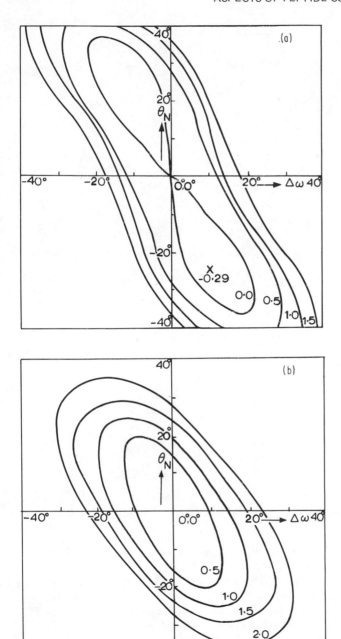

Figure 9. (a) Energy contours (in kcal/mole) for *N*-methylacetamide, calculated by the CNDO/2 method. The conformation marked X has the lowest energy. [Redrawn from data in (12)]. (b) Similar to (a), but obtained with the INDO method.

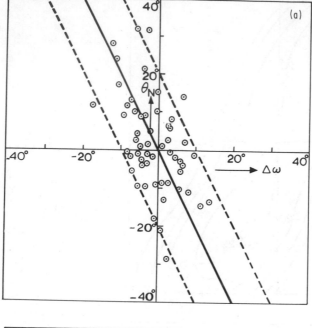

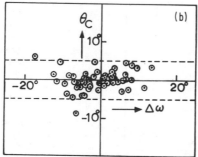

Figure 10. (a) The θ_N-$\Delta\omega$ relationship experimentally observed in crystals of simple peptides and related compounds (open circles). [Data are reproduced from (14)]. Note that all the data lie along the line $\theta_N = -2\Delta\omega$. (b) The θ_C-$\Delta\omega$ relationship experimentally observed in crystals of simple peptides and related compounds (open circles). [Data are reproduced from (14)]. Note that all the data lie close to the $\Delta\omega$ axis and are much less scattered than in (a).

provides perhaps the best evidence for the essential validity of the theoretical deductions obtained from quantum chemistry. Similar calculations by the INDO method have also been performed (14) for formamide. The energy dependence on ω calculated along the line $\theta_N = -2\Delta\omega$ is shown in Fig. 12, together with the experimental results obtained (50) by microwave spectroscopy. Here again, the theoretical and the experimental results are similar,

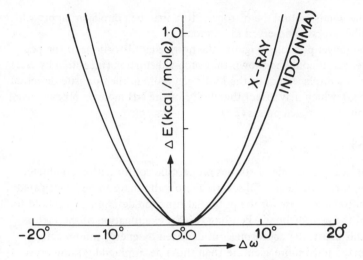

Figure 11. Variation of the energy of *N*-methylacetamide with $\Delta\omega$ along the line $\theta_N = -2\Delta\omega$, calculated by the INDO method. The experimental curve deduced from observed data in crystal structures is shown for comparison.

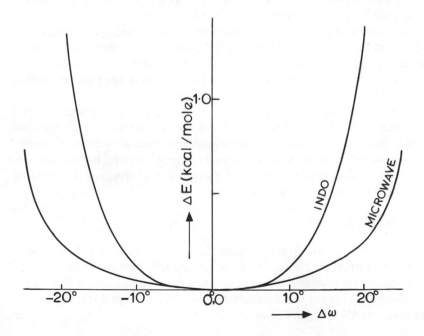

Figure 12. Similar to Fig. 11, for formamide. The two curves were obtained from theory (INDO) and from experiment (microwave studies).

showing that the bonds around the nitrogen atom may be nonplanar, approaching the pyramidal structure observed in ammonia.

Thus, it is no longer possible to ignore the nonplanar distortions of the peptide unit. Moreover, it could be shown that such a distortion (measured by $\Delta\omega$) is associated with a displacement of the hydrogen atom in the opposite direction (measured by θ_N), which may affect the ability of the NH moiety to be involved in the formation of hydrogen bonds (51).

CONCLUSIONS

Today, after nearly a decade of studies on peptide conformation conducted in various laboratories, there are still several factors that have to be investigated. This was shown to be the case for the potential function describing $V(\psi)$ and for the planarity of the peptide bond. Pending further examination of the factors mentioned in this paper, the conformations deduced by energy minimization cannot be expected to be more accurate than those determined by x-ray crystallography, as has sometimes been claimed (20).

Moreover, it is our belief that in conformational energy calculations minimization should be performed not only with respect to the parameters ϕ_j and ψ_j, but also with respect to two new parameters, namely, $\Delta\omega_j$ and θ_{Nj}. Devising the program for such computations does not involve any great difficulty, although considerable time might be required for the calculations. It is also necessary to have atomic models in which the nonplanar distortion of the peptide unit can be incorporated. Such a device might easily be incorporated in the Kendrew models, although its introduction into the space-filling models would probably be more difficult.

In the future, simple crystal structures will continue to play an important role in establishing the validity of potential functions, because the number of variables involved can be reduced and also because such structures are well specified and can be accurately determined experimentally. Therefore, we advocate that more studies like the ones described herein be pursued if refinements of the potential functions now in use are to be achieved.

ACKNOWLEDGMENTS

The author would like to acknowledge the great assistance of A.S. Kolaskar in preparing this paper. He also thanks V.S.R. Rao, C. Ramakrishnan, and K. Venkatesan for providing some unpublished data for inclusion in this paper. He is also grateful to the U.S. Public Health Service for Grant AM15964 in Bangalore and AM11493 in Chicago.

REFERENCES

1. G.N. Ramachandran, in *Conformation of Biological Molecules and Polymers,* E.D. Bergman and B. Pullman, eds., Israel Academy of Science and Humanities, Jerusalem, 1973.
2. G.N. Ramachandran and V. Sasisekharan, *Advan. Protein Chem.,* **23**, 283 (1968).
3. D.E. Williams, *J. Chem. Phys.,* **45**, 3770 (1966).
4. H.A. Scheraga, *Advan. Phys. Org. Chem.,* **6**, 103 (1968).
5. F.A. Momany, G. Vanderkooi, and H.A. Scheraga, *Proc. Nat. Acad. Sci. U.S.,* **61**, 429 (1968).
6. A.I. Kitaigorodsky, in *Advances in Structure Research by Diffraction Methods,* Vol. 3, R. Brill and R. Mason, eds., Pergamon Press, Oxford, 1970, p. 173.
7. D.A. Brant and P.J. Flory, *J. Am. Chem. Soc.,* **87**, 2791 (1965).
8. B. Maigret, B. Pullman, and N. Dreyfus, *J. Theor. Biol.,* **26**, 321 (1970).
9. F.M. Pohl, *Nature New Biol.,* **234**, 277 (1971).
10. F. Jordan, *J. Theor. Biol.,* **41**, 375 (1973).
11. R.A. Scott and H.A. Scheraga, *J. Chem. Phys.,* **45**, 2091 (1966).
12. G.N. Ramachandran, A.V. Lakshminarayanan, and A.S. Kolaskar, *Biochim. Biophys. Acta,* **303**, 8 (1973).
13. G.N. Ramachandran and A.S. Kolaskar, *Biochim. Biophys. Acta,* **303**, 385 (1973).
14. A.S. Kolaskar, A.V. Lakshminarayanan, K.P. Sarathy, and G.N. Ramachandran, in preparation.
15. T. Shimanouchi, *Discuss. Far. Soc.,* **49**, 60 (1970).
16. A.V. Lakshminarayanan, V. Sasisekharan, and G.N. Ramachandran, in *Conformation of Biopolymers,* Vol. 1, G.N. Ramachandran, ed., Academic Press, New York, 1967, p. 61.
17. V. Sasisekharan, in *Conformation of Biological Molecules and Polymers,* E.D. Bergman and B. Pullman, eds., Israel Academy of Sciences and Humanities, Jerusalem, 1973.
18. A.S. Kolaskar, K.P. Sarathy, V. Sasisekharan, and G.N. Ramachandran, in preparation.
19. J.A. Pople and L. Radom, in *Conformation of Biological Molecules and Polymers,* E.D. Bergman and B. Pullman, eds., Israel Academy of Sciences and Humanities, Jerusalem, 1973.
20. P.K. Ponnuswamy, R.F. McGuire, and H.A. Scheraga, *Int. J. Peptide and Protein Res.,* **5**, 73 (1973).
21. R. Chandrasekaran, A.V. Lakshminarayanan, U.V. Pandya, and G.N. Ramachandran, *Biochim. Biophys. Acta,* **303**, 14 (1973).
22. A.S. Kolaskar, personal communication.
23. C. Ramakrishnan and G. Manjula, personal communication.
24. G.N. Ramachandran, K.P. Sarathy, and A.S. Kolaskar, *Z. Kristallogr.,* **138**, 299 (1973).
25. G.N. Ramachandran, K.P. Sarathy, and A.S. Kolaskar, *Z. Natureforschung,* **28**, 643 (1973).
26. E.G. Cox, D.W.J. Cruickshank, and J.A.S. Smith, *Proc. Roy. Soc. (London),* **A247**, 1 (1958).
27. G.E. Bacon, N.A. Curry, and S.A. Wilson, *Proc. Roy. Soc. (London),* **A279**, 98 (1964).

28. A.S. Kolaskar, in preparation.
29. G.J. Piermarini, A.D. Mighell, C.E. Weir, and S. Block, *Science,* **165,** 1250 (1969).
30. R. Balasubramanian, R. Chidambaram, and G.N. Ramachandran, *Biochim. Biophys. Acta,* **221,** 196 (1970).
31. M. Dentini, P. De Santis, S. Morosetti, and P. Piantanida, *Z. Kristallogr.,* **136,** 305 (1972).
32. R.F. McGuire, F.A. Momany, and H.A. Scheraga, *J. Phys. Chem.,* **76,** 375 (1972).
33. V.S.R. Rao and A.R. Srinivasan, in preparation.
34. V.S.R. Rao and A.R. Srinivasan, personal communication.
35. D.A. Brant and P.J. Flory, *J. Am. Chem. Soc.,* **87,** 2788 (1965).
36. H. Fujita, A. Teramoto, T. Yamashita, K. Okita, and S. Ikeda, *Biopolymers,* **4,** 781 (1966).
37. M. Terbojevich, E. Peggion, A. Cosani, G. D'Este, and E. Scoffone, *Eur. Polym. J.,* **3,** 681 (1967).
38. W.L. Mattice and Jung-TehLo, *Macromolecules,* **5,** 734 (1972).
39. E. Benedetti, P. Corradini, M. Goodman, and C. Pedone, *Proc. Nat. Acad. Sci. U.S.,* **62,** 650 (1969).
40. P.R. Andrews, *Biopolymers,* **10,** 2253 (1971).
41. C. Pedone, E. Benedetti, A. Immirzi, and G. Allegra, *J. Am. Chem. Soc.,* **92,** 3549 (1970).
42. P. Ganis, G. Avitabile, S. Migdal, and M. Goodman, *J. Am. Chem. Soc.,* **93,** 3328 (1971).
43. M. Levitt and S. Lifson, *J. Mol. Biol.,* **46,** 269 (1969).
44. F.K. Winkler and J.D. Dunitz, *J. Mol. Biol.,* **59,** 169 (1971).
45. A.S.N. Murthy, K. Guradath Rao, and C.N.R. Rao, *J. Am. Chem. Soc.,* **92,** 3544 (1970).
46. M. Pericaudet and A. Pullman, *Int. J. Peptide and Protein Res.,* **5,** 99 (1973).
47. L. Isbrandt, W.C.T. Tung, and M.T. Rogers, *J. Mag. Resonance,* **9,** 461 (1973).
48. J. Smolikova, A. Vitek, and K. Blaha, *Collection Czechoslov. Chem. Commun.,* **38,** 548 (1973).
49. K. Venkatesan, personal communication.
50. E. Hirota, R. Sugisaki, C.J. Nielsen, and G.O. Sorensen, *J. Mol. Struct.,* in press.
51. S. Arnott and A.J. Wonacott, *J. Mol. Biol.,* **21,** 371 (1966).

THE AMIDE HYDROGEN BOND IN ENERGY FUNCTIONS FOR PEPTIDES AND PROTEINS

A.T. HAGLER and S. LIFSON, Department of Chemical Physics, The Weizmann Institute of Science, Rehovot, Israel, and **E. HULER,** Soreq Nuclear Research Centre, Yavne, Israel

SYNOPSIS: The force field derived in previous papers (1, 2) to describe the intermolecular interactions in amide crystals is examined in terms of the geometric parameters that determine the NH$\cdots$OC hydrogen-bond interaction. The potential is seen to be fairly insensitive to the θ_{HOC} angle. The much stronger dependence on θ_{NHO} is attributed to the increased repulsion between the N and O atoms as θ_{HNO} increases. The geometric dependences of various hydrogen-bond potentials used in conformational analysis are analyzed and compared.

INTRODUCTION

In previous papers (1, 2) a force field for intermolecular interactions in amides was derived by a particular technique of least-squares optimization of energy parameters to fit observed crystal structures (3), together with heats of sublimation of amide crystals and dipole moments of some amides. This force field is intended to be part of a consistent force field for use on biological macromolecules as well as on other systems. The hydrogen bond was found to be adequately represented by partial charges placed on the atoms C, O, H_N, and N of the amide group and by nonbonded interactions of the Lennard–Jones type between C, O, and N. The least-squares analysis indicated that the van der Waals radius of the amide hydrogen, H_N, was small (less than 2.0 Å), and because the nitrogen diameter was ~4 Å, it could be neglected without affecting the goodness of fit. This situation was attributed to a significant withdrawal of electron density from the hydrogen by the highly electronegative nitrogen, and it is consistent with the results of calculations of the electron distribution of XH groups (4) showing that the "size" of the hydrogen is a function of the electronegativity of X. More recently, neutron diffraction studies (5) have also suggested that a hydrogen atom bonded to an electronegative atom has a van der Waals radius less than or equal to 2 Å. The small radius results in a short-range repulsion between the hydrogen and oxygen in the N–H$\cdots$O bond, allowing the oxygen to get close enough to the hydrogen to exert a large attractive (negative) electrostatic force. The importance of this phenomenon for the formation of the hydrogen bond has been shown by previous quantum mechanical calculations (6-8).

In this paper we shall describe the behavior of the N–H$\cdots$O hydrogen-bond energy as a function of the $H_N\cdots$O distance and of the $H_N\cdots$O=C and H–$H_N\cdots$O angles for the potential derived in (1). The energy dependence on geometry calculated from this potential is compared with the dependence of other hydrogen-bond functions commonly used in conformational analysis, as well as with some qualitative concepts of the hydrogen bond.

THE HYDROGEN BOND

The geometric relation between the NH and OC groups forming the hydrogen bond is determined by four degrees of freedom apart from the covalent bond lengths (N–H and O=C). These may be taken as the $H_N \cdots O$ distance, the supplementary N–$H_N \cdots O$ angle (θ_{NHO}), the supplementary $H_N \cdots O$=C angle (θ_{HOC}), and the rotation about the H$\cdots$O "bond." Thus, a full description of the dependence of the hydrogen-bond energy on its geometry requires a four-dimensional surface. In order to gain insight into some properties of this surface [calculated with the potential derived in (1)], we shall present here the dependence of the energy of these systems on some of the aforementioned variables, which are commonly thought to be important to the description of the hydrogen bond.

The O$\cdots$H distance. The total energy of the system NH$\cdots$OC, as well as the energies of its two components—the nonbonded (6-12 potential) and the electrostatic interactions—is plotted in Fig. 1 as a function of the H$\cdots$O distance. The values of θ_{NHO} and θ_{HOC} are fixed at 0° and 60°, respectively. This is the approximate conformation often encountered in amide crystals, in general, and in particular, in those containing centrosymmetric hydrogen-bonded rings. The minimum energy along this curve is -2.51 kcal and it occurs at an O$\cdots$H distance of 2.05 Å. The interaction at short distances (r_{OH} less than ~1.7 Å) is dominated by the repulsive part of the nonbonded potential. As r_{OH} increases beyond the minimum, the energy is determined basically by the electrostatic forces. It increases more gradually with distance than is assumed in current concepts of the hydrogen bond, and in this respect, our force field constitutes a departure from force fields commonly used in conformational calculations (9–11).

The distance dependence of the $NH_2 \cdots OC$ system in a planar conformation for the same values of θ_{NHO} and θ_{HOC} has also been considered. The general features are quite similar, but the energy minimum is somewhat less (-3.1 kcal) and r_{OH} at the minimum is somewhat shorter (2.01 Å).

As pointed out above, the special nature of the hydrogen bond has been attributed to the small radius of the hydrogen, that is, to the falling off of the exchange repulsion at small distances from the hydrogen (6-8), when it is bonded to an electronegative atom. As can be seen from Fig. 1, there is no significant nonbonded repulsion until the oxygen approaches within approximately 3 Å of the nitrogen (i.e., within 2 Å of the hydrogen). As can be seen from the curves, the attractive energy of the electrostatic interaction increases significantly at distances of r_{OH} = 2.5 Å and less. Thus, the hydrogen bond is formed by the combined effect of the short contact distance between H_N and O, and the appreciable electrostatic energy of attraction at this distance. The picture would be similar if charge transfer were included as part of the hydrogen-bond

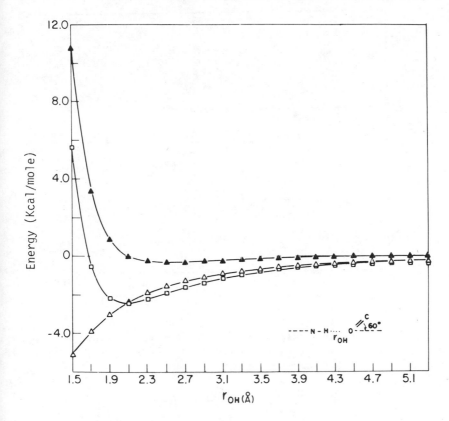

Figure 1. The interaction energy of N–H and O=C groups as a function of r_{OH}, calculated with the 6–12 potential of (1): □ total energy; △ electrostatic contribution; ▲ nonbonded contribution.

interaction. The strength of the charge-transfer interaction falls off rapidly as a function of O···H distance (12–14) and, thus, can become significant only if the oxygen and hydrogen atoms are allowed to come rather close to each other.

The H···O=C angle. The angular dependence of the hydrogen-bond energy is a subject of controversy (15–17), but there is a general consensus that the angular dependence on θ_{HOC} is weak, if it exists at all (10, 15, 18). In Fig. 2 we have plotted the total energy of the N–H···O=C system, as well as the non-bonded and electrostatic contributions, as a function of θ_{HOC} with $r_{OH} = 2.1$ Å and $\theta_{NHO} = 0$. The energy is indeed independent of θ_{HOC} within 60° to 70°. As θ_{HOC} increases beyond this point, the carbonyl carbon comes into repulsive contact with the nitrogen. Even at 90°, the energy is approximately –1 kcal/mole, compared with approximately –3 kcal/mole at the minimum. This minimum corresponds to a linear arrangement of the N=H···O=C atoms. The

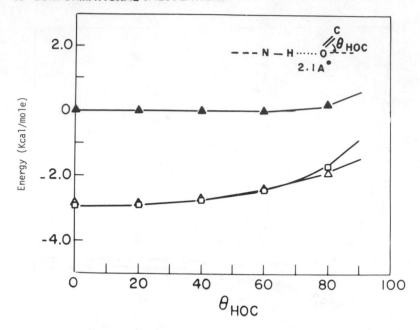

Figure 2. The interaction energy of N–H and C=O groups as a function of θ_{HOC} for $r_{O\cdots H}$ = 2.1 Å and θ_{NHO} = 0° [potential 6–12 of (1)]: □ total energy; △ electrostatic contribution; ▲ nonbonded contribution.

geometry at the minimum has also been a subject of controversy, and both functions in which the minimum occurs at θ_{HOC} = 0° and θ_{HOC} = 60° have been used in conformational analysis (9, 10, 19). The latter corresponds to a geometry in which the N–H is linear with one of the oxygen lone-pair orbitals. Although the crystal data fit well (1, 2) with the geometric dependence evident in Fig. 2, there are some indications that the lone pairs may have some importance (2), for example, in the case of formamide.

The N–H···O angle. The proponents of angular dependence of the hydrogen-bond energy believe the energy is very sensitive to the NHO angle (20). Brant (19), for example, considers the hydrogen bond to be completely broken if the NHO angle deviates from linearity by 40°.

Fig. 3 gives the energy of the NH···O=C system as a function of θ_{NHO} for θ_{HOC} = 0 at r_{OH} = 2.1 Å. The energy is seen to be much more sensitive to the NH···O angle than it is to the H···O=C angle. In fact, it increases rapidly as θ_{NHO} increases beyond 20° to 30°. When θ_{NHO} equals 60°, the total energy is already slightly positive. The difference in the behavior effected by the two angles is easily explained in terms of Figs. 2 and 3. The N and O atoms are in repulsive contact even at θ_{NHO} = 0 and their repulsion increases as θ_{NHO} increases. On the other hand, the increase in energy with respect to the HOC angle involves mainly a decrease in favorable electrostatic energy, which varies less than the nonbonded repulsion. Only when θ_{HOC} is large (~70°) does the

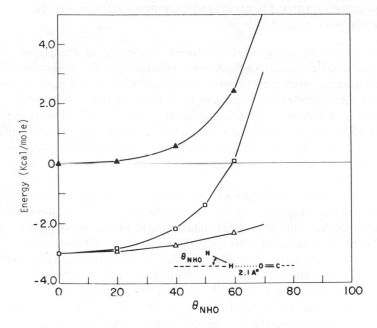

Figure 3. The interaction energy between the N–H and C=O groups as a function of θ_{NHO} for $r_{O\cdots H}$ = 2.1 Å and θ_{HOC} = 0° [potential 6–12 of (1)]: □ total energy; △ electrostatic contribution; ▲ nonbonded contribution.

carbon, which is initially more than 4 Å from the nitrogen, come into repulsive contact with it. Thus, only at this angle does the energy begin to increase sharply.

In conclusion, the general picture that emerges from this potential basically supports previous concepts of the sensitivity of electrostatic energy to these angles. *Rather than being due to some explicit angular dependence in a hydrogen-bond energy function, however, this angular dependence arises naturally from the nonbonded (short-range) repulsion and the other interactions discussed above.*

Many of the hydrogen-bond geometries observed in crystals have an HOC angle of approximately 120° (θ_{HOC} = 60°) (16, 17, 21), and as mentioned above, both 0° and 60° have often been taken as the location of the energy minimum in various hydrogen-bond potential functions. Generally, what has been said about the sensitivity of electrostatic energy with respect to the NHO angle applies in this case, but it is noteworthy that in this common situation in which θ_{HOC} is about 60°, the linear arrangement of NH$\cdots$O is no longer the most favorable one. Rather, the energy minimum occurs at θ_{NHO} = ~10°, mainly due to the electrostatic interaction between the nitrogen and the carbonyl carbon.

COMPARISON OF HYDROGEN-BOND ENERGY FUNCTIONS USED IN CONFORMATIONAL ANALYSIS

A primary purpose in calculating the conformational energy of a biological macromolecule is to determine its equilibrium conformation. The minimum-energy (equilibrium) conformation depends on the interplay of all the inter-actions within the molecule (viz., dispersion, short-range repulsion, electrostatic interaction, etc.). These interactions are particularly strong in the NH$\cdots$OC system, and therefore the hydrogen bond is one of the most important interactions in biological systems. Due to the uncertainty about its functional form, various hydrogen-bond potentials have been proposed to describe the NH$\cdots$O=C inter-action (16, 19, 22).

Because of the importance of the hydrogen-bond in determining peptide and protein conformations, it is of interest to examine the behavior of the proposed potentials as functions of the geometric parameters used to describe the hydrogen bond. Doing this often gives insight into the origin of characteristics of a conformation calculated with a particular potential function. It should be noted that for a meaningful comparison, the *total* potential for the NH$\cdots$OC inter-action (i.e., electrostatic, nonbonded, and where applicable, explicit hydrogen-bond functions) must be considered. In Fig. 4 the total energy of the N$-$H$\cdots$O=C interaction is plotted as a function of the O$\cdots$H distance (N$-$H$\cdots$O=C linear) for various potentials. As can be seen from this figure, there are wide variations among the characteristics predicted with these poten-tials. In order to compare the potentials in more detail we calculated some of their properties as functions of r_{OH} and θ_{NHO} (for both $\theta_{HOC} = 0°$ and $\theta_{HOC} = 60°$) and of θ_{HOC} and θ_{HNO} (see Table 1). The potentials are charac-terized for each variable by the position of the minimum, the energy at the mini-mum, and the second derivative at the minimum (which is a measure of the steepness). Another measure of the steepness is presented, namely, the value of the variable for which the energy is 50% smaller in absolute value (i.e., less nega-tive) than its value at the minimum.

The large variation among the different representations of the N$-$H$\cdots$O=C interaction is evident in Table 1. For example, as a function of r_{OH} for $\theta_{HOC} = 0°$, the minimum energy varies from -6.05 kcal/mole (Liquori *et al.*) to -2.14 kcal/mole (Popov *et al.*). The position of the minimum also varies from potential to potential, from $r_{OH} = 1.67$ Å (Brant) to $r_{OH} = 2.18$ Å (Popov *et al.*). Inter-estingly, the experimental O$\cdots$N distance in amide crystals varies from 2.85 Å to 3.00 Å (1, 16, 17) (corresponding to O$\cdots$H contact distances of approxi-mately 1.85 Å to 2.00 Å). Potential functions whose minimum is at r_{OH} con-siderably less than 1.85 Å probably would not be able to account for the observed contact distance in most amide crystals. The steepness of the poten-tials also varies significantly as measured by the second derivatives of the equi-librium distance, which vary from 14 kcal/Å^2 (Hagler *et al.*) to 150 kcal/Å^2 (Ramachandran *et al.*).

The minimum in θ_{HOC} also varies to some extent, as seen in section 3 of Table 1. The Brant, the modified Lippincott–Schroeder, and the Liquori poten-tials have explicit angular dependence; the angular dependence of the other

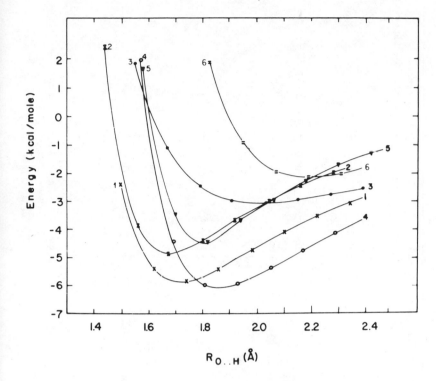

Figure 4. The total interaction energy of the N−H and C=O groups as a function of r_{OH} for the different potential functions used in conformational analysis: (1) Ooi et al.(24); (2) Brant (10); (3) Hagler et al. (1); (4) Liquori (27); (5) Ramachandran et al. (25, 26); (6) Popov et al. (23).

Table 1.   Some characteristics of the NH···OH interactions for different configurations[a]

Configuration		Hagler et al. (1)	Popov et al. (23)	Ooi et al. (24)	Brant (10)	Ramachandran et al. (25, 26)	Liquori (27)
1. N–H···O=C	r_{OH}(min)	2.01 Å	2.18	1.71	1.67	1.80	1.86
	$V_{min}(r_{OH})$	−3.05 kcal/mole	−2.14	−5.86	−4.85	−4.46	−6.05
	$V_{rr}(r_{OH})^c$	14 kcal/Å² mole	25	70	91	150	65
	r_{OH} at $V_{min}/2$	3.04 Å	2.72	2.38	2.16	2.18	2.54
2. N–H···O—C 60°	r_{OH}(min)	2.05 Å	2.18	1.63	1.69	1.80	1.99
	$V_{min}(r_{OH})$	−2.51 kcal/mole	−2.16	−7.27	−4.67	−4.36	−2.71
	$V_{rr}(r_{OH})^c$	13 kcal/Å² mole	24	117	85	147	24
	r_{OH} at $V_{min}/2$	3.02 Å	2.73	2.16	2.17	2.19	2.72
3.[b] N–H···O—C θ_{HOC}	θ_{HOC}(min)	0°	0.	54	0	0.	0
	$V_{min}(\theta_{HOC})$	−3.05 kcal/mole	−2.14	−7.03	−4.85	−4.46	−6.05
	$V_{\theta\theta}(\theta_{HOC})^d$	0.8 kcal/deg² mole	0.07	10	0.05	0.006	7.2
	θ_{HOC} at $V_{min}/2$	82°	99°	86°	85°	93°	54°
4.[b] θ_{NHO} N—H···O=C	θ_{NHO}(min)	0°	0	0	0	0	0
	$V_{min}(\theta_{NHO})$	−3.05 kcal/mole	−2.14	−5.86	−4.85	−4.46	−6.05

Configuration		Hagler et al. (1)	Popov et al. (23)	Ooi et al. (24)	Brant (10)	Ramachandran et al. (25, 26)	Liquori (27)
4.[b] cont.	$V_{\theta\theta}(\theta_{NHO})^d$	3.4 kcal/deg^2	0.1	4.8	22	1.85	8.2
	θ_{NHO} at $V_{min}/2$	43.5°	80	41.5	26.5	28	49
5.[b] θ_{NHO} N—H···O=C 60°	$\theta_{NHO}(min)$	9°	58°	27°	0	0	32°
	$V_{min}(\theta_{NHO})$	−2.55	−2.44	−7.44	−4.67	−4.36	−3.27
	$V_{\theta\theta}(\theta_{NHO})^d$	4.4 kcal/deg^2	2.3	10	24	3	7.8
	θ_{NHO} at $V_{min}/2$	43°	90	52	25	28	61.5
6.[b] H / N—θHNO ···O=C	$\theta_{HNO}(min)$	0°	0	0	0	0	0
	$V_{min}(\theta_{HNO})$	−3.05 kcal/mole	−2.14	−5.86	−4.85	−4.46	−6.05
	$V_{\theta\theta}(\theta_{HNO})^d$	9 kcal/deg^2	0.5	14	58.3	4	8.1
	θ_{HNO} at $V_{min}/2$	40°	50.5	47	18	22	44

[a]The four rows for each configuration represent, respectively, the value of the geometric variable for which the energy is minimum in the given configuration, the minimum energy, the partial second derivative with respect to the variable, the value of the variable at $V = V_{min}/2$.

[b]In sections 3, 4 and 5, r_{OH} for every potential was set at $r_{OH}(min)$ of section 1. In section 6, r_{ON} was set at $r_{OH}(min) + 1$ Å.

[c]V_{rr} is an abbreviation for $\partial^2 V/\partial r^2$.

[d]$V_{\theta\theta}$ is an abbreviation for $\partial^2 V/\partial\theta^2$.

potentials arises implicitly from atomic interactions. The second derivatives show that the potentials are fairly insensitive to this angle, in agreement with previously held qualitative concepts (10, 15, 18), with the exception of the Liquori and the modified Lippincott–Schroeder potentials. In fact, Dentini *et al.* (28) found that the Liquori potential does not account satisfactorily for the crystal structure of *N*-methylacetamide because of the steepness in the functional dependence on θ_{HOC}. After minimization, angle $\theta_{HOC} = 0°$ (as opposed to 60° observed in the crystal), and the potential has to be modified by putting the dipoles on the lone-pair orbitals of oxygen (28).

The dependence on θ_{NHO} is generally thought to be greater than the HOC angular dependence (19, 20), as noted above. The difference is reflected in the data in the two last rows of sections 4 and 5 of Table 1. As mentioned, this feature is not necessarily due to an explicit angular dependence, as in the cases of the potentials of Brant (10), Ramachandran *et al.* (25, 26) and Liquori (27), but arises naturally from electrostatic and nonbonded interactions. It is interesting, in this regard, that in most potentials, θ_{NHO} at the minimum is not 0° when θ_{HOC} is 60°.

Hitherto we have considered the system N–H$\cdots$O=C in isolation and found that the force fields used to represent the energy of the system as a function of its geometry differ greatly. The system does not exist in isolation, however, so it is of interest to compare the various potentials in terms of real molecules. Here we shall present the calculation of some properties of formamide, oxamide, and urea. These molecules were chosen because they are among the smallest molecules that include the amide group and therefore the contribution of other chemical groups to the intermolecular energy is minimal.

The properties calculated are the energy and structure of the hydrogen-bonded cyclic dimers of oxamide and formamide (see Fig. 5) and the lattice energies of the experimentally determined crystal structures of all three molecules. Table 2 shows the equilibrium distance $r_{OH}(\min)$ and the equilibrium energy of the cyclic dimers calculated for the various potentials. A comparison of Table 2 and the first two sections of Table 1 reveals an interesting trend. All potentials except that of Liquori (27) give a somewhat smaller $r_{OH}(\min)$ in the ring, compared with the isolated N–H$\cdots$O=C group, due to the compression effect of intermolecular interactions. Such a compression is known to occur in the crystal phases and causes the contact distances to be smaller than the distance at the minimum of the nonbonded-interaction pair potential (29, 30, 1). In the dipole-dipole interaction potential of Liquori, the compression effect is counteracted by the unfavorable orientation of the dipoles in the ring.

It would be interesting to compare the calculated properties of the ring pairs with gas-phase experiments, but these figures are not available. It may be assumed, however, that the $r_{OH}(\min)$ for the hydrogen-bonded dimers and for the hydrogen-bonded rings in the corresponding crystals are quite similar, because the compression effect is largely due to nearest-neighbor intermolecular interactions. This assumption is supported by our results for the dimers and for the corresponding crystals, as may be seen by comparing Table 1 (section 2) and

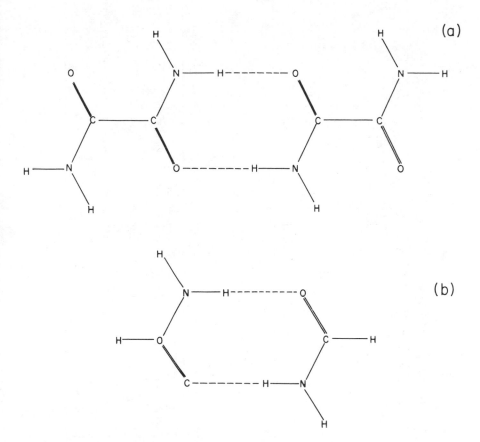

Figure 5. Hydrogen-bonded rings of (a) oxamide and (b) formamide.

Table 2 of this paper with Table 3 in (2) and the corresponding experimental
data. The experimental r_{OH} values are 1.95 Å in both oxamide and formamide.
According to our potential, the difference between the experimental r_{OH} and
the calculated r_{OH}(min), in the dimers as well as in the crystals, is less than
0.03 Å; the difference between r_{OH}(min) of the isolated NH$\cdots$OC group and
the dimers is 0.12 Å. In all other potentials, the compression effect is smaller,
and the calculated r_{OH}(min) for the oxamide and formamide crystals is expected
to be very close to the r_{OH}(min) of the dimer rings given in Table 2. Therefore,
the r_{OH}(min) in Table 2 may be substituted for the r_{OH}(min) of the crystals
that would be predicted by the various potentials.

Table 3 shows the lattice energies of oxamide, formamide, and urea, calculated with the various potentials. The last column gives the experimental lattice
energies obtained from measured heats of sublimation, partly corrected for the
vibrational effect in the crystal phase and rotations, translations, and pressure

Table 2.   Equilibrium O···H distances r_{OH}(min) and energies of formation V(min) of hydrogen-bonded dimer rings of oxamide and formamide

		Hagler et al. (1)	Popov et al. (23)	Ooi et al. (24)	Brant (10)	Ramachandran et al. (25, 26)	Liquori (27)
Oxamide	r_{OH}(min)	1.93	2.15	1.61	1.67	1.78	2.02
	V(min)	−9.43	−6.15	−19.32	−11.82	−18.41	−4.21
Formamide	r_{OH}(min)	1.93	2.15	1.62	1.67	1.79	2.01
	V(min)	−10.7	−6.28	−20.72	−11.90	−18.52	−4.42

Table 3.   Lattice energies (in kcal/mole) for experimental crystal structures calculated with various potentials and derived from heats of sublimation

	Hagler et al. (1)	Popov et al. (23)	Ooi et al. (24)	Brant (10)	Ramachandran et al. (25, 26)	Liquori (27)	Experimental Value
Oxamide	−25.6	−22.76	−25.1	−31.3	−45.83	−20.53	−28.2 (33)
Formamide	−16.5	−10.59	−15.2	−17.2	−23.84	− 9.41	−17.5 (34)
Urea	−22.2	−18.24	−20.4	−22.2	−31.2	−14.46	−22.2 (35)

in the gas phase (1). The inaccuracy of the experimental lattice energy is estimated to be 1 to 2 kcal/mole, mainly due to the rough estimate of the contributions of lattice vibrations. The energies were calculated for the experimental crystal structure and not for the calculated minimum-energy crystal structure. The lattice energies of the calculated equilibrium structures are thus lower than the corresponding values in Table 3, to the extent that the calculated structures deviate from the experimental ones.

Tables 2 and 3 together serve as some indication of the value of the various potentials in reproducing the observed properties of amides, particularly in the crystal phase. It should be pointed out, however, that our potential has been derived by optimizing its parameters with respect to crystal data, including those calculated here. Agreement with these data, therefore, is expected to be better in our case than in other cases, *ceteris paribus*. On the other hand, experimental data on the heats of sublimation and structures of amide crystals provide a most reliable basis for choosing, optimizing, and testing empirical energy functions for the amide hydrogen bond, and that is why they were chosen as the basis for our studies of the amide hydrogen bond.

We hope that the detailed comparison of different potentials will stimulate an interest in reconciling current analytic descriptions of the hydrogen bond with each other and with experimental observations.

REFERENCES

1.   A.T. Hagler, E. Huler, and S. Lifson, *J. Am. Chem. Soc.,* in press.
2.   A.T. Hagler and S. Lifson, *J. Am. Chem. Soc.,* in press.
3.   A.T. Hagler and S. Lifson, *Acta Cryst.,* in press.
4.   R.F.W. Bader, I. Keavany, and P.E. Cade, *J. Chem. Phys.,* 47, 3381 (1967).
5.   M.S. Lehman, J.L. Verbist, W.C. Hamilton, and T.F. Koetzle, *J. Chem. Soc.* (Perkin II), 133 (1973). W.H. Baur, *Acta Cryst.,* B28, 1456 (1972).
6.   C.A. Coulson, *Research,* London, 10, 149 (1957).
7.   J.N. Murrel, *Chem. Brit.,* 5, 107 (1969).
8.   S. Bratoz, *Adv. Quant. Chem.,* 3, 209 (1967).
9.   H.A. Scheraga, *Adv. Phys. Org. Chem.,* 6, 103 (1968).
10.   D.A. Brant, *Macromolecules,* 1, 291 (1968).
11.   R.F. McGuire, F.A. Momany, and H.A. Scheraga, *J. Phys. Chem.,* 76, 375 (1972).
12.   P.A. Kollman and L.C. Allen, *Chem. Rev.,* 72, 283 (1972).
13.   C.A. Coulson and U. Danielson, *Arkiv. Physik,* 8, 245 (1954).
14.   M. Dreyfus and A. Pullman, *Theoret. Chim. Acta* (Berlin), 19, 20 (1970).
15.   J. Donohue, in *Structural Chemistry and Molecular Biology,* A. Rich and W. Davidson, eds., W.H. Freeman, San Francisco, 1968.
16.   G.C. Pimentel and A.L. McClellan, *The Hydrogen Bond,* W.H. Freeman, San Francisco, 1960.
17.   L. Leiserowitz and G.M.J. Schmidt, *J. Chem. Soc.,* (A), 2732 (1969).
18.   See, for example, (16), p. 267.

19.   D.A. Brant, *Ann. Rev. Biophys, & Bioeng.*, **1**, 369 (1972).
20.   See, for example, (16), p. 265, and (19).
21.   See also Table 1 of (1).
22.   A.T. Hagler and S. Lifson in *The Proteins,* submitted.
23.   E.M. Popov, V.G. Dashevskii, G.M. Lipkind, and S.F. Arkhipova, *Mol. Biol.*, **2**, 491 (1968).
24.   T. Ooi, R.A. Scott, G. Vanderkoi, and H.A. Scheraga, *J. Chem. Phys.*, **46**, 4410 (1967).
25.   R. Balasubramanian, R. Chidambaram and G.N. Ramachandran, *Biochim. Biophys. Acta,* **221**, 196 (1970).
26.   G.N. Ramachandran and V. Sasisekharan, *Adv. Prot. Chem.*, **23**, 283 (1968).
27.   A.M. Liquori in *Symmetry and Function of Biological Systems at the Macromolecular Level,* A. Engstrom and B. Strandbey, eds., Wiley, New York, 1969, p. 101.
28.   M. Dentini, P. De Santis, S. Morosetti and P. Piantanida, *Zeit. Kristall.*, **136**, 305 (1972).
29.   A.I. Kitaigorodsky, *Tetrahedron,* **14**, 230 (1961).
30.   D.A. Brant, W.G. Miller, and P.J. Flory, *J. Mol. Biol.*, **23**, 47 (1967).
31.   E.M. Ayerst and J.R.C. Duke, *Acta Cryst.*, **7**, 588 (1954).
32.   T. Ladell and B. Post, *Acta Cryst.*, **7**, 559 (1954).
33.   R.S. Bradley and T.G. Cleasby, *J. Chem. Soc.*, 1681 (1953). E.P. Egan, Jr., Z.T. Wakefield, and T.D. Farr, *J. Chem. Eng. Data,* **10**, 138 (1965).
34.   G. Somen and J. Coops, *Rec. Trav. Chim.*, **84**, 985 (1968). A. Bander and H. Günthord, *Helv. Chim. Acta,* **41**, 670 (1958).
35.   K. Suzuki, S. Oniski, T. Koide, and S. Seki, *Bull. Chem. Soc. Japan,* **29**, 127 (1956).

POLY(AMINO ACIDS), INTERATOMIC ENERGIES, AND PROTEIN FOLDING

HAROLD A. SCHERAGA, Biophysics Department, Weizmann Institute of
Science, Rehovot, Israel

SYNOPSIS: There are two aspects to the computations of the conformations of
polypeptides and proteins from a knowledge of interatomic interactions. First,
it is necessary to develop a framework of procedures (including generation of
chain conformations, energy minimization, etc.) and, second, numerical values
must be assigned to the parameters of the empirical potential energy functions
used in the computations. Progress is being made in both areas, using both small
molecules and macromolecules such as synthetic poly(amino acids) as models for
the dependence of conformation on interatomic interactions. The experience
gained from such model systems has been applied to the refinement of x-ray
structures of proteins, to the calculation of the structure of a protein from the
known structure of a homologous one, and to the calculation of the stable conformations of intermolecular complexes involving proteins. A remaining difficulty to be overcome, before the three-dimensional structure of a protein can be
predicted from its amino acid sequence, is the multiple-minimum problem,
which is being attacked with prediction algorithms based on short- and medium-
range interactions, followed by energy minimization.

INTRODUCTION

A combined theoretical and experimental approach is being used to determine
how interatomic interactions dictate the folding of proteins and their reactions
with other small and large molecules. The theoretical work involves the use of
empirical potential energy functions to describe the interatomic interactions (1).
The experimental studies have served two purposes. First, they have provided
the data (on gas-phase, solution, and crystal structures of model compounds)
from which the nature of these potential functions and their parameters have
been deduced. Second, solutions and crystals (in most cases, fibers and films) of
synthetic homopolymers and copolymers of amino acids have served as excellent
experimental systems (2, 3) for the study of different conformational forms
(e.g., α helix, extended structure, and random coil) and, at the same time, as
useful models for the application of phenomenological and molecular theories
of cooperative transitions from one conformational form to another; for example, synthetic poly(amino acids) and polypeptides have been used to test the
theoretical predictions of the conformational properties of noncyclic (4–7) and
cyclic (8–14) oligomeric peptides, the screw senses of a α-helical homopoly-
(amino acids) in solution (15) and in the solid state (16, 17), the phenomenological parameters σ and s (18–21) of the helix-coil transition (22–24) and corresponding parameters (25) for the helix-helix transition of poly(L-proline) (26),
the tendencies of amino acids to adopt helical, extended, chain-reversal and coil

conformations (27), and the properties of the random coil (28–35). These examples of recent work on synthetic homopolymers and copolymers of amino acids (which have served as useful models for experimental tests of potential functions that are then applied in studies of proteins) will be reviewed here; then the application of the information gained from model systems to the refinement of x-ray data on proteins, to the determination of the structures of homologous proteins, to the folding of proteins, and to their interactions with small molecules will be considered.

BASIS OF CALCULATIONS

The theoretical basis for carrying out conformational energy calculations has been reviewed elsewhere (1, 15, 36–44). Essentially, it is assumed that various portions of the conformational space of a polypeptide are sufficiently accessible from each other (although all of this space is not traversed) that the chain may fold into the native conformation of a protein by minimizing the free energy of the system. This hypothesis, implicit in earlier work of Anfinsen (45), has gained support from recent experiments of Hantgan *et al.* (46), who studied the kinetics of anaerobic reoxidation of reduced ribonuclease by a mixture of oxidized and reduced glutathione. The sulfhydryl groups were observed to disappear much more rapidly than activity resumed or various of the physical properties characteristic of the native molecule were regained. An analysis of the products formed at various times during the reoxidation process led to the conclusion (46) that incorrectly paired sulfhydryl species were first oxidized to form disulfide bonds and that these "wrong" bonds then underwent a glutathione-catalyzed rearrangement to those of the native protein. The attainment of the native structure from those species with incorrectly paired cysteine residues indicates that there are no insurmountable barriers on the conformational pathway leading from the open chain to the folded conformation. A model for the pathway of the thermal unfolding of ribonuclease *with its disulfide bonds intact* (which is presumably a pathway for *folding* because the process is reversible) has been proposed (47) as a result of observations on the sequence in which peptide bonds become susceptible to attack by various proteolytic enzymes during the thermal transition. Of course, this is not necessarily the same pathway followed in the formation of the native enzyme from the fully reduced form.

Thus, the hypothesis given is an appropriate basis for calculating the potential energy minimization of the polypeptide. In one approximation in such a calculation, the dihedral angles for rotation about single bonds are the independent variables for generating any arbitrary conformation of the polypeptide, and the potential energy of each conformation is computed to find the conformation of lowest potential energy (1, 15, 36); in some cases, entropy contributions are included to obtain an estimate of the free energy (15). Usually, the bond lengths, bond angles, and planarity of the amide group are fixed (and selected separately for *each* type of amino acid), but these can be varied by introducing appropriate force constants (11, 48). Starting from a given conformation, it is now possible to minimize the total potential energy, including *all* pairwise interatomic interactions, with respect to the dihedral angles, for a protein the size of ribonuclease or lysozyme in a reasonable amount of computer time (49). Such computations are being carried out for the refinement of x-ray structures of proteins and for the determination of the structures of homologous proteins (see later sections), but the multiple-minimum problem (to be discussed later) must be overcome before a protein can be folded into its native structure from an *arbitrary* starting conformation.

The empirical potential energy functions used for such computations are based on pairwise interatomic interactions. The parameters of these functions have been refined (50) by computing the lattice constants and intermolecular binding energies (and some intermolecular-force constants) for some of the crystals shown in Table 1; the functions were then similarly tested (50, 51) on the remaining crystals of Table 1. Approximate procedures for including the effects of hydration are also available (23, 26, 52). A summary of the geometrical parameters, partial atomic charges, nonbonded potentials, hydrogen-bond potentials, and intrinsic torsional potentials [based, in part, on the calculations of crystal properties (50, 51)] and the procedures for applying them to conformational energy calculations on polypeptides, have been presented by Momany *et al.* (53). In this paper, the bond lengths and bond angles are summarized

Table 1. Crystals used for calculations of packing configurations (50, 51)

Aliphatic Hydrocarbons	Amides and Amine
Pentane[a]	Formamide[a]
Hexane[a]	*N*-Methyl acetamide[a]
Octane	Oxamide[a]
	Succinamide
Aromatic Hydrocarbons	Adipamide[a]
Benzene[a]	Suberamide
Anthracene[a]	Methyl amine[a]
Heterocyclic Compounds	Amino Acids and Peptides
Pyrazine[a]	α-Glycine
Bipyridine[a]	β-Glycine
Phenazine	γ-Glycine
Thianthrene[a]	Glycylglycine
	N-Acetylglycine
Carboxylic Acids	L-Alanine
Formic[a]	L-Glutamine
Acetic[a]	L-Glutamic acid
Butyric	L-Aspartic acid
Oxalic	L-Threonine
Succinic[a]	L-Tyrosine ethyl ester
Suberic[a]	Acetyl proline amide
Sebacic	Glycyl-L-asparagine
	Diketopiperazine

[a] These crystals were used in the determination of energy parameters, which were then tested on the remaining crystals.

separately for each amino acid because they seem to be determined by *local* interactions (i.e., between a side chain and its own backbone). Thus, the initial selection of the proper geometry of each amino residue of a polypeptide chain, *may* make it possible to carry out reliable conformational energy calculations with rigid geometry. Other approaches to the formulation of the potential energy of a polypeptide chain are being explored (54).

APPLICATIONS TO SYNTHETIC POLY(AMINO ACIDS) AND POLYPEPTIDES

The procedures referred to at the beginning of the previous section have been tested and applied to a variety of synthetic oligomers, homopolymers, and co-polymers of amino acids. Some of these applications are summarized in this section.

N-acetyl N'-methyl amides of the naturally occurring amino acids. Conformational energy calculations were carried out on *N*-acetyl *N'*-methyl amides of the 20 naturally occurring amino acids (4). The librational entropy (computed on the assumption that the structure undergoes *small* fluctuations around each minimum-energy conformation) was included in the computations and was shown to influence the stability in numerous cases. The computed relative amounts of two dominant species, a five-membered and an equatorial seven-membered hydrogen-bonded ring, agree, in general, with values deduced from IR and NMR measurements on these compounds in nonpolar solvents.

Recent reconsideration of the theoretical results (4) and related experimental data has shown (6) that a third conformation, which has no internal hydrogen bond (see Fig. 1), can be assumed by amino acids other than glycine, especially by amino acids with bulky side chains. Heretofore, this so-called γ conformation and the equatorial seven-membered ring conformation were considered to be the same (4).

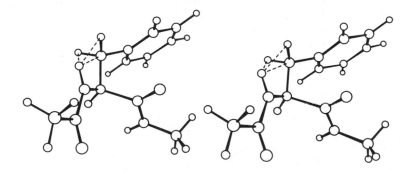

Figure 1. ORTEP stereo diagram of *N*-acetyl *N'*-methyl-L-phenylalanine amide in the γ conformation ($\phi = -60°$, $\psi = 140°$, $\chi_1 = 180°$, $\chi_2 = 80°$) (6). The dashed lines indicate nonbonded interactions that are thought to affect the N–H stretching frequency.

Noncyclic proline oligomers. The conformational preferences of oligomers of L-proline [di-, tri-, tetra-, and penta(L-proline)] were computed (7), allowing for two puckering conformations of the pyrrolidine ring for *cis-trans* isomerization in the peptide group and for rotation about the C_α–C' bond (55). Interactions over the *whole* molecule were taken into account, and the librational entropy was included. The calculated probabilities of occurrence of the *trans* conformation at the *i*th residue (in the *interior* of the chain) are in good agreement with experimental values of Deber *et al.* (56), provided allowances are made for end-group effects. The computed conformations are also in good agreement with those determined for proline trimers and tetramers by x-ray diffraction (57). These results, together with results discussed below provide a test of the geometrical and interaction parameters for the proline residue (e.g., for the different geometry of *cis*- and *trans*-proline).

N-acetyl N'-methyl amides of dipeptides. Calculations are being carried out for a variety of *N*-acetyl *N'*-methyl amides of dipeptides to determine the tendency for short segments of peptide chains to adopt chain-reversal conformations. Thus far, results are available only for Gly-Gly and L-Ala-L-Ala (58), although work is in progress (59) with other dipeptides.

For Gly-Gly and L-Ala-L-Ala, most combinations of single-residue energy minima correspond to local minima on the energy surface of the dipeptide, and the global minima of both dipeptides are simply the combinations of the global minima for each single residue (see Fig. 2). For minimum-energy bend conformations, however, there is a significant departure from additivity; that is, the stable conformations are not simply the pairwise combinations of local minima for single residues. The type-II bend (60, 61) for L-Ala-L-Ala (with a conformation (55) of $\phi_1 = -66°$, $\psi_1 = 110°$, $\phi_2 = 58°$, $\psi_2 = 44°$) (58) has the most favorable interactions between the two amino acid units and is a local minimum of the dipeptide energy surface that does not correspond to a combination of single-residue minima. Ralston and DeCoen (62) have also carried out computations on dipeptides.

N-acetyl N'-methyl amides of tetrapeptides. Chain-reversal conformations have also been examined in the *N*-acetyl *N'*-methyl amides of tetrapeptides. Lewis *et al.* (61) studied the chain reversals found in the native structures of eight proteins. Of the 135 bends located, over 40% did not possess a hydrogen bond between the C=O of residue *i* and the NH of residue *i* + 3. In addition, conformational energy calculations were carried out (61) on three tetrapeptides having amino acid sequences found to occur in the bends of the native structure of α-chymotrypsin and on tetra(L-alanine). The results for the chymotrypsin tetrapeptides indicate that the bends occur not only in the whole molecule, but also in the isolated tetrapeptide; in other words, the observed bends were the conformations of lowest energy (among the limited number of conformations examined) even in the tetrapeptides. Bend conformations of a tetra(L-alanine) sequence, which would not be expected (63) to form a bend, were not the

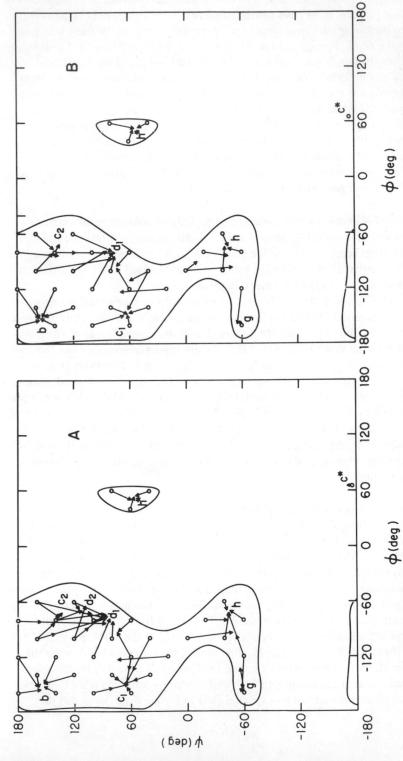

Figure 2. The (ϕ, ψ) space of the first (A) and second (B) single-residue units of the dipeptide L-Ala-L-Ala (58). The energy contour shown is 5 kcal/mole above the global minimum (at point d_1). The circles represent starting conformations for energy minimization. The lower-case letters indicate single-residue minima; however, c_2 and d_2 are not single-residue minima, but form dipeptide minima in combination with particular minima of the other residue. The arrows indicate the direction of movement after one cycle of energy

low-energy conformations after energy minimization (61). The stability of the bends, compared to the stability of bends of other structures, arises principally from side-chain-backbone interactions (e.g., a hydrogen bond between the side chain COO^- of Asp in position $i + 3$ and the backbone NH of the residue in position i) rather than from i to $i + 3$ backbone-backbone hydrogen bonds. This result is consistent with the observation (63) that residues with small polar side chains, such as Ser, Thr, Asp, and Asn, are found frequently in bends, possibly because these residues can interact most strongly with their immediate backbones. Alternatively, water may play a role in stabilizing bends among these polar residues, especially because such bends usually occur on the surface of a protein.

Experiments are in progress (64, 65) to detect bends in N-acetyl N'-methyl amides of tetrapeptides in water. The amino acid sequences used are taken from known bend conformations in α-chymotrypsin and variants thereof. For example, Asp-Lys-Thr-Gly (residues 35–38 of α-chymotrypsin) is in a bend conformation in the native protein. This tetrapeptide, as well as Asp-Lys-Gly-Thr, Lys-Thr-Gly-Asp, and Gly-Thr-Asp-Lys, was examined (64, 65) by NMR (nuclear Overhauser effect) and CD measurements and by conformational energy calculations. From the results of these experiments [interpreted with calculations of rotational strengths made by Woody (66)] and the conformational energy calculations, it appears that a significant fraction of the total population of conformations exhibits a chain reversal in the "native" sequence Asp-Lys-Thr-Gly [and in high-bend-probability (63) sequence variants thereof, viz., Asp-Lys-Gly-Thr and Lys-Thr-Gly-Asp], but not in a low-bend-probability (63) sequence variant (Gly-Thr-Asp-Lys) (65). In the conformational energy calculations (65), we minimized the energies of approximately 80 starting conformations per tetrapeptide (selected on the basis of the single-residue data described above). For the three high-bend-probability sequences, the minimum-energy conformations were bends that were several kcal/mole lower in energy than any other conformation considered. For the low-bend-probability sequence there were five low-energy conformations (including a bend), all within 3 kcal/mole of each other. Thus there seemed to be no strong preference for the latter sequence to adopt a bend conformation, but rather to distribute among at least five different conformations. Results like these, which indicate the possibility of bend formation in tetrapeptide sequences, are applicable in the formulation of algorithms for the folding of proteins.

Random coils. The conformational energy surface of a single residue has been used (28–35) to compute the unperturbed mean square end-to-end distance, $\langle r_0^2 \rangle$, of the random-coil form of a polyamino acid. The results are expressed by the dimensionless characteristic ratio C, which is defined as $\langle r_0^2 \rangle / N\ell^2$, where N is the number of virtual bonds of length ℓ in the polypeptide chain. The use of single-residue energies is justified by experimental evidence (67) that, in the unperturbed state, long-range effects do not contribute to the conformations of some poly(amino acids) [see also Hawkins and Holtzer (68)]. The characteristic ratio can be calculated either by averaging over all single-residue energies (28–30, 33) or by Monte Carlo procedures (31, 32, 34, 35). Thus, a value of ∼9 is obtained for C for a homopoly(amino acid) (but smaller, ∼2, for polyglycine); the value of ∼9 is in good agreement with experimental data on nonglycine homopolymers. In the case of poly(L-proline), the effect of a small number of *cis* peptide groups on the characteristic ratio of form II (*trans*) was investigated (35); the computed value of C, with ∼5% form I (*cis*) included, was in good agreement with the experimental value. The value of C for form II of poly(L-proline) can be reduced considerably by the presence of *cis* residues.

For polypeptide copolymers, C has been found to vary markedly with composition and amino acid sequence (30). For example, the random introduction of glycine residues into poly(L-alanine) led to a monotonic decrease in the value of C from ∼9 to ∼2 as the percentage of glycine increased. The end-to-end distance is also very sensitive to the degree of regularity of L and D residues in a copolymer of L- and D-alanine. Preliminary experimental results on copolymers of glycine and L-glutamic acid and on copolymers of L- and D-glutamic acid tend to confirm the main predictions of the calculations (30).

Cyclic peptides. In carrying out conformational energy calculations on cyclic peptides or on proteins with loops formed by disulfide bonds, it is possible to close such cyclic structures with the aid of a loop-closing potential (69). Alternatively, mathematical conditions can be applied to effect exact ring closure. Recently, procedures have been developed for generating closed rings [with (12) and without (12, 14) symmetry] in polypeptide chains of fixed bond lengths, bond angles, and planar peptide groups. In some calculations, variations in bond lengths, bond angles, and the planarity of the peptide group also have been taken into account (11, 48). These generation procedures, together with energy calculations, have been applied to $cyclo$(-Gly$_3$-Pro$_2$-) (8, 9, 11), cyclohexaglycyl (12), $cyclo$(-Pro-Ser-Gly-)$_2$ (10), $cyclo$(-Pro-Gly-)$_3$ (14), gramicidin S (70), and other cyclic peptides (10). In the case of $cyclo$(-Gly$_3$-Pro$_2$-), librational entropies also were included in the calculations (9).

A useful approach to conformational studies has been the combination of NMR data with conformational energy calculations (10, 12–14, 71–74). Both techniques have given complementary and mutually consistent structural information on many of the cyclic peptides cited above (10, 73, 74).

Helix sense and form. With empirical energy calculations, it is possible to predict the right- or left-handedness of the α-helical form of a large number of homopoly(amino acids) in solution and to identify the energetic factors that determine the helix sense (15, 75).

In addition, the helix sense in the crystal can be computed by taking into account intermolecular as well as intramolecular interactions. Using an earlier set of potential functions, we assessed (16) the energies stabilizing the β structure of poly(L-alanine) and the ω helix of poly(β-benzyl-L-aspartate) in fibers. With the newer functions (50, 51, 53), we extended these computations (17) to assess the relative stabilities of the α- and ω-helical forms of poly[β-(p-chlorobenzyl)-L-aspartate] in fibers. Intermolecular interactions contribute to the increased stability (compared to that of the isolated molecules) of these helices in crystalline fibers. The right-handed α-helical conformation is the most stable one at room temperature, both as an isolated molecule and in the crystal. As the temperature is raised, the entropy contribution from the rotational freedom about the respective helix axes decreases the differences in the free energy of the two helical forms (17). This phenomenon (together with a possible librational entropy contribution from the side chains) may account for the presence (76) of the ω-helical form in fibers at elevated temperatures.

Helix-coil transition. Phenomenological theories have been presented for cooperative conformational transitions in linear biopolymers (18–21). Molecular theories (in which the statistical weight of each ordered or disordered phase is computed from interatomic potentials) also have been developed for these phase transitions [e.g., order ⇌ disorder as in the helix-coil transition or order ⇌ order as in the poly(L-proline) form-I ⇌ form-II interconversion]. The ability to treat the properties of the random coil (in terms of its characteristic ratio or its statistical weight), the helix-coil transition discussed in this section, and the poly(L-proline) I-II interconversion disucssed below provide additional tests of the functions and procedures used in conformational energy calculations.

The Zimm–Bragg parameters σ and s (18) for the thermally induced helix-coil transition in water have been calculated for polyglycine (22, 23), poly(L-alanine) (22, 23), and (only s) for poly(L-valine) (24) from interatomic-interaction energies and the influence of the solvent. For example, in the case of poly(L-valine), three kinds of solvent effects were taken into account, namely, the influence of the solvent on the dielectric constant, the change in hydration (including hydrophobic bonding) around the side-chain methyl groups that accompanies the conformational change, and the binding of water molecules to the free NH and CO groups of residues in the coil state. The entropy loss from the restricted rotational freedom of the side chain in the α helix (in the presence and absence of water) and the binding of water molecules to the free NH and CO groups destabilize the α helix of poly(L-valine); however, the change in hydration around the side-chain methyl groups stabilizes the α helix over the random coil. The s-versus-temperature curve of poly(L-valine) in water [shown in Fig. 3, together with experimental data (77)] exhibits a maximum around 50°C due to the change in hydration around the side-chain methyl groups. Although they do not correspond perfectly, the theoretical curve follows the experimental data in predicting an inverse transition in which the helix becomes more stable as the temperature is raised ($s < 1$ at low temperatures and $s > 1$ at high temperatures).

Although σ and s have been computed from potential functions to test the theoretical treatment, as indicated above, their importance for predicting the likely location of α-helical segments in proteins is so great that an experimental approach is being taken to obtain these parameters for all of the naturally occurring amino acids. Studies are being carried out on the helix-coil transition in random copolymers of a helical host and a low-content guest residue, the latter

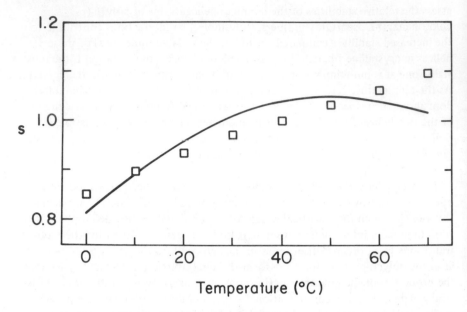

Figure 3. Computed *s*-versus-*T* curve for poly(L-valine) in water (24). The squares are the experimental results (77). [Reprinted with permission from Gō *et al.*, *Macromolecules*, 7 (1974). Copyright by the American Chemical Society.]

being the amino acid whose σ and *s* parameters are to be determined. With these parameters, it is possible to calculate the relative probability of formation of the helix (78) at any temperature during the incipient stages of the folding of a protein (before "globularity" has set in). This information, together with relative probabilities of occurrence of extended and chain-reversal conformations (27, 63), is being applied in attempts to surmount the multiple-minimum problem (see below). Fig. 4 illustrates the *s*-versus-*T* curves (77) thus far determined with the host-guest technique. Similar experiments are in progress with Glu, Gln, Asp, Asn, Tyr, Pro, Lys, and Ile as the guest residues. Helix-breaking residues, such as Gly and Ser, are also those that occur frequently in chain-reversal conformations; that is, helix breaking and chain-reversal forming are equivalent manifestations of the same conformational preferences. The interactions that make Ala a helix former and, say, Gly, Ser, and Asn helix breakers have been discussed elsewhere (4, 23, 42, 79).

The *isothermal* helix-coil transition in an ionizable homopolymer such as poly(L-lysine) can be induced (80) by a change of pH. The properties of this transition at various degrees of ionization in aqueous salt solution were computed by Hesselink *et al.* (34) with the aid of empirical potential energy functions. The free-energy change in the process was considered to be the sum of a contribution (ΔG_0) from the neutral polymer in the aqueous medium and one (ΔG_e) from the electrostatic interactions between the charged side chains. The term ΔG_0 was obtained from the calculations for poly(L-alanine) in water (22,

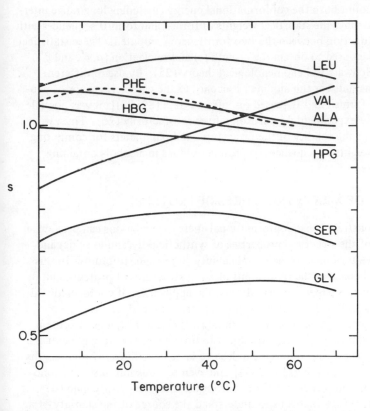

Figure 4. Temperature dependence of s for various amino acid residues in water. [Reprinted with permission from Alter *et al.*, *Macromolecules*, **6**, 564 (1973). Copyright by the American Chemical Society.]

23) and the effect of the lysine side chain was incorporated. To obtain ΔG_e, the partition function for the helix and coil at various degrees of ionization was computed, the coil conformations being generated by a Monte Carlo procedure. A Debye-Hückel screening potential was employed to compute ΔG_e. The computed value of $(\Delta G_0 + \Delta G_e)$ was zero at a degree of ionization of ~50%, compared to an experimental value (81) of ~35%; the discrepancy is thought to arise from an underestimate of the expansion of the coil over its dimensions in a θ solvent. Similar calculations for poly(L-lysine) in aqueous methanol led to the suggestion (34) that the stability of the helical form of this poly(amino acid) in 95% methanol at acid pH is due to ion-pair formation to the extent of about 35%.

Helix-helix transition. The procedures (22–24) used to compute the properties of the helix-coil transition were adapted to treat (26) the helix-helix transition between form I (*cis* peptide bond) and form II (*trans* peptide bond) of poly(L-proline), allowing for two puckering conformations of the pyrrolidine ring.

Taking into account both the conformational energy (including long-range inter-actions) and entropy, the statistical weights of form-I and form-II sequences with and without a junction between the two forms were computed. These statistical weights were then used to obtain the growth (s) and nucleation (σ, β', and β'') parameters of Schwarz's phenomenological theory (25). By use of different equilibrium constants for the binding of alcohol to the carbonyl groups of pep-tide residues in forms I and II [based on infrared data (82, 83)], it was possible to reproduce the experimental transition curves satisfactorily (26). Thus, we have evidence for the validity of the procedures and parameters for computing the statistical weights of sequences of proline residues in alcohol-containing solvent mixtures.

REFINEMENT OF X-RAY STRUCTURES OF PROTEINS

Having demonstrated that conformational energy calculations can provide an understanding of the structural properties of synthetic poly(amino acids) and polypeptides, we turn now to their applicability to protein structures. In this section, we shall consider the refinement of x-ray structures of proteins and, in the remainder of this paper, we shall discuss the application of conformational energy calculations to other aspects of protein structure and reactivity.

The atomic coordinates, obtained from x-ray diffraction studies of protein crystals at a resolution of ~ 2 Å, are subject to uncertainties that may be of the order of 0.5 Å. These uncertainties can lead to large deviations in bond lengths and bond angles [e.g., see Fig. 1 of (84)] and even to omission of some amino acid residues from the structure deduced from the electron-density map (85). If the x-ray structure is to enable us to understand the source of the stability of an enzyme's native conformations or is to be used to investigate the details of enzyme-substrate interactions (86), such as the conformational changes induced by substrates, it is necessary to know the correct (i.e., low-energy) conformation of the enzyme. Because conformational energies can be very sensitive to slight errors in atomic positions, particularly when these errors result in atomic over-laps, the minimization of the conformational energy of a protein should give a structure in which such overlaps are eliminated. If the potential energy func-tions and procedures used are reliable, then the computed structure should be close to the correct one.

Levitt and Lifson (87), Warme et al. (49, 88, 89), and Nishikawa and Ooi (90) have developed refinement procedures to improve the x-ray-determined atomic coordinates of a protein. [A discussion of the similarities and differences among these methods has been presented elsewhere (49, 89).] To conserve computer time and to facilitate computation, the refinement procedure of Warme et al. is carried out in three stages. In stage 1 (88), the dihedral angles of the backbone and side chains of the polypeptide chain are adjusted by a least-squares proce-dure to produce a structure that conforms as closely as possible to the x-ray coordinates, but at the same time, is constrained to have a standard geometry [i.e., fixed bond lengths and bond angles (different for each type of residue) and planar *trans* peptide groups]. These quantities were established from a survey of crystal structures of amino acids and small peptides (53). By the end of stage 1, which is solely a geometric fitting procedure (without the computation of energies), the polypeptide chain has acquired "standard" bond lengths and bond angles and planar *trans* peptide groups, but atomic overlaps have not yet been eliminated. It is relatively easy, however, to formulate the matrix operations to rotate about single bonds during the course of subsequent energy minimization. The subsequent refinement is most economically carried out in two stages (2 and 3). In stage 2 (89), the effort is made to relieve most of the major atomic overlaps by minimizing only a part of the total energy. The function that is minimized in stage 2 includes only nonbonded, hydrogen-bond, torsional, and disulfide loop-closing energies, as well as a "fitting" potential. (In order to conserve computer time, the electrostatic energy was omitted in stage 2, and the minimization was only partly carried out.) In stage 3 (49), the function that is minimized includes all the energy contributions listed above and also the electrostatic energy; in addition, a more refined hydrogen-bond potential (compatible with the electrostatic energy introduced at this stage) is used, and the minimization of the total conformational energy is carried through many more cycles until a minimum in the energy surface is reached. This refinement procedure has been applied to actinomycin D (91), lysozyme (49), and rubredoxin (84), and its application to several other proteins is in progress. In the case of rubredoxin, the crystallographic reliability index (R factor) of the refined structure was computed at 0.37, which is in the range of those obtained for high-resolution protein-structure determinations.

In order to use an x-ray structure of a protein to study its interactions with other small and large molecules, we must know its atomic coordinates with very high accuracy. Thus, *both* the conformational energy and R factor should be minimized. Such a procedure would, presumably, lead to a structure that would have bond lengths and bond angles in the commonly accepted range, would have a low energy, and would match the electron-density map very closely. It is still very difficult, however, to assess the reliability of the resulting coordinates because of the limited resolution presently attainable in x-ray-diffraction studies of proteins.

COMPUTATION OF STRUCTURES OF HOMOLOGOUS PROTEINS

In recent years, it has become apparent, from comparisons of the amino acid sequences of a large number of proteins, that many proteins have homologous sequences and may therefore have similar three-dimensional structures (92–96). The computation of similar helix- and bend-probability profiles for homologous proteins (63, 97) supports this view. The procedures described earlier for the refinement of x-ray structures by conformational energy minimization therefore were adapted to the computation of the structure of a protein based on the x-ray structure of a homologous protein (98). In considering the homologies in the amino acid sequences of related proteins, the maximum extent of homology becomes apparent only by proper alignment of the sequences of the various proteins; that is, in comparing the various sequences, insertions and/or deletions of amino acids must be performed. Such insertions and deletions must be taken into account in computing a structure for the *complete* homologous protein. Recent conformational energy calculations on oligopeptides (4, 61) provide a basis for incorporating these effects, for example, by providing initial side-chain conformations that are compatible with the assigned backbone conformation.

In an early paper on this subject (98), we reported the computation of the structure of bovine α-lactalbumin from that of hen egg-white lysozyme (99). These two proteins have striking sequence homologies (93, 100). Allowing for several deletions and insertions, 47 of the 123 amino acids in α-lactalbumin, including the 8 half-cystine residues, are identically positioned in the two proteins, and 3 other corresponding pairs are Asn-Asp or Gln-Glu differences. Many other corresponding residues are similar in their polar or nonpolar nature. A variety of experimental approaches have revealed many similarities in these two proteins that strongly indicate structural homology; these approaches have been reviewed elsewhere (98). Browne *et al.* (101) constructed a wire skeletal model of α-lactalbumin that closely approximates the known structure of lysozyme (99). Warme *et al.* (98) used the approximate backbone dihedral angles (101) of α-lactalbumin, corrected the sequence in those places where the identities of the amino acids were uncertain at the time of their work, took account of insertions and deletions, and applied the geometrical fitting procedure [stage 1 refinement (88)] to fit the α-lactalbumin sequence to the x-ray coordinates of the backbone atoms of lysozyme and thereby obtained an initial backbone conformation. The side chains were then added, and their initial conformations were obtained from a knowledge of the interrelationship between backbone and side-chain conformation (4). Then the stage-2, energy-refinement procedure (89) was applied. The resulting low-energy conformation, which is similar to that of lysozyme, correlates well with experimental evidence (98). The structure has not yet been carried through the final (49) refinement procedure (stage 3), and it remains to be seen whether the computed structure, *constrained* to be homologous to that of lysozyme, resembles the one that should be determined soon by x-ray diffraction.

This technique is presently being applied to the homologous serine proteases, chymotrypsin, trypsin, elastase, and thrombin and also to several homologous trypsin inhibitors.

PROTEIN FOLDING

The empirical energy functions used to treat conformational problems in synthetic poly(amino acids) and polypeptides, to refine x-ray structures, and to compute structures of homologous proteins are also being applied to determine the folding of a protein from knowledge of its covalent structure. However, although procedures are available for minimizing the conformational energy of a protein the size of ribonuclease or lysozyme (49), the result depends on the starting conformation; that is, many minima exist in the multidimensional conformational energy surface of a protein, and the usual minimization procedures give the minimum in the same potential energy well in which the minimization procedure was started. The nature of this multiple-minimum problem, the limitations of the mathematical approaches thus far applied to its solution, and the more encouraging physical approach to its solution have been reviewed elsewhere (41, 42). The physical approach, which has been used to develop a protein-folding algorithm, makes use of the concept that the conformation of an amino acid residue is determined mostly by short-range interactions and to a lesser degree by medium-range interactions. This approach will be briefly discussed here.

The concept of the dominance of short-range interactions, the history of which is discussed elsewhere (41), is based on the observation (both by conformational energy calculations and by examination of x-ray structures of proteins) that, to a first approximation, the conformational preference of an amino acid residue is determined by the interactions of a side chain and its own backbone. In a second approximation (102), the conformational preferences, based on short-range interactions, are modified to some extent by neighboring amino acid residues on both sides of a given residue in the polypeptide chain. Thus, the central residue of a given sequence of amino acids in a protein has a conformation that is determined, in first approximation, by the nature of the residue itself and, in second approximation, by the nature and conformation of up to four residues on each side of the central residue.

Empirical probabilities can be assigned to each residue to express its preference for adopting one of the five frequently observed conformational states (27), namely, right- and left-handed α-helical conformations (α_R and α_L, respectively), extended conformations (ϵ), and conformations in the bridge regions between α_R and ϵ (ζ_R) and between α_L and ϵ (ζ_L). By multiplying these probabilities together, it is possible to determine the preference of the central residue in a given nonamer sequence to adopt a particular topographical structure if several residues adopt a given conformational state; thus far (27) the topographical structures considered have been right-handed α helix, extended structure, chain-reversal structure, and everything else (designated "coil"). It should be noted that "extended structure" is *not* β structure because the latter term implies information about the association of at least two extended structures in parallel or antiparallel fashion. Many other empirical algorithms for predicting conformational preferences of amino acid residues have been reported, and these have been analyzed (27) in terms of a prediction index P defined as follows:

$$P = N_s Q - 1 \qquad \text{for } 0 \leqslant Q \leqslant \frac{1}{N_s} \tag{1}$$

$$P = \frac{N_s Q - 1}{N_s - 1} \qquad \text{for } \frac{1}{N_s} \leqslant Q \leqslant 1 \tag{2}$$

where

$$Q = \frac{\text{Number of residue states assigned correctly}}{\text{Number of residues in the chain}} \tag{3}$$

that is, the fraction of correctly predicted states; N_s is the number of states possible for each residue in the prediction model. The index P varies from -1 to +1 depending on the value of Q. Thus, if no states are predicted correctly (i.e., $Q = 0$), then $P = -1$; if the fraction of states predicted correctly is the same as expected for a random assignment of states (i.e., $Q = 1/N_s$), then $P = 0$; and if all states are assigned correctly, $P = 1$. For a given value of P, the larger the value of N_s in the model, the more valuable the information provided about the topography of the chain. Several of the prediction algorithms in use yield fairly high values of P (say, ~0.7) when averaged over many proteins. These "successes" have been amply reviewed elsewhere (27, 42) [and a mechanism for protein folding based on these ideas reported (42, 63)] and will not be repeated here. It must be emphasized that the primary purpose of such algorithms is to obtain a starting conformation from which subsequent energy minimization would lead to the correct structure; if successful, this procedure would constitute a solution of the multiple-minimum problem (see below). Without subsequent energy minimization, these algorithms cannot yield a correct protein structure. In this light, a fairly high value of P (but still less than 1) may possibly suffice to provide a suitable initial structure for subsequent energy minimization. This purpose will be better served, however, if the prediction algorithms can be improved so that P approaches 1. Long-range interactions, not yet included *explicitly* in algorithms based on short- and medium-range interactions, are then taken into account in the final stages in which the conformational energy of the total protein is minimized.

The empirical probabilities (which are assigned to each residue and then multiplied together in nonamer segments) were determined from the known structures of eight proteins (27). Thus, in contrast to probabilities obtained from conformational energies on *single* residues (4), the empirical probabilities reflect not only the short- and medium-range interactions present in proteins, but to some extent, the long-range interactions as well.

Focusing attention on nonamer segments (see, for example, Fig. 5), we have a basis for assigning one of four states to each residue in a nonamer in a protein.

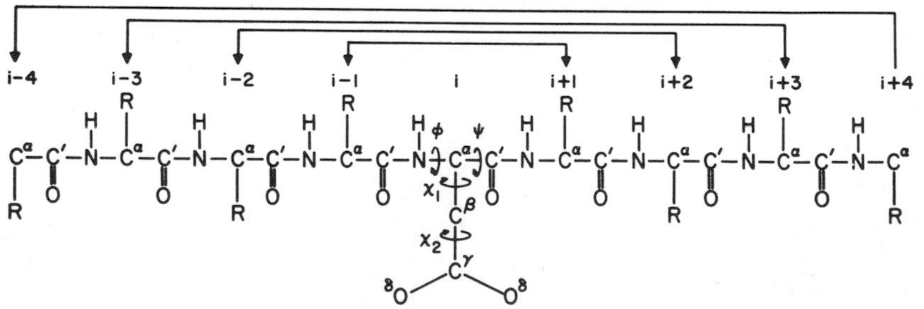

Figure 5. Definition of the central residue i in tri-, penta-, hepta-, and nonapeptides (102). The dihedral angles ϕ, ψ, χ_1, and χ_2 of the central residue are indicated for Asp, as an example.

These assignments are based on conformational energy calculations, on the experimental determinations of σ and s and helix-probability profiles computed therefrom and on the empirical parameters expressing the tendency toward formation of helical, chain-reversal, extended, and coil conformations. The conformational energy of the whole nonamer is minimized with respect to the dihedral angles of only the central residue. Then an overlapping nonamer is selected (shifted by one residue toward the C terminus of the chain), and the four-state model is again applied to all residues except the one whose dihedral angles were varied in the previous step; for this residue, the minimum-energy *dihedral angles* are assigned. The process is illustrated schematically in Fig. 6. Thus, by moving along the entire protein chain (possibly repeating the process several times), we may be able to see whether we are approaching the native conformation by successive approximation. Once the structure begins to resemble that of the native protein, we would be in the correct potential energy well to start the minimization of the conformational energy of the whole protein, not just that of nonamer segments. This procedure is now being tested.

ENZYME-SUBSTRATE INTERACTIONS

Conformational energy calculations may also be used to treat intermolecular complexes by including not only intramolecular but also intermolecular degrees of freedom and interactions. Thus, the complexes involving proteins and ligands, proteins and proteins (e.g., multi-subunit proteins, such as hemoglobin), etc., can be treated. One important application to problems of this type is the use of energy-minimized protein structures to study enzyme-substrate interactions. A start has been made on this problem (86) by treating the bimolecular complex of a rigid enzyme and a flexible substrate. One interesting result of these initial computations was the finding that there are *several* favorable binding sites near the "active-site cleft" of the enzyme; these may contribute to the attraction of the substrate to the active site. Subsequently, the calculations have been extended (103) to allow both the enzyme and substrate to change conformation as the noncovalent Michaelis complex is formed. An example of a computed low-energy structure of a complex of α-chymotrypsin with a tripeptide substrate is shown in Fig. 7. Low-energy structures of the covalent acyl intermediate also have been computed (42, 103). Such calculations should facilitate the interpretation of the wide range of acylation rates of α-chymotrypsin observed (104) when the phenylalanine residue is flanked by different amino acids, which contribute to the binding energy and serve to determine the orientation of the peptide bond to be hydrolyzed with respect to the side chain of Ser 195, which is to be acylated.

SUMMARY

Synthetic poly(amino acids) and polypeptides have been useful model systems for understanding the nature of the interactions within the polypeptide chain and the origin of the conformational preference of each amino acid residue. The empirical conformational energies used to treat these systems are

Step	1	2	3	4	5	6	7	8	9	10	11	12	13 ...
1	β	β	α	α	α	α	α	β	β				
2	β	β	α	α	φ=-52° ψ=-48°	α	α	β	β				
3		β	α	α	φ=-52° ψ=-48°	α	α	β	β	β			
4		β	α	α	φ=-52° ψ=-48°	φ=-51° ψ=-49°	α	β	β	β			
5			α	α	φ=-52° ψ=-48°	φ=-51° ψ=-49°	α	β	β	β	β		
6			α	α	φ=-52° ψ=-48°	φ=-51° ψ=-49°	φ=-52° ψ=-49°	β	β	β	β		

etc.

Figure 6. Schematic illustration of successive steps in the computation of the dihedral angles of a protein (42). The amino acid sequence is specified by the numbers 1, 2, . . . , 13, . . . , and the backbone conformations of each residue are specified *initially* as α, β, and so on, but in terms of the dihedral angles φ and ψ after each computational step. In step 1, one of the four conformational states is assigned to each residue in a nonapeptide. In step 2, the energy of the nonapeptide is minimized with respect to the dihedral angles of residue 5, and so on. [Reprinted with permission from *Current Topics in Biochemistry* (1973), C.B. Anfinsen and A.N. Schechter, eds., New York: Academic Press, 1974.]

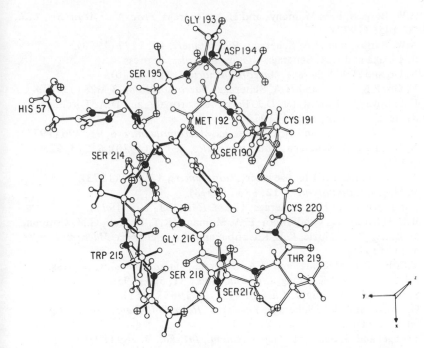

Figure 7. A low-energy conformation of the substrate *N*-acetyl L-alanyl L-alanyl L-phenylalanine *N*-methylamide in the active-site cleft of α-chymotrypsin. The substrate is shown with filled-in line bonds and the enzyme is shown with open line bonds. [Reprinted with permission from *Currents Topics in Biochemistry* (1973), C. B. Anfinsen and A. N. Schecter, ed., Academic Press, New York, 1974.]

applicable to such problems as the refinement of x-ray data on proteins, the computation of the structure of a protein from the known structure of a homologous one, and the computation of the preferred conformations of intermolecular complexes. It is hoped that the information gained from studies of such systems can be used, in the manner indicated in the section on protein folding, or in some other way, to surmount the multiple-minimum problem and thus lead to the prediction of the three-dimensional structure of a protein from a knowledge of its covalent structure.

REFERENCES

1. H.A. Scheraga, *Adv. Phys. Org. Chem.*, **6**, 103 (1968).
2. E. Katchalski, M. Sela, H.I. Silman, and A. Berger, in *The Proteins* (Second Edition), Vol. 2, H. Neurath, ed., Academic Press, New York, 1964, p. 405.
3. N. Lotan, A. Berger, and E. Katchalski, *Ann. Rev. Biochem.*, **41**, 869 (1972).
4. P.N. Lewis, F.A. Momany, and H.A. Scheraga, *Israel J. Chemistry*, **11**, 121 (1973).

5.   A.W. Burgess, F.A. Momany, and H.A. Scheraga, *Proc. Nat. Acad. Sci. U.S.*, **70**, 1456 (1973).

6.   A.W. Burgess and H.A. Scheraga, *Biopolymers*, **12**, 2177 (1973).

7.   S. Tanaka and H.A. Scheraga, *Macromolecules*, in press.

8.   N. Gō and H.A. Scheraga, *Macromolecules*, **3**, 188 (1970).

9.   N. Gō, P.N. Lewis, and H.A. Scheraga, *Macromolecules*, **3**, 628 (1970).

10. F.A. Bovey, A.I. Brewster, D.J. Patel, A.E. Tonelli, and D.A. Torchia, *Accts. Chem. Res.*, **5**, 193 (1972).

11. G.C.C. Niu, N. Gō, and H.A. Scheraga, *Macromolecules*, **6**, 91, 796 (1973).

12. N. Gō and H.A. Scheraga, *Macromolecules*, **3**, 178 (1970); **6**, 273, 525 (1973).

13. A.E. Tonelli and A.I.R. Brewster, *Biopolymers*, **12**, 193 (1973).

14. V. Madison, *Biopolymers*, **12**, 1837 (1973).

15. For a review, see H.A. Scheraga, *Chem. Revs.*, **71**, 195 (1971).

16. R.F. McGuire, G. Vanderkooi, F.A. Momany, R.T. Ingwall, G.M. Crippen, N. Lotan, R.W. Tuttle, K.L. Kashuba, and H.A. Scheraga, *Macromolecules*, **4**, 112 (1971).

17. Y.C. Fu, R.F. McGuire, and H.A. Scheraga, *Macromolecules*, in press.

18. B.H. Zimm and J.K. Bragg, *J. Chem. Phys.*, **31**, 526 (1959).

19. S. Lifson and A. Roig, *J. Chem. Phys.*, **34**, 1963 (1961).

20. D. Poland and H.A. Scheraga, *Theory of Helix-Coil Transitions in Biopolymers*, Academic Press, New York, 1970.

21. J. Engel and G. Schwarz, *Angew. Chem., Int. Ed.*, **9**, 389 (1970).

22. N. Gō, M. Gō, and H.A. Scheraga, *Proc. Nat. Acad. Sci. U.S.*, **59**, 1030 (1968).

23. M. Gō, N. Gō, and H.A. Scheraga, *J. Chem. Phys.*, **52**, 2060 (1970); **54**, 4489 (1971).

24. M. Gō, F.T. Hesselink, N. Gō, and H.A. Scheraga, *Macromolecules*, in press.

25. G. Schwarz, *Biopolymers*, **6**, 873 (1968).

26. S. Tanaka and H.A. Scheraga, *Macromolecules*, submitted.

27. A.W. Burgess, P.K. Ponnuswamy, and H.A. Scheraga, *Israel J. Chemistry*, in press.

28. D.A. Brant and P.J. Flory, *J. Am. Chem. Soc.*, **87**, 2791 (1965).

29. D.A. Brant, W.G. Miller, and P.J. Flory, *J. Mol. Biol.*, **23**, 47 (1967).

30. W.G. Miller, D.A. Brant, and P.J. Flory, *J. Mol. Biol.*, **23**, 67 (1967).

31. H.E. Warvari and R.A. Scott, *J. Chem. Phys.*, **57**, 1146 (1972).

32. S. Tanaka and A. Nakajima, *Macromolecules*, **5**, 708 (1972).

33. W.L. Mattice, K. Nishikawa, and T. Ooi, *Macromolecules*, **6**, 443 (1973).

34. F.T. Hesselink, T. Ooi, and H.A. Scheraga, *Macromolecules*, **6**, 541 (1973).

35. S. Tanaka and H.A. Scheraga, work in progress.

36. G.N. Ramachandran and V. Sasisekharan, *Adv. Prot. Chem.*, **23**, 283 (1968).

37. A.M. Liquori, in *Symmetry and Function of Biological Systems at the Macromolecular Level*, A. Engstrom and B. Strandberg, ed., Nobel Symposium II, Almqvist and Wiksell, Stockholm, 1969, p. 101.

38. G.N. Ramachandran, in (37), p. 79.

39. H.A. Scheraga, in (37), p. 43.

40. H.A. Scheraga, in *Conformation of Biological Molecules and Polymers,* The Jerusalem Symposia on Quantum Chemistry and Biochemistry, **5**, 51 (1973).

41. H.A. Scheraga, *Pure and Applied Chem., 36,* 1 (1973).

42. H.A. Scheraga, in *Current Topics in Biochemistry* (1973), C.B. Anfinsen and A.N. Schechter, New York: Academic Press, 1974, p. 1.

43. B. Pullman, *Adv. Protein Chem.,* in press.

44. A.T. Hagler and S. Lifson, in *The Proteins* (Third Edition), H. Neurath, ed., Academic Press, New York, in press.

45. C.B. Anfinsen, in *New Perspectives in Biology,* M. Sela, ed., Elsevier, Amsterdam, 1964, p. 42.

46. R.R. Hantgan, G.G. Hammes, and H.A. Scheraga, *Biochemistry,* in press.

47. A.W. Burgess, L. Weinstein, D. Gabel, and H.A. Scheraga, *Biochemistry,* submitted.

48. A. Warshel, M. Levitt, and S. Lifson, *J. Mol. Spectroscopy, 33,* 84 (1970).

49. P.K. Warme and H.A. Scheraga, *Biochemistry, 13,* 757 (1974).

50. F.A. Momany, L.M. Carruthers, R.F. McGuire, and H.A. Scheraga, *J. Phys. Chem.,* in press.

51. F.A. Momany, L.M. Carruthers, and H.A. Scheraga, *J. Phys. Chem.,* in press.

52. K.D. Gibson and H.A. Scheraga, *Proc. Nat. Acad. Sci. U.S., 58,* 420 (1967).

53. F.A. Momany, R.F. McGuire, and H.A. Scheraga, *J. Phys. Chem.,* to be submitted.

54. L.L. Shipman, A.W. Burgess, and H.A. Scheraga, work in progress.

55. The nomenclature used here is that in *Biochemistry, 9,* 3471 (1970).

56. C.M. Deber, F.A. Bovey, J.P. Carver, and E.R. Blout, *J. Am. Chem. Soc., 92,* 6191 (1970).

57. G. Kartha, private communication.

58. K. Nishikawa, F.A. Momany, and H.A. Scheraga, *Macromolecules,* submitted.

59. S.S. Zimmerman and H.A. Scheraga, work in progress.

60. C.M. Venkatachalam, *Biopolymers, 6,* 1425 (1968).

61. P.N. Lewis, F.A. Momany, and H.A. Scheraga, *Biochim. Biophys. Acta, 303,* 211 (1973).

62. E. Ralston and J.L. DeCoen, *J. Mol. Biol., 83,* 393 (1974).

63. P.N. Lewis, F.A. Momany, and H.A. Scheraga, *Proc. Nat. Acad. Sci. U.S., 68,* 2293 (1971).

64. H.A. Scheraga, P.N. Lewis, F.A. Momany, P.H. Von Dreele, A.W. Burgess, and J.C. Howard, *Fed. Proc., 32,* 495 (1973).

65. J.C. Howard, A. Ali, H.A. Scheraga, and F.A. Momany, work in progress.

66. R.W. Woody, private communication.

67. D.A. Brant and P.J. Flory, *J. Am. Chem. Soc., 87,* 2788 (1965).

68. R.B. Hawkins and A. Holtzer, *Macromolecules, 5,* 294 (1972).

69. K.D. Gibson and H.A. Scheraga, *Proc. Nat. Acad. Sci. U.S., 58,* 1317 (1967).

70. M. Dygert, N. Gō, and H.A. Scheraga, work in progress.

71. D.N. Silverman and H.A. Scheraga, *Biochemistry, 10,* 1340 (1971).

72.   W.A. Gibbons, G. Nemethy, A. Stern, and L.C. Craig, *Proc. Nat. Acad. Sci. U.S.,* **67**, 239 (1970).
73.   P. DeSantis, R. Rizzo, and G. Ughetto, *Biopolymers,* **11**, 279 (1972).
74.   D. Kotelchuck, H.A. Scheraga, and R. Walter, *Proc. Nat. Acad. Sci. U.S.,* **69**, 3629 (1972).
75.   F.A. Momany, A. Goldman, F.T. Hesselink, and H.A. Scheraga, in preparation.
76.   Y. Takeda, Y. Iitaka, and M. Tsuboi, *J. Mol. Biol.,* **51**, 101 (1970).
77.   J.E. Alter, R.H. Andreatta, G.T. Taylor, and H.A. Scheraga, *Macromolecules,* **6**, 564 (1973).
78.   P.N. Lewis, N. Gō, M. Gō, D. Kotelchuck, and H.A. Scheraga, *Proc. Nat. Acad. Sci. U.S.,* **65**, 810 (1970).
79.   D. Kotelchuck and H.A. Scheraga, *Proc. Nat. Acad. Sci. U.S.,* **61**, 1163 (1968); **62**, 14 (1969).
80.   J. Applequist and P. Doty, in *Polyamino Acids, Polypeptides and Proteins,* M.A. Stahmann, ed., Univ. Wisconsin Press, Madison, Wis., 1962, p. 161.
81.   P.Y. Chou and H.A. Scheraga, *Biopolymers,* **10**, 657 (1971).
82.   H. Strassmair, J. Engel, and S. Knof, *Biopolymers,* **10**, 1759 (1971).
83.   S. Knof, H. Strassmair, J. Engel, M. Rothe, and K.D. Steffen, *Biopolymers,* **11**, 731 (1972).
84.   D. Rasse, P.K. Warme, and H.A. Scheraga, *Proc. Nat. Acad. Sci. U.S.,* in press.
85.   K.D. Hardman and C.F. Ainsworth, *Biochemistry,* **11**, 4910 (1972).
86.   K.E.B. Platzer, F.A. Momany, and H.A. Scheraga, *Int. J. Peptide and Protein Res.,* **4**, 201 (1972).
87.   M. Levitt and S. Lifson, *J. Mol. Biol.,* **46**, 269 (1969).
88.   P.K. Warme, N. Gō, and H.A. Scheraga, *J. Comput. Phys.,* **9**, 303 (1972).
89.   P.K. Warme and H.A. Scheraga, *J. Comput. Phys.,* **12**, 49 (1973).
90.   K. Nishikawa and T. Ooi, *J. Phys. Soc. Japan,* **32**, 1338 (1972).
91.   P.K. Ponnuswamy, R.F. McGuire, and H.A. Scheraga, *Int. J. Peptide and Protein Res.,* **5**, 73 (1973).
92.   B.S. Hartley, *Phil. Trans. Roy. Soc. London,* **B257**, 77 (1970).
93.   K. Brew and P.N. Campbell, *Biochem. J.,* **102**, 258 (1967).
94.   K. Brew, F.J. Castellino, T.C. Vanaman, and R.L. Hill, *J. Biol. Chem.,* **245**, 4570 (1970).
95.   R.E. Dickerson, *J. Molec. Evolution,* **1**, 26 (1971).
96.   K. Nishikawa and T. Ooi, *J. Theor. Biol.,* **43**, 351 (1974).
97.   P.N. Lewis and H.A. Scheraga, *Arch. Biochem. Biophys.,* **144**, 584 (1971).
98.   P.K. Warme, F.A. Momany, S.V. Rumball, R.W. Tuttle, and H.A. Scheraga, *Biochemistry,* **13**, 768 (1974).
99.   C.C.F. Blake, G.A. Mair, A.C.T. North, D.C. Phillips, and V.R. Sarma, *Proc. Roy. Soc.,* **B167**, 365 (1967).
100. K. Brew, T.C. Vanaman, and R.L. Hill, *J. Biol. Chem.,* **242**, 3747 (1967).
101. W.J. Browne, A.C.T. North, D.C. Phillips, K. Brew, T.C. Vanaman, and R.L. Hill, *J. Mol. Biol.,* **42**, 65 (1969).
102. P.K. Ponnuswamy, P.K. Warme, and H.A. Scheraga, *Proc. Nat. Acad. Sci. U.S.,* **70**, 830 (1973).
103. K.E.B. Platzer and H.A. Scheraga, unpublished results.
104. W.K. Baumann, S.A. Bizzozero and H. Dutler, *Eur. J. Biochem.,* **39**, 381 (1973).

STUDIES OF SOME EMPIRICAL ENERGY FUNCTIONS USED IN PEPTIDE CONFORMATIONAL ANALYSIS

A.J. HOPFINGER, Department of Macromolecular Science, Case Western Reserve University, Cleveland, Ohio 44106

SYNOPSIS: Theoretical conformational analysis using empirical potential energy functions has become a useful tool in the physical characterization of peptides. A considerable amount of work remains to be done, however, in order to extend the generality and applicability of this technique. One specific area needing attention is the development and the refinement of these functions. This paper describes work done on the empirical energy functions to characterize torsional rotation about the peptide bond, polypeptide–aqueous medium interactions, and pairwise electrostatic interactions between atoms in molecules.

INTRODUCTION

Application of empirical potential energy functions to determine the conformational energy of polypeptides has been extensive. Theoretical conformational analyses of this nature have provided important information concerning conformational preferences and distributions of conformational states as a function of amino acid sequence. Moreover, the incorporation of theoretical conformational analysis into a comprehensive experimental investigation of molecular structure is proving to be a useful means of identifying and limiting conformational choices [see, for example, (1)]. A considerable amount of work remains to be done in the refinement and generalization of theoretical conformational analysis, however. Scheraga (2) has reported on the problem of energy minimization when many degrees of freedom are involved and on the progress he and his co-workers have made in this area. Lifson *et al.* (3) and Ramachandran (4) also have an active interest in the refinement of potential energy functions, as has our own laboratory. This paper presents the results of our attempt to describe the empirical potential energy functions for torsional rotation about the amide— or imide—bond, polypeptide–aqueous solvent interactions, and pairwise electrostatic interactions between nonchemically bonded atoms.

TORSIONAL ROTATIONS ABOUT THE PEPTIDE BOND

One common feature of conformational analyses carried out until approximately two years ago was the fixation of the torsional rotation (ω) about the amide bond so that the peptide unit would be *trans* planar (i.e., $\omega = 180°$). The basis for this conformational restriction is the high-energy barrier, estimated to be 20 kcal/mole (5), to rotation about this bond. In no way, however, is a large torsional energy required to be associated with a small rotation about the peptide bond (4). Both theoretical and experimental studies, including those of Ramachandran (4), performed on amides and other peptide model compounds suggest that rotational fluctuations of ω up to $15°$ from *trans* or *cis* planarity are energetically possible (5, 6). Furthermore, these and other studies (7, 8) suggest that a functional representation for torsional fluctuations about the peptide bond is

$$U(\omega) = \frac{A}{2}(1 - \cos 2\omega) \qquad A \approx 20\,\text{kcal/mole} \qquad (1)$$

We report here the results of allowing rotation about the peptide bond for poly(L-alanine) and poly(L-tyrosine) in and out of aqueous solution. Extended secondary structures such as the β conformation exhibit only minor conformational sensitivity to torsional rotations about the peptide bond because of the high chain flexibility as a function of the torsional angles ϕ and ψ. Thus, our conformational analyses are restricted to that region in (ϕ,ψ,ω) space where the tightly wound, right-handed α helix is located. The exact position in (ϕ,ψ,ω) space of the energy minimum corresponding to the right-handed α helix varies slightly from one homopolypeptide to another (9).

Fig. 1 shows the vacuum conformational energy maps of poly(L-alanine) in the vicinity of the right-handed α helix for $\omega = 180°$ and ω chosen to minimize the total conformational energy; the energies were calculated using the method, parameters, and functions discussed in (10). It is obvious from these maps that minimizing the total conformational energy with respect to ω markedly increases the conformational flexibility of the polypeptide chain. The effect is similar for poly(L-tyrosine). Table 1 lists the average potential energy $\langle E \rangle$, the entropy $\langle S \rangle$, and the free energy $\langle A \rangle$ for poly(L-alanine) and poly(L-tyrosine) in and out of aqueous solution at $300°K$.

The increase in the value of $\langle S \rangle$ when ω is allowed to vary is a measure of enhanced chain flexibility. The decrease in the value of $\langle E \rangle$ when torsional rotations about the peptide bond are allowed indicates a gain in stabilization energy from nonbonded, electrostatic, and hydrogen-bonding interactions that more than compensates for the expenditure of potential energy in the peptide torsional rotations. The free energy associated with polypeptide-solvent interaction is not particularly sensitive to torsional rotation about the peptide bond, at least for the two examples given here. The free-energy gain accompanying torsional rotations about the peptide bonds appears to be dependent upon the size of the side chains: the larger the side chain, the greater the gain in stabilization free energy. Stereochemical restrictions between backbone and side chain are usually minimized by allowing peptide torsional rotations.

POLYPEPTIDE–AQUEOUS SOLVENT INTERACTIONS

Most theoretical conformational analyses of polypeptides have been based upon the free-space approximation, which is the assumption that the polypeptide

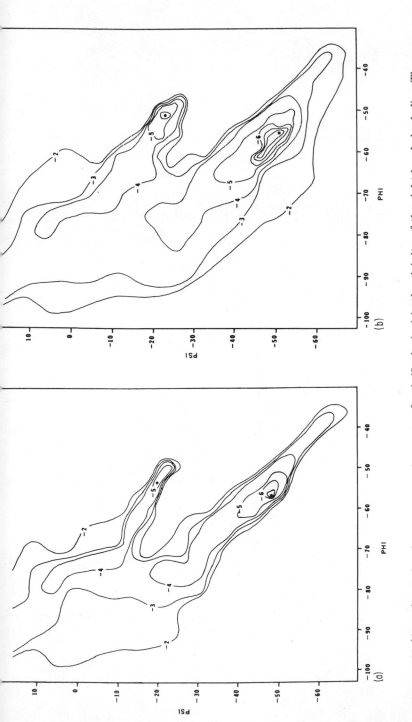

Figure 1. (ϕ, ψ) conformational energy maps, *in vacuo*, of poly(L-alanine) in the vicinity of the right-handed α helix. The energy contours are in kcal/mole/residue. The large dots denote relative minima. (a) $\omega = 180°$; (b) ω chosen to minimize the total conformational energy for each choice of (ϕ, ψ).

Table 1. Values of the average potential energy $\langle E \rangle^a$, entropy $\langle S \rangle$, and free energy $\langle A \rangle$ for poly(L-alanine) and poly(L-tyrosine) in the vicinity of the right-handed α helix[b] (T = 300°K)

Polymer	Molecule	Conformation	$\langle E \rangle$ (kcal/mole/residue)	$\langle S \rangle$ (e.u./residue)	$-T \langle S \rangle$ (kcal/mole/residue)	$\langle A \rangle$ (kcal/mole/residue)
Poly(L-alanine)	$\omega = 180°$; fixed planar unit	R.H. α (H_2O)	-6.9	5.4	-1.6	-8.5
		R.H. α (in vacuo)	-7.3	5.5	-1.6	-8.9
	ω allowed to vary	R.H. α (H_2O)	-8.0	6.4	-1.9	-9.9
		R.H. α (in vacuo)	-8.0	6.6	-2.0	-10.0
Poly(L-tyrosine)	$\omega = 180°$; fixed planar unit	R.H. α (H_2O)	-8.7	5.9	-1.8	-10.5
		R.H. α (in vacuo)	-6.1	7.0	-2.1	-8.2
	ω allowed to vary	R.H. α (H_2O)	-9.8	9.6	-2.9	-12.7
		R.H. α (in vacuo)	-7.1	10.2	-3.1	-10.2

[a] A contact distance of 2.10 Å was used for H $\cdots$ H interactions.
[b] The energy was minimized with respect to the side-chain rotation, x_1, for each (ϕ, ψ, ω) triplet in all energy scans.

does not interact with its medium; in other words, the polypeptide is in a vacuum. We have attempted to develop empirical energy functions to describe the free-energy interaction between polypeptides and an aqueous medium. Our solute-solvent interaction model is based on the concept of a hydration shell (11), which has been used in solvation theories for many years to describe the behavior of a solvent molecule near a solute species. Gibson and Scheraga (12) modified the hydration-shell model to apply to the atoms of a solute macromolecule. The model used in our laboratory is a further modification of the basic hydration-shell concept, which differs from the Gibson–Scheraga model in the size and properties of the hydration shells and the criteria for calculating excluded volumes in the hydration shells.

The concept of a hydration shell implies that around each atom of the solute molecule can be centered a characteristic sphere, the size of which is dependent upon the solvent molecule and atom of the solute molecule. A particular change in free energy is associated with the removal of a solvent molecule from the hydration shell. The size of the hydration shell and the shape of the solvent molecule dictate how many solvent molecules can occupy the hydration shell. The sum of the intersections of the van der Waals volumes of the atoms of the solute molecule with the hydration sphere results in an excluded hydration-shell volume, which determines how many solvent molecules are removed from the hydration shell when the solute molecule is in a particular conformation. Thus the hydration shell is sensitive to conformation via excluded hydration-shell volumes.

The hydration-shell model is a four-parameter system in which n is the maximum number of solvent molecules that can occupy the hydration shell, Δf is the free-energy change associated with the removal of one solvent molecule from the hydration shell, R_v is the effective radius of the hydration shell, and V_f is the free volume of packing associated with one solvent molecule in the hydration shell.

Detailed descriptions of the solute molecule–solvent interaction model are available in (10) and (13–15). We have used this model in the conformational analysis of several homopolypeptides. The results of these calculations are summarized below for four homopolypeptides restricted to adopting ordered secondary structures.

In the case of poly(L-alanine), intramolecular energetics governs the choice of the right-handed α helix as the preferred conformation both in and out of an aqueous medium. However, peptide–aqueous medium interactions promote the extended β and PPII conformations over the α-helical structure. The value of $\Delta A = 0.4$ kcal/mole/residue in going from vacuum to aqueous solution for the right-handed α helix is consistent with the insolubility of poly(L-alanine) in water. *Trans*-poly(L-proline) has a moderately favorable free-energy interaction with the aqueous medium that is consistent with the relatively high solubility of this polymer in water. The hydrophobic interactions between the side chains and aqueous solvent dominate the molecular thermodynamics of poly(L-valine) to the extent that this biopolymer has a net repulsive interaction free energy with water in all conformations. The difference between ΔA of neutral poly(L-lysine) and charged poly(L-lysine) ranges from 6 to 7 kcal/mole/residue (9), demonstrating that the free-energy interaction that takes place between an NH_3^+ group and water is exceedingly favorable. This finding carries over to all other charged groups of both signs. The demonstration of the highly favorable interactions between charged polar groups and an aqueous medium is consistent with the finding that the polar groups are found on the surfaces of globular proteins crystallized from an aqueous solvent. [For detailed comparative molecular thermodynamic data on the examples given above, see (9).]

ELECTROSTATIC POTENTIAL

The prevalent method of computing the pairwise electrostatic potential energy between two nonchemically bonded atoms in the conformational analysis of polypeptides is to use Coulomb's law:

$$\phi_{ij} = \frac{kq_i q_j}{\epsilon r_{ij}} \tag{2}$$

where ϕ_{ij} is the potential energy, k is a conversion constant, q_i and q_j are the partial charges on atoms i and j, respectively, ϵ is the dielectric constant, and r_{ij} is the interaction distance. Considerable effort (16) has gone into developing methods of accurately calculating the partial charges. There is little in the literature, however, concerning the reasonableness of two assumptions implicit in the use of Eq. (2). First, atomic charge is distributed over space and not fixed at a point, as is assumed by Coulomb's law. Miller and Rai (17) and Scheraga et al. (18) independently have recognized the spatial distribution of charge on the ionized groups of the side chains of charged poly(L-glutamic acid) (the COO^- group) and charged poly(L-lysine) (the NH_3^+ group), respectively. Each group adopted a screened electrostatic potential function to describe interactions involving the charged groups in their conformational studies.

The second assumption made in using Coulomb's law is that the medium separating the pair of interacting charges is homogeneous with respect to interactions with the electric fields generated by each of the charges. This is the requirement for adopting a dielectric constant in Coulomb's law. Most pairwise partial-charge interactions between nonchemically bonded species in a molecule take place through a medium containing polarizable, partially charged atoms. This medium is heterogeneous with respect to electrical fields and, thus, the conditions needed for adoption of a dielectric constant are not satisfied.

We are presently studying the pairwise interaction between point charges in which the separating medium contains polarizable point charges. Our hope is to replace ϵ in Eq. (2) with some $\epsilon\,(r_{ij}, i, j)$ that yields different dielectric values as a function of interaction distance, r_{ij}, and type of pair species (i,j). The functional form of $\epsilon\,(r_{ij}, i,j)$ will come from our model calculations. Our present approach is to selectively place polarizable point charges between two interacting point charges subject to the constraints that no polarizable point charge is more than 10 Å from either of the interacting charges i and j and that the separation distances of all point charges are consistent with general stereochemical rules. The total electrostatic potential energy for a pair interaction separated by a medium containing polarizable point charges is given by:

$$\phi'_{ij} = \frac{kq_i q_j}{r_{ij}} + \sum_{1}^{\nu} \frac{km_1 m_1}{2\alpha_1} + \sum_{1}^{\nu} k \frac{q_i(m_1 r_{i1})}{r_{i1}^3} + \frac{q_j(m_1 r_{j1})}{r_{j1}^3} \qquad (3)$$

where

$$m_1 = \alpha_1 E_1 \qquad (4)$$

and

$$E_1 = \Sigma_n \left[-\frac{\nabla(m_n r_{1n})}{r_{1n}^3} + \frac{q_n r_{1n}}{r_{1n}^3} \right] \qquad (5)$$

Equations (4) and (5) allow the induced dipoles m_1 to be determined. Those values are then substituted into Eq. (3) and the value of ϕ'_{ij} is computed. The α_1 values are the atomic coefficients of polarizability (19), ν is the number of atoms between i and j, the 1 index indicates that the sum is over only the ν set of atoms, and the n index indicates the sum is over the ν set of atoms and atoms i and j. The value of $\epsilon(r_{ij}, i,j)$ can be found by equating (2) and (3) and solving for $\epsilon(r_{ij}, i,j)$:

$$\epsilon(r_{ij}, i,j) = \frac{kq_i q_j}{\phi'_{ij} r_{ij}}$$

Table 2 lists values of $\epsilon(r_{ij}, i,j)$ as a function of r_{ij} for two interacting carbonyl oxygens separated by a medium containing two methyl groups. The methyl groups were positioned as near as stereochemically possible to the oxygens. The value of $\epsilon(r_{ij}, i,j)$ increases in a nearly linear fashion until r_{ij} becomes approximately equal to the separation distance needed to accommodate the two methyl groups directly between the oxygens. At this distance, $\epsilon(r_{ij}, i,j)$ reaches a maximum value and then decreases very gradually at longer separation distances. Clearly, much more work needs be done in the study of the electrostatic potential energy, and we feel that the work reported here represents a reasonable starting point for such investigation.

Table 2. ϵ (r_{ij}, i,j) as a function of r_{ij} for two interacting carbonyl oxygens, $q_i = q_j = -0.463$ AU, separated by a medium containing two methyl groups, $\alpha_1 = 2.05$ A3[a]

r_{ij}	ϵ (r_{ij}, i,j)	r_{ij}	ϵ (r_{ij}, i,j)
3.0 A	1.83	8.0 A	4.91
3.5	2.19	8.5	5.11
4.0	2.33	9.0	5.18
4.5	2.78	9.5	5.26
5.0	3.21	10.0	5.32
5.5	3.64	10.5	5.35
6.0	3.95	11.0	5.31
6.5	4.15	11.5	5.28
7.0	4.33	12.0	5.22
7.5	4.68		

[a] The van der Waals radius of the oxygen atom chosen was 1.35 A and that picked for the methyl group was 2.05 A.

REFERENCES

1. M. Goodman, F. Chen, C. Gilon, R. Ingwall, D. Nissen, and M. Palumbo, in *Peptides, Polypeptides, and Proteins,* Proceedings of the Rehovot Symposium 1974, New York: Wiley(Interscience), 1974.
2. H. Scheraga, *ibid.*
3. A.T. Hagler, S. Lifson, and E. Huler, *ibid.*
4. G.N. Ramachandran, *ibid.*
5. A.S.N. Murthy, K.G. Rao, and C.N.R. Rao, *J. Am. Chem. Soc.,* **92**, 925 (1970).
6. P.R. Andrews, *Biopolymers,* **10**, 2253 (1971).
7. S. Krimm and C.M. Venkatachalam, *Proc. Nat. Acad. Sci. U.S.,* **68**, 2468 (1971).
8. F.K. Winkler and J.D. Dunitz, *J. Mol. Biol.,* **59**, 169 (1971).
9. T. Ooi, R.A. Scott, G. Vanderkooi, and H.A. Scheraga, *J. Chem. Phys.,* **46**, 4410 (1967).
10. A.J. Hopfinger, *Conformational Properties of Macromolecules,* New York: Academic Press, 1973.
11. E.A. Moelwyn-Hughes, *Physical Chemistry,* New York: Pergamon Press, 1957.
12. K.D. Gibson and H.A. Scheraga, *Proc. Nat. Acad. Sci. U.S.,* **58**, 420 (1967).
13. A.J. Hopfinger, *Macromolecules,* **4**, 731 (1971).
14. K.H. Forsythe and A.J. Hopfinger, *Macromolecules,* **6**, 423 (1973).
15. H.J.R. Weintraub and A.J. Hopfinger, *J. Theor. Biol.,* **41**, 53 (1973).
16. J.F. Yan, F.A. Momany, R. Hoffmann, and H.A. Scheraga, *J. Phys. Chem.,* **74**, 420 (1970).
17. J.H. Rai and W.G. Miller, *Biopolymers,* **12**, 845 (1973).
18. F.T. Hesselink, T. Ooi, and H.A. Scheraga, *Macromolecules,* **6**, 541 (1973).
19. K.S. Pitzer, in *Advances in Chemical Physics,* Vol. 2, J. Prigogine, ed., New York: Wiley (Interscience), 1959, p. 59.

EFFECTS OF METHYLATION ON THE ENERGETICALLY PREFERRED HELICAL CONFORMATIONS OF POLYPEPTIDES

ANTONY W. BURGESS, YVONNE PATERSON, and **SYDNEY J. LEACH,**
Russell Grimwade School of Biochemistry, University of Melbourne, Parkville, Victoria, Australia, 3052

SYNOPSIS: Polysarcosine and poly(α-amino isobutyric acid) have been used as models to study the effects of methylation on the conformations of polypeptide helices and coils. The residues per turn (n) and helix pitch (h) of *cis* polypeptide helices have been calculated and plotted as a function of the dihedral angles (ϕ, ψ). The (ϕ, ψ) potential-energy surfaces for the *cis* and *trans* isomers of N_α-acetylsarcosine N,N-dimethylamide are presented. The energy calculations were extended to *cis* and *trans* oligomers of sarcosine; of the six helical conformations of minimum potential energy, the two most energetically preferred had *cis* peptide bonds. Similar calculations and measurements were made for poly(α-amino isobutyric acid).

INTRODUCTION

In recent years much research has been devoted to the understanding of both ordered and random conformations of polypeptides. Most attention has been given to the polypeptides of those amino acids that occur in proteins, but there are several other naturally occurring amino acids that have interesting conformational properties. In particular, two methylated amino acids, namely, sarcosine (N-methylglycine) and α-amino isobutyric acid (α-methylalanine) occur in cyclic antibiotics (1–5) where they assume important conformational roles. The theoretical (6–8) and experimental (9–11) studies on polysarcosine and equivalent model di- and tripeptides indicate considerable conformational restrictions due to steric overlaps between the N-methyl group and other atoms within the sarcosyl residue. Methylation of the peptide nitrogen introduces additional conformational isomerism such that both the *cis* ($\omega = 0°$) and *trans* ($\omega = 180°$) forms occur in the polypeptide chain (9, 10). Methylation at the C_α atom also restricts the conformations available to the polypeptide chain. The effect of a single C_β atom is well documented both theoretically (12, 13) and experimentally (14). The conformational restrictions due to *two* C_β atoms [as found in poly(α-amino isobutyric acid)], however, have been studied only briefly (5, 15). In this paper, the energetically most stable regular conformations of *cis* and *trans* polysarcosines are described and the results compared with the preferred conformations for N_α-acetyl sarcosine N,N-dimethylamide. The random-coil characteristics determined both experimentally and theoretically are described in another publication (16).

THEORETICAL AND EXPERIMENTAL METHODS

Computation of conformational energies. For N_α-acetylsarcosine N,N-dimethylamide and N_α-acetyl-α-amino isobutyryl N-methylamide the computations were carried out over the whole range of ϕ and ψ, but ω was held fixed at $0°$ or $180°$ for the former and at $180°$ for the latter. The geometry of each residue was adapted from x-ray-crystallographic data. For the sarcosine residue the structure of N,N-diphenylacetamide (17) seemed the most appropriate; the bond angles and lengths were very similar to Pauling–Corey geometry (18), except that the peptide bond length was 1.37 Å. The crystallographic structure for α-amino isobutyric acid (19) indicated that Pauling–Corey geometry was appropriate for this residue and this geometry was also assumed for the peptide bond configuration. The geometry used for each residue is shown in Fig. 1.

An empirical energy algorithm was used to estimate the relative potential energy for each (ϕ, ψ) conformation of the N_α-acetyl amino acid N-methylamides. The scheme used was very similar to that described by Gibson and Scheraga (20), but the partial electrostatic charges of Yan *et al.* (21) and an effective dielectric constant of 3.0 were used. The nonbonded energy between atoms was calculated using a Lennard–Jones 6-12 nonbonding potential by the method of Gibson and Scheraga. The alkyl hydrogen atoms were not treated explicitly, however, but were regarded as part of an enlarged spherical carbon atom. The

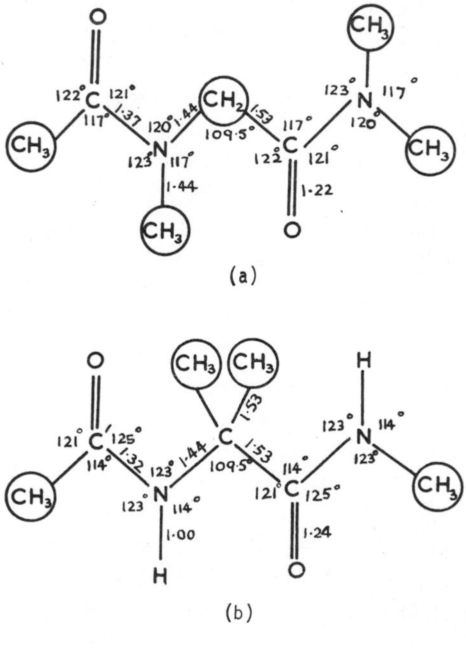

(a)

(b)

Figure 1. Molecular geometry of (a) N_α-acetylsarcosine N,N-dimethylamide and (b) N_α-acetyl-α-amino isobutyryl N-methylamide assumed for all computations on polysarcosine and poly(α-amino isobutyric acid), respectively.

Lennard–Jones parameters used for the atoms of the peptide bond were the same as those used by Gibson and Scheraga, but the parameters for the enlarged carbon atoms were slightly different because they were recalculated from a more extensive compilation of refractive index and density data for hydrocarbons (20).

Dimensions of regular helical chains. Two parameters commonly used to describe regular polypeptide helices (i.e., helices in which (ϕ, ψ) pairs are identical for each residue in the chain) are the number of residues per turn in the helix (n) and the axial translation in Å per residue along the helix (h). Using the method outlined by Leach *et al.* (23) and based on general equations set down by Miyazawa (24), n and h were calculated for *cis* and *trans* polysarcosine throughout the whole range of ϕ and ψ values. (The results for the *trans* polysarcosine helix were almost identical to the values for a similar calculation (25) on a helix with Pauling–Corey geometry.

RESULTS AND DISCUSSION

Conformational energy calculations. Many of the properties of a polypeptide with *trans* peptide bonds are determined by the interatomic forces within each residue of the chain (14, 23, 25, 26). It is possible, therefore, to arrive at a first approximation to the most likely regular helical conformations, as well as to the random-coil properties (16), from the conformational energy surface for the individual residues of these polymers. The effect of methylation on the conformational freedom of *trans* polypeptides is well illustrated by a comparison of the (ϕ, ψ) conformational energy diagrams of N_α-acetyl-*trans*-sarcosine N,N-dimethylamide (6, 8) and N_α-acetyl-α-amino isobutyryl N-methylamide (5), shown in Figs. 2(a) and (b). The addition of a C_β methyl group or an N-methyl

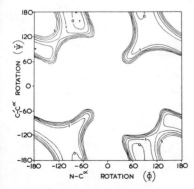
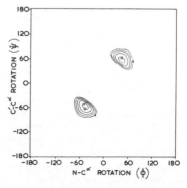

Figure 2. Conformational energy diagrams as a function of ϕ and ψ for (a) N_α-acetyl-*trans*-sarcosine-N,N-dimethylamide ($\omega = 180°$) and (b) N_α-acetyl-α-amino isobutyryl-N-methylamide ($\omega = 180°$). The energy contour values are relative to that of the lowest-energy contour. The locations of one set of local energy minima are shown by dots.

group to the glycyl residue reduces the area of low-energy conformations from over 50% to less than 20% (within 5 kcal/mole of the global minimum). This is true regardless of whether the *cis* or *trans* isomer of the *N*-methyl–substituted peptide is considered. The addition of still another methyl group either at the C_β position or on the peptide bond of the alanine residue (27, 28) reduces the low-energy areas to less than 5% of the total (ϕ, ψ) area. The allowed (i.e., low-energy) regions for these methyl-substituted peptides are quite different; therefore their properties would be expected to differ. The positions and relative energies for the local energy minima for N_α-acetylsarcosine *N,N*-dimethylamide (*cis*, $\omega = 0°$; *trans*, $\omega = 180°$) and N_α-acetyl-α-amino isobutyryl *N*-methylamide are given in Table 1.

The most stable conformations for the sarcosyl residue appear to be *trans*. This result is supported by NMR studies of a similar model compound: "The *trans* conformer is preferred by about 2.5:1 in DMSO-d_6 .... The *trans/cis* ratio varies somewhat in other solvents ... but the *trans* is always strongly preferred" (11). Similar theoretical results for the sarcosyl residue have been reported by Conti and De Santis (8) using slightly different geometry and different empirical energy algorithms. No experimental evidence on the (ϕ, ψ) conformations for these peptides is yet available.

Table 1. Dihedral angles and relative energies for the preferred conformations of the sarcosyl and α-amino isobutyryl model compounds

Residue	Dihedral Angles[a]			Relative Energies[b] (kcal/mole)
	ϕ	ψ	ω[c]	
Sarcosyl	$-70°$	$160°$	$180°$	0.00
	$-80°$	$-180°$	$180°$	0.14
	$-170°$	$90°$	$180°$	0.39
	$-150°$	$80°$	$0°$	0.63
	$-90°$	$-110°$	$0°$	0.65
	$-80°$	$150°$	$0°$	0.80
α-Amino isobutyryl	$-50°$	$-50°$	$180°$	0.00

[a] The conformational energy maps for both model compounds are centrosymmetrical and each of the preferred conformations (ϕ, ψ) listed has a counterpart $(-\phi, -\psi)$, which also describes a conformation of minimum energy.

[b] The conformational energy of the global minimum for each energy map was set to zero; the other minima are expressed relatively.

[c] Although both the sarcosyl and α-amino isobutyryl model compounds have two peptide bonds, *cis* and *trans* isomerism about both bonds need not be considered. Thus, in the case of the sarcosyl model compound, the *cis* and *trans* isomers of the right-hand peptide group are identical. The presence of a free N–H group in the α-amino isobutyryl model compound, however, makes the *trans* conformation the most probable one.

The effect of extending N_α-acetylsarcosine N,N-dimethylamide to the octa-peptide N_α-acetyl(sarcosyl)$_8$ N,N-dimethylamide for both *cis* ($\omega = 0°$) and *trans* ($\omega = 180°$) isomers is shown in Figs. 3(a) and (b), and the relative energies of the local minima of the octapeptides are given in Table 2. It can be seen that the conformational energy maps for *trans*-sarcosine peptides change little when the chain is extended. There is, however, a small area in the vicinity of $(-50°,-90°)$ that is disallowed for oligomers of more than four residues. In this respect, our *trans*-polysarcosine maps differ from those presented by Tanaka and Nakajima (6) and Mattice [Fig. 3(c) in (3)]. Their polysarcosine maps are very similar to

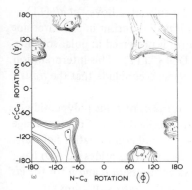 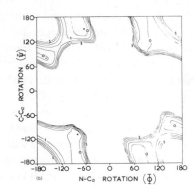

Figure 3. Conformational-energy diagrams as a function of ϕ and ψ for regular (a) *cis*-polysarcosine and (b) *trans*-polysarcosine helices.

Table 2. Characteristics of stable helical conformations of polysarcosine

Desig-nation	Dihedral Angle[a]			h (Å)	n	Sense of Helix	Relative Energy (kcal/mole per Residue)
	ϕ	ψ	ω				
trans I	$-80°$	$-110°$	$+180°$	-1.60	-5.10	Left	1.51
trans II	$-160°$	$+80°$	$+180°$	$+3.04$	-3.05	Right	1.88
trans III	$-60°$	$+150°$	$+180°$	$+2.81$	$+3.34$	Left	1.64
cis I	$-90°$	$-110°$	$0°$	-2.71	$+3.25$	Right	1.96
cis II	$-160°$	$+90°$	$0°$	-1.67	-3.43	Left	0.99
cis III	$-90°$	$+150°$	$0°$	$+1.38$	-3.35	Right	0.00

[a]The characteristics of only half of the possible helices of minimum potential energy for *cis* and *trans* polypeptides are described. The remaining helices are stereoisomers arising from the twofold centrosymmetry of the conformation map and occur at $(-\phi,-\psi)$.

our maps for N_α-acetylsarcosine N,N-dimethylamide. The conformational energy diagrams for regular *cis* sarcosine peptides change significantly on extending N_α-acetylsarcosine N,N-dimethylamide to the regular *cis* polypeptide. For the *cis* polymer, unlike the *trans* polypeptide, conformations in the region around $(-120°, 130°)$ are now disallowed.

The energy values of the preferred conformations for *trans* polysarcosines are similar to those for the *trans* model compound. However, the conformational energies of the most stable *cis* polysarcosines are significantly lower than those of the *cis* model compound. Thus, in a polymer of at least four residues the *cis* conformation of the peptide bond is energetically preferred.

It should be noted that because the effect of solvation has not been included in these calculations, these results are most applicable to regular polysarcosine helices in the solid phase, in films, and in nonpolar solvents. Bovey *et al.* (11) have deduced from NMR measurements that *cis-trans* isomerism in polysarcosine varies markedly with solvent. While the *trans* form is preferred in polar solvents, "the *cis* conformation is strongly preferred in $CDCl_3$" (11). This interpretation, however, has been questioned by Sisido *et al.* (29) who conclude that the *trans* conformer predominates in $CDCl_3$.

The pitch (h) and the number of residues per turn (n) of *trans* polypeptides as a function of all possible (ϕ, ψ) values have been described (25). Frequent usage of these maps has produced a familiarity with the overall conformations of *trans* polypeptides. The geometry of the corresponding *cis* polypeptides ($\omega = 0$) has not yet been described, however. Fig. 4(a) shows contours of n and h for the *cis* structures plotted on a (ϕ, ψ) map; the contours for the *trans* structures are shown in Fig. 4(b) for comparison. Both sets of calculations are based on the amide geometry appropriate to sarcosyl peptides (17).

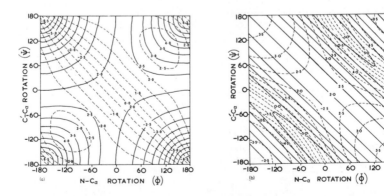

Figure 4. Contour maps showing the axial translation per residue (– –) and number of residues per turn (—) for (a) *cis* poly helices and (b) *trans* poly helices, using the geometry of Fig. 1(a).

All helices described by the area around the contour $h = 0$ are disallowed. For example, the center of the map, where $\phi \cong \psi \cong 0°$, describes flat, narrow helices in which h approaches zero and $n \cong 2$. Such helices would be excluded by severe atomic overlaps between adjacent turns. At the corners of the map, where $\phi \cong \psi \cong \pm 180°$, the helices are flat, wide spirals in which h is again zero and $n > 6$; severe steric restrictions again exclude such structures. The remaining disallowed areas are between $\phi \cong \psi \cong 0°$ and $\phi \cong \psi \cong \pm 60°$, where the distances (nh) between adjacent turns of the helices are again too short to avoid atomic contacts between turns. Each of the six sterically permitted areas [see Fig. 3(a)] contains one structure of minimum potential energy. The values of n, h, and helix chirality of all the preferred conformations of both *cis* and *trans* polysarcosines are shown in Table 2.

The most stable conformation of the *trans*-polysarcosine molecule (*trans* I, Table 2) is a rather broad, left-handed helix with a small axial translation per residue. The atoms are closely packed, and even a small increase in ϕ ($\sim 10°$) causes severe interatomic overlaps to occur between the N-methyl of the ith residue and the carbonyl oxygen atom of the $(i + 4)$th residue. In this lowest-energy conformation the methyl groups on adjacent peptide-bond nitrogen atoms are separated by 3.95 Å – the minimum in the Lennard–Jones potential for two interacting methyl groups (20). This feature is one of the major stabilizing forces for this helix. The steric proximity of the methyl on the ith peptide nitrogen and the carbonyl oxygen of the $(i + 3)$th and $(i + 5)$th residues also contributes to the stability of this helix. Fig. 5(a) shows this conformation. There are two other minima on the conformational energy map for *trans* polysarcosine [see Table 2 and Fig. 3(a)]. The conformation designated *trans* II is an extended right-handed helix in which the methyl groups are directed in towards the helical axis and lie within the arc of the polypeptide backbone. *Trans* III is also extended, but it is left-handed and the methyl groups are directed away from the helix axis. The atomic packing in these two helices does not provide as much stabilization energy as in the case of the *trans*-I helix.

The two most stable *cis* polysarcosine helices, designated *cis* II and *cis* III in Table 2, have lower conformational energies than the *trans*-I helix. The most stable of these two structures (*cis* III), illustrated in Fig. 5(b), is a right-handed helix with 3.4 residues per turn and an axial translation per residue of 1.4 Å – similar to the n and h values for the *trans* α helix described by Pauling and Corey (18, 23). The second most stable structure (*cis* II) has a similar overall geometry ($n = 3.4, h = 1.7$), but is left-handed. The major stabilizing forces in the *cis*-II and *cis*-III helices include van der Waals contacts between the N-methyl of the ith residue and the carbonyl oxygen of the $(i + 1)$th residue. The *cis*-III helix is further stabilized by other close contacts, $C_{\alpha,i} \cdots O_{i + 2}$, $C_{\alpha,i} \cdots O_{i + 3}$, and $C_{\alpha,i} \cdots N$-methyl$_{i + 2}$. The *cis*-I helix (Table 2) is an extended helix that has few van der Waals contacts to provide strong stabilizing forces. Fig. 5(c) is a molecular representation of this structure.

It could be predicted from this theoretical study that the *trans* conformation would be preferred for N_α-acetylsarcosine N,N-dimethylamide and that polymers of sarcosine should favor the *cis* conformation.

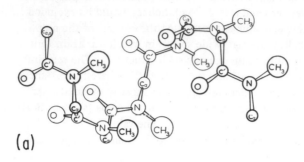

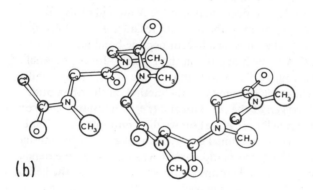

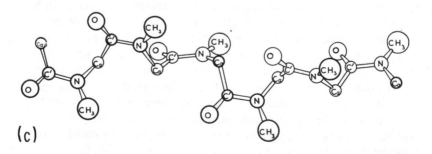

Figure 5. Molecular drawings of regular polysarcosine helices corresponding to three of the six minima in the conformational energy diagrams shown in Fig. 3. The view shown has its axis in the plane of the paper. $C_{\alpha,0}$ denotes the *first* α-carbon atom in the polymer chain at the N terminus; six amino acid residues are shown for each helix: (a) *trans* I (–80°, –110°), (b) *cis* III (–90°, +150°), (c) *cis* I (–90°, –110°).

There are only two regions of the conformation map available to N_α-acetyl-α-amino isobutyryl N-methylamide. These small areas surround the positions on the map corresponding to the left- and right-handed α helices [Table 1 and Fig. 2(b)]. The severely restricted conformational space available to each residue suggests that the random-coil characteristics of the poly(amino acid) would be unusual; these characteristics are reported elsewhere (16).

The obligatory α-helical character of the α-amino isobutyryl residue would make it a useful tool in helix-coil transition studies. The introduction of such residues into biologically active peptides or proteins may also allow specific conformations to be engineered. Taking this concept further, methylation at the C_α position might be combined with the proline residue to give N_α-acetyl-L-α-methylproline N-methylamide, in which the (ϕ, ψ) energy surface should be restricted to a *single* region, namely, the right-handed α-helical conformation. The D isomer would be restricted to the left-handed α-helical conformation.

ACKNOWLEDGMENTS

Support from a Commonwealth Scientific and Industrial Research Organization postgraduate studentship (A.W. Burgess) and from the Australian Research Grants Committee (S.J. Leach) is gratefully acknowledged. The authors also thank the Centre for Computer Services of Melbourne University for assistance and the Computing Center of the National Institutes of Health, Bethesda, Md., for the use of ADAGE display facilities.

REFERENCES

1. P. Mueller and D.O. Rudin, *Biochem. Biophys. Res. Commun.*, **26**, 398 (1967).
2. G.W. Kenner and R.C. Sheppard, *Nature* (London), **181**, 48 (1958).
3. H.M. Sobell, S.C. Jain, T.D. Sakore, G. Ponticello, and C.E. Norman, *Cold Spring Harbor Symposia on Quantitative Biology*, **36**, 263 (1971).
4. A.I. McMullen, D.I. Marlborough, and P.M. Bayley, *FEBS Lett.*, **16**, 278 (1971).
5. A.W. Burgess and S.J. Leach, *Biopolymers*, **12**, 2599 (1973).
6. S. Tanaka and A. Nakajima, *Poly. J.* (Japan), **2**, 717 (1971).
7. A.W. Burgess and S.J. Leach, *Proc. Aust. Biochem. Soc.*, **4**, 12 (1971).
8. F. Conti and P. De Santis, *Biopolymers*, **10**, 2581 (1971).
9. J.H. Fessler and A.G. Ogston, *Trans. Faraday Soc.*, **47**, 667 (1951).
10. G.D. Fasman and E.R. Blout, *Biopolymers*, **1**, 99 (1963).
11. F.A. Bovey, J.J. Ryan, and F.P. Hood, *Macromolecules*, **1**, 305 (1968).
12. S.J. Leach, G. Nemethy and H.A. Scheraga, *Biopolymers*, **4**, 369 (1966).
13. G.N. Ramachandran, C. Ramakrishnan, and V. Sasisekharan, *J. Mol. Biol.*, **7**, 95 (1963).
14. H.A. Scheraga, *Chem. Rev.*, **71**, 195 (1971).
15. A. Eliott, *Proc. Roy. Soc.* (London), **A226**, 408 (1954).
16. A.W. Burgess, Y. Paterson, and S.J. Leach, *J. Polymer Sci.*, in press.
17. W.R. Krigbaum, R.-J. Roe, and J.D. Woods, *Acta Cryst.*, **B24**, 1304 (1968).

18.   R.B. Corey and L. Pauling, *Proc. Roy. Soc.* (London), **B141**, 10 (1953).
19.   S. Hirokawa, S. Kuribayashi, and I. Nitta, *Bull. Chem. Soc.* (Japan), **25**, 192 (1952).
20.   K.D. Gibson and H.A. Scheraga, *Proc. Nat. Acad. Sci. U.S.*, **58**, 420 (1967).
21.   J.F. Yan, G. Vanderkooi, and H.A. Scheraga, *J. Chem. Phys.*, **49**, 2713 (1968).
22.   A.W. Burgess, Ph.D. Thesis, University of Melbourne, Victoria, Australia.
23.   S.J. Leach, G. Nemethy, and H.A. Scheraga, *Biopolymers*, **4**, 887 (1966).
24.   T. Miyazawa, *J. Poly. Sci.*, **55**, 215 (1961).
25.   G.N. Ramachandran and V. Sasisekharan, *Adv. Prot. Chem.*, **23**, 283 (1968).
26.   D.A. Brant and P.J. Flory, *J. Am. Chem. Soc.*, **87**, 2791 (1965).
27.   J.E. Mark and M. Goodman, *Biopolymers*, **5**, 809 (1967).
28.   A.M. Liquori and P. De Santis, *Biopolymers*, **5**, 815 (1967).
29.   M. Sisido, Y. Imanishi, and T. Higashimura, *Biopolymers*, **11**, 399 (1972).
30.   W.L. Mattice, *Macromolecules*, **6**, 855 (1973).

GUIDES TO THE EVALUATION OF PEPTIDE CONFORMATION

VINCENT S. MADISON, Department of Biological Chemistry, Harvard Medical School, Boston, Mass. 02115

SYNOPSIS: A method is described for interfacing theoretical and experimental results in the determination of solution conformations of peptides. The role of dipolar interactions in stabilizing γ turns and β turns, which are evident in cyclic hexapeptides, is explored. Although dipolar interactions are sufficient to stabilize γ turns in $cyclo(\text{-Pro-Gly-})_3$, additional forces are necessary for the stabilization of β turns in $cyclo(\text{-Gly-Pro-Gly-})_2$.

INTRODUCTION

In the study of solution conformers of peptides, judicious interfacing of theoretical and experimental methods is a powerful tool for the identification of conformational details and leads to characterization of the fundamental molecular forces that determine the preferred conformational state. For example, the consistent-force-field method developed by Lifson *et al.* (1) is a mathematical link between theory and experiment that has been successfully applied to small molecules. A less mathematically precise method that is applicable to larger molecules is summarized. First, a qualitative conformational distribution is computed from simplified potential functions. Second, within the low-energy regions near each local minimum, theoretical values are obtained for molecular properties, including hypo- and hyperchromism, circular dichroism (CD) spectra, dipole moment, proton-proton coupling constants, and chemical shifts induced by paramagnetic ions. Third, utilizing the theoretical results of the first two steps and empirically established correlations, an actual conformational distribution is derived from experimental absorption, CD, nuclear magnetic resonance (NMR), and infrared (IR) spectra; to determine conformational details and to avoid the pitfalls of circular reasoning, complementary information provided by the different types of measurements is *essential*. Finally, having identified the solution conformer, the adequacy of the original potential function, as well as possible additional terms to improve its accuracy, can be evaluated.

Synthetic cyclic hexapeptides are excellent models for the study of conformational determinants. Cyclization and steric constraints allow a complete exploration of conformational space. Symmetric sequences in cyclic peptides permit amplification (by repetition) of subtle free-energy differences. Of course, cyclic molecules avoid the complications of end effects. In addition, the synthetic cyclic peptides have been shown to mimic properties of biological molecules such as cation binding (2) and the formation of hydrogen-bonded turns (3). Two classes of these turn structures have been found in proteins (4). One class, the γ turn, is stabilized by a hydrogen bond between residues i and $i + 2$. In the other class, the β turn, the hydrogen bond is between residues i and $i + 3$ (5).

In this communication the arduous task of determining solution conformers is short-circuited. Rather, the role of dipole-dipole interactions in stabilizing β turns and γ turns (which have been substantiated experimentally) in two representative cyclic hexapeptides is discussed. The proline-containing peptides *cyclo*(-L-prolyl-glycyl-)$_3$ [hereafter *cyclo*(-Pro-Gly-)$_3$] and *cyclo*(-glycyl-L-prolyl-glycyl-)$_2$ [hereafter *cyclo*(-Gly-Pro-Gly-)$_2$], which are thought to contain γ turns and β turns, respectively, are considered.

METHODS OF COMPUTATION

For *cyclo*(-Gly-Pro-Gly-)$_2$ the glycine residues were assumed to have standard Pauling-Corey geometry (6), and the proline residues were taken from the crystal structure of Leu-Pro-Gly (7). The peptide units were fixed in the planar *trans* conformation (8). Dihedral angles for C_2-symmetric cyclic structures were generated by the method of Gō and Scheraga (9). [They have applied this method to the study (10) of *cyclo*(-Gly-)$_6$.] Within the symmetry unit, Gly$_1$-Pro-Gly$_2$, three dihedral angles are independent—(ϕ,ψ)Gly$_1$ and ψPro; for the remaining two dihedral angles—(ϕ,ψ)Gly$_2$—there may be 0, 1, or 2 solutions that give a cyclic structure. For the cyclic structures, the intramolecular nonbonded potential energy was evaluated as the sum of a Lennard–Jones 6–12 term with parameters from Karplus and Lifson (1b) and a coulombic term with monopole charges (11) calculated from Smyth's bond moments (12). For potentially hydrogen-bonded atoms, the Lennard–Jones parameters were adjusted so that the minimum occurred at an H$\cdots$O distance of 1.8 Å. No explicit hydrogen-bonding potential was incorporated. The following abbreviations are used: TPE for total intramolecular potential energy, ESE for electrostatic potential energy, NBE for nonbonded van der Waals potential energy, and ϵ for the dielectric constant.

Similar methods utilized for calculations on *cyclo*(-Pro-Gly-)$_3$ have been reported elsewhere (13).

RESULTS AND DISCUSSION

The solution conformers of the cyclic hexapeptide *cyclo*(-Pro-Gly-)$_3$ have been thoroughly investigated by ^{1}H and ^{13}C NMR (2b, 14), as well as by computed intramolecular potential energies and theoretical CD spectra (13). Experimental CD spectra and additional NMR data that will allow completion of the conformational analysis are forthcoming (15).

In nonpolar solvents, such as chloroform and dioxane, *cyclo*(-Pro-Gly-)$_3$ takes up a C_3-symmetric conformer that is stabilized by three $1 \leftarrow 3$ hydrogen bonds, γ turns (Fig. 1). This observed conformer is near the global minimum in the computed intramolecular potential energy. The electrostatic energy of dipole-dipole interactions contributes significantly to the stabilization of the γ-turn conformer (see Fig. 2 near ψPro = 70°). The intramolecular $1 \leftarrow 3$ hydrogen bonds are not observed in polar solvents (such as water and alcohols), which attenuate electrostatic interactions.

Figure 1. Photograph of Corey-Pauling-Koltun model of γ-turn conformer of *cyclo*(-Pro-Gly-)$_3$. Note that although each of the peptide units is planar, the glycine N—H, which is C-terminal to proline, deviates significantly from the plane of the peptide unit that is N-terminal to proline. The dihedral angles of the pictured conformer are given in Table 1.

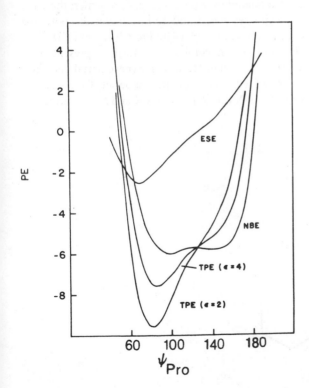

Figure 2. Intramolecular potential energies (kcal/mole) computed for C_3-symmetric conformers of *cyclo*(-Pro-Gly-)$_3$ [from (13)]. For each ψ_{Pro}, the single independent variable, only one of the two solutions for the dependent dihedral angles yields a sterically reasonable structure. The ESE was computed for a dielectric constant $\epsilon = 4$. For the two TPE curves, the values of ϵ are indicated in parentheses.

In order to explore further the role of dipolar interactions in stabilizing the γ-turn conformation, the ESE between individual groups was computed for the array of atoms shown in Fig. 3. Cartesian coordinates for these atoms were computed for the *cyclo*(-Pro-Gly-)$_3$ conformer (Table 1) that best matches both theoretical and experimental results (15). Using atomic charges (Set 2, Table 2), we computed the ESE *in vacuo* via Coulomb's law. The results (Table 3) show that interactions between the N–H and C=O atoms within a single glycine residue and those of residues 1 and 3 both stabilize the conformation. The atoms shown in Fig. 3 contribute almost all of the ESE to the molecule; each of the remaining polar groups, the proline carbonyls, is well separated from other polar groups. When the γ turns are broken, both the $1 \leftarrow 3$ and the $1 \rightarrow 1$ interactions are disrupted so that the ESE increases dramatically (Fig. 2).

A second type of structure, the β turn (Fig. 4), has been postulated for *cyclo*(-Gly-Pro-Gly-)$_2$ and related molecules (3). It has been shown conclusively that the N–H of the glycine preceding proline (residues 1 and 4 in Fig. 4) is shielded from the solvent and that the amide proton's chemical shift is relatively insensitive to temperature (16). These facts, as well as the $J_{N\alpha}$ (HNC$_\alpha$H) coupling constants, are nicely accommodated by a model (Fig. 4) in which there are two hydrogen bonds between glycine residues 1 and 4. Intramolecular potential energies for all C_2-symmetric conformers of *cyclo*(-Gly-Pro-Gly-)$_2$ and CD spectra for the low-energy regions have been computed and will be presented elsewhere. These theoretical results, in conjunction with experimental data from L.G. Pease, are being used to derive further conformational details for a series of cyclic hexapeptides of the type *cyclo*(-X-Pro-Y-)$_2$, where X and Y are various amino acids.

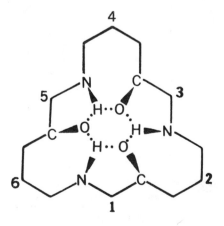

Figure 3. Schematic representation of γ-turn conformer of *cyclo*(-Pro-Gly-)$_3$ (see Fig. 1 and Table 1). The glycine atoms shown were used to compute ESE. Dotted lines indicate possible hydrogen-bonding interactions. The residue numbering system is indicated by the numerals near the C$_\alpha$'s.

Table 1. Selected conformers of cyclic hexapeptides

Molecule	Conformer[a]	Symmetry	(ϕ,ψ,ω)			NBE	ESE[a]
			Gly_1	Pro	Gly_2	(kcal/mole)	
cyclo(-Pro-Gly-)$_3$	γ turn	C$_3$	(170°,−162°,180°)	(−80°,70°,180°)	—	−4.0	−12.4
cyclo(-Gly-Pro-Gly-)$_2$	β turn	C$_2$	(−160°,180°,180°)	(−68°,100°,180°)	(67°,40°,180°)	−1.7	−4.5
cyclo(-Gly-Pro-Gly-)$_2$	Local minimum energy	C$_2$	(160°,180°,180°)	(−68°,100°,180°)	(65°,74°,180°)	−6.1	−0.7

[a]These conformers were selected: (1) γ turn, best match with theoretical and experimental data (15); (2) β turn, a relatively low-energy conformer with 1→4 and 4→1 hydrogen bonds; (3) local minimum energy, a local-minimum-energy conformer near the β turns, but without 1→4 hydrogen bonds.

[b]The electrostatic energy *in vacuo* ($\epsilon = 1$).

Table 2. Static atomic charges

Atom	Charge[a]	
	Set 1[b]	Set 2
C'	0.42	0.39
O	–0.39	–0.39
N (tertiary)	–0.10	0.00
N (secondary)	–0.34	–0.27
H (N)	0.27	0.27

[a]The charges are in fractions of the electronic charge.

[b]There are additional charges on other atoms so that the molecule is electrically neutral. Charge Set 1 was used in calculating ESE of the entire molecule. Charge Set 2 was used in calculating ESE for the more limited atomic arrays (see Figs. 3 and 5).

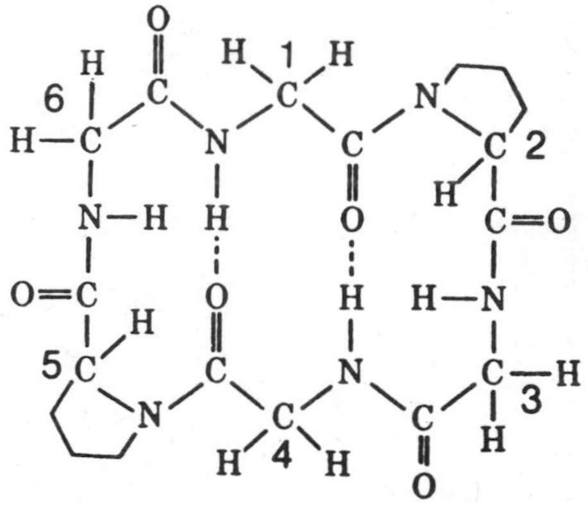

Figure 4. Diagram of proposed β-turn conformer of *cyclo*(-Gly-Pro-Gly-)$_2$ (16). The residue numbering system is indicated.

The role of dipolar interactions in stabilizing β turns was investigated for an idealized planar array (Fig. 5). Both amino acid residues have $(\phi,\psi) = (180°, 180°)$. The second residue was generated from the first by a rotation of 180° about the line through C_α that is perpendicular to the plane of the atoms shown in Fig. 5, followed by translation to give the desired hydrogen bond (O–H) distance. The results show that attractive and repulsive terms decrease simultaneously with hydrogen bond distance, so that the net effect on the on the molecular ESE is smaller that would otherwise be expected.

For the computed C_2-symmetric *cyclo*(-Gly-Pro-Gly-)$_2$ conformers (Table 1 and unpublished results), cyclization and steric constraints prevent coplanarity of the N–H and C=O groups of residues 1 and 4, thus reducing the ESE of the

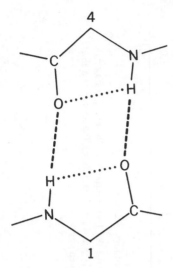

Fig. 5. Idealized representation of hydrogen-bonded atoms in a cyclic hexapeptide β turn. All atoms are coplanar. The distance between the oxygen atom of
residue 1 and the hydrogen of residue 4 is 2.30 Å in this figure.

$1 \rightarrow 4$ and $4 \rightarrow 1$ interactions. In fact, for the cyclic hexapeptide, the electrostatic stabilization is less for a hydrogen-bonded conformer than for a neighbor
ing conformer without $1 \rightarrow 4$ and $4 \rightarrow 1$ hydrogen bonds (compare the β-turn
and local-minimum-energy conformers in Table 3). In $cyclo$(-Gly-Pro-Gly-)$_2$,
changes in the ESE are dominated by changes in the $1 \rightarrow 1$, $1 \leftarrow 3$, and $2 \rightarrow 4$
(rather than $1 \rightarrow 4$) type interactions.

The ESE is not sufficient to stabilize β-turn conformers of $cyclo$(-Gly-Pro-
Gly-)$_2$ relative to other conformational states. Formation of $1 \rightarrow 4$ and $4 \rightarrow 1$
hydrogen bonds leads to an increase in nonbonded van der Waals energy in C_2-
symmetric conformers of $cyclo$(-Gly-Pro-Gly-)$_2$ and also in $cyclo$(-Gly-)$_6$ (10).
In order for the β-turn conformers to be favored in cyclic hexapeptides, their
higher van der Waals energy must be compensated or reduced. Compensation
energy could arise from a chemical-bonding component of the hydrogen-bonding
potential. The van der Waals energy could probably be reduced by allowing
small variations of bond angles and/or deviations from peptide unit planarity.
An alternative would be to relax the C_2-symmetry requirement so that only one
$1 \rightarrow 4$ (or $4 \rightarrow 1$) hydrogen bond is formed at a time. Work is in progress (17)
that will be useful in identifying the forces that stabilize the various conformational states of $cyclo$(-X-Pro-Y-)$_2$ peptides.

CONCLUSION

Intramolecular ESE is sufficient to stabilize the γ-turn conformation of $cyclo$-
(-Pro-Gly-)$_3$. In contrast, the ESE does not seem to contribute significantly to
the stabilization of β turns in $cyclo$(-Gly-Pro-Gly-)$_2$. Yet the enigma remains

Table 3. Electrostatic energies of hydrogen-bonded arrays

Molecule	Conformer	Type of Interaction	O···H Distance (Å)	Electrostatic Energies[a] (kcal/mole)		
				Single Interaction	Repulsion of Array[b]	Net in Molecule[c]
cyclo(-Pro-Gly-)$_3$	γ turn	1→1	2.24	-2.7	–	-8.1
		1→3	2.02	-2.9	3.3	-5.4
Planar array	β turn	1→1	2.22	-2.8	–	-5.6
		1→4	1.85	-4.3	2.8	-5.8
		1→4	2.00	-3.6	2.6	-4.6
		1→4	2.30	-2.8	2.0	-3.6
cyclo(-Gly-Pro-Gly-)$_2$	β turn	1→1	2.26	-2.7	–	-5.4
		1→4	2.34	-1.3	2.0	-0.6
cyclo(-Gly-Pro-Gly-)$_2$	Local minimum energy	1→1	2.26	-2.7	–	-5.4
		1→4	3.26	-1.1	1.1	-1.0

[a]All electrostatic energies were calculated *in vacuo* ($\epsilon = 1$).

[b]Electrostatic repulsion due to the close proximity to the first hydrogen bond of additional ones (see Figs. 3 and 5).

[c]The net ESE in the molecule from the arrays of atoms shown in Figs. 3 and 5 was calculated: (1) for the 1→1 interactions, the energy of a single interaction was multiplied by the number of interactions of this type in the molecule; (2) for the 1←3 and 1→4 interactions, the energy of a single interaction was multiplied by the number of interactions followed by subtraction of the repulsion energy.

that β-turn conformers seem to be stable in polar solvents, such as water, but γ turns (which have the lower computed intramolecular energy) are stable only in nonpolar solvents, such as dioxane. One factor that may favor β-turn stability is that the 1 → 4 type hydrogen bonds are much more nearly coplanar and linear than the 1 ← 3 type hydrogen bonds in γ turns.

ACKNOWLEDGMENTS

Thanks are due my colleagues, Lila G. Pease and Charles M. Deber, whose experimental work and incisive comments have increased the relevance of this work. I also wish to thank Elkan R. Blout for the impetus and support he has given this project. The U.S. Public Health Service provided financial support through a postdoctoral fellowship and Grant AM-07300.

REFERENCES

1. (a) A. Warshel, M. Levitt, and S. Lifson, *J. Mol. Spec.*, **33**, 84 (1970).
   (b) S. Karplus and S. Lifson, *Biopolymers*, **10**, 1973 (1971).
2. (a) B.F. Gisin and R.B. Merrifield, *J. Am. Chem. Soc.*, **94**, 6165 (1972).
   (b) C.M. Deber, D.A. Torchia, S.C.K. Wong, and E.R. Blout, *Proc. Nat. Acad. Sci. U.S.*, **69**, 1825 (1972). (c) T. Wieland, in *Chemistry and Biology of Peptides*, J. Meienhofer, ed., Ann Arbor Science Publishers, Ann Arbor, Mich., 1972, p. 377. (d) V.T. Ivanov, A.V. Evstratov, L.V. Sumskaya, E.I. Melnik, T.S. Chumburidze, S.L. Portnova, T.A. Balashova, and Yu.A. Ovchinnikov, *FEBS Letters*, **36**, 65 (1973).
3. (a) I.L. Karle and J. Karle, *Acta Cryst.*, **16**, 969 (1963). (b) R. Schwyzer and U. Ludescher, *Helv. Chim. Acta*, **52**, 2033 (1969). (c) K.D. Kopple, A. Go, R.H. Logan, Jr., and J. Savrda, *J. Am. Chem. Soc.*, **94**, 973 (1972). (d) D.A. Torchia, S.C.K. Wong, C.M. Deber, and E.R. Blout, *J. Am. Chem. Soc.*, **94**, 616 (1972).
4. (a) B.W. Matthews, *Macromolecules*, **5**, 818 (1972). (b) G. Nemethy and M. Printz, *Macromolecules*, **5**, 755 (1972). (c) J.L. Crawford, W.N. Lipscomb, and C.G. Schellman, *Proc. Nat. Acad. Sci. U.S.*, **70**, 538 (1973).
5. (a) C.M. Venkatachalam, *Biopolymers*, **6**, 1425 (1968).
   (b) R. Chandrasekaran, A.V. Lakshminarayanan, U.V. Pandya, and G.N. Ramachandran, *Biochim. Biophys. Acta*, **303**, 14 (1973).
6. L. Pauling, *The Nature of the Chemical Bond* (Third Edition), Cornell University Press, Ithaca, N.Y., 1960, pp. 281, 498.
7. Y.C. Leung and R.E. Marsh, *Acta Cryst.*, **11**, 17 (1958).
8. IUPAC-IUB Commission on Biochemical Nomenclature, *Biochemistry*, **9**, 3471 (1970).
9. N. Gō and H.A. Scheraga, *Macromolecules*, **6**, 273 (1973).
10. N. Gō and H.A. Scheraga, *Macromolecules*, **6**, 525 (1973).
11. P.M. Bayley, E.B. Nielsen, and J.A. Schellman, *J. Phys. Chem.*, **73**, 228 (1969).
12. C.P. Smyth, *Dielectric Behavior and Structure*, McGraw-Hill, New York, 1955, pp. 244–245.

13. V. Madison, *Biopolymers,* **12**, 1837 (1973).
14. C.M. Deber, E.R. Blout, D.A. Torchia, D.E. Dorman, and F.A. Bovey, in *Chemistry and Biology of Peptides,* J. Meienhofer, ed., Ann Arbor Science Publishers, Ann Arbor, Mich., 1972, p. 39.
15. V. Madison, M. Atreyi, C.M. Deber, and E.R. Blout, Cyclic Peptides. IX. Conformations of a Synthetic Ion-Binding Cyclic Peptide, *Cyclo-*(Pro-Gly)$_3$, from Circular Dichroism and ^{1}H and ^{13}C Nuclear Magnetic Resonance, submitted to *J. Am. Chem. Soc.*
16. L.G. Pease, C.M. Deber, and E.R.Blout, *J. Am. Chem. Soc.,* **95**, 258 (1973).
17. L.G. Pease, unpublished data.

ON THE NATURE OF THE BINDING OF HEXA-*N*-ACETYLGLUCOSAMINE SUBSTRATE TO LYSOZYME

MICHAEL LEVITT, Department of Chemical Physics, Weizmann Institute of Science, Rehovot, Israel, and MRC Laboratory of Molecular Biology, Cambridge, England

SYNOPSIS: Minimization of an empirical energy function has been used to study the binding of hexa-*N*-acetylglucosamine substrate to lysozyme. The problems associated with nonconvergent and convergent energy minimization, selection of flexible and rigid parts of the complex, and choice of a suitable enzyme conformation are detailed. In the various refined conformations of the bound substrate, all sugar rings have the full-chair conformation. A more detailed study of the binding of artificially locked substrates to the D subsite shows that there is no nonbonded or purely steric preference for a half-chair sugar ring in that subsite. This result differs from the finding of Blake *et al.* (1) that short nonbonded contacts force the D sugar ring toward a half-chair conformation. A consequence of these calculations is the speculation that for lysozyme binding energy contributes to catalysis by increasing the electrostatic potential energy of charged side chains in the active site region rather than by subjecting the substrate to mechanical or geometric strain.

INTRODUCTION

There are some basic problems associated with the study of the conformation of enzyme-substrate complexes. Good substrates do not usually form stable complexes with the enzyme because they bind, react chemically, and the products then rapidly leave the active site. Poor substrates or inhibitors that sometimes form stable complexes with the enzyme may not be bound in exactly the same way as good substrates. The conformation of bound substrates and of transition states in the catalytic process must be inferred from the conformation of stable complexes of inhibitors, indirect chemical measurements, and careful model building. Energy minimization can be used to refine the proposed model of the complex because the energy of the enzyme and substrate is minimal when the substrate is bound. It may seem paradoxical that the enzyme-substrate complex is at a free-energy minimum for, if that is the case, how does the reaction proceed? In fact, initial binding usually occurs in a fast equilibrium step that leads to a relatively stable complex. At this point, the system cannot pass out of the potential well, over a barrier, and on to the final state consisting of the products until it receives sufficient kinetic energy.

In 1967 Blake *et al.* (1) used x-ray crystallography to determine the conformation of the complex between lysozyme and an inhibitor, tri-NAG (tri-*N*-acetyl-chitotriose, the $\beta(1-4)$ linked trimer of *N*-acetylglucosamine). They then built three more NAG residues into the active site of lysozyme to give a very plausible model for the binding of the hexa-NAG substrate. Of the six substrate residues labeled A to F, that occupy six subsites, A to F, in the cleft of lysozyme, the conformation of the first three residues, A to C, was determined by x-ray crystallography and the conformation of the second three, D to F, was determined by model building. Furthermore, Blake *et al.* (1) found that the saccharide residue bound in the D subsite makes bad contacts with the enzyme and that these contacts can be relieved by distorting the D saccharide ring from a full-chair toward a half-chair conformation. Phillips (2) proposed a reasonable mechanism for the cleavage of saccharide residues by lysozyme in which residue 35 of lysozyme (glutamic acid), acting by general acid catalysis, transfers a hydrogen to the glycosidic oxygen (O4), thereby joining substrate residues D and E. The bond between atoms C1 and O4 of residues D and E breaks. In the transition state that results there is a positive charge on atom C1 of residue D. This positive charge, known as a carbonium ion, is stabilized by the negatively charged lysozyme residue, aspartic acid 52, until it can combine with a hydroxyl radical donated by a water molecule. The carbonium ion is stabilized also by sharing its charge with the ring oxygen (O5) of the D residue, thereby making the D ring torsion angle θ_{C1-O5} planar. The distortion of sugar residue D toward a half-chair, reduces the energy difference between the bound substrate and transition state, making the cleavage of the substrate easier (3).

Energy minimization was used to study the binding of hexa-NAG to lysozyme. Certain features of the accepted mode of binding of this substrate are not supported by this work. In particular, the ring of the D substrate residue need not take up a half-chair conformation when bound to lysozyme.

METHODS

The energy of the lysozyme-substrate complex is expressed as an empirical function of all bond lengths, bond angles, torsion angles, and nonbonded interatomic distances. Nonbonded interactions are calculated between atoms separated by more than three bonds and closer than a maximum cut-off distance. Hydrogen atoms and atomic partial changes are not explicitly included in these calculations, but their effect is reproduced by the special nonbonded potentials used. These potentials are abnormally strongly repulsive at short distances to compensate for the van der Waals repulsion between the missing hydrogens. They also reproduce accurately the energy and directionality of hydrogen bonds (4), the main effect of atomic partial charges. Strong electrostatic forces between groups carrying a substantial net charge have not been treated adequately, but in a protein these groups are neutralized by the solvent. These special potentials have been used previously to refine the x-ray conformation of lysozyme (5).

The —CC— torsion potential is also used for the —CO— torsion angles in the substrate, because the lone-pair electrons of the divalent oxygen may give the same sort of steric torsion barrier as the two hydrogen atoms bound to a tetravalent carbon, —CH_2—.

Binding of the substrate to the enzyme involves formation of a stable complex that must be at a free-energy minimum. The energy of the system is minimized in two ways. One method, the nonconvergent method of steepest descents, is used to change the Cartesian coordinates of all the atoms in the enzyme-substrate complex. This method, which is known as *nonconvergent energy refinement*, moves atoms slightly in order to relieve strain, but it does not cause the large movements that may be needed for a favorable conformation. A second procedure, the convergent Davidon method (6) is used to change the Cartesian coordinates of a small number of atoms in the presence of the rest of the molecule. The enzyme is held fixed while overlapping zones of the substrate consisting of two saccharide residues are minimized sequentially, starting with the F residue. Thus, the energy of residues E and F is minimized first, that of E and D second, and so on until the whole substrate has been energy-refined. When the overlapping-zone-refinement procedure is repeated a second time, no substrate residue moves appreciably, indicating that the zone-minimization procedure has converged. As the Davidon method used for zone refinement converges, it can cause the large atomic movements that may be needed for a favorable conformation. Nevertheless, zone minimization has a disadvantage in that if atoms in two different zones interact strongly and make many nonbonded contacts, then the conformation generated by zone minimization depends on which zone is refined first. For example, if the atoms of zone P are refined first, while atoms in zone Q are fixed, then zone P will move to optimize the contacts with zone Q. Then, if the atoms of zone Q are moved while those of zone P are fixed, atoms in zone Q will hardly move, because the fit between zones P and Q has already been optimized by moving atoms of P. As a general rule, when two zones of atoms that interact strongly are energy-refined, the zone with less accurate conformation should be refined first. Zone energy refinement can be used to fit a flexible substrate to a rigid enzyme, but the treatment of substrate-induced changes of the enzyme conformation, which may make the substrate bind better, is somewhat problematic.

RESULTS

The conformations of the lysozyme–hexa-NAG complex generated by energy minimization for this study are as follows:

1. LN conformation. Complex of lysozyme and hexa-NAG that have not been energy-refined together. The lysozyme coordinates were real-space-refined and energy-refined [set ER5D from (5)]. The hexa-NAG coordinates were measured from a wire model by Phillips and coworkers.

2. ELN conformation. Complex of lysozyme and hexa-NAG that have been energy-refined together. The nonconvergent energy minimization method was used to move all the atoms of the complex and eliminate bad contacts between lysozyme and the substrate.

3. LNA conformation. Complex of lysozyme and hexa-NAG not energy-refined together, apart from four side chains that were refined by the convergent method from several starting conformations to accommodate the initial substrate conformation. The side chains were Asn 44 (χ_1 changed from $-80°$ to $180°$), Asp 101 (χ_1 changed from $-80°$ to $-160°$), Val 109 (χ_1 changed from $66°$ to $-50°$), and Arg 114 (many changes).

4. DLN conformation. Complex with the substrate energy-refined to fit the LN lysozyme conformation. The atoms of the substrate were moved by convergent energy minimization to optimize the fit between a rigid enzyme and a flexible substrate.

5. DELN conformation. Complex with the substrate convergently energy-refined to fit the ELN lysozyme conformation.

6. DLNA conformation. Complex with the substrate convergently energy-refined to fit the LNA lysozyme conformation.

Table 1 shows the movements of enzyme atoms caused by nonconvergent minimization of the energy of the whole enzyme-substrate complex. All the large movements relieve close contacts between lysozyme and hexa-NAG. It is encouraging that the initial fit is good and that so few residues have to move to accommodate the substrate. The lysozyme coordinates were extensively refined by real-space refinement (7) and then by energy refinement (5). The substrate coordinates were measured from a wire model by Phillips and coworkers. The changes in the conformation of lysozyme that were caused by nonconvergent energy refinement are not nearly as extensive as the changes observed by Blake *et al.* (1) in the electron density difference map between the tri-NAG complex and native lysozyme. In particular, the ring of tryptophan 62, which moves 0.75 Å in the difference map, does not move more than 0.1 Å after nonconvergent energy refinement. This discrepancy arises from the fact that nonconvergent energy refinement moves the atoms only to eliminate bad contacts between nonbonded atoms, but not to make favorable contacts. Proper binding of the substrate can be obtained only by convergent energy minimization, which can move a few atoms to make favorable contacts. Here the enzyme is considered rigid and the substrate flexible; convergent energy minimization is used to refine coordinates of the substrate without moving the enzyme at all. Problems arise when choosing a suitable conformation for the rigid enzyme because side chains of the real-space–refined lysozyme conformation that are not clearly seen in the electron density map may have been positioned incorrectly so that they block part of the active site. Unless these side chains are moved first, the substrate may not be able to bind properly. Other parts of the lysozyme conformation

Table 1. Shifts of lysozyme atoms and close contacts relieved after nonconvergent energy refinement of the hexa-NAG complex of lysozyme

Atom Shifted[a]	Shift (Å)[b]	Initial Close Contact[c]
Asn 44, $N^{\delta 2}$	0.5	E,N2; E,O7; E,C7
$O^{\delta 1}$	0.3	E,O7
C^{γ}	0.3	E,O7; E,C7
C^{β}	0.3	–
Asp 52, $O^{\delta 2}$	0.9	D,C1; D,C5; D,O5
Gln 57, O	0.6	D,C6
Trp 108, $C^{\epsilon 2}$	0.3	–
$N^{\epsilon 1}$	0.4	C,C8
$C^{\delta 1}$	0.3	C,C8

[a] The designation of enzyme atoms follows the accepted convention (Kendrew *et al.*, 1970). Substrate and residue designations are shown in Fig. 1. For example, Gln 57,O refers to the peptide oxygen of the enzyme residue glutamine 57; D,C6 refers to atom C6 of substrate residue D.

[b] Shifts smaller than 0.25 Å are not given.

[c] Close contacts are closer than the extreme close-contact distance given in (5).

may make bad contacts with the substrate because the substrate was positioned badly before refinement. When the energy of the whole complex is minimized, lysozyme atoms that are wrongly placed and thus block the active site will move; the lysozyme atoms that make bad contacts with the badly positioned substrate also will move. It is difficult to move lysozyme atoms in the first case, but not in the second. For these reasons three rigid lysozyme conformations are used throughout this work: (1) the best conformation of lysozyme determined by extensive refinement; (2) the conformation generated by nonconvergent energy refinement of (1) together with the hexa-NAG substrate; and (3) the conformation generated by convergent energy refinement of selected side chains of (1) in the presence of the bound substrate. In conformation (1) certain side chains may block the active site, whereas in conformation (2) certain atoms may have been moved incorrectly because the substrate was badly positioned before refinement.

The root-mean-square shifts of the atoms of each substrate residue for minimization from the different starting conformations are given in Table 2. Convergent energy refinement moves the substrate 1 Å for each starting conformation, whereas the nonconvergent refinement causes much smaller movements (0.2 Å). In view of the substantial movement caused by convergent refinement, it is important to check whether the substrate may not have moved from the active site.

Table 2. Root-mean-square shifts[a] of the atoms of each hexa-NAG residue

Conformation	Root-Mean-Square Atom Shift (Å)						
	A	B	C	D	E	F	All
ELN	0.1	0.1	0.2	0.2	0.4	0.1	0.2
DELN	1.1	0.4	0.5	0.7	0.6	1.8	1.0
DLN	1.1	0.5	0.6	0.9	1.0	1.6	1.0
DLNA	0.8	0.6	0.6	1.0	0.9	1.6	1.0

[a]The shift of an atom is $[(x - x')^2 + (y - y')^2 + (z - z')]^{1/2}$, where x' is the x coordinate of the atom in the LN conformation, and x is the x coordinate of the same atom in the particular conformation.

The terminal residues A and F move about 0.5 Å into the cleft and residues B to E move out slightly (<0.4 Å), except for DLNA in which case the D ring moves out by 0.7 Å; the rearrangements of side chains in the DLNA conformation may have altered the geometry of the D subsite.

The naming of the atoms in NAG units and the residues of lysozyme that are near the substrate in the bound complex are shown in Fig. 1. Stereoscopic drawings of some of the conformations studied in this work are given in Fig. 2. Hydrogen bonds between lysozyme and hexa-NAG substrates are listed in Table 3. Several hydrogen bonds proposed by Blake et al. (1) are absent in all the conformations studied here. Thus, it may be concluded that the extensive real-space and energy refinement of the lysozyme coordinates moved certain side chains of residues such as Arg 114 and Asp 101 from the position shown in the drawing of Blake et al. (1). Because these residues are far from the center of the active site, the differences are unimportant. Hydrogen bonds to the D saccharide change if lysozyme is first refined to accommodate the substrate: Atom O6 of D hydrogen-bonds to Glu 57, O in DLN, and to Ala 107, O in DELN.

Binding energies of each substrate residue are given in Table 4. The binding energies of the initial conformation LN are very high due to bad contacts. Nonconvergent energy minimization moves all the atoms of the whole complex to give conformation ELN with reasonable binding energies. Subsequent convergent energy minimization, which moves the substrate alone, gives a conformation DELN with still lower binding energies. When the substrate is fitted to the unrelaxed enzyme (starting conformation, LN), the binding energies of the resulting conformation, DLN, are low but not as low as for conformation DELN. The internal strain energies of the convergently refined substrate residues are similar; no residue is particularly strained, and residues at the ends of chains are less strained than those in the middle of the chain. In the convergently refined complexes, all substrate residues bind equally well except for residue B, which binds less well. Blake et al. (1) observed that the binding of tri-NAG changes the conformation of lysozyme, the largest change being the movement of the indole ring of Trp 62 onto the saccharide ring of residue B. Such a movement,

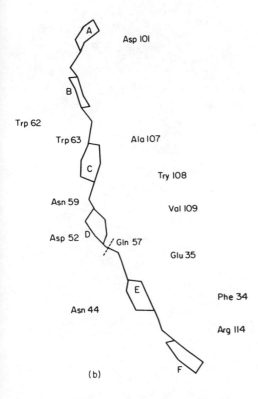

(a)

(b)

Figure 1. (a) An NAG residue. Successive residues are linked through $\beta(1 \to 4)$ bonds. The reducing terminal is shown on the right. In the α anomer, oxygen O1 is connected to C1 in the axial rather than the equatorial position (relative to the plane of the ring). (b) The residues that are close to the bound substrate. The bond between substrate residues D and E that is crossed by a dashed line is cleaved.

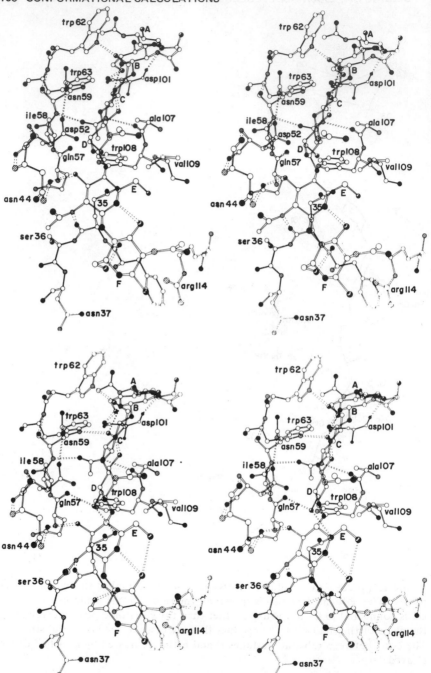

Figure 2. Stereoscopic drawings of some of the lysozyme-substrate conformations studied in this work. All of the lysozyme residues that interact with the substrate are included. Hydrogen bonds, which were selected by the criteria given in Table 3, are represented by dashed lines. Carbon is represented by O, nitrogen by ◉, carbonyl and carboxyl oxygens by ●, and hydroxyl oxygen by ◎.

Table 3. Hydrogen bonds between lysozyme and hexa-NAG in different conformations

Atoms that Hydrogen-Bond[a]	LN	ELN	DLN	DELN	Model[b]
$A,N \cdots Asp\ 101,O^{\delta 2}$	−	−	−	−	+
$A,O4 \cdots Asp\ 101,O$	$+,\wp$	$+,\wp$	+	+	−
$B,O3 \cdots A,O5$	+	+	+	+	+
$B,O6 \cdots Asp\ 101,O^{\delta 2}$	+	+	$+,\wp$	$+,\wp$	+
$C,N2 \cdots Ala\ 107,O$	+	+	+	+	+
$C,O3 \cdots B,O5$	+	+	+	+	+
$Asn\ 59,N \cdots C,O7$	+	+	+	+	+
$Try\ 62,N^\epsilon \cdots C,O6$	+	+	$+,\wp$	+	+
$Try\ 63,N^\epsilon \cdots C,O3$	−	−	$+,\wp$	−	+
$D,O3 \cdots C,O5$	$+,\wp$	$+,\wp$	+	+	+
$D,O6 \cdots Glu\ 35,O^{\epsilon 2}$	−	−	+	−	−
$D,O6 \cdots Gln\ 57,O$	−	−	+	−	+
$D,O6 \cdots Ala\ 107,O$	−	−	−	+	−
$E,N2 \cdots Glu\ 35,O$	+	−	+	−	+
$E,O3 \cdots Gln\ 57,O^{\epsilon 1}$	+	+	+	+	+
$Asn\ 44,N^{\delta 2} \cdots E,O7$	−	−	−	−	+
$F,O3 \cdots E,O4$	−	−	−	−	+
$F,O3 \cdots E,O5$	+	+	−	+	+
$F,O3 \cdots E,O6$	−	−	+	+	−
$F,O6 \cdots Phe\ 34,O$	+	+	+	+	+
$F,O6 \cdots Glu\ 35,O$	−	−	+	+	+
$Arg\ 114,N^{\eta 1} \cdots F,O7$	−	−	+	+	−
$Arg\ 114,N^{\eta 1} \cdots F,O5$	−	−	−	−	+
$Arg\ 114,N^{\eta 2} \cdots F,O6$	−	−	−	−	+
$Asn\ 37,N^\delta \cdots F,O6$	−	−	−	−	+

[a] The hydrogen-bond donor is written first. For all hydrogen bonds the deviation from linearity is $<35°$ (measured by the angle $180° - \widehat{NH} \cdots O$); the angle at the acceptor is $>80°$ (measured by the angle $N \cdots \widehat{OC}$); the donor-acceptor separation is <3.3 Å for bonds marked $\wp$ and <3.05 Å for all others. The hydrogen atom of hydroxyl groups can rotate freely to optimize each hydrogen bond. A hydrogen bond is indicated by "+"; the absence of a hydrogen bond is noted "−." The atom designations are explained in footnote a to Table 1.
[b] The hydrogen bonds of the model were taken from Fig. 19 in Blake *et al.* (1).

which would lower the binding energy in subsite B, did not occur in the nonconvergent energy refinement that gave conformation ELN. When convergent energy refinement was used on Trp 62 in the DLN or DELN conformation, the indole ring did not move and the binding energy of substrate residue B did not decrease. When convergent energy refinement was used on Trp 62 in the ELN conformation, the six-membered part of the indole ring moved 0.6 Å to close the cleft, and the binding energy of substrate residue B dropped by 1.4 kcal/mole. The indole ring could not move in the DLN and DELN conformations because convergent energy refinement had already fitted the B residue into the wider cleft.

Table 4. Binding energies computed for each substrate residue in the different enzyme-substrate conformations

Residue	Energy (kcal/mole)					
	NAG6[a]	LN	ELN	DLN	DELN	DLNA
Total energy[b]						
A	−7	0	−11	−16	−18	−19
B	−10	15	−12	−11	−13	−15
C	−8	36	−16	−16	−21	−18
D	−4	468	−9	−15	−21	−16
E	−7	5191	−17	−14	−20	−23
F	−7	7	−9	−16	−20	−19
Internal strain energy[c]						
A	2	14	3	2	2	2
B	2	18	5	6	5	4
C	4	18	6	6	5	4
D	6	16	8	5	6	5
E	5	15	7	6	6	6
F	3	17	4	3	2	2

[a] NAG6 is the isolated hexa-NAG molecule not bound to lysozyme, but still in an extended conformation.
[b] The strain energy plus the nonbonded energy contributions from interactions within the particular sugar residue, from interactions with atoms in other sugar residues, and from interactions with the enzyme.
[c] The sum of bond, bond-angle, and torsion-angle strain energies.

This clearly indicates the problem caused by zone refinement of two strongly interacting zones of atoms: the zone that is refined first moves, preventing the subsequent movement of the other zone.

The ring torsion angles of the different D saccharide conformations are given in Table 5. In each case this ring is close to a full-chair conformation, as are the other five saccharide rings. Convergent energy refinement does not force the sugar ring of residue D toward a half-chair conformation, but that does not prove that a half-chair would not bind more strongly: There might be an energy barrier separating the full-chair and half-chair conformations, or the full-chair conformation may just be at a local minimum. It is of central importance whether steric factors enhance the binding of a half-chair sugar ring, so this line of investigation was pursued further. An artificial potential was introduced to constrain the torsion angle, $\theta_{C1\text{-}O5}$, of the D sugar to a set value between −60° (full chair) and 0° (half-chair). For each value of the angle constraint, the convergent energy minimization was used to fit residues E and D, then D and C to the enzyme, and finally the binding energy of this refined D residue was calculated. Fig. 3 shows the variation of the calculated D sugar binding energy and its components for different locked ring conformations between full-chair and

Table 5. Ring torsion angles of saccharide residue D

Conformation	Torsion Angle					
	C1-C2	C2-C3	C3-C4	C4-C5	C5-O5	O5-C1
Half-chair[a]	50°	−60°	50°	0°	−30°	0°
LN	39°	−52°	59°	−55°	46°	−37°
ELN	43°	−53°	58°	−55°	49°	−43°
DLN	53°	−56°	53°	−51°	54°	−54°
DELN	52°	−58°	57°	−53°	52°	−51°
Half-chair Intermediate[b]	38°	−63°	64°	−38°	12°	−12°

[a] Estimated from the drawing on p. 89 of Phillips (2).
[b] Created by changing the ring torsion potential before convergent minimization from the LN conformation (like DLN) as follows: θ_{C1-O5} is pushed toward 0° with a force constant of 20 kcal/mole radian (2) and the θ_{O5-C5} torsion barrier is reduced from 2.8 kcal/mole to zero.

half-chair. The total binding energy increases as ring-bond and torsion angles are strained, but the nonbonded energy is constant, showing that the active site has no purely steric preference for the half-chair sugar ring. Of course, any change in charge distribution that results, for example, from formation of a carbonium ion transition state could cause better binding through electrostatic interactions with aspartic acid 52. Because the three different lysozyme conformations (LN, ELN, and LNA) gave such similar results, any effects due to misplaced side chains or shifted parts of the enzyme cannot be important.

DISCUSSION AND CONCLUSIONS

Current calculations strongly suggest that the hexa-NAG substrate can bind to lysozyme with all the sugar rings in the unstrained, full-chair conformation. In particular, tests with substrates having artificially distorted D ring conformation show that subsite D has no purely steric or van der Waals preference for a half-chair ring. Good fits of the substrate to the active site were obtained for three different enzyme conformations: the real-space and energy refined (5, 7) x-ray conformation, the conformation for which the energy has been refined to accommodate the substrate, and the conformation in which four important side chains were rearranged to fit the substrate. The results of these calculations are similar, showing that the problems that, in principle, may occur in zone refinements with a rigid enzyme (see "Methods") were not very significant for lysozyme. lysozyme.

Blake et al. (1) found that Glu 35 and Asp 52 are closest to the cleaved bond of bound hexa-NAG. In the energy-refined conformations, $O^{\epsilon 1}$ of Glu 35 is within 3.2 Å of atom O4 of residue E, and $O^{\delta 2}$ of Asp 52 makes a close contact with atom C1 of residue D (3.0 Å) even after the coordinates have been

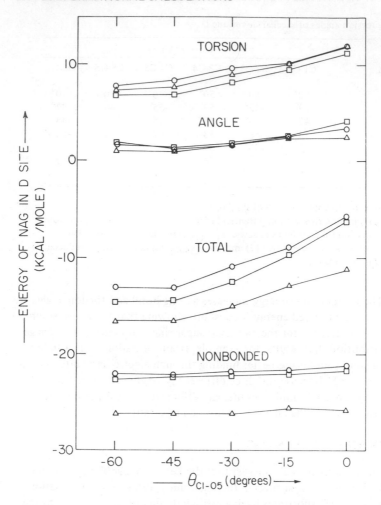

Figure 3. The dependence of the different energy contributions of the D sugar ring on the torsion angle θ_{C1-O5} for various artificially locked hexa-NAG substrates bound to the three different enzyme conformations. $\theta_{C1-O5} = -60°$ for a full chair and $0°$ for a half-chair. $\triangle$ is DELN; $\circ$ is DLN; and $\square$ is DLNA.

extensively refined to eliminate close contacts. We have taken advantage of the results of Blake *et al.* (1) by using their proposed conformation for sugar rings D, E, and F as a starting point for our energy minimization. In one respect the results do not agree with those of Blake *et al.* (1). Thus the D saccharide has moved a little out of the cleft, and its conformation changed by 1 Å, so that it is now a full chair. These differences are relatively small, however, and our results must therefore be considered a refinement of the rigid-wire model originally proposed by Blake *et al.* (1).

Binding studies (8, 9) estimate that binding to subsite D is unfavorable by about 3 kcal/mole. In particular, the binding constant for NAG-NAM-NAG is 2.8×10^5 per mole (9.8 kcal/mole), and for NAG-NAM-NAG-NAM it is 2.1×10^3 per mole (6.9 kcal/mole). Current calculations (see Table 4) indicate, however, that the nonbonded interaction of a NAG residue in subsite D is as favorable as at subsites C and E. Part of this discrepancy could be due to the expulsion of tightly bound water molecules from subsite D, a process that was not included in our calculations. Water molecules have been located by Phillips and coworkers (10), who fitted peaks in the isomorphous density map, and also by Moult and coworkers (11), who fitted peaks in a suitably weighted difference map using calculated phases. In both cases, two or three water molecules, which are within 3.5 Å of carboxyl oxygen $O^{\delta 2}$ of Asp 52, must be displaced when sugar residue D is bound. Because of its low pK ($\sim$4.5) (9), the carboxyl group of Asp 52 is charged at the pH used for the binding experiments (pH 5.4). Expelling two water molecules that are tightly bound to Asp 52 may require several kcal/mole. If so, binding would be less favorable to the D site than to the C site.

Other tetrasaccharides, such as $(NAG)_4$, $(NAG)_3$-XYL (12) (XYL is a xylose residue), and $(NAG)_3$-LAC (13) [$(NAG)_4$ with the reducing terminal sugar oxidized to the lactone], all bind more strongly than $(NAG-NAM)_2$ by 3 to 5 kcal/mole. If $(NAG)_4$ does not bind in subsites A to D (8), in contrast to other tetrasaccharides, good D subsite binding would seem to depend on either not having a primary alcohol group (C6, O6) on the sugar [as in $(NAG)_3$-XYL] or on having a half-chairlike ring conformation [as in $(NAG)_3$-LAC]. This proposal supports the suggestion of Blake *et al.* (1) that overcrowding at C6 forces the D sugar towards a half-chair. An alternative explanation is that all of these tetra-saccharides bind in subsites A to D, with less favorable D-site binding for NAM residues. The latter explanation receives strong support from results of Pollock (14) who found that NAG-NAM-NAG-NAG (which must occupy sites A to D because of the NAM residue) binds more strongly than $(NAG-NAM)_2$ by 2.6 kcal/mole. In summary, D-subsite binding for NAG is about 3 kcal/mole less favorable than expected from energy calculations, probably because two water molecules must be removed from atom $O^{\delta 2}$ of Asp 52; for NAM, the binding is even less favorable (by an additional 2.6 kcal/mole) due to steric interference of the lactyl side group. This problem is currently under study by energy minimization of the different tetrasaccharide conformations.

The catalytic mechanism of lysozyme proposed by Phillips (2) can be modified slightly to be consistent with all of the above observations. First, the hexa-NAG substrate binds as suggested by Phillips (2), but the D-residue ring is not distorted into a half-chair. Water molecules that are expelled by the substrate disturb the solvation shell of the negatively charged Asp 52,$O^{\delta 2}$, and the electro-static potential energy of the Asp 52 carboxyl group is raised. Second, concerted action of Glu 35,$O^{\epsilon 1}$ and Asp 52,$O^{\delta 2}$ cleaves the bond between atom C1 of D and O4 of E. Atom O4 of E is protonated by the glutamic acid. The carbonium ion on atom C1 of D is promoted by the aspartic acid because the electrostatic potential energy of the negatively charged aspartic acid is reduced when the positively charged carbonium ion is formed. The carbonium ion shares its charge with the D-ring oxygen, O5, and forms a resonating bond that forces the D-residue ring toward a half-chair conformation. The electrostatic interaction between the carbonium ion and Asp 52 increases the binding of the transition state relative to that of the substrate. Finally, a water molecule donates a hydroxyl radical to the carbonium ion, and the cleaved substrate dissociation from lysozyme is assisted by the strong tendency of water molecules to bind in subsites D and E. The main differences between this mechanism and that of Phillips (2) are that the binding energy of residue D is used to increase the electrostatic energy of Asp 52 rather than to distort the ring of the D residue and that subsite D has little preference for the half-chair conformation, which forms only because the carbonium ion on C1 shares its charge with O5.

In 1946 Pauling (3) suggested that the drop in energy difference between the bound substrate and transition state needed for catalytic rate enhancement was due to increased affinity of the enzyme for the transition state. (This proposal does not mean that the transition state complex has lower energy than the substrate complex or that the transition state complex is at an energy minimum.) Because the binding of a substrate to an enzyme is mediated by nonbonded van der Waals forces, which fall off rapidly with separation, and because both enzyme and substrate are flexible, it is difficult to generate large distorting forces upon binding. In the distortion of a full chair to a half-chair, one atom moves about 0.5 Å and the strain energy rises by ~5 kcal/mole; such a distortion probably requires too great a force to be caused by van der Waals binding forces. In general, the conformational changes between substrate and transition state are small (<1 Å), but because catalysis involves electronic rearrangements, the differences between the atomic partial charges of a substrate and the transition state can be substantial. This suggests that transition states may often be stabilized by electrostatic rather than by van der Waals steric effects.

ACKNOWLEDGMENT

I wish to thank D.C. Phillips for the coordinates of hexa-NAG.

REFERENCES

1. C.C.F. Blake, L.N. Johnson, G.A. Mair, A.C.T. North, D.C. Phillips, and V.R. Sarma, *Proc. Roy. Soc.* (London), **B167**, 378 (1967).
2. D.C. Phillips, *Scientific American,* **215(5)**, 78 (1966).
3. L. Pauling, *Chem. Eng. News,* **24**, 1375 (1946).
4. M. Levitt, Ph.D. Thesis, University of Cambridge (1972).
5. M. Levitt, *J. Mol. Biol.,* **82**, 393 (1974).
6. W.C. Davidon, *A.E.C. Research and Development Report,* **ANL-5990** (rev.) (1959).

7. R. Diamond, *J. Mol. Biol.,* **82**, 371 (1974).
8. D.M. Chipman and N. Sharon, *Science,* **165**, 454 (1969).
9. T. Imoto, L.N. Johnson, A.C.T. North, D.C. Phillips, and J.A. Rupley, in *The Enzymes,* **Vol. 7**, P.D. Boyer, ed., Academic Press, New York, 1972, p. 665.
10. D.C. Phillips, personal communication.
11. A. Yonath, personal communication.
12. P.V. Eikeren and D.M. Chipman, *J. Am. Chem. Soc.,* **94**, 4788 (1972).
13. I.I. Secemski and G.E. Lienhard, *J. Am. Chem. Soc.,* **93**, 3549 (1971).
14. J.J. Pollock, Ph.D. Thesis, The Weizmann Institute of Science (1969).

PREDICTION OF PROTEIN CONFORMATION: CONSEQUENCES AND ASPIRATIONS

GERALD D. FASMAN* and **PETER Y. CHOU,** Graduate Department of Biochemistry, Brandeis University, Waltham, Mass. 02154

SYNOPSIS: The authors' predictive model for the secondary structure of globular proteins [*Biochemistry*, **13**, 211 (1974)] is discussed in relation to its applicability to biochemical–molecular biology problems. The efficacy of prediction for conformational problems is discussed under the following headings: A Guide to X-Ray–Crystallographic Studies; Segments with Potential for Conformational Changes; Rationale for Amino Acid Substitutions in Peptide Synthesis of Biologically Active Polypeptides; Recognition of Homologous Conformations in Analagous Proteins from Various Species with Sequence Differences; Conformational Dependence of Protein Binding Sites to Membranes, Nucleic Acids, and so on; and Understanding the Loss of Biological Activity Through Enzyme Cleavage.

INTRODUCTION

Structure-function relationships have become one of the basic tenets of modern biochemistry. The conformational determinacy of biological activity is now well established, and the unique three-dimensional structures of many proteins have been elucidated by x-ray–crystallographic studies during the past twenty years. An intriguing ancillary problem of recent years has been to unravel and understand the factors that dictate the formation of these tertiary structures. Anfinsen (1–6) has been instrumental in focusing attention on the hypothesis that the primary structure, that is, the sequence, contains all the necessary information for the folding of polypeptide chains into their biologically active conformation. He has further elaborated the proposition of Kendrew (7) that the goal of predicting tertiary structure from primary sequence alone is not far from attainment.

Many studies have utilized theoretical treatments of the energetics of polypeptide folding (8–10). Scheraga's (11) work is illustrative of the increasing sophistication in this approach. The formidable problems inherent in this approach, however, preclude the possibility of calculating the complete three-dimensional structure in the immediate future (13). Therefore, more direct attempts have been made to predict protein conformation from the primary sequence (14–23). These empirical approaches are gaining in accuracy and the ease of application makes them exceedingly attractive.

The impetus for the prediction of protein conformation comes from studies on poly(α-amino acids). The availability of these model proteins is a tribute to the many years of brilliant work by Katchalski (Katzir) and colleagues (24).

---

*Author to whom correspondence should be addressed.

The helix-coil theories of polypeptides (25–27) utilized two parameters: σ, the cooperation factor for helix initiation, and s, the equilibrium constant for converting a coil residue to a helical state at the end of a long helix (25). These parameters have been investigated for many poly(α-amino acids) [for review see (11, 23)], for which a series of relative formation and stability of α helices has been established. Recent analysis of the 20 naturally occurring amino acids yielded a set of conformational parameters, P_α, P_β, P_c, and P_t, based on the frequency of occurrence of each amino acid residue in the α, β, β-turn, and coil conformations in 15 proteins (23) whose structures were determined by x-ray crystallography. A comparison of the experimental Zimm–Bragg helix growth parameter s from poly(α-amino acids) and the P_α values obtained from proteins revealed excellent agreement. It was found, for example, that Leu has the largest measured s value. The $P_{i\alpha}$ for this amino acid is also the largest. Thus Leu is shown to have the strongest tendency for α-helix formation.

This series of conformational parameters for α, β, and β-turn conformations prompted the development of a new approach for predicting protein conformation from the amino acid sequence (23). Previously, the amino acids were placed in strict categories whereby a particular residue was labeled a helix former, helix breaker, or indifferent. It is now acknowledged that an amino acid may have a high preference for helix formation, but can accommodate itself with equal facility to the β conformation. Other residues are α or β formers only. Still others are found most frequently in β turns. Classification by these structures accounts for 80% of all residues, leaving only 20% in unstructured regions, the so-called *random regions*. With these parameters and several rules, it was possible to correctly predict 80% of all residues in the helical, β-sheet, and coil conformations in the 19 proteins evaluated. The "permissibility" of amino acid replacement in homologous sequences now can be more easily rationalized. A permissible replacement is one in which residues of approximately equal P value are exchanged; the replacement of a defective mutant may alter the $\langle P \rangle$ values of a segment, thereby altering its conformation.

Pancreatic trypsin inhibitor, a 58-residue polypeptide, was used to demonstrate this new predictive method (23c). Fig. 1 is a schematic diagram of the helical, β-sheet, and reverse β-turn predicted regions for this protein. Eighty-seven percent of the helical residues and 95% of the β residues are predicted correctly, and 86% of total residues are correctly identified as α, β, or coil when compared to the x-ray–determined structure of Huber *et al.* (27). As a result of this study, the sequences of a protein having the potential to undergo conformational changes could be predicted. Thus, if a sequence has both $\langle P_\alpha \rangle$ and $\langle P_\beta \rangle$ values greater than one, it has the potential to undergo a conformational change, under appropriate conditions (e.g., change of environment, temperature, ionic strength, etc.). Another consequence of the study is perhaps a new method to elaborate on the mechanism of protein folding, either during synthesis or upon renaturation. Lewis *et al.* (18b) proposed that regions of high helical probability help to direct the folding of polypeptide chains to yield the final native protein, with β bends playing an important role in the folding mechanism, as had been suggested originally by Perutz (29). It was not stipulated, however, which

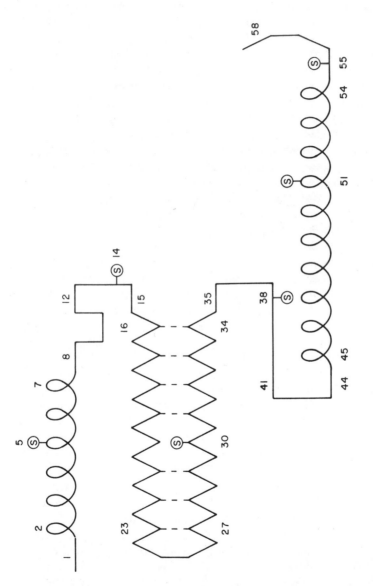

Figure 1. Schematic diagram of helical, β sheet and reverse β-turns predicted in pancreatic trypsin inhibitor. Residues are represented in their respective conformational state: ℓ, helical; Λ, β sheet; −, coil. Chain reversals denote β-turn tetrapeptides. Hydrogen bonding between the antiparallel β sheets is represented by dashed lines. Conformational boundary residues are numbered, as are the six Cys residues indicated by s. Each helical loop represents a single helical residue. [Reprinted with permission from Chou and Fasman, *Biochemistry*, **13**, 222 (1974). Copyright by the American Chemical Society.]

section of the protein would fold first. Using the computed conformational parameters P_α and P_β it is possible to compare the various helical and β segments in each protein in terms of their $\langle P_\alpha \rangle$ and $\langle P_\beta \rangle$ values. There are indications that helix initiation proceeds faster than β formation (30), and it is postulated that the first fold in a protein occurs at the helix region with the highest helical potential, that is, the largest $\langle P_\alpha \rangle$ value, the next fold near the helix region with the second largest $\langle P_\alpha \rangle$ value, and so on (23b). The probable folding sequence for myoglobin has been predicted by this approach (23b).

Likewise, the proposed folding mechanism might apply to the folding of the chain during synthesis on the ribosome. Thus, the chain would grow until a center of nucleation was produced, at which time helix formation would occur and propagate in both directions until terminated by the appropriate tetrapeptide sequence. This mechanism would support the suggestion (31) that the polypeptide chain may progressively assume a three-dimensional conformation similar or identical to its conformation in the completed protein molecule. It is assumed that the information necessary for folding is provided through short-range interactions (i.e., single-residue information represented by P_α and P_β) and medium-range interactions (i.e., neighboring-residue information represented by $\langle P_\alpha \rangle$ and $\langle P_\beta \rangle$), rather than long-range interactions. The accuracy of this predictive model supports the mechanism.

Conformation prediction by itself can be fascinating, but it also has great potential applications in many areas of biochemistry. The main theme of this paper is the utility of protein conformational prediction.

A GUIDE TO X-RAY–CRYSTALLOGRAPHIC STUDIES

In the past, both x-ray crystallographers and protein chemists have lacked confidence in the predictive technique as an aid in delineation of electron-density maps. It has frequently been demonstrated, however, that at low resolution (~4 Å), x-ray electron-density maps frequently omit areas of conformation that are recognized at high resolution (~2 Å). Prediction could aid in the recognition of such areas. For example, refinement of 1.85 Å (32) added helical residues to the original structure of carp myogen (33). Several of these regions had been predicted before the final structure emerged (23c). Likewise, staphylococcal nuclease regions 88–94 and 111–115 were predicted to be β sheets, in contradiction to the x-ray findings of Arnone et al. (34), but more recent x-ray studies have shown that residues 87–94 and 109–112 indeed are in the β conformation (35). Other examples of this type of correlation can be found in the original paper (23c).

This method should permit quicker resolution of electron-density maps in the x-ray–crystallographic studies of proteins. The general secondary structures of proteins whose conformations have not yet been scrutinized by means of x-ray crystallography can be determined via the predictive technique. As is evident from the *Atlas of Protein Sequence and Structure* (35), there are hundreds of known sequences for which secondary structures are still lacking. The ability to predict these structures offers those interested in structure-function relationships a valuable tool.

SEGMENTS WITH POTENTIAL FOR CONFORMATIONAL CHANGES

Such predictions allow us to focus on certain domains of a protein if con-formational changes are known to occur in response to a variation in conditions. For example, alterations in solvent are known to cause changes in conformation and such changes can be located by the predictive method. Many regions in proteins have both helical and β-forming potentials, and one conformation is usually preferred, depending on the environmental conditions. Although only 2% helix is found in the native structure of concanavalin A (37) in x-ray-crystallographic studies (38, 39), it has been shown (37) that higher helicity can be induced in the native protein with 2-chloroethanol. The circular dichroism spectra showed 55% helicity in 70% chloroethanol. The predictive scheme (23c) had a total of 47% of its residues in 13 regions with α potential, although many of these had still higher β potentials. Similarly, elastase (40) was shown to be only 7% α helix (41) by x-ray analysis, but circular dichroism studies (40) showed approximately 35% helical content in sodium dodecylsulfate. In 15 seg-ments there are 79 potential helical residues, which account for 33% helicity.

The predicted conformation of glucagon (42), a 29-amino-acid hormone, illustrates the delicate balance of this polypeptide between two major con-formational states. In Table 1 the $\langle P_\alpha \rangle$ and $\langle P_\beta \rangle$ values for glucagon are given. The structural sensitivity of this hormone is caused by residues 19–27, which have α-helical potential ($\langle P_\alpha \rangle = 1.19$), as well as β-sheet potential ($\langle P_\beta \rangle = 1.25$). Two conformational states are predicted for glucagon (Fig. 2). In predicted form (a), residues 5–10 form a β-sheet region and residues 19–27 form an α-helical region (31% α, 21% β). Circular dichroism spectra of glucagon (43) at 12.6 mg/ml support this prediction. In predicted form (b), both regions (resi-dues 5–10 and residues 19–27) are β sheets (0% α, 52% β) in agreement with the infrared spectral evidence that glucagon gels and fibrils have a predominant β-sheet conformation (44). Glucagon may possess tertiary structure caused by the predicted three reverse β turns at residues 2–5, 10–13, and 15–18. This con-formational change can be brought about by a change in concentration of the

Table 1. Conformational prediction for glucagon: $\langle P_\alpha \rangle$, $\langle P_\beta \rangle$, $\langle P_t \rangle$, and P_t values computed for helical, β-sheet, β-turn, and random-coil regions

Predicted	$\langle P_\alpha \rangle$	$\langle P_\beta \rangle$		
1–4 c	0.93	0.87		
5–10 β	0.86	1.08		
11–18 c	0.90	0.91		
19–27 β	1.19	1.25		
28–29 c	0.78	0.93		
Predicted β Turns	$P_t \times 10^{-4b}$	$\langle P_t \rangle$	$\langle P_\alpha \rangle$	$\langle P_\beta \rangle$
2–5	0.96	1.20	0.83	0.99
10–13	0.88	1.27	0.77	1.01
15–18	0.33	1.21	0.84	0.83

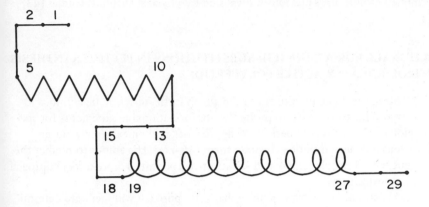

(a) Glucagon Solution : 31 % α Helix
                        21 % β Sheet

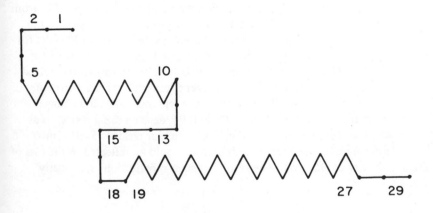

(b) Glucagon Gel : 52 % β Sheet

Figure 2. Schematic diagram of the secondary structure predicted in glucagon. Residues are represented in helical (ℓ), β-sheet (∧) and coil (−) conformations. Chain reversals denote β-turn tetrapeptides. Region 19–27 has both helical ($\langle P_\alpha \rangle$ = 1.19) and β-sheet potential ($\langle P_\beta \rangle$ = 1.25) so that the predicted conformation (a) is 31% helix and 21% β sheet and conformation (b) is 52% β sheet.

hormone in solution, as previously suggested by physical chemical studies (43, 44).

RATIONALE FOR AMINO ACID SUBSTITUTIONS IN PEPTIDE SYNTHESIS OF BIOLOGICALLY ACTIVE POLYPEPTIDES

The literature abounds with studies in which intricate synthetic amino acid substitutions have been used to probe the conformational requirements for specific biological activity. At best, these approaches rely on extremely clever deductions and deep intuition to guide them (45-47). The ability to predict the consequences of certain amino acid substitutions permits a more direct approach in such investigations.

It has been suggested, for example, that this approach will facilitate determination of the active conformation for glucagon binding to the receptor site on the membrane (42). The conformation of region 19-27 is delicately balanced between the α and β conformations, so it is predicted that replacement of one or two residues of high β potential in this region with strong α formers would lock the conformation in the α-helical structure. If the conformation necessary for binding to the membrane is the β conformation, then this homolog of glucagon would preclude binding. Replacement of Glu with Val 23 would raise $\langle P_\alpha \rangle$ from 1.19 to 1.23 and lower $\langle P_\beta \rangle$ from 1.25 to 1.09 for segment 19-27, so that the helical conformation would predominate. A double substitution of Phe-Val for Glu-Glu in residues 22 and 23 would result in $\langle P_\alpha \rangle = 1.28$ and $\langle P_\beta \rangle = 0.98$ for region 19-27. This modification not only would further strengthen the helical potential of this region, but it would also prevent formation of the β structure because $\langle P_\beta \rangle < 1$. Substituting Phe 6 for Glu in β region 5-10 would change $\langle P_\alpha \rangle$ from 0.86 to 0.93 and $\langle P_\beta \rangle$ from 1.08 to 0.91, causing this β region to assume a random-coil conformation. If the β structure at residues 5-10 is involved in the receptor binding site, such a modification would be reflected in the loss of biological activity of glucagon. The results of such substitutions are eagerly awaited.

RECOGNITION OF REGIONS OF HOMOLOGOUS CONFORMATIONS IN ANALOGOUS PROTEINS FROM VARIOUS SPECIES WITH SEQUENCE DIFFERENCES

A. C. Paladini and his associates (45) have carried out a comparative study of several growth hormones and ovine prolactin. The pituitary growth hormones of several animal species (HGH, human growth hormone; BGH, bovine growth hormone; OGH, ovine growth hormone; EGH, equine growth hormone), human placental lactogen (HPL) and ovine prolactin are closely related proteins with interesting biological properties. All growth hormones stimulate weight increases, but only HGH exhibits slight lactogenic action. HPL and ovine prolactin are preferentially lactogenic hormones although HPL does promote a small amount of growth. Furthermore, these hormones show species specificity in their action.

Sequence comparisons of these two types of biological activity have been unsuccessful (46), but utilizing the predictive procedure, Paladini *et al.* (45) were able to demonstrate the existence in HGH of a central region with considerable antiparallel β-sheet structure that is analogous to a less extensive β region in BGH, OGH, and EGH. A similar conformation was conspicuously absent in HPL and in ovine prolactin; HPL has only shorter segments of β sheet in common with HGH and ovine prolactin has none. Paladini *et al.* found it tempting to associate the similar, larger β-sheet areas in HGH, BGH, OGH, and EGH with growth activity and to correlate other zones in common to HGH, HPL, and ovine prolactin to the lactogenic action of these molecules. These predictions incorporated a host of experimental data.

CONFORMATIONAL DEPENDENCE OF PROTEIN BINDING SITES TO MEMBRANES, NUCLEIC ACIDS, AND SO ON

The conformational dependence of binding to receptor sites was discussed briefly in the above section on amino acid substitutions for glucagon. The binding of the *lac* repressor to the *lac* operator of *E. coli* is an excellent example of a specific and tight interaction between a protein and DNA. The amino acid sequence of the *lac* repressor has recently been elucidated (47). This protein of 347 residues (MW 38,000–40,000) binds as a tetramer (48) (MW 148,800) to the operator. The DNA operator, which also has been sequenced recently, (49) is a double-stranded 24-base-long sequence. By means of enzymatic cleavage, it has been demonstrated that the 59 N-terminal residues of the *lac* repressor contain the binding site for DNA (50). Circular dichroism studies (51). indicate that the native *lac* repressor contains 40% helical and 42% β-sheet conformations and the trypsin-resistant core (residues 60–324) is 16% helix and 54% β sheet. The predicted conformation (52) of the *lac* repressor, presented in Fig. 3, shows 37% helix, 35% β sheet; the core contains 29% helix and 41% β sheet. The sharp reduction in helicity in the trypsinized *lac* repressor could be due to the loss of two long α-helical regions, predicted to be residues 26–45 and 328–344.

Extensive β sheets (8 strands) predicted in the 215–324 region may be involved in the stabilization of the tetramer found both in the *lac* repressor and in the core. It is interesting to note that the β sheets are almost devoid of charged residues, so that the *lac* repressor has an extremely hydrophobic nucleus. On the other hand, the predicted β turns, which are found predominantly on the surface of proteins, consist of 50% charged, polar residues (Ser and Thr). Hence, the observation (53) that polar and nonpolar residues occupy, respectively, the inside and outside of proteins is also borne out in the predicted structure of the *lac* repressor.

Residues 17–33 were previously predicted to be helical and proposed to bind in the major groove of DNA (54). The present prediction (Fig. 3) shows that region 17–24 is a β sheet with $\langle P_\beta \rangle = 1.29$ and $\langle P_\alpha \rangle = 0.95$ and that region 26–45 is helical with $\langle P_\alpha \rangle = 1.22$ and $\langle P_\beta \rangle = 0.94$. At this time it is difficult to assess which of these two segments is chiefly responsible for binding or whether both conformations play an essential role in the interaction of the *lac* repressor with the operator.

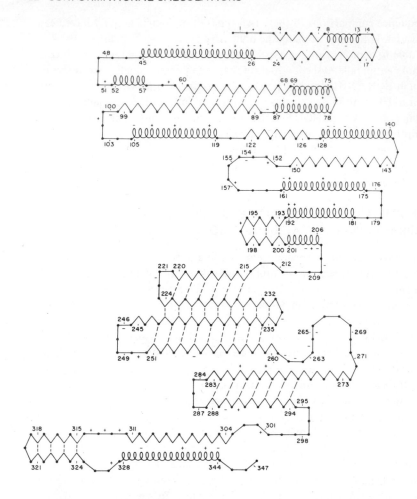

Figure 3. Schematic diagram of the secondary structure predicted for *lac* repressor. There are 37% helical (ℓ), 35% β-sheet (∧), and 28% coil (−) residues. Chain reversals denote β-turn tetrapeptides, which make up approximately 20% of the amino acids of the protein. Adjacent β turns occur with high probability at regions 152–157, 260–265 and 265–269. Kinks along the chain are depicted at 212–215, 271–272 (Pro-Pro), 284–287, 301–304, and 325–328 because these β turns are not predicted to have 180° chain reversal. Possible hydrogen bonding between β sheets is represented by dashed lines. The positions of the charged residues are indicated, and conformational boundary residues are numbered. Each residue is drawn to scale according to its conformation along the polypeptide chain: helical residue, 1.5 Å; β-sheet residue, 3.47 Å; coil residue, 3.62 Å. It should be noted that the $C_{\alpha,1}$–$C_{\alpha,4}$ distance for β turns in the present scale is 7.2 Å, not less than 6 Å as usually found in proteins. This was done so that the conformational boundary residues could be numbered with clarity. Hence the vertical dimension should be reduced from 108 Å to 90 Å, and this length should be even smaller upon tertiary folding. The horizontal length of the molecule is approximately 75 Å.

There are numerous *lac*-repressor mutants (55) that allow a single amino acid substitution to eliminate repression *in vivo*. Examination of the Ser 16-to-Pro 16, Thr 19-to-Ala 19, and Ala 53-to-Val 53 mutations by our predictive method reveals that the β turn at 14–17 has shifted to 15–18, the β region 17–24 acquires a slight helical potential ($\langle P_\alpha \rangle$ = 0.95 to 1.03), and the helical region 52–57 flips to a β conformation ($\langle P_\alpha \rangle$ = 1.29 to 1.24, $\langle P_\beta \rangle$ = 1.21 to 1.33). Hence the present model of the *lac* repressor can be utilized to show that conformational differences in these mutants could alter the operator binding activity.

UNDERSTANDING THE LOSS OF BIOLOGICAL ACTIVITY THROUGH ENZYME CLEAVAGE

The sequences essential for biological activity can be monitored by the progressive loss of segments of a protein through enzyme cleavage. It is now possible to analyze biological activity in terms of these segments, which may or may not have an independent conformational state. Thus certain cleavages may be of little conformational consequence and result, perhaps, in no loss of activity. Similarly, certain cleavages can be expected to remove domains of specific conformation with concomitant loss of activity.

CONCLUSIONS

The ability to predict the secondary structure of proteins offers a new method of attack on many problems in molecular biology and biochemistry. Biological activity involving conformational dependence can now be approached in a more direct, more rational manner. Thus, the loss of activity by mutants perhaps can be understood in terms of the conformational consequences of mutation. In addition, conformational transitions that may be important, say, in membrane binding, can be located in the sequence and these areas can be further investigated. One appealing factor in this new predictive method is its simplicity. Without resort to complex computer programs an individual can make an educated guess as to the conformation of a protein.

ACKNOWLEDGMENT

We gratefully acknowledge the support, in part by grants, from the U.S. Public Health Service (GM17533), National Science Foundation (GB 29204X), American Cancer Society (P-577), and the American Heart Association (71-1111).

REFERENCES

1.   (a) M. Sela, F.H. White, and C.B. Anfinsen, *Science,* **125**, 691 (1957). (b) C.B. Anfinsen, E. Haber, M. Sela, and F.H. White, Jr., *Proc. Nat. Acad. Sci. U.S.,* **47**, 1309 (1961).
2.   C.B. Anfinsen, *New Perspectives in Biology,* **4**, 42 (1964).
3.   C.B. Anfinsen, *The Harvey Lectures,* **61**, 96 (1967).

4. C.B. Anfinsen, *Developmental Biol.* Supp., **2**, 1 (1968).
5. C.B. Anfinsen, *Biochem. J.*, **128**, 737 (1972).
6. C.B. Anfinsen, *Les Prix Nobel en 1972*, Nobel Foundation, 1973.
7. J. Kendrew, *Pontificiae Academiae Scientiarum Scripta Varia*, **22**, 449 (1962).
8. D.A. Brant and P.J. Flory, *J. Am. Chem. Soc.*, **87**, 663 (1965).
9. A.M. Liquori, *Ciba Found. Symp. Principles Bio. Mol. Org.*, 40 (1966).
10. (a) O.B. Ptitsyn, *J. Mol. Biol.*, **42**, 501 (1969). (b) O.B. Ptitsyn and A.V. Finkelstein, *Biofizika*, **15**, 757 (1970).
11. H. Scheraga, *Chem. Rev.*, **71**, 195 (1971).
12. G.N. Ramachandran, C.M. Venkatchalam, and S. Krimm, *Biophys. J.*, **6**, 849 (1966).
13. P.J. Flory, *Polymerization in Biological Systems,* Ciba Foundation Symposium 7, Elsevier, Amsterdam, 1972, p. 294.
14. A.V. Guzzo, *Biophys. J.*, **5**, 809 (1965).
15. J.W. Prothero, *Biophys. J.*, **6**, 367 (1966).
16. B.W. Low, F.M. Lovell, and A.D. Rudko, *Proc. Nat. Acad. Sci. U.S.*, **60**, 1519 (1968).
17. M. Schiffer and A.B. Edmundson, *Biophys. J.*, **7**, 131 (1967).
18. H.A. Scheraga (see Review #11). (a) D. Kotelchuck and H.A. Scheraga, *Proc. Nat. Acad. Sci. U.S.*, **62**, 14 (1969). (b) P.N. Lewis, N. Gō, M. Gō, D. Kotelchuck, and H.A. Scheraga, *Proc. Nat. Acad. Sci. U.S.*, **65**, 810 (1970). (c) P.N. Lewis, F.A. Momany, and H.A. Scheraga, *Proc. Nat. Acad. Sci. U.S.*, **68**, 2293 (1971). (d) P.N. Lewis and H.A. Scheraga, *Arch. Biochem. Biophys.*, **144**, 576 (1971).
19. E.A. Kabat and T.T. Wu, (a) *Biopolymers,* **12**, 751 (1973). (b) *Proc. Nat. Acad. Sci. U.S.*, **70**, 1473 (1973). (c) T.T. Wu and E. Kabat, *Proc. Nat. Acad. Sci. U.S.*, **68**, 1501 (1971). (d) *J. Mol. Biol.*, **75**, 13 (1973).
20. B. Robson, R.H. Pain, *J. Mol. Biol.*, **58**, 237 (1971).
21. (a) O.B. Ptitsyn and A.V. Finkelstein, *Biofizika*, **15**, 757 (1970). (b) A.V. Finkelstein and O.B. Ptitsyn, *J. Mol. Biol.*, **62**, 613 (1971).
22. K. Nagano, *J. Mol. Biol.*, **75**, 401 (1973).
23. P.Y. Chou and G.D. Fasman, (a) *J. Mol. Biol.*, **74**, 263 (1973). (b) *Biochemistry,* **13**, 211 (1974). (c) *Biochemistry,* **13**, 222 (1974).
24. (a) E. Katchalski, *Adv. Protein Chem.*, **6**, 123 (1951). (b) E. Katchalski and M. Sela, *Adv. Protein Chem.*, **13**, 243 (1958). (c) M. Sela and E. Katchalski, *Adv. Protein Chem.*, **14**, 391 (1959).
25. (a) B.H. Zimm and J.K. Bragg, *J. Chem. Phys.*, **31**, 526 (1959). (b) B.H. Zimm and S.A. Rice, *Mol. Phys.*, **3**, 391 (1960).
26. S. Lifson and A. Roig, *J. Chem. Phys.*, **34**, 1963 (1961).
27. J. Applequist, *J. Chem. Phys.*, **38**, 934 (1963).
28. R. Huber, D. Kukla, A. Rühlmann, and W. Steigemann, *Cold Spring Harbor Symp. Quant. Biol.*, **36**, 141 (1972).
29. M. Perutz, Dunham Lectures, Harvard Medical School, Boston, 1963.
30. C.R. Snell and G.D. Fasman, *Biochemistry,* **13**, 1017 (1973).
31. D.C. Phillips, *Proc. Nat. Acad. Sci. U.S.*, **57**, 484 (1967).
32. R.H. Kretsinger and C.E. Nockolds, *J. Biol. Chem.*, **248**, 3313 (1973).
33. R.H. Kretsinger and C.E. Nockolds, G.J. Coffie and R.A. Bradshaw, *Cold Spring Harbor Symposium Quant. Biol.*, **36**, 217 (1972).

34.   A. Arnone, C.J. Bier, F.A. Cotton, V.W. Day, E.E. Hazen, Jr., D.C.
       Richardson, and J.S. Richardson, and (in part) A. Yonath, *J. Biol. Chem.*
       246, 2302 (1971).
35.   (a) F.A. Cotton, C.J. Bier, V.W. Day, E.E. Hazen, Jr., and S. Larsen, *Cold
       Spring Harbor Symp. Quant. Biol.*, **36**, 243 (1972). (b) E.E. Hazen, per-
       sonal communication.
36.   M.O. Dayhoff, ed., *Atlas of Protein Sequence and Structure*, Vol. 5,
       National Biomedical Research Foundation, Washington, D.C., 1972, and
       Supplement I, 1973.
37.   W.D. McCubbin, K.O. Kawa, and C.M. Kay, *Biochem. Biophys, Res.*
       Commun., **43**, 666 (1971).
38.   G.M. Edelman, B.A. Cunningham, G.N. Reeke, Jr., J.W. Becker, M.J.
       Waxdal, and J.C. Wang, *Proc. Nat. Acad. Sci. U.S.*, **69**, 2580 (1972).
39.   K.D. Hardman and C.F. Ainsworth, *Biochemistry*, **8**, 4108 (1972).
40.   L. Visser and E.R. Blout, *Biochemistry*, **10**, 743 (1971).
41.   D.M. Shotton and H.C. Watson, *Nature* (London), **225**, 811 (1970).
42.   P.Y. Chou and G.D. Fasman, to be submitted.
43.   P.A. Srere and G.C. Brooks, *Arch. Biochem. Biophys.*, **129**, 708 (1969).
44.   (a) W.B. Gratzer, E. Bailey, and G.H. Beaven, *Biochem. Biophys. Res.*
       Commun., **28**, 914 (1967). (b) W.B. Gratzer, G.H. Beaven, H.W.E. Rattle,
       and E.M. Bradbury, *Eur. J. Biochem.*, **3**, 276 (1968). (c) G.H. Beaven,
       W.B. Gratzer, and H.G. Davies, *Eur. J. Biochem.*, **11**, 37 (1969). (d) R.M.
       Epand, *Can. J. Biochem.*, **49**, 166 (1971).
45.   A.C. Paladini, J.A. Santome, and J.M. Dellacha, a personal communication.
46.   T.A. Bewley, J.S. Dixon, and C.H. Li, *Int. J. Peptide Protein Research*, **4**,
       281 (1972).
47.   K. Beyreuther, K. Adler, N. Geisler and A. Klemm, *Proc. Nat. Acad. Sci.*
       U.S., **70**, 3576 (1973).
48.   B. Müller-Hill, K. Beyreuther, and W. Gilbert, L. Grossman and K. Moldave,
       eds., *Methods in Enzymology*, Vol. 21, Academic Press, New York, 1971,
       p. 483.
49.   W. Gilbert and A. Maxam, *Proc. Nat. Acad. Sci. U.S.*, **70**, 3581 (1972).
50.   K. Weber, T. Platt, D. Ganem, H.H. Miller, *Proc. Nat. Acad. Sci. U.S.*, **69**,
       3624 (1972).
51.   P.Y. Chou, A.J. Adler, and G.D. Fasman, *Fed. Proc.*, in press.
52.   P.Y. Chou and G.D. Fasman, to be submitted.
53.   J.C. Kendrew, *Brookhaven Symp. Biol.*, **15**, 216 (1962).
54.   K. Adler, K. Beyreuther, E. Fanning, N. Geisler, B. Groyenborn, A. Klemm,
       B. Müller-Hill, M. Pfal, and A. Schmitz, *Nature*, **237**, 322 (1972).
55.   K. Weber, T. Platt, D. Ganem, and J.H. Miller, *Proc. Nat. Acad. Aci. U.S.*,
       69, 3624 (1972).

PART II.   POLYPEPTIDE AND PROTEIN CONFORMATION

CONFORMATIONAL STUDIES OF POLYPEPTIDES AND POLYDEPSIPEPTIDES

MURRAY GOODMAN, FU CHEN, CHAIM GILON, RICHARD INGWALL, DIETMAR NISSEN, and MANLIO PALUMBO, Department of Chemistry, University of California (San Diego), La Jolla, Calif. 92037

SYNOPSIS: The background for the modern conformational analysis of polypeptides is presented. In this paper we discuss our recent studies on high-molecular-weight and model compounds related to N-methylalanine and β-amino butyric acid. We also include a description of the preparation of alternating high-molecular-weight polydepsipeptides and model compounds. A conformational analysis of these structures is developed based upon circular dichroism, nuclear magnetic resonance, and infrared spectroscopy. Results from the experimental techniques are discussed in terms of conformational energy calculations.

INTRODUCTION

More than twenty years have passed since Ephraim Katchalski and his associates (1, 2) turned their attention to the preparation and study of synthetic polypeptides. Their work on the synthesis of α-amino acid N-carboxy anhydrides led to the emergence of the field that is a main focus of this volume. Independently, Blout (3), Doty (4), and Bamford (5), commenced systematic investigations of the polymerization mechanism of α-amino acid N-carboxy anhydrides. Through their efforts, it became possible to prepare high-molecular-weight polypeptides by suitable strong-base initiation in aprotic solvents (6); low-molecular-weight materials were synthesized by primary-amine initiation (5). In the early 1960s, our research group undertook a study of the mechanism of strong-base-initiated polymerization of α-amino acid N-carboxy anhydrides (7). We were able to demonstrate that these polymerizations proceed via an "active monomer" mechanism (8, 9). Initiator studies by Scoffone (10), Peggion (11) and their associates confirmed our findings.

During the years immediately following Katchalski's pioneering work, he and many other researchers undertook the conformational analysis of these materials (12, 13). Pauling realized that high-molecular-weight poly(γ-benzyl-L-glutamate) can form fibers possessing an α-helical conformation (14, 15). Bamford (16), Blout (6), and Doty (17), demonstrated that helicity could be maintained under specific conditions in solution.

Various spectroscopic techniques have been applied to the study of these materials in solution and in the solid state (13). Derived constants for optical rotatory dispersion initially proposed by Moffitt (18, 19) provided some insight into the conformational characteristics of polypeptides in solution. Much more substantial measurements on polypeptides in solution came from the development in the 1960s of highly sensitive ultraviolet recording, optical rotatory

126

dispersion (ORD), and circular dichroism (CD) instruments (20). Holzwarth and Doty (21), for example, clearly showed that CD could be used to assign helical, random, and β conformations to polypeptides and some proteins in solution (Fig. 1). Ultraviolet absorption spectroscopy (UV) below 200 nm confirmed these results (17); see Fig. 2. Infrared spectroscopy (IR) also was employed in assigning the conformations of polypeptides. Shifts of the absorption bands for the amide groups and infrared dichroism studies were carried out by many, including Blout and Miyazawa (22, 23). They were able to relate group frequencies in many polypeptide systems to specific conformations. High-resolution nuclear magnetic resonance spectroscopy (NMR) also is a most important technique by which to study polypeptide conformations (24). Under appropriate conditions, structural assignments can be made on the basis of chemical shifts and coupling constants.

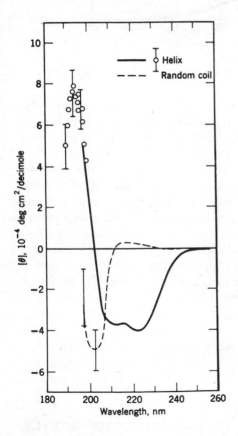

Figure 1. CD spectrum of poly(α-glutamic acid) in 0.1 M NaF at pH 4.3 (helix) and pH 7.6 (random coil) (21).

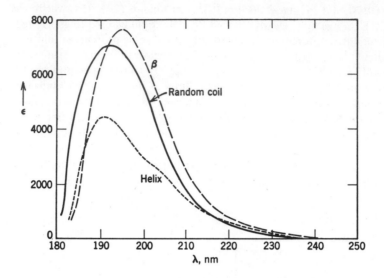

Figure 2. Absorption spectra of poly(L-lysine) in aqueous solution (17).

No analysis of the study of polypeptide conformations can be complete without mention of the statistical mechanical treatment of the transitions between different conformations in a given polypeptide system and the calculations of preferred conformations by semiempirical methods. Zimm and Bragg (25), Gibbs and DiMarzio (26), and Lifson and Roig (27) developed the theoretical models for helix-coil transitions. Other groups followed and refined the initial discoveries. Experimental proof that these transitions indeed occur with various synthetic polypeptides and proteins came almost simultaneously with the appearance of the initial theories of transition. The work of Ramachandran (28), Liquori (29), Scheraga (30), Lifson (31), Flory (32), and others (33) has added substantially to the development of the approach using contributions to potential energy to deduce preferred conformations. Detailed treatments of these semiempirical energy approaches are presented in this volume by Hagler and Lifson, Scheraga, Hopfinger, and Ramachandran.

Work in our laboratory is dependent upon all of the techniques noted above. Included are various syntheses, spectroscopic characteristics of conformation based upon CD, UV, IR, and NMR, helix-coil transitions, and semiempirical conformational energy calculations, all of which are fitted together to present a substantial conformational analysis.

POLY(THIAZOLIDINE-4-CARBOXYLIC ACID) AND POLY(N-METHYL-L-ALANINE)

Helical structures are common features of proteins, nucleic acids, polysaccharides and many synthetic polypeptides. The study of the helical structure of polypeptides in the solid state and in solution is of great significance because it leads to a clear interpretation of results obtained from conformational studies of fibrous and globular proteins. Pauling and Corey (14) proposed that the α-helix is one of the more stable secondary structures for folded polypeptide chains. Some of their basic assumptions are that the peptide group is a *trans* planar conformation and that each peptide C=O and N—H group is hydrogen-bonded with a nearly linear disposition of the NH···O=C bond. However, polypeptides such as poly(L-proline) and poly(3-hydroxyproline), in which the α-amino nitrogen atom is disubstituted, have no amide hydrogen available for hydrogen bonding. If such chains are capable of maintaining fixed helical structures, they must be stabilized by factors other than hydrogen bonds.

Poly(L-proline) was first synthesized by Katchalski *et al.* (1). Because hydrogen bonding is impossible, any stable configuration must be due to rotational restrictions of bonds along the polyproline chains. Katchalski *et al.* observed that poly(L-proline) can assume two distinct helical forms in solution. The structures of solid-state forms corresponding to the two helical forms were established later from x-ray–diffraction studies (34). The two stable forms are poly(L-proline) I, a right-handed helix with a residue translation of 1.85 Å and with all peptide bonds in the *cis* configuration, and poly(L-proline) II, a left-handed helix with a residue translation of 3.12 Å and with all peptide bonds in the *trans* configuration. A thorough study of the reversible cooperative transformation of forms I and II by ORD, CD, NMR and so on (35) indicates that it is due to a *cis-trans* isomerization of the peptide bond.

Poly(L-proline) is constrained by the five-membered pyrrolidine ring. To ascertain whether the rigid five-membered ring can contribute to the stabilization of the polymer chains, we synthesized poly(*N*-methyl-L-alanine), an acyclic analog of poly(L-proline), by amine initiation of *N*-methylalanine *N*-carboxy anhydride.

To investigate further the nature of mutarotation exhibited by *N*-disubstituted polypeptides, we synthesized poly[(*S*)-thiazolidine-4-carboxylic acid], a cyclic analog of poly(L-proline) in which the γ-methylene has been replaced by sulfur. The NMR spectrum of poly[(*S*)-thiazolidine-4-carboxylic acid] remains unchanged after alteration of the solvent from chloroform-d_1 to trifluoroacetic acid (36, 37). We interpret the NMR spectrum to be consistent with an all-*trans* structure for the polypeptide. The CD results are similar to those obtained for poly(L-proline) II. Using CD and NMR spectra, we were not able to detect mutarotation. From energy calculations, we found that the all-*trans* conformation for poly[(*S*)-thiazolidine-4-carboxylic acid] is more stable than the all-*cis* conformation by 5 kcal/mole of peptide unit (38). This compares with a difference between *trans*- and *cis*-poly(L-proline) of about 2 kcal/mole per peptide unit.

Poly(*N*-methyl-L-alanine) can be considered as poly(L-proline) in which the five-membered ring has been opened by the removal of the central methylene group, thereby permitting some rotational freedom about the N—C_α bond. Because the amide nitrogen is alkylated, hydrogen bonding is impossible. To study the conformation of poly(*N*-methyl-L-alanine) we used NMR, CD, ORD, and UV spectroscopy. A model compound for the polymer, *N*-acetyl-*N*-methyl-L-alanine methyl ester, was synthesized. Fig. 3 shows NMR spectra of both poly-(*N*-methyl-L-alanine) and the model compound. In the latter, the *N*-methyl

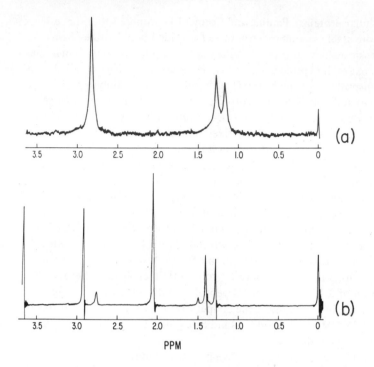

Figure 3. 220 MHz NMR spectrum in methylene chloride of (a) poly(N-methyl-L-alanine) and (b) N-acetyl-N-methyl-L-alanine methyl ester.

resonance appears as a large peak at 3.0 ppm and a smaller peak at 2.8 ppm, and the α-methyl group appears as two doublets near 1.5 ppm. The appearance of two sets of peaks indicates that this compound exists in both the *cis* and *trans* conformations. In the spectrum of the polymer, only one N-methyl and one α-methyl resonance are observed, and they are in the positions corresponding to the *trans* conformation. Thus, unlike the model compound, the polymer has an all-*trans* conformation. This feature was found also in methylene chloride and trifluoroethanol (TFE). Optical rotatory dispersion of poly(N-methyl-L-alanine) in TFE shows a negative Cotton effect with a trough at 233 nm ($[m'] =$ $-15,000°$) and a peak at 202 nm ($[m'] = +32,000°$). Such chromophoric behavior is indicative of a helical conformation.

The helical conformation was further substantiated by a marked hypochromism ($\epsilon = 5600$ at 201 nm) in the UV absorption spectrum, a characteristic of strongly interacting chromophores in a helix. The CD spectrum of the polymer (Fig. 4) shows a very broad, negative band centered at 223 nm, followed by a sharper positive band at 192 nm. This spectrum was interpreted to be similar to the CD spectrum of the α helix, that is, a negative n-π^* band followed by a positive π-π^* band. The first two bands are so close together that two separate minima cannot be distinguished. The spectrum of the model compound shows only a weak n-π^* band, followed by the beginning of a negative π-π^* band. The above CD assignments were recently confirmed (39) by calculations of the

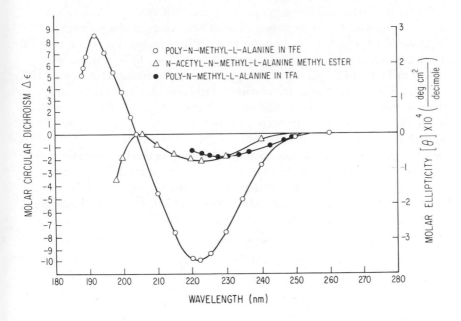

Figure 4. CD spectrum of poly(*N*-methyl-L-alanine) in TFA and TFE and *N*-acetyl-*N*-methyl alanine methyl ester in TFE.

optical activity of poly(*N*-methyl-L-alanine), giving further support to the idea of a helical conformation; if the polymer were a random coil, such a marked contrast between the CD spectrum of the model and that of the polymer would not be expected. These results provided substantial evidence that chains of this polypeptide are helical with *trans* amide bonds in some helix-supporting solvents, such as methylene chloride (CH_2Cl_2) and TFE. Energy calculations carried out by Mark and Goodman (40), Conti and De Santis (41) indicate that the preferred conformation of this chain is either a right-handed, approximately threefold helix or a slightly distorted, left-handed α helix.

Recent theoretical treatments (42), as well as optical activity (39), suggest that poly(*N*-methyl-L-alanine) exists preferentially as a right-handed helix with $(\phi, \psi) = (30°, 250°)$, in agreement with the structure proposed by Mark and Goodman (40):

Splitting of the α- and N-methyl peaks in the NMR spectra of this polymer has been attributed to the formation of dyad conformers in the presence of trifluoroacetic acid (TFA) (41). In order to investigate further the conformational changes induced by TFA, spectral changes of poly(N-methyl-L-alanine) were studied in TFA/CDCl$_3$ mixed solvent systems made by adding TFA to the CDCl$_3$ solution. These spectra at different solvent ratios are shown in Fig. 5. In 1% TFA, the spectrum is similar to that observed in CD$_2$Cl$_2$ and CDCl$_3$. In addition to the α- and N-methyl resonances, which appear at 1.39 and 3.1 ppm, respectively, a small peak appears around 2.8 ppm. The area of this peak increases and the spectrum of the polymer changes drastically as the TFA concentration is increased. However, the difference between the chemical shifts of the two sets of N-methyl resonances remains unchanged (0.3 ppm) at different TFA concentrations. In 100% TFA, the NMR spectra show two sets of N-methyl multiplets and a complex α-methyl multiplet.

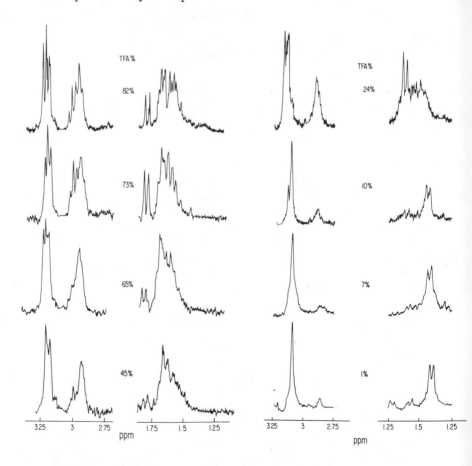

Figure 5. Partial 220 MHz spectrum (32 CAT scans accumulated) of the α-methyl region (about 1.4–1.8 ppm) and N-methyl region (*trans*, 3–3.2 ppm; *cis*, 2.8–3 ppm) of poly(N-methyl-L-alanine) in TFA/CDCl$_3$ at 22°C and 0.1% concentration (w/v).

Spectral changes in poly(N-methyl-DL-alanine) in the TFA/CDCl$_3$ system were found to be very similar to those observed in poly(N-methyl-L-alanine). The same results were also observed when these two polypeptides were studied in the TFA/CD$_2$Cl$_2$ or the TFA/TFE solvent systems.

The conformational change of poly(N-methyl-L-alanine) in TFA was also studied using CD. In TFE CD data show a helix CD pattern, as reported above. When TFA is added to TFE solutions, the CD absorption band decreases. In 100% TFA, the CD shows a broad band at 229 nm with a molar circular dichroism ($\Delta\epsilon$) of –2 compared to –10 in TFE (Fig. 4). When the TFA is evaporated and the polypeptide is redissolved in TFE, it exhibits the same CD spectra as originally observed in TFE. The molecular weight of this polypeptide, before and after treatment with TFA, was the same. Apparently, then, no degradation of the polymer chain occurs during the TFA treatment.

Experimental results for poly(N-methyl-L-alanine) shown by NMR and CD clearly indicate that there is a conformational change after addition of TFA to the solution. The polymer exists predominantly as an ordered helical structure with *trans* amide bonds in helix-supporting solvents. As TFA is added to the solution, a new resonance appears at 0.3 ppm upfield from the original *trans* N-methyl resonance, which corresponds to the polymer segment in the *cis* configuration exhibited by the model compound. Thus, two major sets are observed for these two N-methyl multiplets, which correspond to the polymer in *cis* and *trans* conformations. The drastic change in the CD spectrum confirms the NMR results in that the polymer exhibits a conformational change as the TFA content of solutions in helix-supporting solvents is increased.

From our investigations, we conclude that in TFA there is a partial *trans-cis* peptide-bond mutarotation, which converts the polymer into a disordered state. The observed effects are very similar to those observed for poly(L-proline) in concentrated salt solutions.

Because the model compound, N-acetyl-N-methyl-L-alanine methyl ester, exists in both *trans* and *cis* forms and the polymer shows only *trans* structure in helix-supporting solvents, we are synthesizing the N-methyl-L-alanine oligomers in order to understand the dependence of the *trans-cis* transition on the number of residues in the chain. Preliminary results show that N-methyl-L-alanine oligomers [N-acetyl-(N-methyl-L-alanine)$_n$ methyl ester] still exhibit both *trans* and *cis* configurations at $n = 3$. We speculate that the acetyl N-methyl-L-alanyl tetramer will exhibit structure in which only *trans* peptide units exist.

POLY[(S)-β-AMINO BUTYRIC ACID]

Poly [(S)-β-amino butyric acid] is a homolog of poly(L-alanine) (43, 44). The latter polymer exhibits an α-helical form both in solution and as a film (45).

Because of the additional methylene group, poly [(S)-β-amino butyric acid] is expected to have different conformational properties than poly(L-alanine):

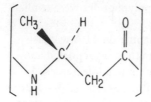

We have studied the conformations of poly [(S)-β-amino butyric acid] in solution by CD and UV spectroscopy (46).

The CD spectrum of poly [(S)-β-amino butyric acid] in hexafluoroisopropanol (HFIP) (Fig. 6) shows a trough at 216 nm and a positive peak at 197 nm with molar ellipticities of –3.2 × 10^3 and 6.6 × 10^3 (deg cm³/d mole), respectively.

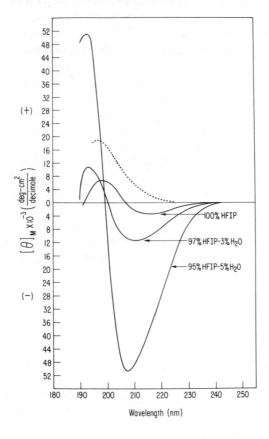

Figure 6. Circular dichroism of poly [(S)-β-amino butyric acid] in hexafluoroisopropanol, in hexafluoroisopropanol-water mixtures, and N-acetyl (S)-β-amino butyric-N'-methyl amide in hexafluoroisopropanol.

The crossover occurs at 207 nm. The model compound, N-acetyl (S)-β-amino butyric-N'-methyl amide, shows only an intense positive band near 197 nm; there is no negative band. The spectral differences can be interpreted to indicate a conformational difference for a residue in the polymer compared to the model compound. The overall shape of the CD spectrum of poly $[(S)$-β-amino butyric acid] in HFIP is very similar to that reported for several polypeptides in a β conformation (20).

To ascertain the solvent effect on poly $[(S)$-β-amino butyric acid], films were cast from HFIP solutions. The CD spectra of these films (Fig. 7) show a trough centered at 212 nm and a positive peak near 190 nm. Compared to the solution spectra, these spectra are shifted 6–7 nm towards the blue; the spectral patterns are similar, however. The β conformation with the antiparallel chains has been confirmed for the polymer film by Schmidt (44) by means of x-ray and polarized infrared dichroism. This evidence further supports our CD assignments.

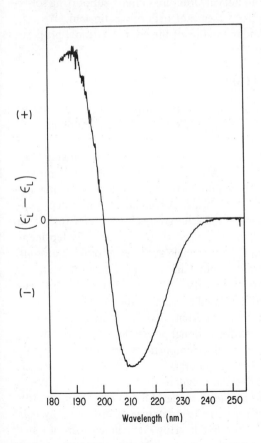

Figure 7. Circular dichroism of poly $[(S)$-β-amino butyric acid] film cast from hexafluoroisopropanol.

The ultraviolet spectrum of poly [(S)-β-amino butyric acid] in HFIP shows an absorption maximum at 192 nm [ϵ_{max} = 9180 ± 5% (M^{-1} cm^{-1})]. The addition of water shifts the maximum to a longer wavelength, near 197 nm. The overall spectrum is similar to that of the β structure of poly(L-serine) (47).

The β conformation of poly [(S)-β-amino butyric acid] in HFIP is established from the UV and CD studies described above. Adding a trace amount of water to HFIP enhances the formation of the β conformation, whereas poly(L-alanine) in the same solvent forms a distorted α helix. The presence of methanesulfonic acid disrupts the β conformation of poly [(S)-β-amino butyric acid]. This effect indicates that the additional α-methylene group of poly [(S)-β-amino butyric acid] leads to a preference for a β conformation in solution and in the solid state.

We intend to synthesize the N-methylated poly [(S)-β-amino butyric acid] to compare its conformation with that of our previously reported poly(N-methyl-L-alanine), which exhibits an ordered helical structure in helix-supporting solvents and a disordered structure in the presence of trifluoroacetic acid. We expect these studies to enhance our understanding of the conformational properties of polypeptides as well as poly(β-amides).

ALTERNATING POLYDEPSIPEPTIDES

We have recently initiated an investigation of the synthesis and conformational characteristics of polydepsipeptides comprised of alternating α-amino and α-hydroxy acids (48, 49).

Our interest in polydepsipeptides arises from the close structural analogy between these molecules and polypeptides. Also, we believe that they will prove to be useful models of the conformational properties of naturally occurring depsipeptides such as valinomycin and the enniatins. A segment of a polydepsipeptide chain is shown in Fig. 8. The amide and ester groups both prefer the planar *trans* configuration. The skeletal geometries of polypeptides and polydepsipeptides are nearly identical except that the bond angle at the ester oxygen atom is 10 degrees smaller than the corresponding bond angle at the amide nitrogen. The steric and dipolar factors that determine the conformational energy of peptides and depsipeptides also are closely related because of the similarity of their structural units. There are differences in important features of these structural units, however. Comparison of the conformational properties of polypeptides and polydepsipeptides allows assessment of the importance of such features in determining the chain conformation of these polymers. In particular, hydrogen-bonding characteristics of polypeptides and polydepsipeptides differ, because every second amide NH group in the former is replaced by an ester oxygen atom in the latter. This replacement primarily affects the nature of ordered polydepsipeptide structures stabilized by hydrogen bonds. However, dimensions of randomly coiling chains also are affected by elimination of certain steric interactions that involve the replaced amide hydrogen atom.

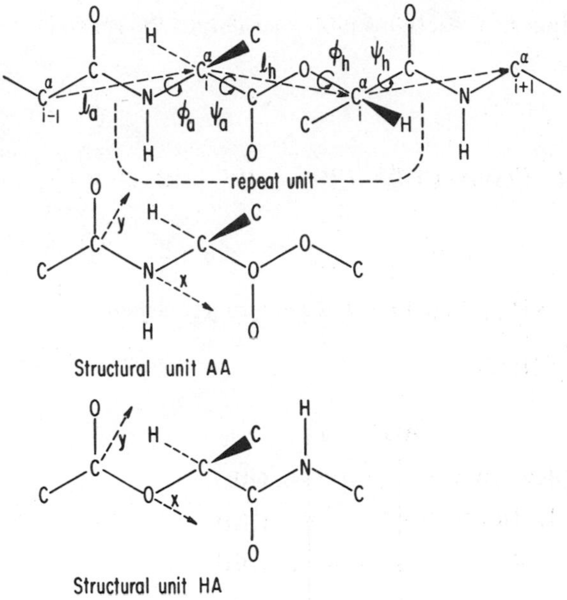

Structural unit AA

Structural unit HA

Figure 8. A segment of a polydepsipeptide chain showing a repeat unit with the torsional angles ϕ_a, ψ_a and ϕ_h, ψ_h and showing the structural units AA and HA.

Synthesis. High-molecular-weight polydepsipeptides were obtained by polymerization of the monomer

I. $NH_2-CHR-CO-O-CHR'-CO-O-$$-Cl$

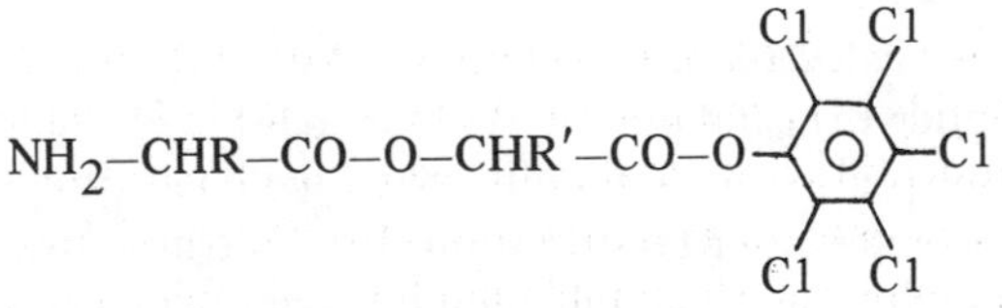

where $R = -CH_3, -CH(CH_3)_2$, $R' = -CH_3$ (48).
    Solution polymerization of the type-I monomer led predominantly to the cyclization product. Solid-state polymerization of monomer-I hydrochloride or trifluoroacetate gave a high-molecular-weight polymer in about 20% yield. The polymer did not show the 1792 cm^{-1} band in the IR of the PCP ester. Strong bands at 3330 cm^{-1} (NH), 1750 cm^{-1} (ester carbonyl), 1670 cm^{-1} (amide I), and 1536 cm^{-1} (amide II) were observed. A molecular weight of 100,000, corresponding to a degree of polymerization of 580, was estimated for poly(Val-Lac) from its intrinsic viscosity (in dichloroacetic acid) of 0.54 dl/g compared to the molecular weight-viscosity relationships determined for poly(γ-benzyl-L-glutamate) (4) and for poly(L-lactic acid) (50). In the polymerization of HCl-Val-Lac-OPCP, a sublimation product was isolated and fractionated. The three resulting compounds were purified and characterized: the starting material (HCl-Val-Lac-OPCP), pentachlorophenol, and *cyclo*(-Val-Lac-). These compounds were characterized by elemental analysis, m.p., t.l.c., IR, and NMR.

To minimize the cyclization side reaction a monomer unit of the general formula

II. 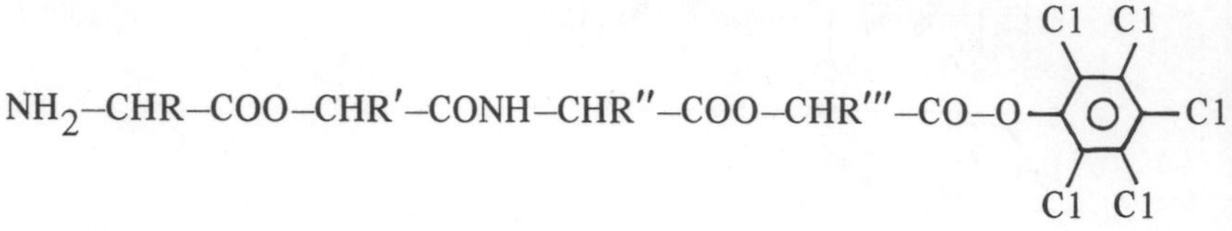 $NH_2-CHR-COO-CHR'-CONH-CHR''-COO-CHR'''-CO-O-\underset{\substack{Cl \quad Cl \\ Cl \quad Cl}}{\boxed{O}}-Cl$

where $R = R' = R'' = R''' = -CH_3$ was synthesized according to Scheme I.

SCHEME I

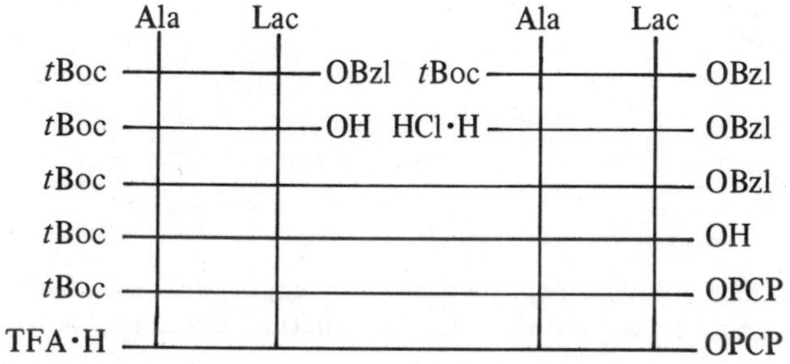

	Ala	Lac		Ala	Lac	
tBoc		OBzl	tBoc			OBzl
tBoc		OH	HCl·H			OBzl
tBoc						OBzl
tBoc						OH
tBoc						OPCP
TFA·H						OPCP

This monomer was polymerized in high yield both in solution and in the solid state.

Conformational calculations and analysis. With procedures developed for treating polypeptide (51) and poly(α-hydroxy acid) (52) chains, conformational energies of depsipeptide structural units were calculated from semiempirical potential functions representing contributions from intrinsic torsional potentials and repulsive, London and Coulombic nonbonded interactions between atom pairs. The potential functions employed were those suggested by Brant, Miller, and Flory (51) for peptides and by Brant, Tonelli, and Flory (52) for α-hydroxy ester structural units. The hydrogen-bond interaction was included, according to the procedure of Brant (53).

Energy contour diagrams for the Ala structural unit, AA, and for the Lac structural unit, HA, are shown in Figs. 9 and 10, respectively. The energy data given were employed to calculate the conformational characteristics of randomly coiling polydepsipeptides. They also served as a basis for a conformational analysis of ordered depsipeptide chains. The results for randomly coiling polydepsipeptides are expressed as characteristic ratios $C_x = \langle r^2 \rangle_0 / x\ell^2$ of $\langle r^2 \rangle_0$, the mean-square unperturbed end-to-end distance to the product of x, the number of α-amino and α-hydroxy acid structural units per chain, and ℓ^2, the square of the average structural unit length. Asymptotic characteristic ratios C_∞ calculated for selected randomly coiling polydepsipeptides are presented in Table 1.

Table 1.   Theoretical asymptotic characteristic ratios C_∞ of several polydepsipeptides

Depsipeptide	C_∞
Poly(glycine-glycolic acid)	2.27
Poly(L-alanine-L-lactic acid)	3.80
Poly(L-alanine-D-lactic acid)	1.55
Poly(D-alanine-L-lactic acid)	1.55

The conformational characteristics of randomly coiling polydepsipeptides reported in Table 1 have important features in common with those reported for polypeptides (51) and for poly(L-lactic acid) (52). In particular, the random-chain dimensions of poly(glycine) and poly(glycine-glycolic acid) are nearly identical. Because of the absence of a suitable solvent, it has not been possible to compare the theoretical characteristic ratio of poly(glycine) to the appropriate experimental quantity. It is our belief that the solubility characteristics of poly-(glycine-glycolic acid) are less restrictive. The chain end-to-end distance of randomly coiling poly(L-alanine-L-lactic acid) is sensitive to energy differences in the allowed regions of Figs. 9 and 10. These regions are closely related to those

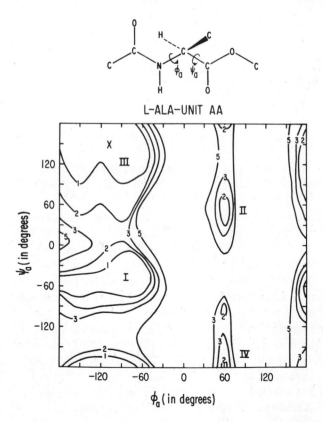

Figure 9. Conformational energy map for Ala structural unit AA.

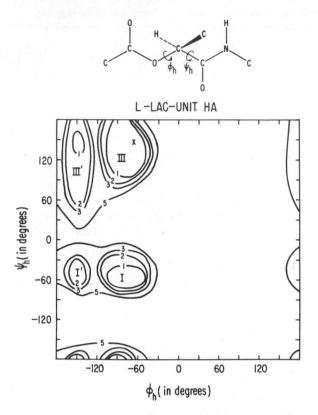

Figure 10. Conformational energy map for Lac structural unit HA.

for polypeptide chains. Our synthesis of high-molecular-weight polydepsipeptides will allow comparison of the theoretical calculations with experimental results.

The conformational energies of ordered depsipeptide chains in which each repeat unit conformation is identical were also calculated. Two low-energy helical structures with 10- and 13-membered hydrogen-bonded rings, respectively, were discovered. The conformational characteristics and total energy of an interior repeat unit of the two minimum-energy helical structures are presented in Table 2. The conformation of the lowest-energy helix of Table 2 is very similar to that of the standard right-handed polypeptide α helix.

In a preliminary experimental conformational analysis of poly(Val-Lac) CD and NMR studies were performed in trifluoroethanol (TFE) and TFE:acid mixtures. In TFE the polymer exhibits a small negative Cotton effect at 237 nm with $[\theta]_m = -2 \times 10^3$, a crossover at 225 nm, two positive bands at 215 nm ($[\theta]_m = 6 \times 10^3$), and 204 nm ($[\theta]_m = 5 \times 10^3$) and a very strong negative band at about 192 nm. Addition of sulfuric acid up to 18% does not cause any significant change in the 237 nm and 215 nm bands, but it does decrease the intensities of the 204 nm and 192 nm bands. Further addition of sulfuric acid causes a red shift associated with a decrease of intensity of the 215 nm band.

Table 2. Conformational characteristics and energy of two minimum-energy polydepsipeptide helices

| Helix[a] | Dihedral Angles | | | | n[b] | d[c] | H Bonding | V_1 (kcal/mole) |
	ϕ_a	ψ_a	ϕ_h	ψ_h				
10	51°	–94°	–144°	30°	1.78	4.50 Å	$(CONH)2^{d} \rightarrow (CONH)$	1.84
13	–65°	–35°	–63°	–47°	1.81	2.89 Å	$(COO)3^{d} \rightarrow (CONH)$	–7.91

[a]Characterized by the number of atoms, 10 or 13, linked in the hydrogen-bonded rings of each helix.

[b]Number of repeat units required for a complete turn about the helix axis.

[c]Displacement along the helix axis per repeat unit.

[d]Number of intervening α-carbon atoms between the indicated hydrogen-bonded groups.

Nuclear magnetic resonance results were similar. Upon addition of 15% trifluoroacetic acid (TFA) to the $CDCl_3$-polymer solutions, the α-CH peaks of the lactyl residue at 5.04 ppm (quartet) and the valyl residue at 4.20 ppm (triplet), were shifted downfield by 0.49 ppm and 0.28 ppm, respectively.

Both the CD and NMR results suggest that there is an acid-induced conformational change of poly(Val-Lac) in these two solvent systems. Considering the similarity of polypeptides and polydepsipeptides, it is likely that this conformational change is from an ordered form in trifluoroethanol and $CDCl_3$ to a disordered form in these solvents with acid.

Infrared spectra of poly(Val-Lac) in dioxane and chloroform solutions and as a film and KBr pellet were studied. The group frequencies are tabulated in Table 3.

The semiempirical conformational energy calculations suggest two low-energy helical structures with 10- and 13-membered hydrogen-bonded rings, respectively. The R = 10 helix, which is characterized by a hydrogen bond between two consecutive amide groups is expected to be more stable in moderately polar solvents because the amide and especially the ester groups can be solvated. Thus, the two bands at 1695 cm^{-1} and 1640 cm^{-1} in dioxane are probably due to a split amide-I band resulting from hydrogen-bond interactions between neighboring amide groups in the R = 10 helical form. The R = 13 helix is characterized by a

Table 3. Group frequencies of poly(Val-Lac) determined on a Perkin–Elmer 180 Spectrophotometer

Spec. No.	Sample	νNH (cm^{-1})	ν-C(=O)–O (cm^{-1})	Amide I (cm^{-1})	Amide II (cm^{-1})
1	Chloroform solution[a,b]	3390 (vs) 3310 (vs)	1732 (vs)	1665 (vs)	1540 (vs)
2	Dioxane solution[a,c]	3580 (vs) 3510 (vs) 3324 (s)	1740 (vs)	1695 (vs) 1640 (s)	1530 (s)
3	Film[d]	3390 (shoulder) 3310 (s)	1740 (vs)	1665 (vs)	1540 (vs)
4	KBr pellet	3430 (shoulder) 3300 (s)	1750 (vs)	1670 (s)	1536 (s)

[a]Spectrum of a 1 mg/ml solution was taken in a 0.5 mm KBr cavity cell with solvent as a reference.

[b]Redissolving a film cast from dioxane in chloroform gave spectrum 1.

[c]Redissolving a film cast from chloroform in dioxane gave spectrum 2.

[d]The same spectrum was obtained for films cast from dioxane or chloroform.

hydrogen bond between an amide and an ester of the following depsipeptide unit. This form is expected in nonpolar solvents in which the polar amide and ester groups are partially protected from the solvent by the helix side chains. Thus the 1665 cm^{-1} and 1540 cm^{-1} bands in chloroform may arise from the $R = 13$ helix form.

Nuclear magnetic resonance spectral parameters determined at 220 MHz for poly(Val-Lac) at a 3% concentration in CDCl$_3$ are presented in Table 4. Upon addition of 15% TFA, there was a 0.49 ppm downfield shift of the lactyl α-CH peak and a 0.28 ppm downfield shift of the valyl α-CH peak.

The CD and NMR results suggest that there is an acid-induced conformational change of poly(Val-Lac) in these two solvent systems. In view of the similarity of polypeptides and polydepsipeptides, this conformational change is likely to be from an ordered form in TFE and CDCl$_3$ to a disordered form in these solvents with acid.

Results from IR and CD studies suggest the existence of two helical structures. We speculate that these structures could be the two lowest-energy ordered forms calculated for these polydepsipeptides, and our efforts are now aimed at confirmation of these deductions.

Table 4.   220 MHz NMR data for poly(Val-Lac)

Assignment	PPM from Internal TMS	Peak Shape	Relative Intensities
NH	7.29	Doublet	1
α-CH, lactyl	5.04	Quartet	1
α-CH, valyl	4.20	Triplet	1
β-CH, valyl	2.32	Multiplet	1
α-CH$_3$, lactyl	1.57	Doublet	3
α-CH$_3$, valyl	1.05	Triplet	6

REFERENCES

1.   A. Berger, J. Kurtz, and E. Katchalski, *J. Am. Chem. Soc.*, **76**, 5552 (1954).
2.   E. Katchalski and M. Sela, *Advances in Protein Chem.*, **13**, 243 (1958).
3.   E.R. Blout and R.H. Karlson, *J. Am. Chem. Soc.*, **78**, 941 (1956).
4.   P. Doty, J.H. Bradbury, A.M. Holtzer, and E.R. Blout, *J. Am. Chem. Soc.*, **76**, 4493 (1954).
5.   C.H. Bamford and H. Block, in *Polyamino Acids, Polypeptides, and Proteins*, M.A. Stahmann, ed., Madison, Wis.: University of Wisconsin Press, 1962, p. 65.
6.   E.R. Blout, *ibid.*, pp. 3 and 79.
7.   M. Goodman and U. Arnon, *J. Am. Chem. Soc.*, **86**, 3384 (1964).
8.   M. Goodman and J. Hutchison, *J. Am. Chem. Soc.*, **88**, 3627 (1966).
9.   M. Goodman, L.B. Jacobsberg, and N.S. Choi, *Israel J. Chem.*, **9**, 125 (1971).
10.   M. Terbojevich, G. Pissiolo, E. Peggion, A. Cosani, and E. Scoffone, *J. Am. Chem. Soc.*, **89**, 2733 (1967).

11. E. Peggion, M. Terbojevich, A. Cosani, and C. Columbini, *J. Am. Chem. Soc.,* **88**, 3630 (1966).
12. E. Katchalski, M. Sela, H.I. Silman, and A. Berger, in *The Proteins,* H. Neurath, ed., Vol. 2, New York: Academic Press, 1964, p. 406.
13. G.D. Fasman, ed., *Poly-α-Amino Acids,* New York: Marcel Dekker, 1967.
14. L. Pauling and R.B. Corey, *Proc. Nat. Acad. Sci. U.S.,* **37**, 729 (1951).
15. M.F. Perutz, *Nature,* **161**, 1053 (1951).
16. C.H. Bamford, L. Brown, A. Elliott, W.E. Hanby, and L.F. Trotter, *Nature,* **169**, 351 (1952).
17. K. Rosenheck and P. Doty, *Proc. Nat. Acad. Sci. U.S.,* **41**, 1775 (1961).
18. W. Moffitt, *Proc. Nat. Acad. Sci. U.S.,* **42**, 731 (1956).
19. W. Moffitt and A. Moscowitz, *J. Chem. Phys.,* **30**, 648 (1959).
20. M. Goodman, A.S. Verdini, N.S. Choi, and Y. Masuda, in *Topics in Stereochemistry,* E.L. Eliel, and N.L. Allinger, eds., New York: Wiley(Interscience), 1970, p. 69.
21. G. Holzwarth and P. Doty, *J. Am. Chem. Soc.,* **87**, 218 (1965).
22. T. Miyazawa and E.R. Blout, *J. Am. Chem. Soc.,* **83**, 712 (1962).
23. T. Miyazawa, in *Poly-α-Amino Acids,* G.D. Fasman, ed., New York: Marcel Dekker, 1967, p. 69.
24. F.A. Bovey, *Polymer Conformation and Configuration,* New York: Academic Press, 1969, p. 68.
25. B. Zimm and J. Bragg, *J. Chem. Phys.,* **31**, 526 (1959).
26. J.H. Gibbs and E.A. DiMarzio, *J. Chem. Phys.,* **28**, 1247 (1958); *Ibid.,* **30**, 271 (1959).
27. S. Lifson and A. Roig, *J. Chem. Phys.,* **34**, 1963 (1961).
28. G.N. Ramachandran, C. Ramakrishnan, and V. Sasisekharan, in *Aspects of Protein Structure,* G.N. Ramachandran, ed., New York: Academic Press, 1963, p. 121.
29. A.M. Liquori, *J. Poly. Sci.,* **C12**, 209 (1966).
30. R.A. Scott and H.A. Scheraga, *J. Chem. Phys.,* **45**, 2091 (1966).
31. S. Lifson, *J. Chem. Phys.,* **40**, 3705 (1964).
32. P.J. Flory, *Statistical Mechanics of Chain Molecules,* New York: Wiley-(Interscience), 1969, p. 24.
33. M. Goodman and N.S. Choi, in *Peptides,* E. Bricas, ed., Amsterdam: North-Holland, 1968, p. 1.
34. L. Mandelkern, in *Poly-α-Amino Acids,* G.D. Fasman, ed., New York: Marcel Dekker, 1967, p. 675.
35. P.H. Von Hippel and T. Schleich in *Biological Macromolecules,* Vol. 2, S.N. Timasheff and G.D. Fasman, eds., New York: Marcel Dekker, 1969, p. 571.
36. M. Goodman, K.C. Su, and G.C.C. Niu, *J. Am. Chem. Soc.,* **92**, 5220 (1970).
37. M. Goodman and K.C. Su, *Biopolymers,* **11**, 1773 (1972).
38. M. Goodman, G.C.C. Niu and K.C. Su, *J. Am. Chem. Soc.,* **92**, 5219 (1970).
39. V. Madison and J. Schellman, *Biopolymers,* **11**, 1041 (1972).
40. J.E. Mark and M. Goodman, *Biopolymers,* **5**, 809 (1967).
41. F. Conti and P. De Santis, *Biopolymers,* **10**, 2581 (1971).
42. W.L. Mattice, *Macromolecules,* **6**, 855 (1973).
43. R. Graf, *Angew. Chem.,* **80**, 179 (1968); *Angew. Chem. Int. Ed.,* **1**, 172, (1968).

44.   E. Schmidt, *Angew. Makromol. Chem.,* **14**, 185 (1970).
45.   J.R. Parrish and E.R. Blout, *Biopolymers,* **11**, 1001 (1972).
46.   F. Chen, G. Lepore, and M. Goodman, *Macromolecules,* in press.
47.   F. Quadrifoglio and D.W. Urry, *J. Am. Chem. Soc.,* **90**, 2760 (1968).
48.   M. Goodman, C. Gilon, M. Palumbo, R.T. Ingwall, *Israel J. Chem.,* in press.
49.   R.T. Ingwall and M. Goodman, *Macromolecules,* submitted.
50.   A.E. Tonelli and P.J. Flory, *Macromolecules,* **2**, 225 (1969).
51.   D.A. Brant, W.G. Miller, and P.J. Flory, *J. Mol. Biol.,* **23**, 47 (1967).
52.   D.A. Brant, A.E. Tonelli, and P.J. Flory, *Macromolecules,* **2**, 228 (1969).
53.   D.A. Brant, *Macromolecules,* **1**, 291 (1968).

X-RAY STUDIES OF POLY(AMINO ACIDS) AS STRUCTURAL MODELS FOR PROTEINS

WOLFIE TRAUB, Department of Structural Chemistry, The Weizmann Institute of Science, Rehovot, Israel

SYNOPSIS: This paper is a review of the role of x-ray studies of poly(amino acids) in elucidating the structures of native proteins. Homopolypeptides are used as models of the regular polypeptide backbone conformations found in fibrous and globular proteins, but they are not good models for specific chain assemblies. These aspects of structure can be studied with poly(amino acids) of ordered amino acid sequence. Compounds of this type are excellent models for silk fibroin and collagen and might possibly prove useful also in structural investigations of multistranded α-helical fibrous proteins, nucleoprotamines, and elastin.

It is suggested that an understanding of the correlation of sequence and structure in globular proteins might be furthered by accurate single-crystal x-ray analyses of short oligopeptides likely to form stable bend conformations. In addition, more precise dimensions of regular chain conformations, including α helices and β-pleated sheets, might be obtained from single crystals of poly(amino acids) of specified molecular weight.

INTRODUCTION

Interest in poly(α-amino acids) has been motivated primarily by their usefulness as simple synthetic models of proteins, yet the role of poly(amino acids) as models logically implies an eventual shift of interest toward the greater complexities of naturally occurring proteins. I have been asked to reassess the value of x-ray structural investigations of poly(amino acids) and, indeed, to consider whether the usefulness of such studies has been largely exhausted. Frankly, I am inclined to answer this question with a qualified "yes." Whereas ten and even five years ago my own research was concerned mainly with poly(amino acids), it is now completely committed to determining structures of "real" proteins. Such a shift of interest is a common trend and reflects both the belief that the main structural features of such simple models already have been determined and also the greatly improved possibilities of studying protein structures directly. Certainly, there are some areas of incompleteness in our understanding of poly-(amino acid) structures and new physiochemical or biological applications may provoke further structural investigations, but I feel that further major advances in this field require the isolation and characterization of new proteins to be modeled or new technology that will provide answers to new kinds of questions. Before speculating about future prospects, however, it is appropriate to consider the past contributions, as well as the limitations, of poly(amino acid) studies in the elucidation of protein structure.

CONFORMATIONS OF POLY(AMINO ACIDS)

The first major achievement of the indirect approach to protein-structure determination was to establish the dimensions of the amide group from a series of single-crystal structural analyses of small peptides and amino acids (1). These dimensions, which have been slightly revised in the light of more extensive and precise peptide structure analyses (2, 3), were used by Pauling and Corey in their investigations of possible regular extended polypeptide conformations that might be stabilized by systematic hydrogen bonding. These studies led them to propose the α helix (4, 5) and the β-pleated sheet (6, 7), as well as several conformations (8) that generally have been forgotten. Two forms of pleated sheets were considered in which neighboring polypeptide chains were either parallel or antiparallel. Both were said to have twofold screw axes, but the repeat distances were estimated to be 6.5 Å, for the parallel case, and 7.0 Å, for the antiparallel. The α helix incorporated an axial translation of 1.50 Å and a rotation of 100° per residue and could be built in either a right-handed or a left-handed form, although the latter was considered less stable because of a rather short contact (2.64 Å) between the β-carbon atom and the carbonyl oxygen of the same residue.

Experimental support for the existence of the α helix came from the observation of a 1.5 Å meridional reflection, corresponding to the residue translation, in the x-ray-diffraction pattern of poly(γ-benzyl-L-glutamate) fibers (9) and from the appearance in the poly(γ-methyl-L-glutamate) pattern of strong layer lines (10), which were shown to occur at the spacings expected for an α helix (11). A more detailed structural analysis made of the α form of poly(L-alanine) showed a particularly rich x-ray pattern and only a single methyl group as a side chain. It was shown that the x-ray intensities could be easily explained with right-handed α helices with a random-chain sense (12); this structure was later subjected to quantitative refinement (13, 14).

Poly(amino acids) were also found to have x-ray fiber patterns indicative of β-pleated-sheet structure, and some, including poly(L-alanine) and poly(γ-methyl-L-glutamate), were obtained in both α-helical and β forms (8, 15). Most of the β x-ray patterns indicate repeat distances intermediate between those expected for parallel and antiparallel chains, although they are generally closer to the latter; that of polyglycine I is, in fact, 7.0 Å (16). The only β structure that has been analyzed in detail is that of β-poly(L-alanine), for which an antiparallel pleated-sheet structure, with neighboring sheets randomly displaced, has been proposed (17).

Poly(amino acids) have been found to exist in several other conformations. Poly(β-benzyl-L-aspartate), for example, forms left-handed helices (18, 19) that, in the solid state, can conform either to α-helical dimensions or to those of the rather similar ω helix, with a translation of 1.3 Å and rotation of 90° per residue (20). The latter form, which may well be related to the occurrence of bulky groups on the β-carbon atom, has been reported also for poly(S-benzylthio-L-cysteine) (21). Another kind of left-handed helix, with 3.1 Å translation and 120° rotation per residue, occurs in poly(L-proline) II (22, 23), poly(L-hydroxyproline) A (24) and polyglycine II (25, 26). Poly(L-proline) also exists in a right-handed helical form that has an unusual *cis* configuration about all peptide bonds (27, 28).

X-ray studies of solvated poly(amino acids) in liquid crystals indicate that their conformations do not vary appreciably over wide ranges of solvation (29-31). Sharp solvent-induced transitions between different conformations occur in a number of cases (31-37), but no previously unobserved conformations have been discovered.

MODELS FOR FIBROUS PROTEINS

It has proved much easier to determine the detailed structures of poly(amino acids) than to determine those of natural proteins. Structural analyses from x-ray fiber patterns, such as those of poly(amino acids) or fibrous proteins, generally proceed through the following stages: measurement of helical parameters and unit-cell dimensions from the x-ray pattern; consideration of possible molecular models consistent with the helical parameters, any other relevant physicochemical data, and appropriate bond lengths, bond angles, and van der Waals distances; consideration of how various possible molecular conformations could fit into the unit cell; and comparison of observed x-ray intensities with those calculated for possible structures. After choosing an approximate conformation and mode of packing, it is possible to refine several parameters in terms of minimum discrepancies between observed and calculated x-ray intensities (13, 14, 17). For these procedures, poly(amino acids) have several advantages over fibrous proteins: They have simple, known amino acid sequences, they pack more frequently in small well-defined unit cells, and their x-ray patterns are unlikely to be complicated by the effects of long-range distortions and extraneous components. Hence, interatomic distances and x-ray intensities corresponding to possible conformations can be calculated and, by these criteria, the structure can be established with far less ambiguity than is possible in direct studies of fibrous proteins. Corresponding features of the x-ray patterns or other physicochemical observations of the proteins then can be interpreted by analogy with the poly(amino acids).

How well do the various poly(amino acid) conformations represent the structures of fibrous proteins? The α helix, β-pleated sheet and polyproline-II helix, respectively, have been used as models for the α keratin, β keratin, and collagen conformations—the three main types of fibrous protein structures classified by Astbury according to their high-angle x-ray patterns (38). These different types of poly(amino acids) and fibrous proteins do indeed have similar conformations, but the greater simplicity of the homopolypeptides limits their relevance as models of protein structure.

Thus, various observations on α keratin and myosin, including the occurrence of a 1.5 Å meridional reflection, indicate that they contain α helices, presumably similar in conformation to those of α-polyalanine. These fibrous proteins, however, also show a strong 5.1 Å meridional reflection not yet observed in x-ray patterns of synthetic poly(amino acids); it apparently arises from the specific assembly of the protein α helices to form coiled coils (39–45).

It appears likely also that the structures of some α-helical fibrous proteins, notably paramyosin (41, 42), are partially determined by side-chain interactions. Such interactions also appear to occur in the structures of poly(amino acids) with large side chains, the x-ray patterns of which show additional periodicities to those due to the α-helical polypeptide backbone (35, 46, 47).

β-polyalanine appears to be a very good model for Tussah silk fibroin, which exhibits an almost identical x-ray pattern (17, 48) and, in fact, contains 39% alanine, as well as 24% glycine and 11% serine (49). The structure reportedly consists of antiparallel pleated sheets, equally spaced, but randomly displaced in the interchain direction (17). However, the x-ray pattern of Bombyx mori silk fibroin, which contains 44% glycine, 29% alanine, and 12% serine (50), is significantly different (51). Sequence analyses indicate that, in most of this protein, glycine residues alternate with other amino acid residues, and it was found that the synthetic polydipeptide poly(Ala-Gly) shows an x-ray pattern similar to that of this silk. The structure of poly(Ala-Gly), and apparently also of Bombyx mori, consists of antiparallel pleated sheets with alternate narrow intersheet spacings that accommodate only glycine residues and wider spacings that accommodate side chains of other residues (52). The conformational specificity of the poly(Ala-Gly) sequence has been emphasized by studies of a series of ordered copolymers of alanine and glycine, which have been found to form a variety of β and non-β structures (53, 54, 55). However, an even closer resemblance to the x-ray pattern of Bombyx mori is shown by the polyhexapeptide poly(Ala-Gly-Ala-Gly-Ser-Gly), which corresponds to a commonly occurring sequence in this silk (50, 56).

Poly(amino acids) of ordered sequence have played an invaluable role in elucidating the structure of collagen (57). Although it was established as long ago as 1955 (58) that the collagen molecule consists of three helical polypeptide chains coiled about a common axis and held together by NH · · · OC hydrogen bonds, controversy continued for many years as to the detailed conformation and the number and mode of interchain hydrogen bonds. Because of the rather poor x-ray pattern, which does not show a unit cell, and limited knowledge of the sequence, decisive criteria that would permit a clear choice of structure from direct studies of collagen proved very elusive, and attention was directed to model compounds.

The threefold left-handed helices formed by homopolypeptides of proline (22, 23), hydroxyproline (24), or glycine (25, 26) do indeed resemble individual strands of the collagen molecule, but they do not form coiled coils. However, various polytripeptides that, like collagen, have glycine as every third residue and proline or hydroxyproline as one or both of the imino acids show x-ray patterns with all of the main features of that of collagen, including the 2.9 Å meridional spacing indicative of a ropelike triple helix (59–64). A detailed and systematic structural analysis of poly(Gly-Pro-Pro) has led to an unambiguous conformation that has been refined within narrow limits (63). Similar analyses of the polyhexapeptides poly(Gly-Ala-Pro-Gly-Pro-Pro), poly(Gly-Pro-Ala-Gly-Pro-Pro), poly(Gly-Ala-Pro-Gly Pro-Ala) and poly(Gly-Ala-Ala-Gly-Pro-Pro), chosen for their variety of potentially H-bonding NH positions, all revealed conformations close to that found for poly(Gly-Pro-Pro) (64, 65). Furthermore, the most detailed collagen x-ray pattern, that of stretched rat-tail tendon, shows good agreement with that calculated from the poly(Gly-Pro-Pro) conformation (66, 67).

Studies of poly(amino acids) also have served to illuminate the structural significance of sequence regularities in collagen. The failure of poly(Ala-Pro-Pro), poly(Ala-Hyp-Hyp), and random copolymers of glycine and imino acids (62, 68) to form collagenlike structures accords with a position for every third residue near the axis of the triple helix where there is room only for glycine, the smallest amino acid. Furthermore, the poly(Gly-Pro-Pro) conformation indicates severe steric restrictions for phenylalanine, tyrosine, and leucine residues in position Y of the poly(Gly-X-Y) sequence, but stabilizing hydrogen-bonding interactions for hydroxyproline, threonine, glutamine, lysine, and arginine residues in this position (69, 70). Apparently, appreciable amounts of conformationally restrictive imino acid residues are required to prevent collagen from assuming energetically more favored α-helical or β-pleated-sheet conformations, such as those found for poly(Gly-Ala-Ala) (55) and poly[Gly-Ala-Glu(OEt)] (71). Polytripeptides of the form poly(Gly-amino-imino) have been found to form triple helices less readily than those of the form poly(Gly-imino-amino), possibly because of different interactions with water (57, 72).

It is hoped that the use of ordered-sequence poly(amino acid) models will be extended to the investigation of several other structural problems concerning fibrous proteins. It appears that α helices can form stable associations through the juxtaposition of sides rich in hydrophobic residues (73) and it is likely also that the coiled coils of α helices in keratin and myosin may be stabilized in this way. Because α helices have 3.6 residues per turn, poly(amino acids) in which similar side chains are separated by three or four positions in the amino acid sequence could form helices in which the polar and hydrophobic groups are lined up on different sides. Sequences of the form poly(A_2B), poly(A_3B), poly(A_2BA_3B), or poly($A_2B_2A_2B$) may provide models for coiled coils (74). These sequences might be chosen also with a view to studying the influence of side-chain interactions (42, 47).

Nucleoprotamines have been modeled by complexes of DNA with the basic polypeptides poly(L-lysine) and poly(L-arginine). X-ray patterns of nucleoprotamine and polylysine-DNA bear a closer resemblance to each other than to the x-ray pattern of polyarginine-DNA (75), a rather surprising finding in view of the very high arginine content of protamines (76). Studies of molecular models indicate that, in contrast to polylysine, polyarginine cannot wind smoothly along long stretches of the DNA small groove with hydrogen bonding between the basic side chains and the negatively charged phosphate groups (75). In fact, protamine sequences have been found to include groups of arginines (commonly four) separated by a few nonbasic residues (76) and, therefore, a polyhexapeptide of the form poly(Arg_4X_2) may be a better model for protamines.

Recently it has been shown that the amino acid sequence of elastin includes several extensive runs of repeating oligopeptides, and it has been suggested that these form β bends that are arranged in regular helical structures (77, 78). This interesting proposal could be tested by x-ray studies of poly(amino acids) with the appropriate sequences.

MODELS FOR GLOBULAR PROTEINS

Globular proteins have far less regular conformations than fibrous proteins. The first globular protein structures to be determined, those of myoglobin and hemoglobin, are more than 70% helix (79, 80), a fortunate feature that greatly facilitated their elucidation. With few exceptions, however, the more than 20 protein structures determined since are appreciably less than 50% helix (81). Most of these helical regions resemble the right-handed α helix of poly(L-alanine), but there are variations resembling the 3_{10} (79, 82, 84–90), α_{II} (83, 84, 88, 91) and π (79, 82) right-handed helices, which have not been observed in poly(amino acids). Apart from two short segments in pancreatic trypsin inhibitor (92) that resemble the polyproline-II conformation, no left-handed helical regions have been observed in globular proteins. Considerable regions of both parallel and antiparallel β-pleated sheets have been found also in proteins. They generally exhibit a right-handed twist when viewed along the polypeptide chain (93, 94). In addition, globular proteins contain many sequences that do not fold into regular conformations, but instead form various bends and loops that largely serve to determine the tertiary structure of the molecule. In a few cases, residues with a *cis* configuration about the peptide bond are reportedly involved in such bends (90, 91, 95).

It seems to be well established that the equilibrium tertiary structures of proteins are a direct consequence of their amino acid sequences, but that this code is degenerate in the sense that quite different sequences can give rise to essentially the same tertiary structure. Much effort has been directed toward the elucidation of this relationship by calculating minimum-energy conformations for protein sequences and by seeking empirical correlations between sequence and conformation. From these studies has emerged a body of rules as to what combinations of amino acid residues occur in helices (73, 96, 97) pleated sheets (97), and the various bend conformations (98, 99, 100). These rules have been used with increasing success to predict which regions of proteins conform to these three categories (101), although they do not yet distinguish between different kinds of helices or bends, nor do they predict interactions between residues far apart in the linear sequence.

The development and evaluation of such methods could be restricted to some extent by the limited accuracy of protein structure determinations, which, despite improved resolution and refinement (102, 103), are much less reliable than structure analyses of small molecules. Therefore it might be useful to undertake single-crystal structural analyses of tetra-, penta-, and hexapeptides involved in bend conformations to determine them more precisely than is possible in analyses of protein structure, where such regions are generally the least well defined. One proteinlike bend conformation already has been found in oligopeptides (104, 105). A series of such studies might indicate also whether bend conformations are firmly fixed by the local sequences or are appreciably influenced by neighboring regions. The identification of highly stable locally determined conformations might greatly facilitate procedures for calculating the conformations of more extended regions.

The prospects of obtaining single crystals of oligopeptides with 3_{10}-, α_{II}-, or π-helical conformations appear to be far less promising. These conformations are generally confined to the final turn of an α helix and would probably not be stable if these sequences were isolated from the adjacent regions of the protein.

It may be possible, however, to perform single-crystal structural analyses of oligopeptides in more stable regular conformations. In recent years several collagenlike polytripeptides of specified molecular weight have been synthesized (106, 107) and it has been proven possible to grow single crystals of (Pro-Pro-Gly)$_{10}$ (108) in which the chains are associated to form triple-stranded structures (109). Similar syntheses and crystallizations of poly(amino acids) of uniform lengths with sequences appropriate to β-pleated sheets, α helices, or even α-helical coiled coils can be envisaged.

The potential precision of an x-ray structure determination can be gauged from the number of observed reflections per atom in the asymmetric unit or, more correctly, per parameter to be determined. In the high resolution (1.5 Å) analysis of rubredoxin, a small protein molecule with only 400 atoms, some 5000 reflections were used (102), whereas in the most precise poly(amino acid) fiber structural analysis 61 reflections were used to locate 5 atoms in the residue of α-poly(L-alanine) (13, 14). These are the best cases, and generally protein fiber structural analyses utilize far fewer reflections per atom. In contrast, single-crystal structural analyses of small molecules, such as tetrapeptides, may utilize as many as 100 reflections per atom, and for the (Pro-Pro-Gly)$_{10}$ crystals the ratio is 44 to 1 (109). Thus, with suitable single crystals, it may be possible to achieve far more precise determinations of bend conformations as well as of torsion angles, hydrogen bond dimensions, and the degree of planarity of the peptide group in α helices and β-pleated sheets.

REFERENCES

1. R.B. Corey and L. Pauling, *Proc. Roy. Soc.* (London), **B141**, 10 (1953).
2. R.E. Marsh and J. Donohue, *Adv. Protein Chem.*, **22**, 235 (1967).
3. G.N. Ramachandran and A.S. Kolaskar, *Biochim. Biophys. Acta,* **303**, 385 (1973).
4. L. Pauling, R.B. Corey, and H.R. Branson, *Proc. Nat. Acad. Sci. U.S.,* **37**, 205 (1951).
5. L. Pauling and R.B. Corey, *Proc. Nat. Acad. Sci. U.S.,* **37**, 235 (1951).
6. L. Pauling and R.B. Corey, *Proc. Nat. Acad. Sci. U.S.,* **37**, 729 (1951).
7. L. Pauling and R.B. Corey, *Proc. Nat. Acad. Sci. U.S.,* **39**, 253 (1953).
8. C.H. Bamford, A. Elliott, and W.E. Hanby, *Synthetic Polypeptides,* Academic Press, New York, 1956.
9. M.F. Perutz, *Nature* (London), **167**, 1053 (1951).
10. C.H. Bamford, L. Brown, A. Elliott, W.E. Hanby, and I.F. Trotter, *Nature* (London), **169**, 357 (1952).
11. W. Cochran, F.H.C. Crick, and V. Vand, *Acta Cryst.*, **5**, 581 (1952).
12. A. Elliott and B.R. Malcolm, *Proc. Roy. Soc.* (London), **A249**, 30 (1959).
13. S. Arnott and A.J. Wonacott, *J. Mol. Biol.*, **21**, 371 (1966).
14. S. Arnott and S.D. Dover, *J. Mol. Biol.*, **30**, 209 (1967).
15. A. Elliott, in *Poly-α-Amino Acids,* G.D. Fasman, ed., Marcel Dekker, New York, 1967.
16. C.H. Bamford, L. Brown, A. Elliott, W.E. Hanby, and I.E. Trotter, *Nature* (London), **171**, 1149 (1953).
17. S. Arnott, S.D. Dover, and A. Elliott, *J. Mol. Biol.*, **30**, 201 (1967).
18. R.H. Karlson, K.S. Norland, G.D. Fasman, and E.R. Blout, *J. Am. Chem. Soc.*, **82**, 2268 (1960).
19. E.M. Bradbury, A.R. Downie, A. Elliott, and W.E. Hanby, *Proc. Roy. Soc.* (London), **A259**, 110 (1960).
20. E.M. Bradbury, L. Brown, A.R. Downie, A. Elliott, R.D.B. Fraser, and W.E. Hanby, *J. Mol. Biol.*, **5**, 230 (1962).
21. R.D.B. Fraser, T.P. MacRae, and I.W. Stapleton, *Nature* (London), **193**, 573 (1962).
22. P.M. Cowan and S. McGavin, *Nature* (London), **176**, 501 (1955).
23. V. Sasisekharan, *Acta Cryst.*, **12**, 897 (1959).
24. V. Sasisekharan, *Acta Cryst.*, **12**, 903 (1959).
25. F.H.C. Crick and A. Rich, *Nature* (London), **176**, 780 (1955).
26. G.N. Ramachandran, V. Sasisekharan, and C. Ramakrishnan, *Biochim. Biophys. Acta,* **112**, 168 (1966).
27. W. Traub and U. Shmueli, in *Aspects of Protein Structure,* G.N. Ramachandran, ed., Academic Press, New York, 1963, p. 81.
28. W. Traub and U. Shmueli, *Nature* (London), **198**, 1165 (1963).
29. D.A.D. Parry and A. Elliott, *Nature* (London), **206**, 616 (1965).
30. P. Saludjian and V. Luzzati, in *Poly-α-Amino Acids,* G.D. Fasman, ed., Marcel Dekker, New York, 1967, p. 157.
31. W. Traub, U. Shmueli, M. Suwalsky, and A. Yonath, in *Conformation of Biopolymers,* G.N. Ramachandran, ed., Academic Press, New York, 1967, p. 449.

32. V. Sasisekharan, *J. Polymer Sci.,* **47**, 373 (1960).
33. U. Shmueli and W. Traub, *J. Mol. Biol.,* **12**, 205 (1965).
34. W. Traub and A. Yonath, *J. Mol. Biol.,* **25**, 351 (1967).
35. M. Suwalsky and W. Traub, *Biopolymers,* **11**, 623 (1972).
36. M. Suwalsky and L. de la Hoz, *Biopolymers,* **12**, 1997 (1973).
37. Y. Mitsui, *Biopolymers,* **12**, 1781 (1973).
38. W.T. Astbury, *Trans. Faraday Soc.,* **34**, 378 (1938).
39. F.H.C. Crick, *Nature* (London), **170**, 882 (1952).
40. L. Pauling and R.B. Corey, *Nature* (London), **171**, 59 (1953).
41. C. Cohen and K.C. Holmes, *J. Mol. Biol.,* **6**, 423 (1963).
42. A. Elliott, J. Lowy, D.A.D. Parry, and P.J. Vibert, *Nature,* **218**, 656 (1968).
43. R.D.B. Fraser and T.P. MacRae, *J. Mol. Biol.,* **3**, 640 (1961).
44. R.D.B. Fraser, T.P. MacRae, and A. Miller, *J. Mol. Biol.,* **14**, 432 (1965).
45. R.D.B. Fraser and T.P. MacRae, *Polymer,* **14**, 61 (1973).
46. Y. Takeda, Y. Iitaka, and M. Tsubio, *J. Mol. Biol.,* **51**, 101 (1970).
47. J.M. Squire and A. Elliott, *J. Mol. Biol.,* **65**, 291 (1972).
48. R. Marsh, R.B. Corey, and L. Pauling, *Acta Cryst.,* **8**, 710 (1955).
49. W.M. Schroeder and L.M. Kay, *J. Am. Chem. Soc.,* **77**, 3408 (1955).
50. F. Lucas, J.T.B. Shaw, and S.G. Smith, *Biochem. J.,* **66**, 468 (1957).
51. J.O. Warwicker, *J. Mol. Biol.,* **2**, 350 (1960).
52. R.D.B. Fraser, T.P. MacRae, F.H.C. Stewart, and E. Suzuki, *J. Mol. Biol.,* **11**, 706 (1965).
53. A. Brack and G. Spach, *Biopolymers,* **11**, 563 (1972).
54. A. Brack, A. Caille, and G. Spach, *Monatshefte für Chemie,* **103**, 1604 (1972).
55. B.B. Doyle, W. Traub, G.P. Lorenzi, F.R. Brown, and E.R. Blout, *J. Mol. Biol.,* **51**, 47 (1970).
56. R.D.B. Fraser, T.P. MacRae, and F.H.C. Stewart, *J. Mol. Biol.,* **19**, 580 (1966).
57. W. Traub and K.A. Piez, *Adv. Protein Chem.,* **25**, 243 (1971).
58. G.N. Ramachandran and G. Kartha, *Nature* (London), **176**, 593 (1955).
59. V.N. Rogulenkova, M.I. Millionova, and N.S. Andreeva, *J. Mol. Biol.,* **9**, 253 (1964).
60. W. Traub and A. Yonath, *J. Mol. Biol.,* **16**, 404 (1966).
61. N.S. Andreeva, N.G. Esipova, M.I. Millionova, V.N. Rogulenkova, and V.A. Shibnev, in *Conformation of Biopolymers,* G.N. Ramachandran, ed., Academic Press, New York, 1967, p. 469.
62. D.M. Segal, *J. Mol. Biol.,* **43**, 497 (1969).
63. A. Yonath and W. Traub, *J. Mol. Biol.,* **43**, 461 (1969).
64. D.M. Segal, W. Traub, and A. Yonath, *J. Mol. Biol.,* **43**, 519 (1969).
65. W. Traub, A. Yonath, and D.M. Segal, *Nature* (London), **221**, 914 (1969).
66. W. Traub and G. Salem, *Acta Cryst.,* **A28**, S38 (1972).
67. G. Salem, M.Sc. Thesis, The Weizmann Institute of Science (1973).
68. J. Engel, J. Kurtz, E. Katchalski, and A. Berger, *J. Mol. Biol.,* **17**, 255 (1966).
69. W. Traub, *Israel J. Chem.,* in press.
70. G. Salem and W. Traub, in preparation.

71. J.M. Anderson, W.B. Rippon, and A.G. Walton, *Biochem. Biophys. Res. Commun.*, **39**, 802 (1970).
72. B.B. Doyle, W. Traub, G.P. Lorenzi, and E.R. Blout, *Biochemistry*, **10**, 3052 (1971).
73. M. Schiffer and A.B. Edmundson, *Biophys. J.*, **7**, 121 (1967).
74. A.G. Walton and K.P. Schodt, in *Peptides, Polypeptides, and Proteins*, Proceedings of the Rehovot Symposium 1974, Wiley(Interscience), New York, 1974.
75. M. Suwalsky and W. Traub, *Biopolymers*, **11**, 2223 (1972).
76. M.O. Dayhoff, *Atlas of Protein Sequence and Structure*, Vol. 5, National Biochemical Research Foundation, Silver Spring, Md., 1972.
77. W.R. Gray, L.B. Sandberg, and J.A. Foster, *Nature* (London), **246**, 461 (1973).
78. D.W. Urry and T. Ohnishi, in *Peptides, Polypeptides, and Proteins,* Proceedings of the Rehovot Symposium 1974, Wiley(Interscience), New York, 1974.
79. H.C. Watson, *Progr. in Stereochemistry*, **4**, 299 (1969).
80. M.F. Perutz, H. Muirhead, J.M. Cox, and L.C.G. Goaman, *Nature* (London), **219**, 131 (1968).
81. A. Liljas and M.G. Rossmann, in press.
82. R.E. Dickerson and I. Geis, *The Structure and Function of Proteins*, Harper & Row, New York, 1969.
83. G. Nemethy, D.C. Phillips, S.J. Leach, and H.A. Scheraga, *Nature* (London), **214**, 363 (1967).
84. A. Liljas, K.K. Kannan, P.C. Bergsten, I. Waara, K. Fridborg, B. Strandberg, U. Carlborn, L. Jarup, S. Lovgren, and M. Petef, *Nature New Biol.*, **235**, 131 (1972).
85. J.J. Birktoft and D.M. Blow, *J. Mol. Biol.*, **68**, 187 (1972).
86. F.S. Mathes, M. Levine, and P. Argos, *J. Mol. Biol.*, **64**, 449 (1972).
87. T.L. Blundell, G. Dodson, D. Hodgkin, and D. Mercola, *Adv. Protein Chem.*, **26**, 279 (1972).
88. T. Imoto, L.N. Johnson, A.C.T. North, D.C. Phillips, and J.A. Rupley, in *The Enzymes,* Third Edition, Vol. 7, P.D. Boyer, ed., Academic Press, New York, 1972, p. 665.
89. R.G. Kretsinger and C.E. Nockolds, *J. Biol. Chem.*, **248**, 3313 (1973).
90. H.W. Wyckoff, D. Tsernoglou, A.W. Hanson, J.R. Knox, B. Lee, and F.M. Richards, *J. Biol. Chem.*, **245**, 305 (1970).
91. F.A. Quiocho and W.N. Lipscomb, *Adv. Protein Chem.*, **25**, 1 (1971).
92. R.H. Huber, D. Kukla, A. Ruhlmann, and W. Steigemann, *Cold Spring Harbor Symp. on Quant. Biol.*, **36**, 141 (1971).
93. G.N. Ramachandran and V. Sasisekharan, *Adv. Protein Chem.*, **23**, 283 (1968).
94. C. Chothia, *J. Mol. Biol.*, **75**, 295 (1973).
95. C.S. Wright, R.A. Alden, and J. Kraut, *Nature* (London), **221**, 235 (1969).
96. P.N. Lewis, N. Go, M. Go, D. Kotelchuck, and H.A. Scheraga, *Proc. Nat. Acad. Sci. U.S.*, **65**, 810 (1970).
97. P.Y. Chou and G.D. Fasman, *Biochemistry*, **13**, 211 (1974).
98. C.M. Venkatachalam, *Biopolymers*, **6**, 1425 (1968).

99. J.L. Crawford, W.N. Lipscomb, and C.G. Schellman, *Proc. Nat. Acad. Sci. U.S.,* **70**, 538 (1973).
100. P.N. Lewis, F.A. Momany, and H.A. Scheraga, *Biochim. Biophys. Acta,* **303**, 211 (1973).
101. P.Y. Chou and G.D. Fasman, *Biochemistry,* **13**, 222 (1974).
102. K.D. Watenpaugh, L.C. Sieker, J.R. Herriott, and L.H. Jensen, *Cold Spring Harbor Symp. on Quant. Biol.,* **36**, 359 (1971).
103. R. Diamond, *Acta Cryst.,* **A27**, 436 (1971).
104. T. Ueki, T. Ashida, M. Kakudo, Y. Sasada, and Y. Katsube, *Acta Cryst.,* **B25**, 1840 (1969).
105. T. Ueki, S. Bando, T. Ashida, and M. Kakudo, *Acta Cryst.,* **B27**, 2219 (1971).
106. S. Sakakibara, Y. Kishida, Y. Kikuchi, R. Sakai, and K. Kakiuchi, *Bull. Chem. Soc. Japan,* **41**, 1273 (1968).
107. S. Sakakibara, K. Inouye, K. Shudo, K. Kishida, Y. Kobayashi, and D.J. Prockop, *Biochim. Biophys. Acta,* **303**, 198 (1973).
108. S. Sakakibara, Y. Kishida, K. Okuyama, N. Tanaka, T. Ashida, and M. Kakudo, *J. Mol. Biol.,* **65**, 371 (1972).
109. K. Okuyama, N. Tanaka, T. Ashida, M. Kakudo, S. Sakakibara, and Y. Kishida, *J. Mol. Biol.,* **72**, 571 (1972).

PSEUDOPOLYELECTROLYTE POLY(AMINO ACIDS) IN THE LITHIUM CHLORIDE–TRIFLUOROETHANOL SOLVENT SYSTEM

NOAH LOTAN, Department of Biophysics, The Weizmann Institute of Science, Rehovot, Israel

SYNOPSIS: In the lithium chloride–methanol solvent system, polyelectrolyte properties have been previously observed in otherwise neutral poly(amino acids). Such behavior is due to the particular state of solvation of the salt. The pseudopolyelectrolyte effect is even more pronounced in trifluoroethanol (TFE), as expected from the known properties of the lithium chloride–TFE solvent system.

INTRODUCTION

Conformational studies carried out in the past on a variety of poly(amino acids) have provided valuable information about the fundamental properties of these compounds and, consequently, about some of the conformation-related properties of natural proteins (1). In this respect, polyelectrolyte properties and the effects of salts have aroused considerable interest, and much effort has been dedicated to their investigation, particularly in aqueous solutions (2). Nonaqueous solvents on the other hand, have received only limited attention (3-6). More recently, however, the interest of biophysicists has been stirred by the observation that some of these solvents can minimize successfully the environmental conditions encountered in highly aggregated systems, such as the cell membranes (7, 8).

In a previous communication (9), the effects of lithium chloride on the conformation of a polypeptide in methanol were described. It was shown that, between 2 and 4 M salt, a helix-coil transition takes place, possibly due to the binding of Li^+ to the polypeptide and the strong solvation of the Cl counterions. At higher LiCl concentrations, more lithium ions are bound, and the polypeptide acquires the character of a polyelectrolyte. If such a mechanism is indeed operative, we would expect LiCl to be more effective (i.e., the transition would take place at lower salt concentration) in solvents that, compared to methanol, solvate the anion even more strongly and/or the cation to a smaller extent. Such solvents are the fluoroalcohols (10), and a study was undertaken to discover the effects of LiCl on the conformation of polypeptides in 2,2,2-trifluoroethanol (TFE). The results of this study, reported herein, amply confirm the conclusions drawn previously (9).

EXPERIMENTAL SECTION

Materials. Lithium chloride (certified grade, Fisher Scientific Co.) was extensively dried and subsequently stored in vacuum over P_2O_5. 2,2,2-Trifluoroethanol (TFE) (Uvasol grade, Merck Co.) was used as received.

Poly[N^5-(2-hydroxyethyl)-L-glutamine] (PHEG) was prepared by a procedure described previously (11). The sample used (PHEG-V) has an average molecular weight of 75,000 (degree of polymerization 440. See also "Results"). Poly[N^5-(4-hydroxybutyl)-L-glutamine] (PHBG) was the same sample (PHBG-V) as the one used previously (1). Its molecular weight is 76,000 (degree of polymerization 380. See also "Results").

Poly(γ-methyl-L-glutamate) (PMLG) has an intrinsic viscosity in dichloroacetic acid $(\eta)_{DCA}$ = 0.975 dl/g. Its molecular weight is estimated (12) to be 160,000 (degree of polymerization 1100).

Methods. The molecular weight of the (PHEG and PHBG) samples studied was determined by sedimentation equilibrium ultracentrifugation in water, at 20°C, using a Spinco Model E ultracentrifuge. The initial concentration of the polymer solutions was 2–3 mg/ml, and the sedimentation patterns were recorded by means of the absorption optics of the split-beam photoelectric scanning system, set at 240 nm. The value $\bar{v}$ = 0.79 dl/g was used for the partial specific volume (11), and the average molecular weight was calculated as suggested by Schachman (13).

Solutions for optical measurements were prepared by first dissolving the polymers in TFE and diluting the solutions thus obtained to the required salt concentration by adding concentrated solution of LiCl in TFE; traces of insoluble material were removed by high-speed centrifugation and filtration through Teflon® membrane filters (0.5 μ pore size, type LSWP-01300, Millipore Corp.). In the clean solutions, the polymer concentration was determined by micro-Kjeldahl nitrogen analysis, and the LiCl was titrated with $Hg(ClO_4)_2$, according essentially to a procedure described elsewhere (14).

Circular dichroism (CD) measurements were made at room temperature on a Cary 60 spectropolarimeter with the Model 6001 CD attachment; the polymer solutions (0.4-2 mg/ml) were contained in quartz cells (Optical Cell Co., Brentwood, Md., and Hellma, Germany) of 0.1-1.0 mm pathlength. The molar residue ellipticity values reported [θ] were not corrected for the refractive index of the solution.

Viscosity measurements were taken at 30°C with calibrated Ubbelohde semimicro-dilution-type viscometers (Cannon Instruments Co.), using polymer solutions of 2–10 mg/ml; the intrinsic viscosity values [η] were obtained by extrapolation to infinite polymer dilution.

Results. The molecular weight of the PHBG sample used (see "Materials") differs from the previously reported value (9), which had been estimated from a viscosity–molecular weight relationship established (11) for a structurally related yet different polymer. The new value is more accurate.

Lithium chloride is less soluble in TFE than in methanol and, at room temperature, saturation of the former solvent is attained at about 1.2 M. At such salt concentrations, direct dissolution of the PMLG and PHEG samples under study is difficult. Therefore, measurements were performed after appropriate dilutions of stock solutions of polymer in TFE.

Representative CD spectra of PHEG, PHBG, and PMLG in the LiCl–TFE solvent system are shown in Figs. 1, 2, and 3, respectively. In the absence of salt, minima are observed at 220–222 nm and 207–208 nm, with molar ellipticity values of –30,000 to –40,000 deg cm^2/decimole for the glutamine derivatives, and –44,000 to –50,000 deg cm^2/decimole for PMLG. The crossover point from negative to positive dichroism is located at about 202 nm. These features are typical for poly(α-L-amino acids) that are largely in the right-handed α-helical conformation. The spectrum of PMLG in TFE (Fig. 3) is similar to the one reported previously (15).

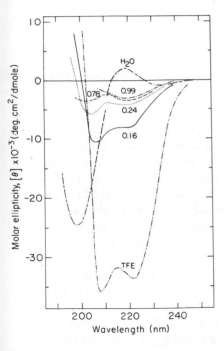

Figure 1. Representative CD spectra of PHEG in the LiCl–TFE solvent system. The molar concentration of salt is indicated. The spectrum of the same polymer in water (20) is included for comparison.

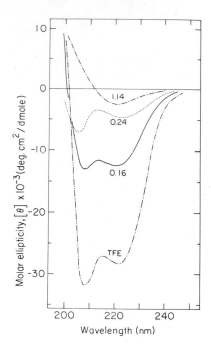

Figure 2. Representative CD spectra of PHBG in the LiCl–TFE solvent system. The molar concentration of salt is indicated.

As the concentration of LiCl is increased, the structure of the polypeptides changes drastically, as indicated by the decrease in the absolute value of the rotatory strength of the 222 nm and 208 nm bands, as well as of the positive dichroism below 200 nm. The 208 nm minimum also is blue-shifted. The transitions, indicated, for example, by the variation of the ellipticity at 222.5 nm, are shown in Fig. 4 for the three polymers investigated. The two glutamine derivatives show similar, sharp changes, with a midpoint at about 0.1 M. For PMLG, on the other hand, the transition is less steep, and requires a higher salt concentration (0.4 M).

At high LiCl concentration (>0.6 M), the CD spectra are rather unusual (Figs. 1, 2, 3). The glutamine derivatives exhibit a minimum at about 220 nm of very low ellipticity ($[\theta]_{min}$ = –2000 to –3000 deg cm^2/decimole); PMLG, on the other hand, exhibits a less-pronounced minimum at 220 nm, and the corresponding band has greater rotatory strength ($[\theta]_{220}$ = –7000 deg cm^2/ decimole).

The conformational changes of PHBG in the LiCl–TFE solvent system were also investigated by viscosity measurements, the results of which are summarized in Fig. 5. A high viscosity ($[\eta]$ = 1.9 dl/g) is observed for the helical polymer in TFE. As the salt concentration is increased, the viscosity drops to a minimum

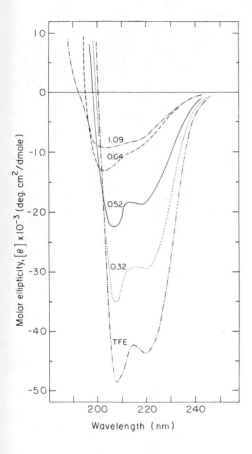

Figure 3. Representative CD spectra of PMLG in the LiCl–TFE solvent system. The molar concentration of salt is indicated.

($[\eta]$ = 0.75 dl/g at 0.15 M LiCl); the transition follows the changes in the CD spectrum. A further increase in salt concentration produces an increase of $[\eta]$ to 1.03 dl/g at 1.2 M LiCl. These changes are almost identical to the ones reported previously (9) for the same polymer in LiCl–methanol solvent system.

DISCUSSION

A study of the PHBG–methanol–LiCl system (9) revealed that as the salt concentration increases, the macromolecule undergoes conformational changes from the α helix to a disordered state and, finally, to an extended chain having polyelectrolyte character. These transformations are the outcome of two parallel processes, namely, the binding of Li$^+$ to the polypeptide and solvation of Cl$^-$ by methanol. It was expected, therefore, that other solvents that enhance one

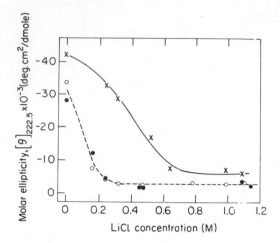

Figure 4. Conformational transitions of polypeptides in the LiCl–TFE solvent system determined by changes in the molar ellipticity at 222.5 nm: x, PMLG; o, PHEG; and ●, PHBG. For the two glutamine derivatives, the experimental data are fitted by the same transition curve (– – –).

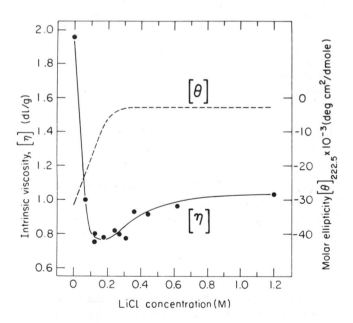

Figure 5. Conformational transitions of PHBG in the LiCl–TFE solvent system determined by changes in the intrinsic viscosity $[\eta]$. For comparison, the changes of $[\theta]_{222.5}$ (taken from Fig. 4) are also included.

or both of these processes would allow the transition to take place at even lower salt concentration. Electrolyte conductance measurements (10) have shown fluorinated alcohols to exhibit extreme anion solvation capacity, so TFE was chosen for the continuation of our studies.

Optical rotatory dispersion measurements have shown PHEG and PHBG to be highly helical in TFE (16), but the addition of LiCl decreases the α-helix content (Figs. 1, 2, and 4). As expected, the transformation takes place at a much lower salt concentration than that required for the same process to occur in methanol (9), the corresponding values being $0.1\,M$ and $3\,M$, respectively. Moreover, the transition in TFE is much sharper than the one occurring in the nonfluorinated solvent.

The α-helical form of PMLG is stable in TFE (15, 17, 18). As in the cases described above, this structure is disrupted by LiCl (Figs. 3, 4), but the conformational transition is less sharp and requires somewhat higher salt concentration (i.e., $0.6\,M$). Unfortunately, PMLG is insoluble in methanol, and therefore, for this polymer, no direct comparison can be made of transformations occurring in the two solvents. However, the very fact that such a transition can also be induced in PMLG, as well as in the hydroxyalkyl glutamine derivatives, indicates that the process has a more general character; that is, it is not restricted to the latter class of polypeptides.

The sharpness of the transitions and the relatively small amount of salt required to induce them suggest that the processes are highly cooperative. Most probably, they are the result of a strong polymer-solvent interaction. The actual occurrence of a Li$^+$-peptide interaction has been substantiated elsewhere [see (9) and references cited therein]. A more detailed description of the complex thus formed can be inferred from recent quantum mechanical studies (19, 20) of low-molecular-weight model compounds. These studies have shown that the complex is of the ion-dipole type, the bonding energy being very high—of the order of magnitude of a chemical linkage. These results strongly support the conclusion that, at high LiCl concentration, the polymers assume a pseudopolyelectrolyte character.

The quantum mechanical calculations mentioned above (19, 20) have also shown that the complex formation is associated with a large Li$^+$-to-amide charge transfer and with a redistribution of the atomic charges of the amide group involved. Such alterations in the electronic structure of the peptide chromophore may well explain the dissimilarity between the CD spectra obtained in the presence of high LiCl concentrations (Figs. 1, 2, 3) (hereafter called "type S") and the ones exhibited by various poly(amino acids) in their disordered conformation (21–23) (hereafter "type C"). PHEG assumes a disordered conformation in water, as indicated by NMR (24) as well as by hydrodynamic measurements (25, 26); the corresponding type-C spectrum (27, 28) is included in Fig. 1. Spectra having the same general features as type S, on the other hand, have been reported (29, 30) for poly(L-proline) and other polypeptides in $6\,M$ aqueous CaCl$_2$ and have also been observed (31) for PHEG and PHBG in the LiCl-methanol solvent system.

The difference between the conformational transitions undergone by the hydroxyalkyl glutamine polymers and by PMLG (Fig. 4) may be due to the difference in the interactions between LiCl and the amino acid–residue side chains, namely, the amide and —OH groups in the former cases and the ester group in the later.

The helix-to-coil-to-extended-chain transitions of PHBG in the LiCl–TFE solvent system are most clearly revealed by measurements of the hydrodynamic properties (Fig. 5). As the solvent composition is changed, the intrinsic viscosity decreases, then increases. Such behavior is typical of polyelectrolytes (32, 33). It was also observed with PHBG in the LiCl–methanol solvent system (1), although at a much higher salt concentration; however, in both cases the drop in the intrinsic viscosity follows closely the corresponding changes in the CD spectra.

It can be concluded that in alcohols, in the presence of LiCl, polypeptides behave as polyelectrolytes and that conformational changes occur at solvent compositions that are related to the anion binding ability of the alcohol. It is expected that similar effects can be obtained with salts containing helix-disrupting anions, provided a cation-solvating solvent is used. Experiments along this line are in progress in our laboratory.

ACKNOWLEDGMENTS

The support of this investigation by a grant from the U.S. Department of Health, Education, and Welfare (Grant no. 06-003) and the technical assistance of Mrs. R. August are gratefully acknowledged.

REFERENCES

1. N. Lotan, A. Berger, and E. Katchalski, *Am. Rev. Biochem.,* **41**, 869 (1972).
2. P.H. von Hippel and Th. Schleich, in S.N. Timasheff and G.D. Fasman, eds., *Structure and Stability of Biological Macromolecules,* Vol. II, M. Dekker, New York, 1967.
3. J.S. Frazen, C. Bobik, and J.B. Harry, *Biopolymers,* **4**, 637 (1966).
4. N. Lotan, M. Bixon, and A. Berger, *Biopolymers,* **5**, 69 (1967).
5. J.H. Bradbury and M.D. Fenn, *J. Mol. Biol.,* **36**, 231 (1968).
6. M. Shiraki and K. Imahori, *Sci. Pap. Coll. Gen. Educ., Univ. Tokyo,* **16**, 215 (1966).
7. D.W. Urry, L. Masotti, and J.R. Krivacic, *Biochem. Biophys. Acta,* **241**, 600 (1971).
8. L. Masotti, D.W. Urry, and J.R. Krivacic, *Biochem. Biophys. Acta,* **266**, 7 (1972).
9. N. Lotan, *J. Phys. Chem.,* **77**, 242 (1973).
10. D.F. Evans, J.A. Nadas, and M.A. Matesich, *J. Phys. Chem.,* **75**, 1708 (1971).
11. N. Lupu-Lotan, A. Yaron, A. Berger, and M. Sela, *Biopolymers,* **3**, 625 (1965).

12. P. Doty, J.H. Bradbury, and A.M. Holtzer, *J. Am. Chem. Soc.,* 78, 947 (1956).
13. H.K. Schachman, *Biochemistry,* 2, 887 (1963).
14. F.W. Cheng, *Microchem. J.,* 3, 537 (1959).
15. J.Y. Cassim and J.T. Yang, *Biopolymers,* 9, 1475 (1970).
16. N. Lotan, A. Yaron, and A. Berger, *Biopolymers,* 4, 365 (1966).
17. M. Goodman and I. Listowski, *J. Am. Chem. Soc.,* 84, 3770 (1962).
18. M. Goodman, I. Listowski, Y. Masuda, and F. Boardman, *Biopolymers,* 1, 33 (1963).
19. D. Balasubramanian, A. Goel, and C.N.R. Rao, *Chem. Phys. Lett.,* 17, 482 (1972).
20. P.V. Kostetsky, V.T. Ivanov, and Yu.A. Ovchinnikov, *FEBS Lett.,* 30, 205 (1973).
21. J.T. Yang, in G.D. Fasman, ed., *Poly-α-Amino Acids,* M. Dekker, New York, 1967.
22. D. Balasubramanian and R.S. Roche, *Chem. Commun.,* 862 (1970).
23. A.J. Adler, N.Y. Greenfield, and G.D. Fasman, in C.H.W. Hirs and S.N. Timasheff, eds., *Methods in Enzymology,* Academic Press, New York, in press.
24. F.J. Joubert, N. Lotan, and H.A. Scheraga, *Biochemistry,* 9, 2197 (1970).
25. W.L. Mattice, J.-T. Lo, and L. Mandelkern, *Macromolecules,* 5, 729 (1972).
26. W.L. Mattice and J.-T. Lo, *Macromolecules,* 5, 734 (1972).
27. A.J. Adler, R. Hoving, J. Potter, M. Wells, and G.D. Fasman, *J. Am. Chem. Soc.,* 90, 4736 (1968).
28. N. Lotan, K. Chen, and R.S. Roche, *Israel J. Chem.,* in press.
29. M.L. Tiffany and S. Krimm, *Biopolymers,* 6, 1767 (1968).
30. M.L. Tiffany and S. Krimm, *Biopolymers,* 8, 347 (1969).
31. N. Lotan, in preparation.
32. P. Doty, A. Wada, J.T. Yang, and E.R. Blout, *J. Polymer Sci.,* 23, 851 (1957).
33. J. Applequist and P. Doty, in M.A. Stahmann, ed., *Polyamino Acids, Polypeptides, and Proteins,* University of Wisconsin Press, Madison, Wis., 1962.

POTENTIOMETRIC AND CIRCULAR DICHROISM STUDIES ON THE COIL-β TRANSITION OF POLY(L-LYSINE)

A. COSANI, M. TERBOJEVICH, L. ROMANIN-JACUR, and **E. PEGGION,**
Institute of Organic Chemistry, University of Padova, Via Marzolo, 1-35100
Padova, Italy

SYNOPSIS: Potentiometric titration experiments were carried out on poly(L-lysine) in 0.1 M potassium chloride in the temperature range $40°$-$80°$C. From the "polyelectrolyte plot" pK_{app} versus α, the free-energy, enthalpy, and entropy changes for the conformational transition from an uncharged random coil to an uncharged β form have been estimated. We found $\Delta H°_c = 130$ cal/mole and $\Delta S°_c = 0.8$ e.u. Circular dichroism (CD) measurements were performed in order to check the polymer conformation during titration. Evidence is presented that at $40°$ and $50°$C in the pH-induced transition the α-helical structure that forms first is transformed subsequently into the β conformation. At temperatures greater than $60°$C the β structure is formed directly because it is thermodynamically impossible for the α-helical conformation to exist.

INTRODUCTION

Poly(L-lysine) [poly(Lys)] in aqueous solution is known to assume either the α-helical structure or the β form, depending on pH and temperature (1–7). At room temperature, the α-helical conformation is stable above pH 10.8. The β conformation results when a polymer solution at pH 10.8 is heated above $40°$C.

Potentiometric studies on the coil-β transition were carried out recently by Hermans and coworkers (8). According to this group, the free-energy and enthalpy changes associated with the formation of an uncharged β structure from an uncharged random coil were –140 cal/mole and 870 cal/mole, respectively, at $25°$C. From direct microcalorimetric experiments, Chou and Scheraga (9) reported an enthalpy change of –1200 cal/mole for the same transition. Between the two sets of experiments there is a difference of 2 kcal/mole, which is unacceptably large.

In this paper, we shall report the results of further potentiometric investigations directed toward obtaining more precise informations on the thermodynamic aspects of the coil-β transition of poly(Lys) in aqueous solution.

EXPERIMENTAL SECTION

Materials. Petroleum ether (Carlo Erba RP), b.p. $40°$-$70°$C, was refluxed over sodium wire, and then distilled. The first 10% of distillate was rejected. Dioxane (Carlo Erba RP) was purified according to the literature (10), using the potassium-anthracene complex. Chloroform (Carlo Erba RP) was dried over $CaCl_2$ and then fractionally distilled. Acetone (Merck puriss.) was kept over Drierite and then fractionally distilled.

N^ϵ-carbobenzoxy-N-carboxy-anhydride (NCA) was prepared from N^ϵ-carbobenzoxy-L-lysine (11) and phosgene according to the literature (12).

Poly(N^ϵ-carbobenzoxy-L-lysine) was prepared by polymerization of the corresponding NCA in dioxane, using sodium methoxide as initiator. The monomer concentration was 3% (by weight) and the monomer-to-initiator molar ratio was 30. At the end of the polymerization, which was determined by infrared spectroscopy, a small portion of the solution was poured into ethyl ether; the polymer precipitated in the form of white fibers, which were isolated by filtration and purified by redissolution in CH_2Cl_2 and precipitation out of petroleum ether. The polymer sample so obtained was characterized by viscosity measurements and elemental analysis. The intrinsic viscosity $[\eta]$ was 1.60 dl/g at 25°C in DMF. (Calculated for $C_{14}H_{18}N_2O_3$: C, 64.1%; H, 6.8%; N, 10.7%. Found: C, 63.8%; H, 6.8%; N, 10.2%.) The remaining larger part of the polymer solution in dioxane was used to obtain poly(Lys).

Poly(Lys)·HCl was obtained by removing the protecting carbobenzoxy groups from poly(N^ϵ-carbobenzoxy lysine), according to the procedure described by Fasman et al. (13). The polymer sample so obtained was characterized by viscometry, ultraviolet (UV) absorption measurements and elemental analysis. The intrinsic viscosity in 0.2 M NaCl was $[\eta]$ = 0.75 dl/g. From the UV spectrum recorded in the 240–270 nm region no trace of residual blocking groups was observed. (Calculated for $C_6H_{13}N_2OCl$: C, 43.8; H, 7.9; N, 17.0. Found: C, 39.62; H, 7.72; N, 15.14. These analytical data are consistent with a water content of 10% by weight.)

METHODS

Potentiometric measurements. Potentiometric measurements were carried out using a Metrohm precision potentiometer Model E. 510, equipped with glass and calomel electrodes. The titration cell was completely dipped in the thermostatted bath and maintained under nitrogen atmosphere. Titrant was added using a Metrohm E 457 microburette. All titrations were carried out in 0.1 M KCl with a polymer concentration of the order of 3×10^{-3} M. The KCl concentration was checked at the beginning and at the end of the titration in order to be sure that there was no significant diffusion of Cl^- ions from the calomel electrode. Substantial variation of ionic strength during the titration would strongly affect the $\Delta G°$ values determined from potentiometric data (14). Titrant solutions of KOH were prepared according to the literature (15) and always handled under nitrogen atmosphere. The pH of a polyelectrolyte solution is given by (16–18)

$$\text{pH} + \log \frac{1-\alpha}{\alpha} = \text{pK}_0 + \frac{0.434}{RT}\left(\frac{\delta G_{\text{ion}}}{\delta \alpha}\right) = \text{pK}_{\text{app}} \tag{1}$$

where α is the fraction of deprotonated amino groups and $\delta G_{ion}/\delta\alpha$ is the variation of electrostatic and conformational free energy resulting from an incremental change, $\delta\alpha$. The standard free-energy change per residue value, $\Delta G°/N$, for the transition from an uncharged coil to an uncharged β structure was calculated according to the literature (16–18) from the usual polyelectrolyte plot of pK_{app} versus α.

CD and UV absorption measurements. A Cary 60 spectropolarimeter equipped with a 6002 CD accessory unit and thermostable cell assembly was used to perform CD measurements. Fused quartz cylindrical cells with Suprasil windows and 0.05 cm optical path were used. A Cary 15 spectrophotometer was used to carry out UV absorption measurements.

RESULTS

Potentiometry. The results of potentiometric titration experiments on poly(Lys) are shown in Fig. 1. Each experimental point is the average of at least five independent measurements, with a standard deviation of 5%. The conformation of the polymer in the transition region after each addition of titrant has been checked by CD measurements. Before evaluating the results in terms of thermodynamic parameters relative to the coil-β transition, several points must be considered. One of the major difficulties encountered by Hermans *et al.* (8) during their experiments was the precipitation of the polymer when the β structure was formed. In order to avoid such an inconvenience, our measurements were carried out with a polymer concentration of 0.5 mg/ml, which is lower than that used by Hermans *et al.* (8), and the same concentration used in the microcalorimetric experiments of Scheraga and Chou (9). At this concentration, the intramolecular cross-β conformation should be preferentially formed (19). In the entire temperature range investigated all solutions remained perfectly clear at least until $\alpha = 0.85$. In a few cases, a very slow precipitation was observed at higher α values. A polymer solution at $\alpha = 1$ heated until transformation into the β structure was complete remained perfectly limpid for more than 1 hr. These findings differ from those of Hermans *et al.* (8), probably because different polymer samples were used in the two sets of experiments. The difference might be attributable to differences in molecular weight and to differing contents of residual blocking groups. As recently shown by Rao and Miller (20), commercial samples of poly(Lys) often contain substantial amounts of residual blocking groups, which can affect the properties of the polymer. As indicated in the experimental section, the poly(Lys) sample used in our work did not show any trace of residual carbobenzoxy blocking groups.

Another important finding is the unusually long time required for the pH to reach the equilibrium value after each addition of titrant in the region of the conformational transition. This behavior has been observed also by Hermans *et al.* (8) and seems to be characteristic of the formation of β structure (21).

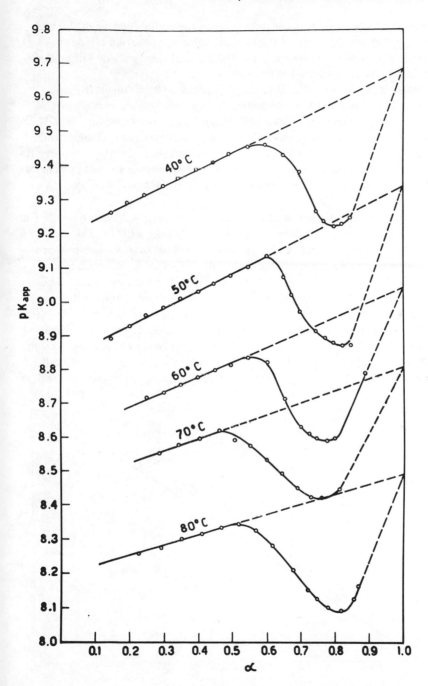

Figure 1. Potentiometric titration curves of poly(Lys) at different temperatures in 0.1 M KCl.

Hermans *et al.* (8) stated that the time to reach the equilibrium was as high as 25 min., but in our experiments the final pH value was observed after a much longer time (in some cases several hrs). The cause of this behavior will be clear from the CD results presented in the next section.

Following previous work of Ciferri (14), Scheraga (9), Hermans (8), and others (22), we made linear extrapolations of the coil titration curves. According to the theory, the extrapolation should not be a straight line, but should exhibit observable curvature, at least at low ionic strength (17). Olander and Holtzer (18) have indeed confirmed this in potentiometric titrations of poly(DL-glutamic acid). On the other hand, titration experiments carried out by Scheraga and Chou (9) did not show any appreciable curvature of the plot of pK_{app} versus α.

The extrapolated pK_0 values at various temperatures are shown in Fig. 2. Our results compare well with those obtained by other workers (9, 14, 23). From the temperature dependence of pK_0 the calculated heat of *dissociation* of protons is 13.4 kcal/mole, which is identical to that reported by Scheraga (9).

From the potentiometric titration curves of Fig. 1 the corresponding ΔG_c° values relative to the conformational transition from an *uncharged* coil to an *uncharged* β structure have been estimated. The limits of error in obtaining free-energy changes from potentiometric titration curves already have been discussed exhaustively by Olander and Holtzer (18). The limit of error is relevant. From the temperature dependence of ΔG_c° (Fig. 3), we calculated ΔH_c° = 130 cal/mole and ΔS_c° = 0.8 e.u. The enthalpy change is positive and about one-sixth of

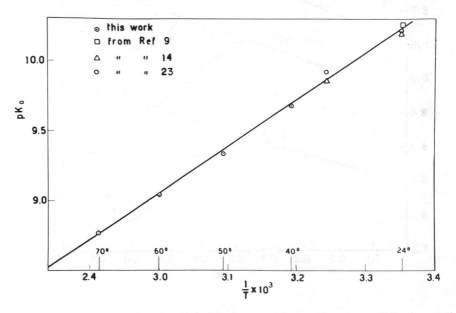

Figure 2. Extrapolated pK_0 values of poly(Lys) as a function of $1/T$.

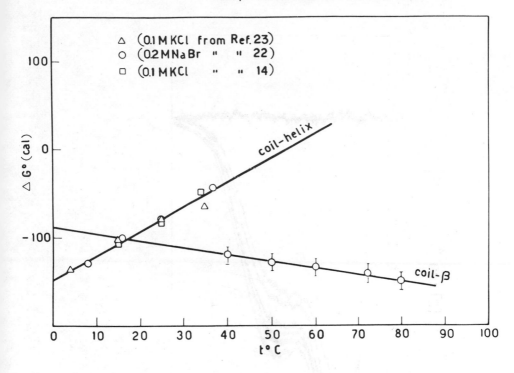

Figure 3. Temperature dependence of ΔG_c° values for the coil-β transition. Data related to the coil-α-helix transition that have been reported in the literature are included for comparison.

that reported by Hermans (8); the discrepancy with the calorimetric result of Scheraga (9) is much more relevant.

CD measurements. In order to establish clearly the conformational state of the polymer at each stage of the transition, CD measurements were carried out on the polymer solutions during the titration, particularly in the transition region. The following preliminary experiment was carried out. To a dilute solution of poly(Lys)·HCl (conc. = 0.5 mg/ml) at 50°C, the amount of 1 N KOH necessary to effect a degree of ionization, α, of ~0.8 (approximately 80% of the ϵ-amino groups are deprotonated) was quickly added. Then the CD pattern was recorded as a function of the time from the addition of titrant. As shown in Fig. 4, 15 min after the addition of the titrant, the polymer conformation consisted of a mixture of α-helical and random-coil structures. By comparison with computed CD curves of poly(Lys) containing various proportions of α helix, β structure and random coil (7), curve 1 of Fig. 4 indicates a helix and coil mixture containing roughly 40% α helix. Subsequently, the polymer underwent a very slow transition from the α-helical form to the β structure, as

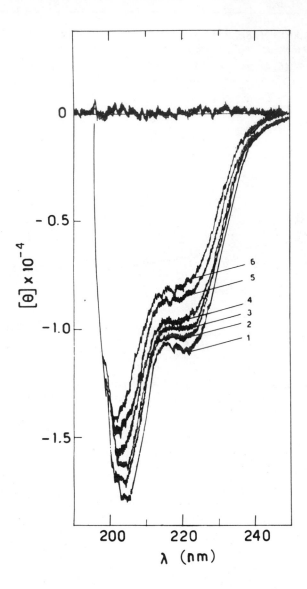

Figure 4. CD spectra of poly(Lys) in 0.1 M KCl at 50°C and $\alpha \cong 0.8$ as a function of the time after addition of titrant: curve 1, after 15 min; curve 2, after 50 min; curve 3, after 90 min; curve 4, after 230 min; curve 5, after 390 min; curve 6, after 530 min.

indicated by the blue shift of the negative CD bands at 222 and 218 nm. The final spectrum (curve 6) is consistent with a mixture of β and random-coil conformations (7). During the α-helix-to-β transition (which took several hours to be complete), the pH of the solution decreased continuously, reaching the final value when no more α helices were present in the system.

When the experiment was repeated at $\alpha \cong 1$, the spectrum recorded immediately after the addition of titrant (Fig. 5) again corresponded to a mixture of α helix and random coil that contains approximately 60% α helix (7). After 40 min, the whole polymer assumed the β conformation, the rate of the transition being much higher than at $\alpha \cong 0.8$. Also in this case, the decreasing pH parallels the formation of the β structure. Finally, at temperatures above 60°C and $\alpha \cong 1$, soon after the addition of titrant, the polymer conformation was almost entirely that of a β structure.

The above results indicate that the amount of time required for the pH to reach an equilibrium value after each addition of titrant is attributable to the slow rate of formation of the β structure. Obviously, reliable thermodynamic data on the coil-β transition of poly(Lys) can be obtained from potentiometric data only if the experimental pH values refer to an equilibrium system containing random coil and β form only. This condition can be achieved only after considerable time, certainly more time than that indicated in previous work. This time is inversely proportional to the temperature.

DISCUSSION AND CONCLUSIONS

The results of our potentiometric experiments indicate that the transition from an uncharged coil to an uncharged β form is *endothermic* and accompanied by a positive change of entropy. Obviously, all thermodynamic parameters relative to the conformational transition, as determined by potentiometric titrations, apply to a polymer system in the presence of solvent. Thus, the calculated values of ΔG_C°, ΔH_C°, and ΔS_C° refer to the transition from a *solvated* random coil to a solvated β structure and therefore contain contributions from eventual changes of solvation associated with the transition. Significantly, ΔS_C° was found to be *positive*, in spite of the fact that the polypeptide chain goes from an unordered state to an ordered one. It is clear that, in this case, the negative contribution to the entropy associated with the disorder-order transition of the peptide backbone is largely overcome by a positive contribution, probably due to a decrease in order in the water structure. This point of view is consistent with the formation of hydrophobic bonds among side chains in the β structure. In fact, it is well known that the formation of hydrophobic bonds is entropy-driven because of the decrease in order in the water structure (24).

At this point it is pertinent to compare the thermodynamic data on the coil-β transition of poly(Lys) with those reported in the literature (14, 22, 23) on the α-helix–coil transition. All data are collected in Fig. 3. The most evident difference is that the coil-β transition is entropy-driven, whereas the α-helix–coil

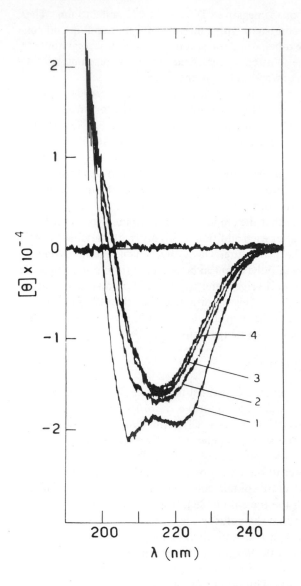

Figure 5. CD spectra of poly(Lys) in 0.1 M KCl at 50°C and $\alpha = 1$ as a function of time after addition of titrant: curve 1, immediately after the addition; curve 2, after 10 min; curve 3, after 22 min; curve 4, after 32 min.

transition is favored by enthalpy factors. In fact, the ΔS_C° associated with the helix–coil transition is *negative*, indicating that the contribution from the disorder-order change of the polymer is predominant and that there is no significant variation in the order of the solvent structure during the transition. It seems reasonable, therefore, to conclude that the stabilizing contribution of hydrophobic bonds is much more important in the β structure than in the α helix.

From the thermodynamic data presented in Fig. 4 it appears that in water, at temperatures higher than 20°C, the β-pleated-sheet conformation is more stable than the α-helical conformation. For instance, at 50°C the β structure is thermodynamically favored by more than 100 cal/mole. In spite of this fact, our CD results showed that at this temperature the deprotonation of the polymer caused the fast formation of the α helix, which subsequently transformed into the β structure. Evidently, the thermodynamically allowed formation of the β structure is disfavored on a kinetic basis. The reason that no intermediate α-helix formation was detected at 60°C is evident from Fig. 3. At 60°C the α-helical structure cannot form for thermodynamic reasons ($\Delta G^\circ > 0$). The formation of the β structure is much more rapid at high temperatures than at low ones. Thus, at temperatures less than 60°C, immediately after the addition of sufficient KOH to deprotonate poly(Lys), the polymer conformation consists almost completely of β structure.

A detailed discussion of the quantitative contribution of individual molecular forces (such as hydrogen bonding, hydrophobic interactions, and so on) to the overall free-energy, enthalpy, and entropy changes observed for the coil-β transition of poly(Lys) is useless in our opinion. In fact, it is not presently possible to measure ΔG_C°, ΔS_C°, and ΔH_C° values with sufficient accuracy or to estimate with acceptable precision the contributions from different, individual molecular forces. In this regard we entirely agree with the point of view expressed by Holtzer and Olander (18), that is, that the conformational stability of a given structure is made up of the sum of different contributions, which are often large in comparison to the overall stability. At the present time, none of them can be calculated or measured with sufficient accuracy to permit a reasonable comparison of calculated and experimental results.

ACKNOWLEDGMENTS

The authors express their gratitude to E. Scoffone for the stimulating discussions during this work. The skillful technical assistance of C. Benvegnu and S. Da Rin Fioretto is gratefully acknowledged.

This work was carried out with the financial support of the Consiglio Nazionale delle Ricerche (CNR).

REFERENCES

1. K. Rosenheck and P. Doty, *Proc. Nat. Acad. Sci. U.S.*, **47**, 1775 (1961).
2. P.K. Sarkar and P. Doty, *Proc. Nat. Acad. Sci. U.S.*, **55**, 981 (1966).
3. B. Davidson, N. Tooney, and G.D. Fasman, *Biochem. Biophys. Res. Commun.*, **23**, 156 (1966).
4. N. Greenfield, B. Davidson, and G.D. Fasman, *Biochemistry*, **6**, 1630 (1967).
5. B. Davidson and G.D. Fasman, *Biochemistry*, **6**, 1616 (1967).
6. R. Townend, T.F. Kumosinski, S.N. Timasheff, G.D. Fasman, and B. Davidson, *Biochem. Biophys. Res. Commun.*, **23**, 163 (1966).
7. N. Greenfield and G.D. Fasman, *Biochemistry*, **8**, 4108 (1969).
8. D. Pedersen, D. Gabriel, and J. Hermans, Jr., *Biopolymers*, **10**, 2133 (1971).
9. P.Y. Chou and H.A. Scheraga, *Biopolymers*, **10**, 657 (1971).
10. A. Cosani, E. Peggion, A.S. Verdini, and M. Terbojevich, *Biopolymers*, **6**, 963 (1968).
11. E. Peggion, A.S. Verdini, A. Cosani, and E. Scoffone, *Macromolecules*, **2**, 170 (1969).
12. M. Sela and A. Berger, *J. Am. Chem. Soc.*, **77**, 1893 (1955).
13. G.D. Fasman, M. Idelson, E.R. Blout, *J. Am. Chem. Soc.*, **83**, 709 (1961).
14. A. Ciferri, D. Puett, L. Rajagh, and J. Hermans, Jr., *Biopolymers*, **6**, 1019 (1968).
15. A. Albert and E.P. Serjeant, *Ionization Constants of Acids and Bases*, Butterworth, London, 1962.
16. B.H. Zimm and S.A. Rice, *Mol. Phys.*, **3**, 391 (1960).
17. M. Nagasawa and A. Holtzer, *J. Am. Chem. Soc.*, **86**, 538 (1964).
18. D.S. Olander and A. Holtzer, *J. Am. Chem. Soc.*, **90**, 4549 (1968).
19. S.Y.C. Wooley and G. Holtzwarth, *Biochemistry*, **9**, 3604 (1970).
20. S.P. Rao and W.G. Miller, *Biopolymers*, **12**, 835 (1973).
21. M. Terbojevich, A. Cosani, E. Peggion, F. Quadrifoglio, V. Crescenzi, *Macromolecules*, **5**, 622 (1972).
22. T.V. Barskaya and O.B. Ptitsyn, *Biopolymers*, **10**, 2181 (1971).
23. J. Hermans, Jr., *J. Phys. Chem.*, **70**, 510 (1966).
24. G. Nemethy and H.A. Scheraga, *J. Phys. Chem.*, **66**, 1773 (1962).

PROTON MAGNETIC RESONANCE SPECTROSCOPY OF SYNTHETIC POLY(α-AMINO ACIDS)

L. PAOLILLO and **P.A. TEMUSSI,** Istituto Chimica, University of Naples, via Mezzocannone 4 and LCFMIB of CNR, Arco Felice, 80134 Naples, Italy, and **E.M. BRADBURY, P.D. CARY, C. CRANE-ROBINSON,** and **P.G. HARTMAN,** Biophysics Laboratories, Physics Department, Portsmouth Polytechnic, Portsmouth PO1 2QG, England

INTRODUCTION

In principle, every nucleus of an amino acid residue can be made to yield a spectrum, so the potential of high-resolution proton NMR for revealing conformational information in polypeptides is considerable and exceeds that of optical methods. The full realization of this potential is difficult, however, due largely to two facts: The increased line widths in polypeptides normally obscure coupling data, and although the chemical shift of a particular nucleus may be highly sensitive to conformational state, it is presently very difficult to calculate the shift values expected from any particular conformation.

Despite these difficulties, some success has been obtained in determining new details of polypeptide conformation. This communication deals with three such investigations concerned with the questions: Why, in partially helical samples, do helix and coil residues frequently give rise to separate resonances (the "double-peak" spectrum) when kinetic data lead us to expect fast exchange? Is there any evidence for strongly preferred side-chain conformations in solution? What is the conformation of racemic DL polypeptides?

THE α-CH DOUBLE-PEAK SPECTRUM

A considerable body of experimental kinetic data, by temperature jump and ultrasonic relaxation techniques (1), and also theoretical studies (2) suggest that at the center of the helix-coil transition, the rate of interconversion of a residue between the helix and coil states is of the order of 10^6 sec^{-1}. However, the observation of separate helix and coil resonances having a shift difference of ~50 Hz (at 100 MHz) suggests a rate that is less than 3×10^2 sec^{-1}. Therein lies the problem for which several explanations have been advanced. Joubert *et al.* (3) proposed that the upfield ("helix") peak is due to unsolvated coil residues and the downfield ("coil") peak is due to solvated coil residues. Thus, the shift difference between the peaks is attributed entirely to interaction with solvent in the coil state, and solvation is postulated to be a slow process because the solvated and unsolvated states are magnetically distinct. The helical residues were postulated to give rise to resonances too broad to be observed. This interpretation has been investigated (4) for poly(γ-benzyl-L-glutamate) (PBLG) in mixtures of deuterated chloroform ($CDCl_3$) and trifluoroacetic acid (TFA) over the complete helix-coil transition region by measurement of the summed area of both peaks relative to internal standards. The total α-CH peak area varied by only eight percent over the complete transition and was of the expected magnitude. It is clear, therefore, that the α-CH resonance of helical residues is almost fully observable under these conditions of measurement.

A second theory of the double-peak phenomenon has been advanced by J.H. Bradbury *et al.* (5) and supported by Tam and Klotz (6). This theory also involves a slow solvation step as the essential process by which two magnetically distinct α-CH proton states are observed. In this case, solvation is postulated to be protonation of the amide group by the acid. The upfield peak is regarded as a composite of unsolvated helix and unsolvated coil in rapid equilibrium (in accord with the kinetic results) and the downfield peak is composed of protonated helix and protonated coil residues, also in rapid equilibrium. Thus, the shift difference between the two peaks is entirely due to solvation, for there is no intrinsic dependence of shift on conformation. Inasmuch as a helix could never maintain a high charge on the peptide groups, the downfield peak must be due almost entirely to protonated coil. In many cases, an extremely good correlation of peak areas is obtained with the helicity determined from b_0 so the contribution of unsolvated (unprotonated) coil to the upfield peak must be slight. Therefore this scheme reduces to: upfield-peak unprotonated helix, downfield-peak completely protonated coil. To test this hypothesis several polypeptides have been studied in dimethylsulfoxide-chloroform. Dimethylsulfoxide (Me_2SO) is a coil-promoting solvent (7) for poly(β-benzyl-L-aspartate) (PBLA) of all molecular weights and for low-molecular-weight PBLG. Typical double-peak spectra were obtained with both polymers, showing that protonation is not required to induce the helix-coil transition or to give a double-peak spectrum.

The explanation of the double-peak spectrum that we favor is that molecular-weight polydispersity in the sample is the primary cause. This explanation assumes that the conclusion drawn from the kinetic measurements is correct; that is, on the NMR time scale there is rapid interconversion of all residues between the helical and coil states. Under specified solvent conditions, the helicity of a polypeptide chain is strongly dependent on the molecular weight, if that is low and the cooperativity is high (as with PBLG). A low-molecular-weight polydisperse sample in the middle of its transition, therefore, will consist in the main of molecules that are either largely helical (and so contribute to the upfield peak) or largely coil (and so contribute to the downfield peak). Thus, a good correlation of the "helix" peak area with b_0 would be obtained for samples having a molecular-weight spread broad enough such that, under any conditions, only a small proportion of the sample is actually in the process of the transition. In samples containing no low-molecular-weight material or in polymers of low cooperativity, all molecules would behave alike and have the same helicity. A single shifting α-CH peak would result. This explanation was first advanced in general terms by Jardetzky (8) and simultaneously fully developed by Ullman (9).

Fig. 1(a) shows typical double-peak spectra. The upfield peak area correlates well with b_0. Polydispersity could be the cause of the double peaks for this sample (R10). Another sample of very similar $\overline{DP}_w$ [Fig. 1(b)] did not show a clear double-peak spectrum, however, and remeasurement at 220 MHz revealed very similar peak shapes throughout the transition. This feature implies a spread of resonances (i.e., a spread of helicities) and not a division of the sample between pure helix and pure coil. If the polydispersity of this sample (S416) were lower than in R10, a spectrum more closely resembling that of a single shifting peak would be expected. Fig. 2 shows the results of fractionating R10 and S416 by precipitation chromatography and bears out the suggestion that R10 is much more disperse than S416. Several fractions obtained from R10 have been examined in detail by ORD and NMR (11). Fig. 3 shows the helix-coil transition of R10 (dark circles) and 3 fractions obtained therefrom. For comparison, the very sharp transition of a high-molecular-weight sample (P17) is included (open circles). At the midpoint of the R10 transition [~50% dichloroacetic acid (DCA)] the 130-mer fraction is still very largely helical and the 50-mer largely coil. Moreover, because a substantial fraction of R10 has a DP below 50 or above 130, it is clear that at 50% DCA the majority of the sample is either largely helix or largely coil. Fig. 4 shows α-CH spectra of a fraction from R10 having $\overline{DP}_w$ close to that of unfractionated R10. The peak shapes are akin to those of unfractionated S416 and very different from the double peaks of unfractionated R10.

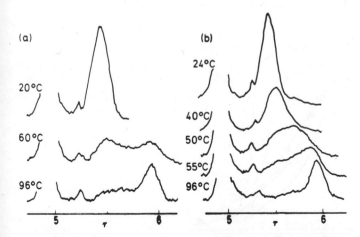

Figure 1. 100 MHz proton spectra, α-CH region, of (a) PBLG, DP = 92 (R10), and (b) PBLG, DP = 100 (S416), both in 20% TFA–80% CDCl$_3$. [Reprinted with permission from Bradbury et al., *Polymer*, **11**, 277 (1970). Copyright by IPC Business Press Ltd.]

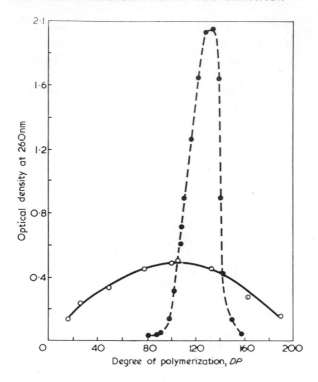

Figure 2. Molecular weight distribution curve for PBLG samples R10 (○) and S416 (●) (degree of polymerization from viscosity measurements).

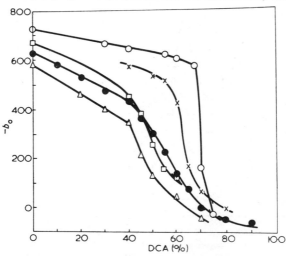

Figure 3. Values of b_0 in DCA/CHCl$_3$ at 22°C for PBLG unfractionated samples P17 (○) (700-mer) and R10 (●) (92-mer). Fractionated samples of R10 are designated H58/59 (✕) (130-mer), H50/51 (□) (80-mer) and H42/43 (△) (50-mer).

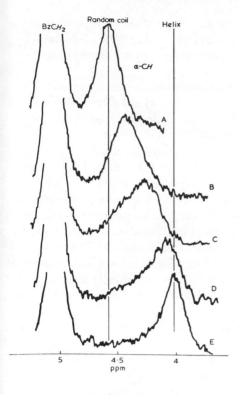

Figure 4. 100 MHz spectra of temperature-induced helix-coil transition of PBLG sample (R10) fraction A20 ($\overline{DP}$ = 110) in 12:88% CDCl₃. (A) 5°C, (B) 16°C, (C) 19°C, (D) 26°C, (E) 38°C.

Furthermore, high-molecular-weight PBLG samples such as P17 (Fig. 3) show a straightforward α-CH spectrum with a single shifting peak as expected, given the polydispersity explanation. We therefore concluded that polydispersity is the major cause of the double-peak spectrum.

Recently Nagayama and Wada (12) have obtained a somewhat better fractionation of PBLG, significantly, at somewhat lower molecular weights ($\overline{DP}_w$ = 45). Their fractionated sample had a value of M_w/M_n = 1.04–1.08, and the NMR spectrum showed what was essentially a single shifting peak. This result has been disputed by Milstein and Ferretti (13) whose fractionated PBLG sample showed $\overline{M}_w/\overline{M}_n$ = 1.06 and yet showed a more or less double-peak spectrum. The resolution of this difference is not fully clear, but it may be related to inaccuracies in estimates of polydispersity from the column-elution profiles. Neither group used a colligative measurement of $\overline{M}_n$ to calculate $\overline{M}_w/\overline{M}_n$, but this ratio may not be very revealing in the present context, because the presence of a high-molecular-weight tail has the stronger influence on this ratio and the presence of a low-molecular-weight tail has the greatest effect (in the polydispersity theory) on the appearance of a double-peak spectrum. We consider the experimental results of Nagayama and Wada (12) and ourselves (10, 11) to be a convincing demonstration that polydispersity is the primary cause of the double-peak spectrum.

A fourth explanation of the double-peak spectrum has been advanced by Ferretti *et al.* (14) (who first observed the double peak) and further elaborated by Miller (15). They propose that nucleation of a helix within a fully coiled molecule is a slow step on the NMR time scale. On this basis, pure coil is seen separately from all other partially helical molecules and the latter are assumed to interconvert rapidly. The predicted spectrum over the helix-coil transition is essentially that of a downfield peak of increasing intensity at the pure-coil position and an upfield peak that moves from the pure-helix position to the coil position. In fact, a suitable choice of the thermodynamic parameters makes it possible even to postulate (13) that the shift of the "helix" peak remains roughly constant at the fully helical position over most of the transition. The critical feature of this slow-helix-nucleation model is that the downfield peak is *pure coil* throughout; that is, it should be of constant shift. Nagayama and Wada's spectra (12) of their unfractionated sample show that this is not the case and that the coil peak approaches the fully coiled position as the helix-coil transition proceeds.

Compared to PBLG, poly(L-alanine) (PLA) has a helix-coil transition of relatively low cooperativity in chloroform-TFA. We checked this fact by comparing (16) the transition in several samples of widely differing molecular weights and found little dependence of helicity on the molecular weight throughout the $CHCl_3$-TFA solubility range from 30 to 100% TFA. In terms of the polydispersity theory, this finding implies that a single shifting α-CH peak should be observed over the helix-coil transition, despite the use of a polydisperse sample. In marked contrast to PBLG, a PLA sample of intermediate molecular weight bears out this expectation. The NMR spectra (Fig. 5) of an intermediate-molecular-weight sample of poly(D-alanine), which, of course, gives the same spectra as PLA, is used to illustrate the transition. The α-CH shift correlates extremely well with b_0, as expected for a peak that is a time average of helix and coil states. We extended this correlation to low TFA values by synthesizing a block PBLG-PLA-PBLG copolymer (17). Fig. 6 shows its NMR spectra. The α-CH shift correlates well with the PLA b_0 value over approximately the same shift range as for PBLG, namely, 4.0 to 4.5 ppm. The spectra of Fig. 5 show the presence of a subsidiary downfield peak (S) that remains largely unchanged in area over the transition. Its shift is just that of poly(DL-alanine) and very-low-molecular-weight PLA; the peak is absent in high-molecular-weight samples (16). The peak can be assigned, therefore, to pure coil and may be due to end residues (16). Thus, the spectrum of PLA appears to fit the Ferretti model of a downfield coil peak and an upfield helix-coil average peak. The situation, however, is quite different from PBLG in that the downfield coil peak (S) of PLA does *not* increase in area over the helix-coil transition. Recently Goodman *et al.* (18) have shown by a study of oligopeptides that the peak can be assigned to three terminal

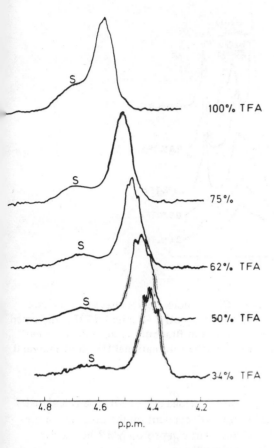

Figure 5. 220 MHz spectra, α-CH region, of poly(D-alanine), η_{sp}/c = 9.04 dl mol^{-1}, in TFA-CDCl$_3$. [Reprinted with permission from Bradbury *et al., Pure and Appl. Chem.,* **36**, 53 (1973). Copyright by the International Union of Pure and Applied Chemistry.]

residues that cannot become helical at all. Therefore the α-CH spectrum of PLA is that of a single helix-coil average peak (as expected for a polymer of low cooperativity) with the addition of an unchanging coil peak due to the presence of end residues that do not exchange at all with the remainder of the molecule.

SIDE-CHAIN ORIENTATION AND HELIX SENSE

The conformations of ester derivatives of poly(L-aspartic acid) depend on the precise nature of the side chain and on the solution conditions (19). Whereas poly(β-benzyl-L-aspartate) (PBLA) and poly(β-methyl-L-aspartate) (PMLA) take up the left-handed (LH) helix in chloroform, the β-ethyl, propyl, and phenethyl

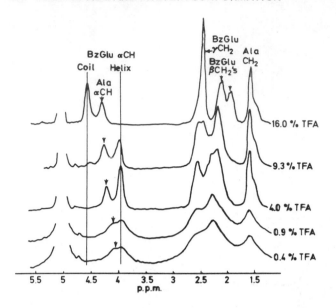

Figure 6. 220 MHz spectra in TFA-CDCl₃ of block poly[benzyl-L-glutamate (39)-L-alanine (46)-benzyl-L-glutamate (33)]. Poly(L-alanine) α-CH is indicated by an arrow. [Reprinted with permission from Bradbury *et al.*, *Rev. Pure and Appl. Chem.*, **36**, 53 (1973). Copyright by the International Union of Pure and Applied Chemistry.]

esters are in the right-handed (RH) form. Scheraga *et al.* (20) have calculated that this delicate balance of helical senses is dependent on the presence of preferred side-chain orientations. To what extent can proton NMR be used to study both helix sense and side-chain orientation?

Fig. 7 shows the spectrum of PBLA in the LH, RH, and coiled forms (21a). The RH form was induced by the inclusion of 10% L-alanine in a random copolymer with PBLA. The NH resonance of the LH form is 0.55 ppm downfield of that in the RH form, whereas the α-CH resonance of the LH form is 0.1 ppm upfield of that in the RH form. This dependence of the main-chain shift values on helix sense has been verified by observing poly(β-methyl-L-aspartate) and poly(β-ethyl-L-aspartate) as examples of LH and RH helices respectively and poly[β-(p-nitrobenzyl)-L-aspartate] as an example of an RH helix (21b). Shift values identical to those of Fig. 7 were obtained, demonstrating that this helix-sense dependence is independent of the nature of the ester group. This result has been used to demonstrate the mixed chirality of a random copolymer of PBLA with 5% L-alanine that showed a b_0 value very close to zero. The dependence of shift on helix sense has been valuable also in analysis of PBLA-PBLG copolymers (22).

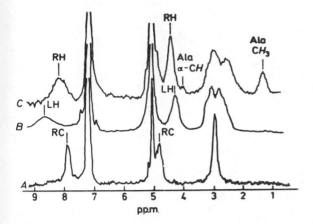

Figure 7.   100 MHz spectra of poly(β-benzyl-L-aspartate):   (a) random-coil (RC) form in 5% TFA–95% CDCl₃; (b) left-handed (LH) helical form in 0.5% TFA–99.5% CDCl₃; (c) right-handed (RH) helical form of poly[β-benzyl-L-aspartate (90)-L alanine (10)] in 0.5% TFA–99.5% CDCl₃. [Reprinted with permission from Bradbury *et al., Rev. Pure and Appl. Chem.,* **36,** 53 (1973). Copyright by the International Union of Pure and Applied Chemistry.]

In general, spin couplings are obscured in polypeptide spectra, but we were able to reveal some splittings in PBLA in chloroform by heating the sample in a sealed tube to 100°C. No changes resulted in the NH, α-CH and β-CH₂ resonances (other than line narrowing) to indicate a conformational change (see Fig. 7), so we judged the polymer to have remained LH-helical.

In Fig. 7 the β-CH₂ resonance of the LH and RH forms at ~3 ppm has the appearance of a broad doublet, the splitting of which ($\triangle$) is 0.37 ppm in the LH form and about 0.55 ppm in the RH form. Closer inspection of PBLA reveals that it is the expected ABX system, in which X is the α-CH proton (see Fig. 8). A shift difference between the two β-CH₂ protons in the helical forms does not necessarily imply restricted rotation about the α-β bond, and so we attempted to analyze the αβ vicinal coupling in terms of the three staggered 60° rotamers, assuming J_{trans} = 13.6 Hz and J_{gauche} = 2.6 Hz. Despite the poorly resolved multiplets, the use of a curve resolver led us to conclude that $J_{\alpha,A}$ = 4 ± 1 Hz and $J_{\alpha,B}$ = 7.4 ± 0.3 Hz. Molecular models indicate that steric hindrance between the main-chain and side-chain carboxyls prohibits the existence of one of the three rotamers, namely, that in which both β hydrogens are gauche to the α hydrogen, and this exclusion implies $J_{\alpha,A} + J_{\alpha,B} = J_{gauche} + J_{trans}$. This is clearly not so and it can be concluded that the regular 60° rotamers are not appropriate. Could *any* single rotamer give rise to the observed couplings? By comparison of the observed Js to a $\cos^2$ variation of J with interbond angle, it becomes clear that no single rotamer can possibly satisfy the observed couplings.

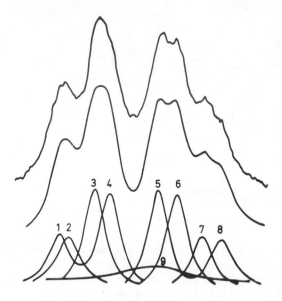

Figure 8. 100 MHz spectrum of the β-CH$_2$ group of poly(β-benzyl-L-aspartate) in CDCl$_3$ at 100°C, together with a curve resolver readout and analysis. [Reprinted with permission from Bradbury *et al.*, *Rev. Pure and Appl. Chem.*, **36**, 53 (1973). Copyright by the International Union of Pure and Applied Chemistry.]

In both of the predicted conformations (20) one vicinal coupling should be 12–13 Hz, the other 1–3.5 Hz. Therefore, there is very considerable freedom of motion about the α-β bond and no strongly preferred single side-chain orientation exists. In fact, it is possible to rationalize the observed couplings in terms of two rotamers (suggested by models) that are distorted away from the staggered 60° rotamers by 10° to 15°, but the available data is insufficient to define the β-CH$_2$ conformation with any precision.

RANDOM DL COPOLYPEPTIDES

The conformations of racemic copolymers of random sequence have been the subject of much study, due largely to the inability of ORD/CD to provide direct information. Until recently, the intensity of the helical amide-V band was the best method of investigation, but this approach is restricted to the solid state. The α-CH or peptide NH region of the NMR spectrum is a more versatile approach now that the chemical shift is known to be conformationally dependent. Bovey *et al.* (23) first showed that a PBDLG sample in chloroform has an α-CH shift of 3.95 ppm (typical of helix conformations) and, moreover, exhibits a double-peak spectrum upon TFA addition. They concluded that the polymer is helical in chloroform. We have studied several PBDLG samples of DPs 170, 159,

100, and 21 in order to assess the dependence of helicity on molecular weight (24). The helicity was assessed from the fraction of α-CH found in the helix peak at 3.95 ppm when the polymer was dissolved in $CDCl_3$–0.5% TFA (TFA used to avoid aggregation). The 170-mer was concluded to be fully helical, the 159- and 100-mers were at least three-quarters helical, and the 21-mer was less than half helical in this solvent.

The dependence of helicity on the solvent used also has been studied using NMR. The α-CH spectra of several racemic poly(benzyl glutamates) in dimethyl-formamide are compared in Fig. 9 with that of the homopolymer. The resonance at ~4.1 ppm clearly can be assigned to helix structure and that at ~4.4 ppm to random coil. It follows that although the 170-mer is fully helical in DMF as well as in chloroform, the 21-mer is fully coiled in DMF. The intermediate-molecular-weight samples are likewise seen to have lower helicity in DMF than in chloroform, and DMF is concluded to be a weaker helicogenic solvent than chloroform. Dimethylsulfoxide is a helicogenic solvent for high-molecular-weight PBLG as evidenced by a b_0 value in the region of –600°. The NMR spectrum of the homopolymer in Me_2SO is very broad, however, no doubt due to aggregation, and the helical α-CH is unfortunately undetectable. All the racemic PBDLG samples gave identical α-CH spectra with a single peak at 4.3 ppm. This peak was assigned to random-coil and not to helix structure because a double-peak spectrum was induced by the addition of chloroform, giving a new single peak at 3.9 ppm in

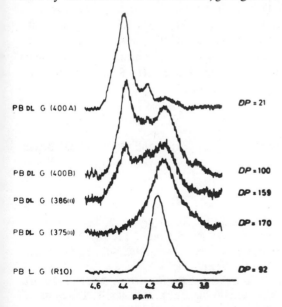

Figure 9. 220 MHz α-CH spectra in DMF of four samples of poly(γ-benzyl-DL-glutamate) having different molecular weights and one sample of poly(γ-benzyl-L-glutamate). [Reprinted from Bradbury *et al., Rev. Pure and Appl. Chem.,* **36,** 53 (1973). Copyright by the International Union of Pure and Applied Chemistry.]

50:50 Me_2SO-$CDCl_3$. This latter peak reliably can be assigned to helix structure because PBLG was fully helical when dissolved in this solvent mixture, did not aggregate, and gave an α-CH at 3.9 ppm. Identical experiments were performed with poly(ϵ-carbobenzoxy-DL-lysine) in Me_2SO. It was likewise concluded to be fully coiled. Therefore Me_2SO is a coil-promoting solvent for these racemic polypeptides.

We have used NMR also in an attempt to judge whether racemic poly(β-benzyl aspartate) is helical in $CDCl_3$ solution, in the same way as PBDLG (24). In chloroform the α-CH resonance of PBDLA is at 4.3 ppm, the same shift as for LH-helical PBLA (see Fig. 7). However, two lines of evidence suggest that this peak cannot be assigned to helix structure: Upon TFA addition there is no evidence of a double-peak spectrum (as observed with both PBDLG and PBLA), and the NH spectrum of DL-aspartate, although broad, is centered at 8.25 ppm with only a weak component at the shift [8.75 ppm; see Fig. 7(b)] characteristic of both types of residue in their preferred helix sense, namely, LH-helical L residues and RH-helical D residues. If PBDLA were helical, these conformations would constitute more than 50% of the sample. It was concluded that the 8.25 ppm and 4.3 ppm peaks must be assigned to coil, which is therefore the predominant conformation of this PBDLA sample (DP = ~60) in chloroform. On a similar basis, a sample of poly(β-methyl-DL-aspartate) of greater DP (~140) was judged to be fully coiled in chloroform.

The above results have shown that the helicity of racemic polypeptides in solution is dependent not only on the nature of the polypeptide, but also on its molecular weight and on the solvent. The ability of the NMR spectrum to estimate the fraction of helix (LH + RH) in a DL copolymer means that in combination with an ORD/CD measurement of the difference (LH – RH), a full conformational analysis into LH, RH, and coil is possible.

ACKNOWLEDGMENTS

The authors acknowledge the continuing support of the SRC of Great Britain and the CNR of Italy.

REFERENCES

1. (a) R. Lumry, R. Legare, and W.G. Miller, *Biopolymers,* **2**, 484 (1964).
    (b) A. Wada, T. Tanaka, and N. Kihara, *Biopolymers,* **11**, 587 (1972).
    (c) A.D. Barksdale and J.E. Stuehr, *J. Am. Chem. Soc.,* **94**, 3334 (1972).
2. G. Schwarz, *J. Mol. Biol.,* **11**, 64 (1965).
3. F.J. Joubert, N. Lotan, and H.A. Scheraga, *Biochemistry,* **9**, 2197 (1970).
4. E.M. Bradbury, P.D. Cary, C. Crane-Robinson, L. Paolillo, T. Tancredi, and P.A. Temussi, *J. Am. Chem. Soc.,* **93**, 5916 (1971).
5. J.H. Bradbury and M.D. Fenn, *Austral. J. Chem.,* **22**, 357 (1969).
6. J.W.O. Tam and I.M. Klotz, *J. Am. Chem. Soc.,* **93**, 1313 (1971).
7. E.M. Bradbury, C. Crane-Robinson, L. Paolillo, and P.A. Temussi, *Polymer,* **14**, 303 (1973).

8.   O. Jardetzky, Third International Conference on Magnetic Resonance in Biological Systems, Warrenton, Virginia (1968).
9.   R. Ullman, *Biopolymers,* **9**, 471 (1970).
10. E.M. Bradbury, C. Crane-Robinson, and H.W.E. Rattle, *Polymer,* **11**, 277 (1970).
11. E.M. Bradbury, C. Crane-Robinson, and P.G. Hartman, *Polymer,* **14**, 543 (1973).
12. K. Nagayama and A. Wada, *Biopolymers,* **12**, 2443 (1973).
13. J. Milstein and J. Ferretti, *Biopolymers,* **12**, 2335 (1973).
14. (a) J.A. Ferretti and B. Ninham, *Macromolecules,* **2**, 30 (1969). (b) J.A. Ferretti, B. Ninham, and V.A. Parsegian, *Macromolecules,* **3**, 34 (1970).
15. W.G. Miller, *Macromolecules,* **6**, 100 (1973).
16. E.M. Bradbury, P.D. Cary, C. Crane-Robinson, and P.G. Hartman, *Rev. Pure and Appl. Chem.,* **36**, 53 (1973).
17. E.M. Bradbury, P.D. Cary, and C. Crane-Robinson, *Macromolecules,* **5**, 581 (1972).
18. M. Goodman, C. Toniolo, and F. Naider, in *Peptides, Polypeptides, and Proteins,* Proceedings of the Rehovot Symposium 1974, Wiley(Interscience), New York, 1974.
19. E.M. Bradbury, B.G. Carpenter, and H. Goldman, *Biopolymers,* **6**, 837 (1968).
20. J.F. Yan, G. Vanderkooi, and H.A. Scheraga, *J. Chem. Phys.,* **49**, 2713 (1968).
21. (a) E.M. Bradbury, B.G. Carpenter, C. Crane-Robinson, and H. Goldman, *Macromolecules,* **4**, 557 (1971). (b) M.H. Loucheux-Lefevre, A. Forchioni, and C. Duflot, *Polymer,* in press.
22. (a) L. Paolillo, P.A. Temussi, E. Trivellone, E.M. Bradbury, and C. Crane-Robinson, *Biopolymers,* **10**, 2555 (1971). (b) L. Paolillo, P.A. Temussi, E.M. Bradbury, and C. Crane-Robinson, *Biopolymers,* **11**, 2043 (1972).
23. F.A. Bovey, J.J. Ryan, G. Spach, and F. Heitz, *Macromolecules,* **4**, 433 (1971).
24. L. Paolillo, P.A. Temussi, E. Trivellone, E.M. Bradbury, and C. Crane-Robinson, *Macromolecules,* **6**, 831 (1973).

CONFORMATIONAL TRANSITIONS OF (BENZYL-ASPARTATE) (BENZYL-GLUTAMATE) COPOLYMERS USING ^{13}C RESONANCE SPECTROSCOPY

E. M. BRADBURY and C. CRANE-ROBINSON, Biophysics Laboratories, Physics Department, Portsmouth Polytechnic, Portsmouth PO, 2QG, England, and **L. PAOLILLO, T. TANCREDI, P. A. TEMUSSI,* and E. TRIVELLONE,** LCFMIB of CNR, Arco Felice, and Istituto Chimico, University of Naples, via Mezzocannone 4, 80134 Naples, Italy

SYNOPSIS: Studies of β-benzyl-aspartate–γ-benzyl-glutamate copolymers show that conformational transitions in solution can be studied easily with ^{13}C resonance spectroscopy. A detailed comparison with previous results obtained by means of ^{1}H resonance spectroscopy indicates that ^{13}C spectroscopy is probably superior in all cases in which two or more different residues of a polypeptide undergo conformational transitions.

INTRODUCTION

The importance of obtaining structural data on protein molecules in solution has stimulated, during the last decade, the search for spectroscopic techniques of increased resolution. There is little doubt, at the moment, that the most promising of all these is nuclear magnetic resonance (NMR) spectroscopy. Some applications of NMR to the study of protein systems (1) have been highly successful, but most of the available quantitative data has been collected in the study of simpler, synthetic polypeptides (2, 3). Most NMR studies of poly(α-amino acids) in solution have dealt with simple homopolymers and have been performed by means of ^{1}H resonance spectroscopy. Only recently has it been possible to extend NMR studies to more realistic model systems such as copolymers (4, 5, 6), which, although they are not nearly as complex as natural polypeptides, are not easily studied by other spectroscopic methods traditionally employed in the polypeptide field.

It is the purpose of this paper to report results obtained recently in the study of benzyl-aspartate–benzyl-glutamate copolymers by means of ^{13}C resonance spectroscopy. This method seems, at present, to be a complement rather than an alternative to ^{1}H resonance spectroscopy in the study of polypeptides, so it is appropriate to review first some results obtained in our laboratories for the same or similar polymers by means of proton magnetic resonance.

*Senior author.

PROTON MAGNETIC RESONANCE STUDIES

Several copolymers of benzyl aspartate and benzyl glutamate were studied by proton magnetic resonance spectroscopy in two solvent systems: chloroform-trifluoroacetic acid (CDCl$_3$-TFA) (4, 5) and chloroform-dimethylsulfoxide (CDCl$_3$-Me$_2$SO-d_6) (6). The choice of residues was motivated by the possibility of observing both left-handed (LH) and right-handed (RH) helical conformations in the same solvent systems. Residues of L-aspartate are known to exist both in LH and RH helices and only small differences in intramolecular and/or inter-molecular forces can cause a change in the helix sense. The "natural" helix sense of benzyl-L-aspartate residues in helicogenic solvents is LH, but random copolymerization with benzyl-L-glutamate (whose homopolymer has a very stable RH helix) can swing the helix sense of the aspartate residues from LH to the "unnatural" RH. Proton magnetic resonance spectroscopy can be used to monitor independently any conformational change affecting the two residues. Of the many NMR parameters that are sensitive to conformational changes, the most useful ones in our studies were the α-CH chemical shift and the line-shape changes. In our copolymers, the helix and coil conformations of both glutamate and aspartate residues gave readily distinguishable α-CH peaks (4, 5). Because previous studies on aspartate polymers (7) had shown that the α-CH chemical shift was also dependent on the helix sense, we were able to follow the individual breakdowns of different helical segments. To illustrate the type of information obtained, Fig. 1 shows the upfield region of the spectra of a random copolymer containing 55% benzyl-L-aspartate and 45% benzyl-L-glutamate in mixtures of TFA-CDCl$_3$. The spectrum of the completely helical polymer (in 0.3% TFA) shows that the chemical shift of the benzyl-D-glutamate α-CH peak is 3.93 ppm (as for the L-glutamate residues in a RH helix) and that the chemical shift of the benzyl-L-aspartate α-CH peak is 4.30 ppm. This last shift corresponds to the LH helical form of the residue [as in pure poly(benzyl-L-aspartate)] because both components, as homopolymers, have LH helix sense. Upon an increase in the fraction of TFA in the solvent mixture, the helical copolymer undergoes transition to the random-coil form. This transition can be monitored by measuring the α-CH peak positions of the component residues because the chemical shifts of the solvated random coils of both residues are approximately 0.5 ppm downfield from the chemical shifts characteristic of the helical forms. The conformational changes of both residues are accompanied by line-shape changes — that are essentially due to the polydispersity of the sample (8, 9, 10). This so-called "double-peak" phenomenon can be very useful as a conformational probe because it reflects small changes in the helical content of the chains in solution. The helix-coil transition of the copolymers can be traced also in the spectra of the side-chain groups. In the helical structure of benzyl-L-aspartate the β-CH$_2$ resonance is coupled to the α-CH resonance and forms the AB part of an ABX system. The multiplet is centered near 3 ppm, with a downfield component at 3.22 ppm and an upfield component at approximately 2.75 ppm overlapping the γ-CH$_2$ resonance of benzyl-L-glutamate. In the coil form, the aspartate β-CH$_2$ is

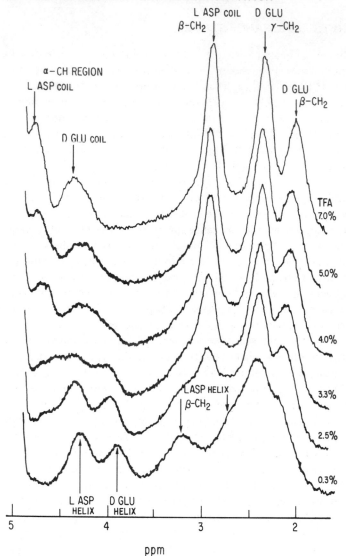

Figure 1. Upfield region of 100 MHz spectra of random poly(β-benzyl-L-aspartate)$_{55}$ (γ-benzyl-D-glutamate)$_{45}$ in CDCl$_3$-TFA mixtures.

centered at 3.05 ppm with a half-height width of only 0.07 ppm, and therefore the transition between the two forms is observable. The β- and γ-CH$_2$ resonances of glutamic residues are centered at 2.2 and 2.4 ppm, respectively, in the helical form. The shift of the γ-CH$_2$ peak remains virtually constant over the transition, but the β-CH$_2$ resonance moves upfield about 0.3 ppm. Thus, although the β-CH$_2$ spectrum is, in fact, complex, the shift of its center of gravity with respect to that of the γ-CH$_2$ can be used to follow the helix-coil transition.

The presence of benzyl-L-glutamate residues, instead of D residues, in the copolymers gives rise to different helix senses of the L-aspartate residues for different compositions. The helix sense of the aspartate component can be obtained easily from the chemical shift of the α-CH resonance. Both in CDCl$_3$-TFA and in CDCl$_3$-Me$_2$SO mixtures, the α-CH proton of the RH helix is at 4.40 ppm and that of the LH helix is at 4.30 ppm.

Aspartate components are RH in copolymers with high L-glutamate content and LH in copolymers with very high L-aspartate content. An inversion of the helix sense from the LH to RH form takes place when the composition varies from 5 to 15% benzyl-L-glutamate.

EXPERIMENTAL SECTION

The samples of poly(γ-benzyl-L-glutamate) (PBLG) and of poly(β-benzyl-L-aspartate) (PBLA) both had a DP of ~100 and were the same as used in a previous study (11). Poly(β-benzyl-L-aspartate)$_{55}$(γ-benzyl-D-glutamate)$_{45}$ was sample 441(5) of Ref. (5). Samples of poly(β-benzyl-L-aspartate)(γ-benzyl-L-glutamate) were all taken from Series 432 of Ref. (5) and 449 of Ref. (4). Polymer concentrations from 10 to 15% w/v were used in 12 mm tubes at 29°C. Two different Varian XL-100-15 spectrometers were used to record all ^{13}C spectra. The internal reference for all chemical shifts in the ^{1}H and ^{13}C spectra was tetramethylsilane (TMS).

^{13}C STUDIES

The applicability of ^{13}C spectroscopy to the study of helix-coil transitions has already been demonstrated by preliminary work on PBLG performed in our laboratories (11) and by subsequent studies in other laboratories (12, 13). More detailed comparisons of ^{1}H studies and ^{13}C studies on several polymers are necessary, however, to determine the relative merits of the two methods. Here we present the results of a ^{13}C investigation on PBLG, PBLA, and some aspartate-glutamate copolymers. Fig. 2 shows the complete ^{13}C spectra of PBLG in several CDCl$_3$-TFA mixtures. All the peaks can be easily assigned (11) according to literature data on model compounds (14). Most of the peak positions are sensitive to the helix-coil transition, the more sensitive being those of carbon atoms in the chain backbone. As in the case of proton spectra (where the α-CH resonance is the most useful for monitoring the transition) the resonance of the α-C is the most sensitive to conformational changes and, presumably, is barely affected by solvation. Accordingly, we chose this resonance to follow in detail the conformational dependence of ^{13}C spectra of polypeptides.

Fig. 3 shows the expanded C$_\alpha$ region of the spectra of Fig. 2, along with the α-CH spectra of the same polymer in identical solvent mixtures. The C$_\alpha$ peak corresponding to the right-handed helix of PBLG is about 3 ppm downfield with respect to the random-coil C$_\alpha$ peak. This relative shift compares favorably with the corresponding shift observed in the proton magnetic resonance spectra, which is of the order of 0.5 ppm. The most interesting feature, however, is the

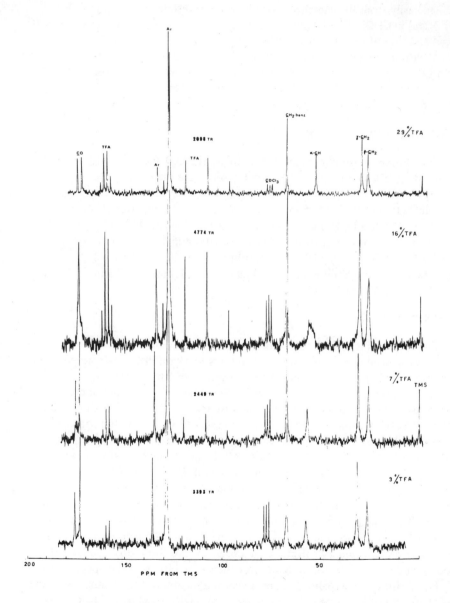

Figure 2. Natural abundance ^{13}C Fourier transform NMR spectra of PBLG in CDCl$_3$-TFA mixtures (5 kHz scale).

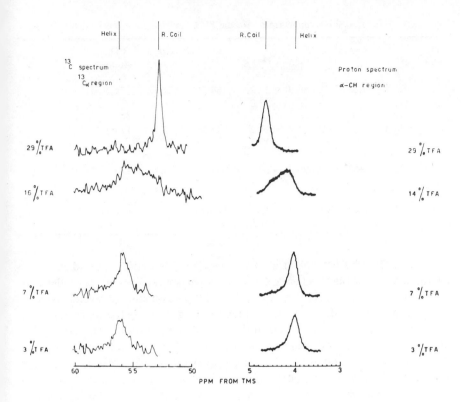

Figure 3. Comparison of the expanded regions of C_α and α-CH NMR spectra of PBLG (S416) in $CDCl_3$-TFA mixtures.

appearance of multiple peaks at the transition midpoint, a phenomenon already observed in proton spectra and attributable essentially to end-group effects and polydispersity (8, 9, 10). An important difference between the proton spectra and the ^{13}C spectra is that in the latter, the broadening of the C_α peak at the transition midpoint is paralleled by similar line-shape changes of all other carbon peaks. Such an effect, in general, is not observed in proton spectra or is masked by the presence of spin-spin couplings. This characteristic of ^{13}C spectra makes it clear that most peaks might be used to follow the transition.

Table 1 summarizes the chemical shift data for PBLG. It can be seen that although the side-chain peaks are barely affected by a change in the solvent system (only the ester CO shows a substantial downfield shift due to TFA solvation), the backbone protons (the C_α and the amide CO) are shifted upfield by the conformational change.

The chemical shift changes of PBLA are quite similar to those of PBLG, but somewhat smaller in magnitude. The only relevant difference is found in the carbonyl region where only one peak is observed for all solvent mixtures. In

Table 1. ^{13}C chemical shifts (ppm) from internal TMS of PBLG solutions in CDCl$_3$-TFA

% TFA (v/v)	C_β	C_γ	C_α	C_{Bzl}	$C_{Ar_{2,6}}$	C_{Ar_1}	CO Ester	CO Amide
3	26.4	31.5	57.2	66.9	127.2 127.5	134.6	171.5	174.0
7	26.3	31.5	57.1	67.4	127.2 127.5	134.3	172.3	174.2
16	26.9	31.5	55	67.9	127.2 127.5	133.9	173	173
29	26.7	30.6	53.6	68.2	127.2 127.5	133.3	173.6	171.5

order to confirm the ^{13}C chemical shift dependence on conformation and to determine whether the C_α chemical shifts of aspartate residues are dependent on the sense of the helix as are the α-CH chemical shifts, we studied several copolymers of benzyl aspartate and benzyl glutamate. Fig. 4 shows the complete spectra of two random copolymers in the helical conformation; both the aspartate and glutamate C_α peaks are clearly distinguishable. The chemical shift changes in the C_α region of poly[(β-benzyl-LAsp)$_{55}$(γ-benzyl-DGlu)$_{45}$] throughout the transition are described in Table 2. Because the helical sense of poly-(β-benzyl-L-aspartate) is the same as poly(γ-benzyl-D-glutamate), the C_α peak positions are the same as those for the corresponding homopolymers. Fig. 5 shows the changes in the C_α region of poly[(β-benzyl-LAsp)$_{50}$(γ-benzyl-LGlu)$_{50}$]. It can be seen that when the polymer is in the helical conformation the L-aspartate C_α peak is displaced 2 ppm downfield with respect to the left-handed C_α peak, that is, 53 ppm compared to 51 ppm. This change can be associated with a change in helix sense because it is known that even small amounts of Glu residues in a random copolymer can force the Asp residues to assume a right-handed helical conformation (see section on proton NMR studies in this paper). Table 3 lists data for several $^{13}C_\alpha$ shifts of this copolymer.

For PBLG in nonaqueous solution, the helix resonance is 3–4 ppm downfield from the coil. Furthermore, this conformationally dependent shift is largely independent of solvent. Similar results have been obtained for three water-soluble polymers, although the shift difference between helix and coil seems somewhat smaller than in organic solvents (see Table 3). The possibility exists, therefore, that this shift difference could be of value in structural studies of proteins in a way that is impossible with proton NMR spectroscopy, in which the α-CH shift is strongly solvent dependent.

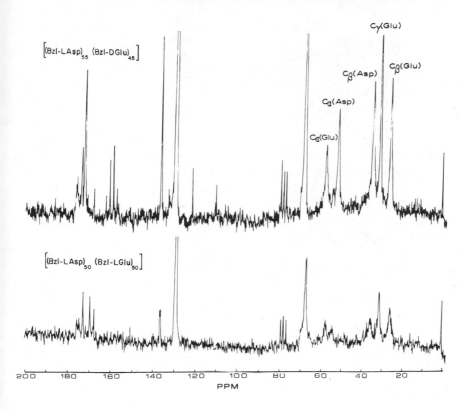

Figure 4. Natural abundance ^{13}C Fourier transform NMR spectra of poly[(β-benzyl-L-aspartate)$_{55}$(γ-benzyl-D-glutamate)$_{45}$] and of poly[(β-benzyl-L-aspartate)$_{50}$(γ-benzyl-L-glutamate)$_{50}$] in helical form (1% TFA–99% CDCl$_3$; 5 kHz scale).

Table 2. ^{13}C chemical shifts (ppm) from internal TMS of poly[(β-benzyl-LAsp)$_{55}$(γ-benzyl-DGlu)$_{45}$] solutions in CDCl$_3$-TFA

% TFA (v/v)	Residue	C_β	C_γ	C_α	C_{Bzl}	$C_{Ar_{2,6}}$	C_{Ar_1}	CO Ester	CO Amide
1	DGlu	25.6	30.7	57.3	65.7	127.5		172.0	175.1
	LAsp	34.1		51.3	65.7	127.5	135.8	170.8	172.0
4	DGlu	25.6	30.6	57.0	66.3	127.6	135.2	173.1	175.3
	LAsp	34.1		51.1	66.3	127.6		171.8	171.8
6	DGlu	25.7	30.6	57 55	67.4	128.2	135.1	173.6	175.2
	LAsp	34.3		50.9	67.4	128.2		171.9	171.9
10	DGlu	26.2	30.5	54.3	68.0	127.8	134.5	174.6	173.3
	LAsp	35.4		50.6	68.0	129.0	134.8	172.0	172.0

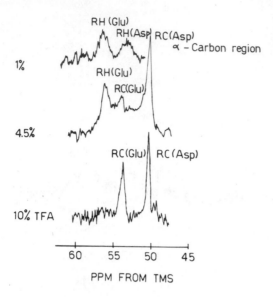

Figure 5. Expanded spectra of the C_α region of poly[(β-benzyl-L-aspartate)$_{50}$-(γ-benzyl-D-glutamate)$_{50}$] [432(2)] in CDCl$_3$-TFA mixtures.

CONCLUSION

The ^{13}C spectra of homo- and copolymers of benzyl aspartate and benzyl glutamate show very clearly that this type of spectroscopy can be used to study conformational transitions of polypeptides in solution. It is interesting to note that all features already observed in ^1H spectra of polypeptides are present also in ^{13}C spectra. The possibility of using more peaks to monitor conformational transitions, however, probably makes ^{13}C spectroscopy superior to ^1H spectroscopy. It is worth noting also that in the ^{13}C spectra of our copolymers, all peaks of the two different residues were clearly spaced. It is hoped that this may be the case even for polymers containing more than two residues.

Table 3. $^{13}C_\alpha$ shifts of several polymers measured with respect to TMS (internal for organic solvents, external for water)

Sample	Solvent	L(or D)Glu or LLys	LAsp
Poly[(Asp)$_{50}$(Glu)$_{50}$]	CDCl$_3$–1% TFA	56.9 helix	53.7 RH helix
Poly [(Asp)$_{55}$(DGlu)$_{45}$]	CDCl$_3$–1% TFA	57.3 helix	51.3 LH helix
Poly(γ-benzyl-L-glutamate)	CDCl$_3$–3% TFA	57.2 helix	
Poly(β-benzyl-L-aspartate)	CDCl$_3$–1% TFA		51.3 LH helix
Poly(γ-benzyl-L-glutamate)	CDCl$_3$–29% TFA	53.6 coil	
Poly(β-benzyl-L-aspartate)	CDCl$_3$–6% TFA		49.5 coil
Poly[(Asp)$_{50}$(Glu)$_{50}$]	CDCl$_3$–10% TFA	54.2 coil	50.5 coil
Poly(γ-benzyl-L-glutamate)	Me$_2$SO	56.2 helix	
Poly(γ-benzyl-DL-glutamate)	Me$_2$SO	51.9 coil	
Poly(γ-benzyl-L-glutamate)	m-CRESOL	57.9 helix	
Poly(γ-benzyl-DL-glutamate)	m-CRESOL	54.1 coil	
Poly(Glu^{42}Lys^{28}Ala^{30})	Water	54.6 coil	
Poly(Glu^{42}Lys^{28}Ala^{30})	Water	56.6 (65% Helix)	
Poly(L-lysine) (16)	Water	53.5 coil	
Poly(L-lysine) (16)	Water	55.1 helix	
Poly(L-glutamic acid)[a]	Water	53.7 coil	
Poly(L-glutamic acid)[a]	Water	55.5 (~80% Helix)	

[a]Literature data (15) corrected from CS$_2$ to TMS.

REFERENCES

1. R.A. Dwek, *NMR in Biochemistry,* Clarendon Press, Oxford, England, 1973.
2. F.A. Bovey, *High Resolution NMR of Macromolecules,* Academic Press, New York, 1972.
3. E.M. Bradbury, C. Crane-Robinson, L. Paolillo, and P.A. Temussi, *J. Am. Chem. Soc.,* **95**, 1683 (1973).
4. L. Paolillo, P.A. Temussi, E. Trivellone, E.M. Bradbury, and C. Crane-Robinson, *Biopolymers,* **10**, 2555 (1971).
5. L. Paolillo, P.A. Temussi, E.M. Bradbury, and C. Crane-Robinson, *Biopolymers,* **11**, 2043 (1972).

6. E.M. Bradbury, C. Crane-Robinson, L. Paolillo, and P.A. Temussi, *Polymer,* **14**, 303 (1973).
7. E.M. Bradbury, B.G. Carpenter, C. Crane-Robinson, and H. Goldman, *Macromolecules,* **4**, 557 (1971).
8. R. Ullman, *Biopolymers,* **9**, 471 (1970).
9. P.A. Temussi and M. Goodman, *Proc. Nat. Acad. Sci. U.S.,* **68**, 1767 (1971).
10. K. Nagayama and A. Wada, *Chem. Phys. Lett.,* **16**, 50 (1972).
11. L. Paolillo, T. Tancredi, P.A. Temussi, E. Trivellone, E.M. Bradbury, and C. Crane-Robinson, *Chem. Commun.,* 335 (1972).
12. G. Boccalon, A.S. Verdini, and G. Giacometti, *J. Am. Chem. Soc.,* **94**, 3639 (1972).
13. A. Allerhand and E. Oldfield, *Biochemistry,* **12**, 3428 (1973).
14. W. Horsley, H. Sternlicht, and J. Cohen, *J. Am. Chem. Soc.,* **92**, 680 (1972).
15. J.R. Lyerla, B.H. Barber, and M.H. Freedman, *Canadian J. Biochem.,* **51**, 460 (1973).
16. H. Saito and I.C.P. Smith, *Arch. Biochem. Biophys.,* **158**, 154 (1973).

REFOLDING OF THE REDUCED PANCREATIC TRYPSIN INHIBITOR

THOMAS E. CREIGHTON, Medical Research Council, Laboratory of Molecular Biology, Hills Road, Cambridge, England

SYNOPSIS: Refolding of the reduced pancreatic trypsin inhibitor has been investigated using thiol-disulfide exchange with several disulfide reagents to regenerate the three disulfide bonds of the native inhibitor. Essentially, quantitative refolding is routinely obtained, and the renatured inhibitor is indistinguishable from the original native inhibitor in several respects.

The thiol-disulfide exchange reaction was rapidly quenched by acidification or by the addition of iodoacetate or iodoacetamide at various times during the refolding reaction, and the species so trapped was separated by gel electrophoresis and by ion-exchange chromatography. Several transient intermediate species with one or two disulfide bonds accumulated to detectable levels; their kinetic roles in the refolding process were investigated and the cysteine residues involved in the disulfide bonds determined.

Implications for the general problem of protein folding are discussed.

INTRODUCTION

Protein folding is a highly cooperative process, due to the relative instability of the intermediate states (1-6), but the identification of such intermediates, the characterization of their kinetic natures, and the elucidation of their conformational properties will be required to define the pathways by which a protein folds and unfolds. Consequently, to achieve these goals the intermediate states must be trapped in a way that overcomes their inherent instability. Of the various interactions within a protein, only disulfide bonds between cysteine residues, due to their redox nature, appear to be of a type that might be stabilized in order to trap intermediate conformations; the usual formation or breakage of a disulfide bond requires the transfer of electrons between the two cysteine residues involved and another appropriate electron donor or acceptor of the environment. It is possible, however, to initiate quickly and to maintain conditions under which disulfide bonds may not be formed or broken.

The thiol-disulfide exchange reaction (7-9)

$$R_1SH + R_2SSR_2 \rightleftharpoons R_1SSR_2 + R_2SH \tag{1}$$

is the most satisfactory method of making or breaking particular disulfide bonds (10). The formation or breakage of a single protein disulfide bond requires two sequential thiol-disulfide exchanges:

$$P_{SH}^{SH} + RSSR \rightleftharpoons P_{SH}^{SSR} + RSH \tag{2}$$

$$P_{SH}^{SSR} \rightleftharpoons P_{S}^{S} + RSH \tag{3}$$

proceeding through the intermediate mixed disulfide between the protein (P) and the reagent (R). For most studies of protein folding, the mixed disulfide intermediate should be minimized. The most feasible minimization is with a cyclic disulfide reagent

$$P_{SH}^{SH} + R_{S}^{S} \rightleftharpoons P_{SH}^{SSRSH} \rightleftharpoons P_{S}^{S} + R_{SH}^{SH} \tag{4}$$

of low redox potential, such as oxidized dithiothreitol (11, 12) (DTT_S^S).[1] The rate of the reaction may be varied with the nature of the reagent, the concentrations of the thiol and disulfide forms of the reagent, and with the pH as the reaction requires the thiolate anion. Consequently, at least four kinetic steps should be distinguishable in protein folding: those involving (1) net disulfide bond formation, (2) net disulfide bond breakage, (3) intramolecular thiol-disulfide exchanges, and (4) other unimolecular conformational transitions not involving disulfide-bond exchange reactions.

The pancreatic trypsin inhibitor is a favorable model for such a study because it has three disulfide bonds (linking Cys-5 and -55, Cys-14 and -38, and Cys-30 and -51) (13, 14), only 58 amino acid residues (13), and a very stable conformation (15, 16) known to very high resolution (17). It binds tightly to the proteases trypsin, chymotrypsin, plasmin, and kallikrein; the kinetics (18–21) and stereochemistry (22) of its interaction with trypsin and chymotrypsin are known in great detail. The primary structures of four proteins homologous with the inhibitor are known (23–25), and considerable progress has been made in the synthetic preparation of analogs of the inhibitor with altered primary structures (26).

A more detailed report of this investigation is being published elsewhere (27).

---

[1] Abbreviations used: DTT_{SH}^{SH} and DTT_S^S, the reduced dithiol and oxidized disulfide forms of dithiothreitol, respectively; $(HOEtS)_2$, hydroxyethyl disulfide, the oxidized form of mercaptoethanol.

RENATURATION OF REDUCED INHIBITOR

The disulfide bonds of the inhibitor are readily reduced by excess thiol reagents, such as DTT_{SH}^{SH}, in 6 M guanidinium-Cl. After removal of the thiol reagent and denaturant by gel filtration, the reduced inhibitor[1] has no detectable ability to inhibit trypsin and is insoluble at concentrations greater than 0.1 mM in nondenaturing solutions at neutral pH. The optical rotatory dispersion studies of Pospisilova *et al.* (28) confirm the conclusion that the reduced inhibitor [like the reduced forms of other proteins that normally contain disulfide bonds (29–31)] has no stable conformation, but exists as a disordered polypeptide chain. Consequently, renaturation of the reduced inhibitor involves refolding of the polypeptide chain.

The reduced inhibitor rapidly forms three disulfide bonds and refolds to the native structure upon the addition of excess disulfide reagent, such as DTT_S^S or $(HOEtS)_2$ (Fig. 1). After a short lag period, the renatured inhibitor appears with approximately pseudo-first-order kinetics; this rate is also first order in disulfide reagent concentration (Fig. 2) and is a function of the nature of the reagent. At

[1]The term inhibitor is used for all forms of the polypeptide chain derived here from the pancreatic trypsin inhibitor; it does not imply that the protein has inhibitory activity; of the forms discussed here, only the fully renatured inhibitor (N) is an inhibitor of trypsin.

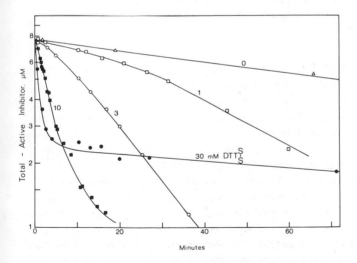

Figure 1. Kinetics of renaturation of reduced inhibitor by DTT_S^S. Reduced inhibitor (8 μM) was renatured at 25° in 0.1 M triethanolamine-HCl buffer, pH 8.5, containing 1 mM EDTA and the indicated mM concentrations of DTT_S^S. Aliquots were assayed for inhibition of trypsin activity (32). (Reprinted with permission from T.E. Creighton, *J. Mol. Biol.*, in press. Copyright by Academic Press.)

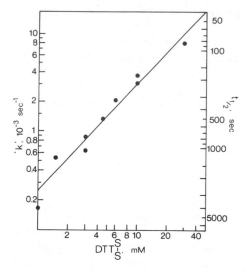

Figure 2. Rate of renaturation of reduced inhibitor versus DTT$_S^S$ concentration. Conditions were as in Fig. 1. The apparent pseudo-first-order rate constant ('k') was estimated from data like that of Fig. 1 by the time required to reach half the final level of trypsin inhibitor activity ($t_{1/2}$). The slope of the line is 1.08 and corresponds to apparent second-order rate constants of 0.25 to 0.33 sec^{-1} M^{-1}. (Reprinted with permission from T.E. Creighton, *J. Mol. Biol.*, in press. Copyright by Academic Press.)

high concentrations of DTT$_S^S$ or with (HOEtS)$_2$, the reaction is distinctly biphasic, a portion of inhibitor refolding relatively slowly.

Low concentrations of DTT$_{SH}^{SH}$ inhibit the rate, but not the extent, of renaturation (Fig. 3), suggesting the occurrence of kinetically significant metastable intermediates on the folding pathway.

Quantitative recovery of renatured inhibitor is routinely obtained. The renatured inhibitor is indistinguishable from native inhibitor in its inhibition of trypsin and chymotrypsin (including its rates of association and dissociation with these two proteases), resistance to digestion by pepsin, ultraviolet absorption spectrum, and polyacrylamide gel electrophoresis (see Fig. 4). Furthermore, diagonal maps (33) prepared with the peptides produced by thermolysin digestion (34) of native and renatured inhibitors were indistinguishable. The disulfide bonds indicated by the diagonal maps consisted of the correct pairs of cysteine residues.

ENTRAPMENT OF MOLECULAR SPECIES DURING RENATURATION

The thiol-disulfide exchange reaction involved in inhibitor renaturation is rapidly quenched by three reagents: HCl to 0.2 M, and iodoacetamide or iodoacetate to 0.1 M. The first reagent greatly decreases the rate of

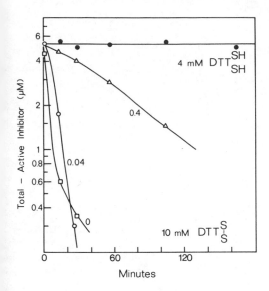

Figure 3. Inhibition by DTT_{SH}^{SH} of renaturation of reduced inhibitor. Reduced inhibitor (5 μM) was renatured in 0.1 M NH$_4$HCO$_3$ (pH 8.7), 1 mM EDTA, 10 mM DTT$_S^S$, with DTT$_{SH}^{SH}$ added to the indicated mM concentrations. After one day, all mixtures had attained full renaturation, with the exception of that containing the highest concentration of DTT$_{SH}^{SH}$, which produced no detectable inhibition of trypsin. (Reprinted with permission from T.E. Creighton, *J. Mol. Biol.*, in press. Copyright by Academic Press.)

thiol-disulfide exchange by preventing ionization of thiol groups, because the reaction requires the thiolate anion (7). It has the advantage of acting quickly, protonation being much more rapid than thiol-disulfide exchange, thereby ensuring that the quenched mixture is an accurate representation of the reaction mixture at the time of quenching. The second and third reagents rapidly alkylate all thiol groups in the renaturation mixture; they have the advantage that the trapped species are relatively stable, having only disulfide bonds and irreversibly blocked thiol groups.

Carboxymethylation with iodoacetate introduces a new acidic group at each cysteine residue not involved in a disulfide bond, so separation of the possible inhibitor species with different numbers of disulfide bonds should be feasible. As expected, the carboxymethylated reduced inhibitor migrates considerably more slowly than that trapped by acid or iodoacetamide in the low pH electrophoresis system of Reisfield *et al.* (35) for basic proteins (Fig. 4). The $-CH_2CONH_2$ groups introduced by iodoacetamide are neutral, as are free cysteine residues at this pH, so it is not surprising that both forms migrate similarly. The electrophoretic mobilities of the native and renatured inhibitors are identical and not altered by the treatment with HCl, iodoacetamide, or iodoacetate, as would be expected for inhibitor with no free cysteine residues.

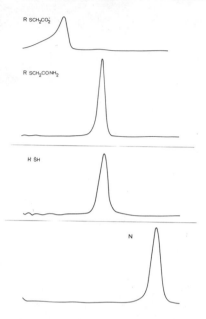

Figure 4. Electrophoretic mobilities of the reduced and renatured inhibitors. The fully reduced inhibitor was reacted with iodoacetate ($R-SCH_2CO_2^-$), iodo-acetamide ($R-SCH_2CONH_2$) and HCl ($R-SH$) with the standard trapping methods. The fully renatured inhibitor (N) had been renatured with 5 mM DTT$_S^S$ for two days. Electrophoresis was at 80 V for 7 hr with the pH 3.8 system for basic proteins (35) in 15% polyacrylamide gels. Densitometer traces were obtained after staining with Coomassie blue. Migration was to the right. (Reprinted with permission from T.E. Creighton, *J. Mol. Biol.*, in press. Copyright by Academic Press.)

However, the native and renatured inhibitors migrate considerably faster than any of the above forms of the reduced inhibitor (Fig. 4), so there must be differences in the conformational properties of the reduced and native inhibitors. This could be a result of altered pK values of the acidic groups, overall shape differences, or altered interactions with the gel matrix. In any case, this electrophoresis system provides a means of distinguishing between the two extreme conformational states of the inhibitor and the possibility of detecting and identifying intermediate states with one or two disulfide bonds.

Examination of samples trapped at varying times during the inhibitor renaturation with 5 mM DTT$_S^S$ shows the disappearance of the reduced inhibitor (R) and appearance of the renatured inhibitor (N) (Fig. 5). In addition, several intermediate species are observed to accumulate and then disappear. Three bands, designated A, B, and C in order of increasing electrophoretic mobility, are resolved in the portions trapped by acid or iodoacetamide, which yield identical electrophoretic profiles. When trapped with iodoacetate, the intermediate species each migrates approximately as predicted for species with single

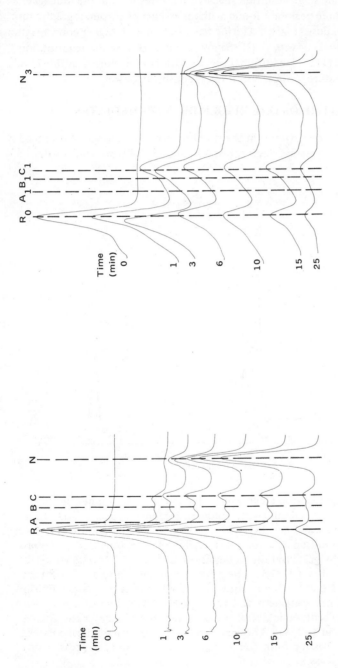

Figure 5. Electrophoretic analysis of the species trapped during renaturation of the reduced inhibitor by 5 mM DTT$_S^S$. Reduced inhibitor (30 μM) was renatured at 25°C in 0.1 M Tris-HCl (pH 8.7), 0.2 M KCl, 1 mM EDTA, and 5 mM DTT$_S^S$ for the indicated periods of time, when portions were treated with iodoacetamide (left) or iodoacetate (right). Electrophoresis was as in Fig. 4. The species R, A, B, C and N are identified on the left; on the right are the *predicted* mobilities of these species with the number of disulfide bonds given by the subscripts. (Reprinted with permission from T.E. Creighton, *J. Mol. Biol.*, in press. Copyright by Academic Press.)

disulfide bonds. There is very little accumulation of inhibitor species with the electrophoretic mobilities expected of species with two disulfide bonds.

Although multiple single-disulfide species are not resolved in the iodoacetate-trapped samples, three peaks are found with ion-exchange chromatography on carboxymethyl cellulose (Fig. 6). During renaturation with higher concentrations of DTT_S^S (e.g., 25 mM) or with $(HOEtS)_2$, where in each case the renaturation reaction is biphasic (Fig. 1), additional species with two or more disulfide bonds accumulate. The nature of these species is under investigation.

MOLECULAR NATURE OF ONE-DISULFIDE INTERMEDIATES

The cysteine residues involved in the disulfide bonds of the species are readily determined with the diagonal electrophoresis method of Brown and Hartley (33).

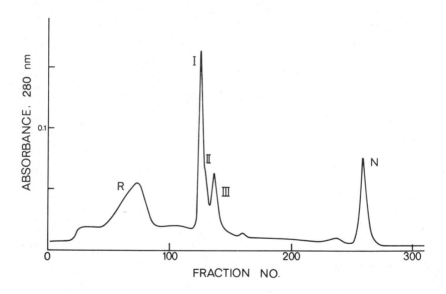

Figure 6. Ion-exchange chromatography of the iodoacetate-trapped species in the renaturation of reduced inhibitor. Reduced inhibitor (30 μM, total of 3.6 μmoles) was renatured with 5 mM DTT_S^S as in Fig. 5. The reaction was quenched after 6 min by the addition of iodoacetate. After an additional 2 min, the protein was recovered by gel filtration on Sephadex G25 in 0.02 M imidazole-HCl buffer, pH 6.2, with 1 mM EDTA. The mixture was applied to a 1.5 × 80 cm column of carboxymethyl cellulose equilibrated with the same buffer. Elution was with a 1.0 liter linear gradient of 0 to 0.7 M NaCl; fractions of 2.6 ml were collected. The fully reduced inhibitor is designated by R; N designates the renatured inhibitor; and 1, 2, and 3 designate the single disulfide intermediates. (Reprinted with permission from T.E. Creighton, *J. Mol. Biol.*, in press. Copyright by Academic Press.)

The protein is proteolytically digested and the peptides separated by electrophoresis under conditions in which the disulfide bonds are maintained intact. Exposure of the peptides to performic acid, which oxidizes the disulfide bonds, converts both Cys residues to cysteic acid and separates the peptides originally linked by each disulfide bond. Any carboxymethyl-Cys and methionine residues are also oxidized to the corresponding sulfones. The oxidized peptides are then subjected to electrophoresis under the same conditions as, but perpendicular to, the first electrophoresis. Peptides unaltered by the oxidation migrate identically in both instances and thus define a diagonal line of peptides. However, the peptides originally linked by a disulfide bond generally migrate differently in the second dimension, due to the intervening cleavage of the disulfide bond and the introduction of a new acidic cysteic acid residue on each fragment. Oxidation of methionine to the sulfone normally does not affect the mobility of a peptide in which it occurs. Oxidation of carboxymethyl-Cys residues markedly alters the pK value of the side-chain carboxyl group, however. If the electrophoresis is carried out at pH 3.5, the mobilities of the carboxymethyl-Cys peptides are altered so that they lie off the diagonal. Several of these often define a "carboxymethyl-Cys diagonal," as shown in Fig. 7(a), which illustrates such a

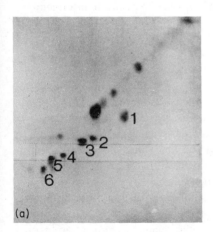 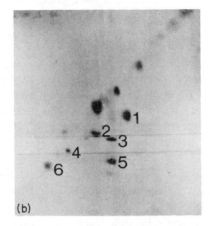

Figure 7. Diagonal maps of (a) carboxymethylated reduced inhibitor and (b) single-disulfide intermediate 1. The proteins were digested with trypsin followed by chymotrypsin. The peptides were separated by electrophoresis at pH 3.5 in the horizontal direction (anode at the right). After exposure to performic acid, the electrophoresis was repeated in the vertical direction (anode at the top). The diagonal maps were stained with ninhydrin-cadmium acetate. The numbered peptides were identified from their amino acid compositions to represent the following residues of the inhibitor primary structure: peptide 1, residues 36–39; 2, 1–15; 3, 47–53; 4, 5–15; 5, 27–33; 6, 54–58. (Reprinted with permission from T.E. Creighton, *J. Mol. Biol.,* in press. Copyright by Academic Press.)

diagonal map prepared of fully reduced inhibitor carboxymethylated with iodo[^{14}C] acetate. Six peptides lie off the diagonal, defining the carboxymethyl-Cys diagonal. Radioautography indicates that all the ^{14}C present was in these peptides, which accounts for all six Cys residues of the inhibitor.

Diagonal maps prepared at pH 3.5 are thus immensely useful in this investigation, as they allow simultaneous determination of the Cys residues carboxymethylated and of those involved in disulfide bonds. Such a diagonal map of fraction 1 of the carboxymethylated single-disulfide intermediates isolated as in Fig. 6 is shown in Fig. 7(b). Comparison with the diagonal map of fully reduced inhibitor indicates that a single pair of peptides are absent from the carboxymethyl-Cys diagonal, but are in related positions off both diagonals, indicating they initially had been joined by a disulfide bond. The electrophoretic mobilities, staining characteristics, and amino acid compositions of these two peptides unambiguously identified them as consisting of residues 47–53 and 27–33, including Cys-30 and -51. The virtual absence of peptides containing these residues from the carboxymethyl-Cys diagonal indicates that fraction 1, which accounts for just over 50% of the single-disulfide intermediates, contains a single species with an intramolecular disulfide bond linking Cys-30 and -51. This disulfide bond is also present in the native inhibitor.

Fraction 2 accounts for approximately 20% of the one-disulfide intermediates. Diagonal maps of the peptides produced by digestion with trypsin followed by thermolysin indicate that 2 is a mixture of two species, one with a disulfide bond linking Cys-30 and -55, the other linking Cys-5 and -51. Similarly, fraction 3, which comprises approximately 25% of the one-disulfide species, consists of essentially a single species with a disulfide bond linking Cys-5 and -30. These three disulfide bonds are not present in the native inhibitor.

It has been found that the same disulfide bonds are present in the single-disulfide intermediates trapped with iodoacetamide (species A, B, and C in Fig. 5). However, only limited preparative separation of these species has been thus far obtained, but the major intermediate species C has been found to have the disulfide bond between Cys-30 and -51, the same as that present in the predominant fraction 1 trapped by iodoacetate. Assignment of the other disulfide bonds to these electrophoretic species is not yet possible.

The single-disulfide species are trapped in similar quantities by acid, iodoacetamide, and iodoacetate, and they accumulate to similar relative levels during renaturation with a variety of disulfide reagents. Consequently, the trapped species are accurate representations of the species accumulating during the refolding process and thus are true intermediates in protein folding. They are not artifacts produced by the disulfide reagents or the trapping reactions.

KINETIC ROLES OF INTERMEDIATE SPECIES

The time-dependent concentrations of the species have been determined electrophoretically during renaturation with three different concentrations of DTT_S^S

(5, 10, and 25 mM) and of $(HOEtS)_2$ (0.09, 0.3, and 0.64 mM). An example of the results obtained by electrophoretic analysis of the species trapped by acid or iodoacetamide during renaturation with 5 mM DTT^S_S is shown in Fig. 8. The re-duced inhibitor (R) disappears in a second-order reaction, being first order with respect to both R and disulfide reagent. The one-disulfide intermediates are always present in constant relative proportions and thus exhibit similar kinetic behavior, which suggests that they are normally in a state of rapid equilibrium with each other via an intramolecular thiol-disulfide exchange reaction. The rates of both appearance and disappearance of the single-disulfide species are proportional to the disulfide reagent concentration. Simulation of the reaction demonstrates that the combined rates of formation of the single-disulfide species fully account for the rate of disappearance of R. Because the rate-limiting step

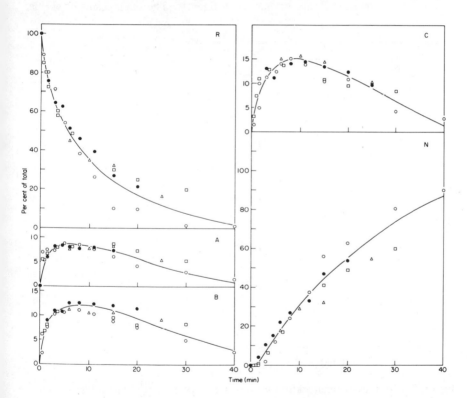

Figure 8. Kinetic behavior of the species trapped during renaturation of reduced inhibitor with 5 mM DTT^S_S. The relative concentrations of the species R, A, B, C, and N estimated from densitometer tracings of electrophoretic analyses as in Fig. 5 are plotted versus time of trapping for four independent experiments. The values indicated ○ were of species trapped with HCl; ●, □, and △ were trapped with iodoacetamide. (Reprinted with permission from T.E. Creighton, *J. Mol. Biol.*, in press. Copyright by Academic Press.)

in disappearance of the one-disulfide species seems to involve disulfide bond formation and because the only species with two or more disulfide bonds that accumulates under these conditions is the fully renatured inhibitor, N (Fig. 5), the rate-limiting step in the disappearance of the single-disulfide intermediate is very probably their conversion to N. Consequently, the one-disulfide intermediates may appear to be directly on the correct folding pathway. Folding pathways independent of the accumulated species are probably not significant. The short lag period in the appearance of N is consistent with this conclusion because it corresponds to the period during which the single-disulfide intermediates are accumulating.

The effect of DTT_{SH}^{SH} (Fig. 3) on the kinetic behavior of various species confirms this conclusion. As shown in Fig. 9, the single-disulfide intermediates accumulate at the normal rate to their normal levels. The single-disulfide species

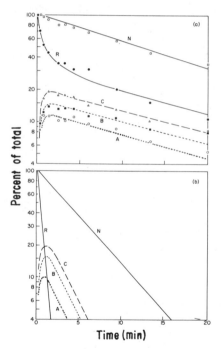

Figure 9. Kinetics of renaturation with addition of DTT_{SH}^{SH}. The relative concentrations of the species R, A, B, C, and N trapped at the indicated times by iodoacetamide were estimated as in Fig. 8 during renaturation with (a) 25 mM DTT_S^S and DTT_{SH}^{SH} added initially to 0.2 mM. The species N is plotted as 100 minus its percent occurrence. (b) A similar plot of renaturation with 25 mM DTT_S^S in the absence of added DTT_{SH}^{SH}, as estimated from six independent experiments. (Reprinted with permission from T.E. Creighton, *J. Mol. Biol.*, in press. Coypright by Academic Press.)

and R, however, disappeared at a common, greatly reduced rate, suggesting that a steady state is approached. The renatured inhibitor (N) appeared at a rate slower than that observed in the absence of added DTT_{SH}^{SH}. The DTT_{SH}^{SH} apparently decreases the net rate of formation of the second disulfide bond, presumably by increasing the rate of the reverse reaction. Virtually every other possible kinetic role for the one-disulfide species would have predicted that their levels of accumulation be decreased or their rates of disappearance increased.

The nature of the disulfide bonds present in the single-disulfide intermediates supports the conclusion that they are in a state of rapid equilibrium. Nearly half the intermediates have disulfide bonds that are not present in the native inhibitor. For these species to be converted to native inhibitor, the disulfide bond must be interchanged at some stage to one of those present in the native inhibitor, and a rapid unimolecular transition most readily accounts for the kinetic behavior of these species.

Because the intermediate with the disulfide bond between Cys-30 and -51 is the major intermediate species and the only one with a detected disulfide bond also present in the native inhibitor, it is likely, but not necessary, that only this species proceeds directly along the further correct folding pathway. Each of the other disulfide bonds found in the single-disulfide intermediates, namely, those between Cys-5 and -30, Cys-30 and -55, and Cys-5 and -51, involve either Cys-30 or Cys-51, so a single intramolecular thiol-disulphide exchange reaction could be sufficient under the renaturation conditions to convert each of these disulfide bonds to one between Cys-30 and -51. It is not surprising then that such species are in rapid equilibrium. The rate-limiting step in their disappearance would be formation of a second disulfide bond in the intermediate with the Cys-30 and -51 disulfide bond.

Formation of such a bond would imply conformational restraints upon the folding process because there is probably a "right" and a "wrong" way to form the Cys-30 to -51 disulfide bond. The conformation of the native inhibitor (17) is somewhat unique in that the NH_2-terminal half of the polypeptide chain passes through the covalent loop defined by the disulfide bond between Cys-30 and -51. This portion of the polypeptide chain probably must be correctly "threaded" before the Cys-30 to -51 disulfide can be incorporated into the native conformation.

Although intermediate states accumulate, the folding reaction of the inhibitor is a thermodynamically cooperative process, all possible intermediate states being unstable relative to R and N. At DTT_S^S to DTT_{SH}^{SH} ratios of between 10 and 10^2, apparently equilibrium mixtures of R and N are obtained in which no other species with one or two disulfide bonds are detected by the above procedures. Consequently, the species observed here are only kinetic intermediates, and the folding of the inhibitor does not appear to be basically inconsistent with the two-state models of reversible folding found for other proteins.

DISULFIDE BONDS AND PROTEIN FOLDING

It was a basic tenet of this investigation that the fundamental principles of protein folding are essentially similar in proteins with and without disulfide bonds in the native conformation and that the interaction between two cysteine residues to form a disulfide bond is analogous to any other attractive side-chain interaction, be it salt-bridge, hydrogen-bond, or hydrophobic interaction. Consequently, which particular disulfide bonds, if any, are formed between otherwise identical cysteine residues will be determined primarily by the proximity of the two cysteine residues, reflecting the conformation of the protein, as determined by the primary and secondary structure and the environment. The disulfide bonds formed during protein folding may then be interpreted as probes of the conformational transitions undergone by the protein, which will reflect the presumed universal principles of protein folding.

IMPLICATIONS FOR PROTEIN FOLDING

A nonrandom spectrum of intermediate conformation is present at the single-disulfide stage of inhibitor folding, and this specificity undoubtedly restricts the additional folding pathways that are kinetically accessible. In particular, over half the molecules have a "correct" disulfide bond, between Cys-30 and -51, so the flux of molecules through the correct pathway must be much greater than that that would occur with a random set of disulfide bonds. The conformations present at the single-disulfide stage of folding may reflect the presence of one type of nucleation center, such as those postulated for this purpose (36–42).

It is important to recognize that the rapid equilibrium between the single disulfide species implies that their relative proportions are principally a product of thermodynamic considerations, not kinetic. That is, the particular single-disulfide intermediates do not accumulate necessarily because they are formed more rapidly than the others, but because they are thermodynamically more stable.

At the present time it is possible only to infer what conformational forces stabilize these single-disulfide intermediates. In the native conformation of the inhibitor (17), there are no atomic contacts between the two segments of residues 24–34 and 42–58, with the exceptions of the Cys-30 to -51 disulfide bond and the proximity of the side chain of Met-52 to the backbone of residues 29 and 30; specific interactions between the residues immediately adjacent to Cys-30 and Cys-51 that occur in the native conformation would not appear to explain the predominance of this disulfide bond in the single-disulfide intermediates.

There is a striking correlation between the occurrence of particular cysteine residues in the disulfide bonds of the monodisulfide intermediates and the presence of the cysteine residues in segments of secondary structure within the native structure, either helices or β sheet. Cys-30 is near the middle of a double-stranded antiparallel β sheet comprising residues 16–36. The other pre-dominant cysteine residue, Cys-51, is in the middle of the only significant α helix (residues 47–56). Cys-5 occurs in a very short helical segment of residues 3–7 that is apparently distorted by the native conformation, and residues 7–10 form one turn of a polyproline-II helix. Cys-55 is at the end of the segment of α helix, and Cys-14 and Cys-38 occur in extended portions of the polypeptide chain.

It seems possible that the proximity of the cysteine residues at the one-disulfide stage of folding is determined primarily by the tendency of segments with helical or β-sheet propensities to aggregate. Interactions between α helices have been proposed for poly(L-alanine) (43–45). Studies by Nagano (46) and by Chou and Fasman (47) suggest that the β structure and helical regions of native inhibitor reflect the propensities of the corresponding amino acid sequences.

Further investigations of the conformations of the trapped intermediates, of the effects on refolding of alteration of the environment and of the inhibitor primary structure (26) and of disulfide bond formation and stability in model compounds should delineate the roles of the various possible conformational forces.

ACKNOWLEDGMENTS

I thank Farbenfabriken Bayer AG and R.C. Sheppard for the purified trypsin inhibitor. This study has benefited from many useful discussions with and much encouragement from D.F. Dyckes and R.C. Sheppard.

REFERENCES

1. R. Lumry, R. Biltonen and J.F. Brandts, *Biopolymers,* **4**, 917 (1966).
2. C. Tanford, *Adv. Protein Chem.,* **23**, 121 (1968).
3. C. Tanford, *Adv. Protein Chem.,* **24**, 1 (1970).
4. W.M. Jackson and J.F. Brandts, *Biochemistry,* **9**, 2294 (1970).
5. F.M. Pohl, *Angew. Chem. Int. Ed.,* **11**, 894 (1972).
6. C. Tanford, K.C. Aune, and I. Ikai, *J. Mol. Biol.,* **73**, 185 (1973).
7. L. Eldjarn and A. Pihl, *J. Biol. Chem.,* **225**, 499 (1957).
8. L. Lumper and H. Zahn, *Adv. Enzymol.,* **27**, 199 (1965).
9. P.C. Jocelyn, *Biochemistry of the SH group,* Academic Press, London, 1972.
10. V.P. Saxena and D.B. Wetlaufer, *Biochemistry,* **9**, 5015 (1970).
11. W.W. Cleland, *Biochemistry,* **3**, 480 (1964).
12. S. Lapanje and J.A. Rupley, *Biochemistry,* **12**, 2370 (1973).
13. B. Kassell and M. Laskowski, Sr., *Biochem. Biophys. Res. Commun.,* **20**, 463 (1965).
14. F.A. Anderer and S. Hornle, *J. Biol. Chem.,* **241**, 1568 (1966).
15. A. Masson and K. Wüthrich, *FEBS Lett.,* **31**, 114 (1973).
16. S. Karplus, G.H. Snyder, and B.D. Sykes, *Biochemistry,* **12**, 1323 (1973).
17. R. Huber, D. Kukla, A. Ruhlmann, O. Epp, and H. Formanek, *Naturwiss.,* **57**, 389 (1970).

18. N.M. Green, *Biochem. J.,* **66**, 407 (1957).
19. J. Putter, *Hoppe-Seyler's Z. Physiol. Chem.,* **348**, 1197 (1967).
20. J.P. Vincent and M. Lazdunski, *Biochemistry,* **11**, 2967 (1972).
21. J.P. Vincent and M. Lazdunski, *Eur. J. Biochem.,* **38**, 365 (1973).
22. A. Ruhlmann, D. Kukla, P. Schwager, K. Bartels, and R. Huber, *J. Mol. Biol.,* **77**, 417 (1973).
23. D. Cechova, V. Svestkova, B. Keil, and F. Sorm, *FEBS Lett.,* **4**, 155 (1969).
24. D.J. Strydom, *Nature New Biol.,* **243**, 88 (1973).
25. H. Takahashi, S. Iwanaga, Y. Hokama, T. Suzuki, and T. Kitagawa, *FEBS Lett.,* **38**, 217 (1974).
26. D.F. Dyckes, T.E. Creighton, and R.C. Sheppard, *Nature* (London), **247**, 202 (1974).
27. T.E. Creighton, *J. Mol. Biol.,* in press.
28. D. Pospisilova, B. Meloun, I. Fric, and F. Sorm, *Coll. Czech. Chem. Commun.,* **32**, 4108 (1967).
29. W.F. Harrington and M. Sela, *Biochim. Biophys. Acta,* **31**, 427 (1959).
30. C.K. Woodward and A. Rosenberg, *Proc. Nat. Acad. Sci. U.S.,* **66**, 1067 (1970).
31. A.M. Tamburro, M. Boccu, and L. Celotti, *Int. J. Protein Res.,* **2**, 157 (1970).
32. B. Kassell, *Methods in Enzymol.,* **19**, 844 (1970).
33. J.R. Brown and B.S. Hartley, *Biochem. J.,* **101**, 214 (1966).
34. B. Kassell and T.W. Wang, in *Proceedings of the International Research Conference on Proteinase Inhibitors,* H. Fritz and H. Tschesche, Ed., Walter de Gruyter, Berlin, 1971, p. 89.
35. R.A. Reisfield, U.J. Lewis, and D.E. Williams, *Nature* (London), **195**, 281 (1962).
36. C. Levinthal, *J. Chim. Phys.,* **65**, 44 (1968).
37. H.F. Epstein, A.N. Schechter, R.F. Chen, and C.B. Anfinsen, *J. Mol. Biol.,* **60**, 499 (1971).
38. P.N. Lewis, F.A. Momany, and H.A. Scheraga, *Proc. Nat. Acad. Sci. U.S.,* **68**, 2293 (1971).
39. H.A. Scheraga, *Chem. Rev.,* **71**, 195 (1971).
40. T.Y. Tsong, R.L. Baldwin, and P. McPhie, *J. Mol. Biol.,* **63**, 453 (1972).
41. C.B. Anfinsen, *Science,* **181**, 223 (1973).
42. D.B. Wetlaufer, *Proc. Nat. Acad. Sci. U.S.,* **70**, 697 (1973).
43. R.T. Ingwall, H.A. Scheraga, N. Lotan, A. Berger, and E. Katchalski, *Biopolymers,* **6**, 331 (1968).
44. D.A.D. Parry and E. Suzuki, *Biopolymers,* **7**, 199 (1969).
45. D.N. Silverman and H.A. Scheraga, *Arch. Biochem. Biophys.,* **153**, 449 (1973).
46. K. Nagano, *J. Mol. Biol.,* **75**, 401 (1973).
47. P.Y. Chou and G.D. Fasman, *Biochemistry,* **13**, 222 (1974).

THE THERMOSTABILITY OF SECONDARY STRUCTURE OF D-GLYCERALDEHYDE-3-PHOSPHATE DEHYDROGENASE FROM Thermus thermophilus CHARACTERIZED BY A NEW METHOD

SHINOBU C. FUJITA and KAZUTOMO IMAHORI, Department of Agricultural Chemistry, Faculty of Agriculture, University of Tokyo, Tokyo, Japan

SYNOPSIS: A new method (the melting-profile method) for characterization of thermostability of proteins is proposed and described. The method essentially consists of continuous observation of a CD signal at a fixed wavelength, as the sample is being heated.

The method was applied to D-glyceraldehyde-3-phosphate dehydrogenase (GAPDH) from an extreme thermophile, *Thermus thermophilus*, and revealed that above around 85°C the enzyme molecule suffers irreversible loss of ordered structure essentially in an all or none manner, with concomitant loss of activity.

A model for the thermal denaturation of the thermophile enzyme was postulated, and the numerical simulation of the melting profile supported the model in all essential aspects studied. Analysis of the model, together with melting-profile experiments of thermophile and rabbit GAPDHs with and without urea, ethanol, or detergents and under various pHs, revealed that the relative thermostability of the thermophile protein need not require any peculiar architecture, yet a 12% difference in one thermodynamic parameter of the homologous proteins could account for a difference in thermostability as large as 40°.

INTRODUCTION

Although there have been various attempts to elucidate the mechanism of thermostability of the enzymes derived from thermophilic organisms (1, 2), knowledge of the mechanism still is largely speculative. Fascinated by the ability of some thermophiles to proliferate at 80°C and persuaded in the belief that careful study of the abnormal always yields clues to an understanding of the normal, we purified D-glyceraldehyde-3-phosphate dehydrogenase (GAPDH) [E.C. 1.2.1.12] from the extreme thermophile *Thermus thermophilus* HB8 and studied its basic properties (3). The thermophile enzyme closely resembles its mesophilic counterparts in molecular weight, tetrameric quaternary structure, overall secondary structure, amino acid composition, and behavior toward sulfhydryl reagents, but it differs appreciably in thermostability, resistance to denaturing agents, and a few other properties.

It should be noted that in our laboratory, as in most others, the activity continuing after heat treatment is taken as the measure of thermostability. The thermostability of activity and that of higher structure, although closely related (the latter is most likely a prerequisite for the former), constitute different aspects of the problem. We have developed a method whereby the thermostability of the higher structure of proteins (and possibly of other biological macromolecules) can be characterized in a convenient manner. Here we report the method as applied to the thermophile GAPDH. The results pertaining to the irreversible thermal denaturation process turned out to be most easily interpreted in terms of a simple model. Numerical simulation of the process supported the model and revealed that relatively slight changes in thermodynamic parameters can give rise to a pronounced difference in the thermostability of the protein.

MATERIALS AND METHODS

The growth of the thermophile *Thermus thermophilus* HB8 (previously classified in the *Flavobacterium* genus), purification and crystallization of GAPDH, estimation of protein concentration, and assay of GAPDH activity were conducted as described in (3). Crystalline holoenzyme preparation, which gives a single band on disc gel electrophoresis, was used throughout. The enzyme preparation had the specific activity of 48 international units per mg protein in the presence of 90 mM NH_4Cl at 25°C.

Reagents. Tris(hydroxymethyl)aminometane (Tris) was purchased from Sigma Chemical Co., and Brij 35 was a product of Atlas Powder Co. Sodium dodecylsulfate (SDS) was recrystallized from Duponol C (4), and dodecyltrimethylammonium chloride was a kind gift from Kao-Atlas Co. Ethyl alcohol, urea, and inorganic salts, all of reagent grade, were obtained from Wako Pure Chemical Industries. Rabbit muscle GAPDH was purchased from Boehringer Mannheim. Redistilled water was used throughout.

Instruments. Circular dichroic studies were made through a Jasco Model J-20 automatic recording spectropolarimeter with a temperature-controlling system employing a Haake bath. The temperature of the sample was monitored through a Takara Type SPD-1D thermistor. A Gilford Model 240 spectrophotometer equipped with a Hitachi Model QPD_{73} recorder was used to assay GAPDH activity. The numerical calculations in the simulation analysis were aided by a MELCOM 7700 computer.

The melting-profile method. The method essentially consists of continuous observation of a CD signal at a fixed wavelength while the sample is being heated at a constant rate. Care was taken to ensure temperature homogeneity within the capped 3 mm quartz sample cell. The temperature of the sample solution was monitored through a thermistor attached through the cap of the cell. A constant heating rate was maintained by raising the temperature of the ethylene glycol circulating through the cell-holder jacket with a Haake constant temperature bath. The CD signal was automatically recorded as a function of time. Such a plot is hereafter referred to as a melting profile and the method as the melting-profile method.

Let $[\theta]/[\theta_0]$ denote the fractional signal intensity, that is, the intensity at any temperature divided by the intensity at room temperature. Then a convenient measure of thermostability of the secondary protein structure to use in comparing different denaturing conditions or different proteins is the temperature $T_{1/2}$ at which half the change in $[\theta]/[\theta_0]$ has occurred. Under carefully controlled experimental conditions the melting profiles were usually reproducible within ±0.5° in terms of $T_{1/2}$. In general, $T_{1/2}$ was expected to shift at different heating rates, but contrary to this expectation, $T_{1/2}$ was found to be quite insensitive to minor changes in the heating rate (see below). The heating rate of 5° per min was adopted throughout the work, unless otherwise stated. All melting profiles reported below were obtained at 219 nm, at which the CD signal of the *T. therm.* GAPDH was most intense.

The method is based on the reasonable assumption that in the temperature range of present interest, the CD spectrum obtained is a function of the chromophore's geometry only (5). Strictly speaking, the data should be corrected for the sample-cell deformation and the volume change of the sample solution accompanying the temperature change. It was considered, however, that such corrections were not warranted in view of the current precision of the method.

The merit of the method lies in the ease with which the thermostability of a protein can be characterized and compared and the direct manner in which the change in the secondary structure can be monitored. The present method is admittedly in need of further refinement, and improvement of the apparatus is under way.

RESULTS

It has been already reported (3) that the CD spectrum exhibited by *T. therm.* GAPDH in the 210–250 nm region has a single minimum at 219 nm with $[\theta]$ = –13,000 [Fig. 1(a)]. Similar profiles indicative of abundant β structure are common to all GAPDHs so far examined (6–8). When the temperature is raised to and held at a subdenaturing temperature (e.g., 75°C), the CD band at 219 nm becomes about 10% less deep, but no qualitative change occurs [Fig. 1(b)]. If the sample is cooled to the room temperature, the 219 nm trough deepens to coincide with the original spectrum. Because no loss of activity occurs at subdenaturing temperatures (3), it may be concluded that the transition in this temperature range is completely reversible and involves relatively minor rearrangement of the secondary structure.

If the temperature is rapidly brought to 86°C and held there, the GAPDH slowly and irreversibly loses the secondary structure. When an irreversibly denaturing sample is scanned along wavelength, the trace gives curves like C in Fig. 1. Curve C is not a true spectrum because the scan was made while the spectrum was changing with time. With this in mind, the curve was interpreted to mean that no significant accumulation of spectrally distinguishable intermediate(s) occurs during this irreversible process. This interpretation, in turn, suggests that the protein molecule unfolds in an all or none manner.

In order to gain more information about the denaturation process, we subjected the thermophile GAPDH to melting-profile experiments. First, the sample was heated at varying rates of heating (Fig. 2). It is readily seen that the

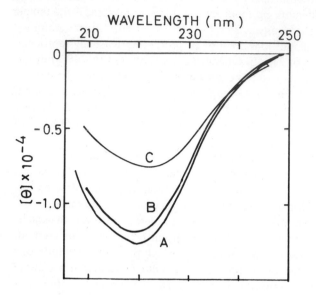

Figure 1. The CD spectrum of *T. therm*. GAPDH in the 210–250 nm region at various temperatures. Curve A, at 25°C; B, 77°C; C, 86°C. Curve C is a scan of the changing spectrum of the irreversibly denaturing sample. The scan was made from right to left at the rate of 5 nm/min. Protein concentration = 0.12 mg/ml in 25 *mM* Tris-HCl buffer (pH 7.5).

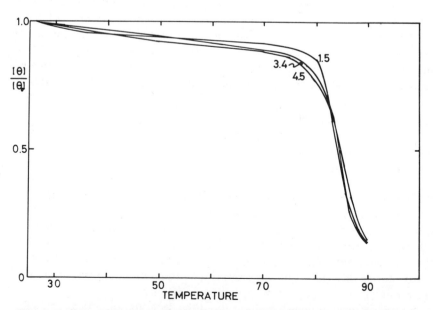

Figure 2. Effect of various heating rates on the melting profile of *T. therm*. GAPDH in 0.2 *M* Tris-HCl buffer (pH 9.3). Rates of temperature elevation in degrees per minute are indicated in the figure. Protein concentration = 0.11 mg/ml.

profile is so insensitive to the variation in the heating rate that the effect of a 10% fluctuation in the latter is quite negligible.

In the next experiment, heating was interrupted at various points along the profile and the sample was rapidly cooled to allow us to examine the reversibility of changes in $[\theta]_{219}$ (Fig. 3). When heating was interrupted before a temperature of 80°C was attained and the sample was cooled to room temperature, $[\theta]/[\theta_0]$ essentially reverted to 1. If heating was interrupted after the sharp transition had set in, however, $[\theta]/[\theta_0]$ remained what it was at the time heating was interrupted. In both cases, the recovered sample yielded the same $T_{1/2}$ as the native system in a second melting-profile experiment. Thus the whole process consisted of two phases, one completely reversible and the other irreversible.

In another experiment, the protein concentration was varied between 0.01 and 0.07 mg/ml. No significant effect on $T_{1/2}$ was observed. The small sample did not permit repetition of the experiment at higher protein concentrations.

In the fourth experiment, a mixture of rabbit and thermophile GAPDHs was heated (Fig. 4). The melting profile of the mixture exhibits two phases of denaturation with $T_{1/2}$ values of 52.5°C and 87.0°C corresponding, respectively, to

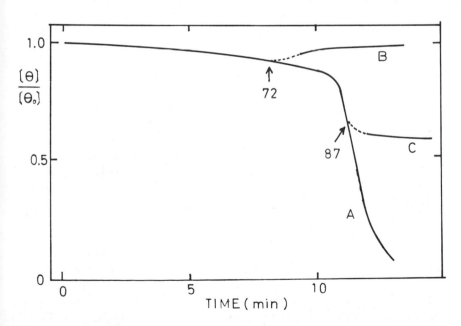

Figure 3. Effect of heating interruption and cooling on the melting profile of *T. therm.* GAPDH in 50 *mM* Tris-HCl buffer (pH 8.5). Heating was interrupted at the point indicated by an arrow at the temperature indicated by the number under the arrow. Lines B and C represent the interrupted profiles; Line A represents an uninterrupted one. Lines B and C were obtained after the sample had been quickly cooled to room temperature. Heating rate = 5° per min; protein concentration = 0.15 mg/ml.

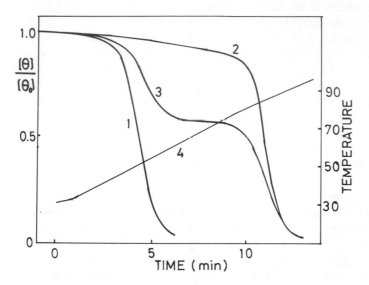

Figure 4. Melting profile of the mixture of rabbit and thermophile GAPDHs in 50 mM Tris-HCl buffer (pH 8.5). (1) Rabbit GAPDH alone (protein concentration = 0.04 mg/ml); $T_{1/2}$ = 52.0°C. (2) Thermophile GAPDH alone (0.05 mg/ml); $T_{1/2}$ = 88.0°C. (3) Rabbit and thermophile GAPDHs (0.04 and 0.05 mg/ml, respectively); 52.5 and 87.0°C. Line 4 shows temperature elevation.

52.0°C for the pure rabbit-enzyme system and 88.0°C for the pure thermophile-enzyme system. The sharpness of the transitions in both proteins was not affected by the presence of another protein. In other words, each species of GAPDH seems to denature independently, neither stabilizing nor destabilizing the other.

In order to find the relation between the stability of the higher structure and that of the activity, we subjected a sample to repeated heating-interruption experiments. After each interruption, the sample was cooled to room temperature and the remaining ellipticity at 219 nm and the ongoing activity were determined. As is readily seen in Fig. 5, a linear relationship exists between the two quantities.

In other melting-profile experiments, effects of buffer concentration, salts, and pH were examined. It was found that the heat stability is not sensitive to changes in salt concentration up to 0.1 M or to variations in pH between 6.4 and 9.2 or to the presence of various inorganic ions. Even phosphate ion, which is a substrate of GAPDH, does not stabilize the enzymes by more than 3 to 5° in $T_{1/2}$. On the other hand, magnesium ion at 0.1 M was destabilizing.

Urea, ethanol, and some detergents have comparable effects on the thermophile and rabbit enzymes (e.g., Fig. 6). As the perturbant concentration was raised, its effect became apparent at lower temperatures and the transition became less sharp. Another notable point is that the effects of perturbing conditions on the

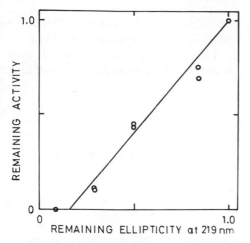

Figure 5. The relation between the continuing activity and the remaining ellipticity of partially denatured systems of *T. therm.* GAPDH. The sample in 50 *mM* Tris-HCl buffer (pH 8.5) was subjected to repeated heating-interruption experiments in the irreversibly denaturing temperature region. After each interruption, two 10 μl aliquots were withdrawn for assay of enzymatic activity. These samples were measured at 25°C and are expressed as fractions of the activity of the original sample. Protein concentration = 75 μg/ml.

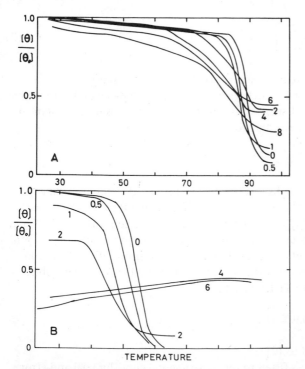

Figure 6. Melting profiles of (A) thermophile and (B) rabbit GAPDHs in urea. The numbers by the profiles refer to the molarity of urea in 50 *mM* Tris-HCl buffer (pH 8.5). Protein concentration = 0.06 mg/ml (A) and 0.09 mg/ml (B).

thermophile enzyme were qualitatively similar to the effects on the rabbit enzyme, except for urea at concentrations exceeding $2\,M$. Thus nonionic Brij 35 at 1% was almost without effect on both enzymes, yet the same concentration of cationic dodecyltrimethylammonium chloride or anionic SDS greatly altered the melting profiles of both proteins.

THEORETICAL CONSIDERATIONS

Based on the results described above, we suggest that the *T. therm.* GAPDH denatures according to the model

$$A \underset{}{\overset{K(T)}{\rightleftharpoons}} B \overset{k(T)}{\rightarrow} D$$

Here A is predominant over B at room temperature: B is stabilized at higher temperatures and in rapid equilibrium with A, but unfolds in an all or none manner to yield D, the irreversibly denatured form. The temperature dependencies of the equilibrium constant $K(T)$ and of the rate constant $k(T)$ are assumed to follow the relations

$$K(T) = \exp\left[CK - \frac{H}{RT}\right] \tag{1}$$

$$k(T) = \exp\left[CV - \frac{E}{RT}\right] \tag{2}$$

where $CK, H, CV,$ and E are parameters and R is the gas constant.

The melting profile of the model system was simulated by assuming that time elapses in discrete steps of 1 sec, during which some molecules in the B state are transferred to the D state according to

$$-\Delta B = k(T)[B]$$

and then the molecules are redistributed between A and B states according to $K(T)$. ([B] here refers to the fraction of molecules in the B state.) Because the minimum value of $[\theta]/[\theta_0]$ in the *reversibly* unfolded region was about 0.9 and its value in the *irreversibly* unfolded region was near zero, it would be appropriate to assume that the ellipticity contributed by the B form is 90% of the contribution of the A form and that the contribution of the D form is negligible. Then the assumed quantity THETA, defined as

$$\text{THETA} = [A] + 0.9 \times [B]$$

should closely correspond to $[\theta]/[\theta_0]$.

The calculated curve was fitted to the experimental curve by trial-and-error adjustments of the four parameters (e.g., Fig. 7). When similar calculations were made with discrete time intervals of 0.1 instead of 1 sec, the difference in terms

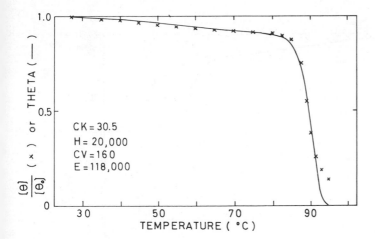

Figure 7. Simulated melting profile compared with the experimental values. The curve represents the melting profile computed with the values of parameters given. Crosses were taken from the experimental profile to illustrate fit.

of $T_{1/2}$ was only 0.15°, which is well within the current level of experimental accuracy of the melting-profile method and justifies the choice of 1 sec in the present analysis.

The simulated melting profile was relatively indifferent to changes in the heating rate. For example, an increase of 10% in the heating rate is reflected by an increase of only 0.21° in $T_{1/2}$. Therefore, a heating rate of 5° per min was adopted throughout the analysis.

The model system was found to mimic the experimental system in its response to heating interruption and cooling (Fig. 8). Thus, if interrupted during the reversible phase, THETA essentially reverts to 1, but if interrupted during the irreversible phase, it remains where it was when heating was stopped. Such close correspondence between the experimental and the model systems, together with the built-in independence of the model system from protein concentration, lead us to consider the model system to be sufficiently representative of the experimental system.

Now the question is: How do the parameters dictate the thermostability of the model system? We are more concerned with the irreversible process $B \rightarrow D$ than with the equilibrium $A \rightleftharpoons B$, so let us examine CV and E. (CK and H are analogous to CV and E, respectively.) The steepness of the transition is mainly dependent upon E. In fact, as a rearrangement of Eq. (2) readily reveals, if the transition is plotted against reciprocal temperature, a change in CV simply causes translational displacement of the melting profile along the $1/T$ axis without affecting the sharpness of the transition (Fig. 9).

What is more remarkable is the fact that a relatively small change of CV from 180 to 160 with E constant boosts $T_{1/2}$ from about 50°C to about 90°C. If E alone increases from 105,000 to 118,000, then $T_{1/2}$ increases from 50°C to 90°C.

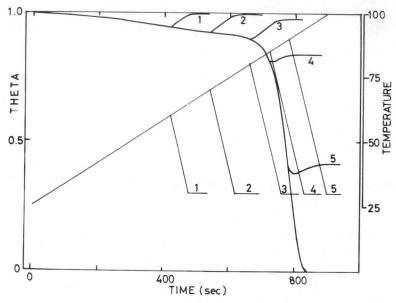

Figure 8. Behavior of the melting profile of the model system upon heating interruption and cooling. The thin lines indicate the temperature change; the thick curves show accompanying changes in THETA.

A 6% decrease in CV coupled with a 6% increase in E gives the same result. Thus, relatively minor changes in the parameters apparently enhance the thermostability of proteins.

What are the parameters CV and E in physical terms? The rate constant $k(T)$ is expressed in terms of the absolute reaction rate theory as

$$k(T) = \kappa \frac{kT}{h} \exp\left[\frac{\Delta S^{\ddagger}}{R} - \frac{\Delta H^{\ddagger}}{RT}\right] \tag{3}$$

with conventional notations of physical quantities (9). Assuming T in the preexponential factor is constant within a limited range of irreversible transition, a comparison of Eqs. (2) and (3) shows that E is proportional to $\Delta H^{\ddagger}$, the enthalpy change of the system in going from the initial to the activated state, and CV is a quantity $\Delta S^{\ddagger}/R + $ const., where $\Delta S^{\ddagger}$ is the corresponding entropy change.

DISCUSSION

After the thermophile GAPDH has been partially denatured in heating-interruption-and-cooling experiments, the system retains the essential features of the original CD pattern and exhibits the same $T_{1/2}$ as the native system. This fact is most simply explained by considering that the partially denatured system contains no appreciable population of partially denatured molecules, but consists of

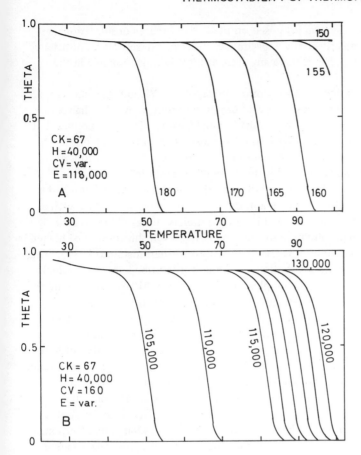

Figure 9. Effect of changing parameters (A) CV and (B) E on the melting profile of the model system. The numbers by the profiles are the values of parameter (A) CV and (B) E. The values of other parameters are given in each figure. In the present analysis (and also in Fig. 7), the equilibrium A ⇌ B has been intentionally displaced in favor of B to highlight the effects of the parameters controlling the irreversible denaturation.

fully native and fully denatured molecules. This point is borne out by the finding that the ongoing activity after heat treatment is linearly related to the remaining ellipticity (Fig. 5). Thus each molecule denatures in an all or none manner; in other words, the forces stabilizing the protein molecule succumb to heat only in a highly cooperative way.

Other experiments have shown that the molecules denature independently, in accordance with our earlier finding (3) that heating at constant temperature irreversibly destroys the activity according to first-order kinetics. The coincidence of the process of irreversible destruction of the secondary structure and activity is attested by the results shown in Fig. 5, as well as by the fact that the two thermodynamic parameters derived from an Arrhenius plot of inactivation rate constants are in good agreement with CV and E used in the simulation experiment illustrated in Fig. 7.

The simple model system proposed on the basis of these considerations was found to behave similarly to the experimental system. No serious inconsistency has been encountered, nor is there anything intuitively objectionable in the model.

It has been shown that relatively minor changes in $\Delta H^{\ddagger}$ and $\Delta S^{\ddagger}$ can cause dramatic shifts in the thermostability of protein molecules. A 12% change in $\Delta H^{\ddagger}$, for example, can cause $T_{1/2}$ to shift as much as 40°. This fact implies that for a mesophilic protein to evolve into a thermostable protein in adaptation to a hotter environment, no unusual devices or gross architectural alterations are necessary. In fact, high solubility in ammonium sulfate solution and abundant β structure are common to all mesophilic and thermophilic GAPDHs studied so far. Thus, there is nothing surprising in the fact that no special feature such as covalent cross linkages (10), sugar (11) or lipid moiety, unusual amino acid residue, increased hydrophobic stabilization (12), or even the absence of higher structure (13) is required in order for a protein to be thermostable (1, 14). This view is supported by the perturbation experiments (described above) in which the thermostabilities of the thermophile and the rabbit GAPDHs were found to be similarly affected by urea, ethanol, and detergents. It should be emphasized that the perturbed melting profiles relative to the unperturbed control curves are essentially the same for the thermophile and the rabbit proteins.

These facts strongly support the view that the two proteins are stabilized by essentially the same kind of forces operating in similar ways, the small difference being quantitative. These considerations strengthen our earlier contention (3) that the relative thermostability of thermophilic proteins arises from subtle differences in the architecture of the molecule, such as efficient hydrophobic packing, properly disposed hydrogen bonds and favorable charge distribution.

As reported earlier (3), there is evidence that the thermophile enzyme is as active at its physiological temperatures of around 70°C as the mesophilic enzymes are at 30°C, although activity assay is not possible above 50°C. So far, no instance of functional impairment in thermophile enzymes has been reported. *T. therm.* phosphofructokinase, for example, exhibits more pronounced allosteric kinetics at 75°C than at 30°C (15). Under optimal conditions (16), at about 70°C, *T. therm.* has been observed to have a doubling time of 20 min. Thus, thermophile enzymes, in general, do not seem to suffer any loss of catalytic or regulatory efficiency in exchange for thermostability, in contrast to what was once suspected (10). If that is the case, the question already raised by Oshima (17) is: Why are enzymes in mesophilic organisms *thermolabile*? It has been taken to be axiomatic that proteins are heat labile. Now it is time to reexamine this concept in light of knowledge gained through the study of thermophile enzymes, "the abnormals."

ACKNOWLEDGMENT

We are deeply grateful to Tairo Oshima for his help and kind collaboration in the culture of the thermophile and extraction of the enzyme.

REFERENCES

1. R. Singleton, Jr., and R.E. Amelunxen, *Bacteriol. Rev.*, **37**, 320 (1973).
2. K. Mizusawa and F. Yoshida, *J. Biol. Chem.*, **248**, 4417 (1973).
    E. Stellwagen, M.M. Cronlund, and L.D. Barnes, *Biochemistry*, **12**, 1552 (1973).
3. S.C. Fujita, T. Oshima, and K. Imahori, submitted to *Eur. J. Biochem.*
4. A.M. Crestfield, K.C. Smith, and F.W. Allen, *J. Biol. Chem.*, **216**, 185 (1955).
5. A. Moscowitz, K.M. Wellman, and C. Djerassi, *Proc. Nat. Acad. Sci. U.S.*, **50**, 799 (1963).
6. D.W. Darnall and T.D. Barela, *Biochim. Biophys. Acta*, **236**, 593 (1971).
7. K. Suzuki and K. Imahori, *J. Biochem.*, **73**, 97 (1973).
8. K. Suzuki and K. Imahori, *J. Biochem.*, **74**, 955 (1973).
9. S. Glasstone, K.J. Laidler, and H. Eyring, *The Theory of Rate Processes*, McGraw-Hill, New York, 1941.
10. T.D. Brock, *Science*, **158**, 1012 (1967).
11. R.E. Amelunxen, *Biochim. Biophys. Acta*, **139**, 24 (1967).
12. Y. Ohta, Y. Ogura, and A. Wada, *J. Biol. Chem.*, **241**, 5919 (1966).
13. G.B. Manning, L.L. Campbell, and R.J. Foster, *J. Biol. Chem.*, **236**, 2958 (1961).
14. B.W. Matthews, J.N. Jansonius, P.M. Colman, B.P. Schoenborn, and D. Dupourque, *Nat. New Biol.*, **238**, 37 (1972).
15. M. Yoshida, T. Oshima, and K. Imahori, *Biochem. Biophys. Res. Commun.*, **43**, 36 (1971). M. Yoshida, *Biochemistry*, **11**, 1087 (1972).
16. T. Oshima and K. Imahori, *J. Gen. Appl. Microbiol.*, **17**, 513 (1971).
17. T. Oshima, *Protein, Nucleic Acid, and Enzyme*, **18**, 454 (1973). In Japanese.

RECURRENCE OF β TURNS IN REPEAT PEPTIDES OF ELASTIN: THE HEXAPEPTIDE ALA-PRO-GLY-VAL-GLY-VAL SEQUENCES AND DERIVATIVES

D.W. URRY and **T. OHNISHI**, Laboratory of Molecular Biophysics and the Cardiovascular Research and Training Center, University of Alabama in Birmingham, Birmingham, Ala. 35294

SYNOPSIS: Proton magnetic resonance studies are reported for the repeating hexapeptide of elastin. Temperature dependence and solvent-mixture dependence of chemical shift were used to delineate solvent-shielded peptide NH protons. The hexapeptide sequence $(Ala-Pro-Gly-Val-Gly-Val)_n$ was studied with $n = 1, 2,$ and 3. The results are interpreted in terms of a β turn involving the Val^4 NH and the Ala^1 carbonyl. In this conformation, Pro^2 and Gly^3 are at the corners of a type-2 β turn, that is, in positions $i + 1$ and $i + 2$. This β turn with Pro^2 and Gly^3 at the corners occurs also in repeating pentapeptides and tetrapeptides of elastin.

The cyclic permutation of Boc-Pro-Gly-Val-Gly-Val-Ala-OMe, which also was studied, gave an interesting *cis-trans* isomerization in which the two states were nearly equally populated. The *cis* isomer resulted in loss of solvent shielding of the Val^4 NH. Coalescence of the peaks near 50°C made it possible to see the averaged and resolved states. The free energy of activation for the *cis-trans* isomerization was calculated to be approximately 16 kcal/mole.

The problem of quantification of the probability of β-turn occurrence is discussed in terms of the standard mole-fraction analysis using the temperature coefficients of chemical shift and the chemical shift change on going from methanol to trifluoroethanol, with gramicidin S as the model system.

INTRODUCTION

Gray and Sandberg and their colleagues have recently reported repeating peptide sequences in tropoelastin (1, 2), the precursor protein of the elastic fiber of ligaments, arterial walls, skin and lungs. They found a tetrapeptide (Gly-Gly-Val-Pro), a pentapeptide (Pro-Gly-Val-Gly-Val), and a hexapeptide (Pro-Gly-Val-Gly-Val-Ala). The tetrapeptide repeats four times in a single sequence, except that Gly^6 is replaced by an alanine and residue 16 is phenylalanine (2). The pentapeptide repeats six times with only one variant; the second glycine is replaced by serine in the fourth repeat (2). The hexapeptide repeats five times, again with a single exception: the first valine is replaced by isoleucine (2). To the extent of the positions of the glycine residues, these sequences had been anticipated in a proposed mechanism for the initiation of calcification by elastin in the arterial wall (3). In this context, it was proposed further that elastin contains β turns and other conformations, such as the π_{LD} helices [now called β helices (4)], that is, conformational features that uniquely require glycines (or D-amino acid residues) in specific sequence relationships to the L-amino acids.

The repeating peptides of elastin have been synthesized in this laboratory and their conformations and interactions are under study. Proton magnetic resonance studies have been utilized to study the elastin peptides in a manner similar to the one we employed in studying the conformations of valinomycin (5-8), gramicidin S (5, 6, 9), gramicidin A (10, 11), oxytocin (12, 13), stendomycin (14), and telomycin (15, 16). The peptide NH protons are delineated by their chemical-shift dependences on temperature and on the methanol-trifluoroethanol solvent mixture, as well as by their rates of H-D exchange. This information, in combination with magnitude of chemical shift and α-CH-NH coupling constants, has been most useful in identifying 10-atom hydrogen-bonded rings, commonly referred to as β turns.

Such studies on the repeat pentamer (Val-Pro-Gly-Val-Gly)$_n$, where $n = 1, 2$, and 3, have been reported and interpreted to indicate a β turn involving the carbonyl of Val[1] and the NH of Val[4] (17, 18). Pro[2] and Gly[3] are at the corners of the β turn at positions $i + 1$ and $i + 2$ [see Fig. 1 (a)]. Study of the tetramer (Val-Pro-Gly-Gly)$_n$ indicates a similar β turn for this repeating peptide (19). The β turn of the tetramer has a hydrogen bond between the carbonyl of Val[1] and the peptide NH of Gly[4] (20). Pro[2] and Gly[3] are again at the corners of the β turn in positions $i + 1$ and $i + 2$ [Fig. 1 (b)].

This paper is a report on proton magnetic resonance studies of the hexapeptide Ala-Pro-Gly-Val-Gly-Val, its oligomers $n = 2$ and 3, and the cyclic permutation Pro-Gly-Val-Gly-Val-Ala. The hexapeptide studies provide an appropriate opportunity to reiterate the meaning of solution proton magnetic resonance studies indicating that β turns are significant conformational features. To say that a β turn is a significant conformational feature in solution does not mean that at a given moment all molecules are locked in the β turn, but rather that a sufficient number of molecules are in the conformation to substantially shield from the solvent the peptide NH in the 10-atom hydrogen-bonded ring (21).

The hexapeptide Boc-Pro-Gly-Val-Gly-Val-Ala-OMe is easily studied because the barrier for interconversion from a β-turn conformation to another conformation is of such a magnitude that at room temperature two conformational states are readily resolved and characterizable, but at 60°C the interconversion is sufficiently fast on the NMR time scale to be seen as an average of the states. The hexapeptide Boc-Ala-Pro-Gly-Val-Gly-Val-OMe, on the other hand, gives averaged properties over the same temperature range.

MATERIALS AND METHODS

Proton magnetic resonance studies were carried out on a Varian 220 MHz spectrometer equipped with an SS-100 computer system and with a tracking

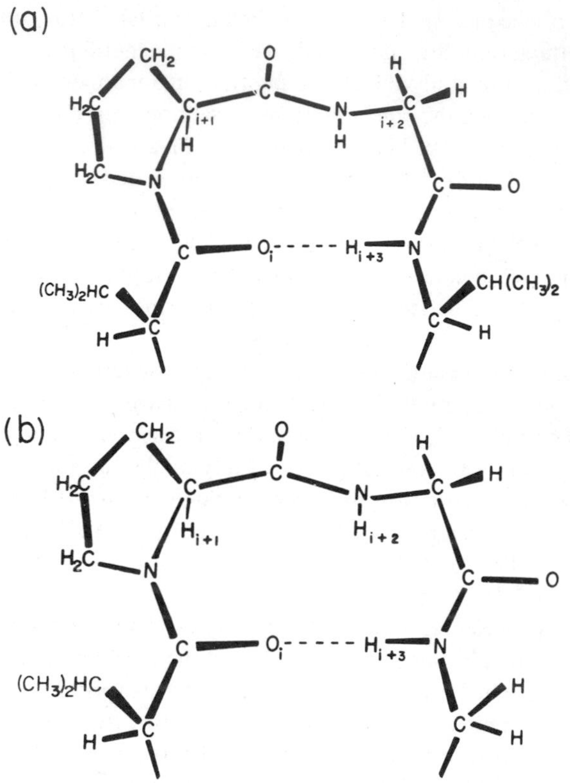

Figure 1 (a). The β turn proposed for the repeating pentapeptide Val-Pro-Gly-Val-Gly of elastin. Residues i and $i + 3$ are valines. Pro is residue $i + 1$ and Gly is $i + 2$. [Reprinted with permission from Urry et al., Biochemistry, **13**, 609 (1974). Copyright by the American Chemical Society.] (b) The β turn proposed for the repeating tetrapeptide Val-Pro-Gly-Gly of elastin. As in the pentapeptide, Pro[2] and Gly[3] are residues $i + 1$ and $i + 2$, and residue i is Val[1]. In position $i + 3$ is Gly[4]. [From Urry and Ohnishi (18).]

frequency-sweep decoupling accessory. The internal reference was tetramethylsilane. Probe temperature was measured to within $\pm 2°C$ using the chemical shifts of methanol or ethylene glycol samples. Spectra were calibrated, particularly on expanded scales, with modulation side bands.

Dimethylsulfoxide-d_6 (Me$_2$SO-d_6, 99.5% and 99.8%) was obtained from Diaprep Corp., Atlanta, Ga., and from Columbia Organic Chemical Co., Columbia, S.C. The source of 2,2,2-trifluoroethanol was Halocarbon Products, Hackensack, N.J. The commercial product was redistilled in glass (b.p. 72°C, 738 mm

Hg) in a Duflon column containing glass beads (lower 3/4) and glass helices (upper 1/4). This column was set for a 10% take. Sodium bicarbonate was added to neutralize traces of acid. A pot residue of 20–30% and the first 10–15% of distillate were discarded.

Spectroquality methanol was obtained from Matheson, Coleman, and Bell.

Characterization of synthetic hexapeptides. Thin-layer chromatography (TLC) was performed on silica gel G, with the following solvent systems: chloroform:methanol:acetic acid (95:15:3, v/v) and n-butanol:acetic acid: pyridine:water (30:6:20:24, v/v).

1. Boc-Ala-Pro-Gly-Val-Gly-Val-OMe: an amorphous powder; Rf (a) 0.68; m.p., 173–185°C. Anal. Calcd. for $C_{28}H_{48}N_6O_9$: C, 54.88%; H, 7.89%; N, 13.71%. Found: C, 54.66%; H, 7.82%; N, 13.72%.

2. HCO-Ala-Pro-Gly-Val-Gly-Val-OMe: An amorphous powder; Rf (b) 0.74. Amino acid analysis: Pro, 1.14; Gly, 2.00; Ala, 1.02; Val, 2.00.

3. HCO-(Ala-Pro-Gly-Val-Gly-Val)$_2$-OMe: Rf (b) 0.70; m.p., 210–212°C. Anal. Calcd. for $C_{46}H_{76}N_{12}O_{14}$ ·2CH$_3$OH: C, 53.12%; H, 7.78%; N, 15.48%. Found: C, 52.94%; H, 7.33%; N, 15.13%. Amino acid analysis: Pro, 1.00; Gly, 2.00; Ala, 1.06; Val, 2.04.

4. HCO-(Ala-Pro-Gly-Val-Gly-Val)$_3$-OMe: Rf (b) 0.67; m.p., 196–199°C. Anal. Calcd. for $C_{68}H_{112}N_{18}O_{20}$·AcOEt: C, 54.38%; H, 7.60%; N, 15.87%. Found: C, 54.74%; H, 7.81%; N, 15.70%. Amino acid analysis: Pro, 1.00; Gly, 1.91; Ala, 1.02; Val, 2.04.

5. HCO-Val-Ala-Pro-Gly-Val-Gly-OMe: Rf (a) 0.2, Rf (b) 0.6; m.p., 253–255°C. Anal. Calcd. for $C_{24}H_{40}N_6O_8$: C, 53.32%; H, 7.46%; N, 15.55%. Found: C, 53.31%; H, 7.35%; N, 15.77%. Amino acid analysis: Pro, 1.05, Gly, 2.00; Ala, 1.09; Val, 2.06.

6. Boc-Pro-Gly-Val-Gly-Val-Ala-OMe: Rf (b) 0.75; m.p., 276–279°C (decomp.). Anal. Calcd. for $C_{28}H_{48}N_6O_9$: C, 54.88%; H, 7.89%; N, 13.72%. Found: C, 54.89%; H, 7.65%; N, 13.46%. Amino acid analysis: Pro, 1.12; Gly, 2.00; Ala, 1.04; Val, 2.16.

RESULTS

The complete proton magnetic resonance spectrum is given in Fig. 2. In general, assignments were achieved by decoupling experiments. The proline assignments were assisted by the studies of McDonald and Phillips (22), and delineation of the valine and glycine resonances was achieved as outlined below.

Comparison of the chemical shifts and their temperature dependence in Me$_2$SO-d_6 of Val-Pro-Gly-Val-Gly with those of Ala-Pro-Gly-Val-Gly-Val allow assignment of the Val^4 NH. In Me$_2$SO-d_6 Val^4 of the pentapeptide exhibits a temperature coefficient in Hz for a 10° increment of –9.68 and an extrapolated 0°C intercept of 7.77 ppm for Boc-Val-Pro-Gly-Val-Gly-OMe, of

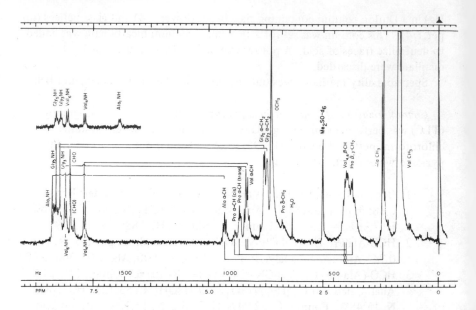

Figure 2. Complete proton magnetic resonance spectrum at 220 MHz of HCO-Ala-Pro-Gly-Val-Gly-Val-OMe in Me_2SO-d_6 at 19°C. Insert is the Boc derivative showing the upfield position of the Ala[1] NH resonance. The downfield peak beside the Pro[2] α-CH resonance near 950 Hz and a satellite formyl, CHO, peak near 1750 Hz indicated that 20% of the molecules are *cis* isomer. The assignments were achieved by decoupling and derivatives. The glycine assignments are tentative.

7.78 ppm for HCO-(Val-Pro-Gly-Val-Gly)$_2$-OMe and of 7.81 ppm for HCO-(Val-Pro-Gly-Val-Gly)$_3$-OMe. In the hexapeptide, one Val NH exhibits a 0°C intercept of 7.75 ppm in the Boc-Ala-Pro-Gly-Val-Gly-Val-OMe, of 7.75 ppm in the HCO-Ala-Pro-Gly-Val-Gly-Val-OMe, of 7.79 ppm in the HCO-(Ala-Pro-Gly-Val-Gly-Val)$_2$-OMe, and of 7.77 ppm in the HCO-(Ala-Pro-Gly-Val-Gly-Val)$_3$-OMe derivatives (see Table 1). The slopes are –9.0, –9.2, –9.7, and –8.8, respectively, for a 10° increment in temperature. The second Val NH exhibits a steeper slope and a larger chemical shift. For Boc-Ala-Pro-Gly-Val-Gly-Val-OMe, the temperature coefficient is –13.4 and the 0°C intercept is 8.18 ppm; similar values are obtained for a single Val NH proton in HCO-(Ala-Pro-Gly-Val-Gly-Val)$_2$-OMe and in HCO-(Ala-Pro-Gly-Val-Gly-Val)$_3$-OMe (Table 1). Thus, Val[4] NH must be the resonance having a 0°C intercept at 7.75 ppm in the hexapeptide and the Val[6] NH resonance, the terminal residue in the hexapeptide, must have 0°C intercepts at 8.18 ppm and 8.19 ppm. The residue with a similar slope and intercept in the dodecamer and in the octadecamer is assigned to Val[12] NH and Val[18] NH, respectively. By elimination, the Val[6] NH of the dodecamer and the Val[6] NH and Val[12] NH of the octadecamer are the resonances with intercepts at 7.86 ppm and 7.87 ppm, respectively.

Table 1. Temperature dependence of peptide proton chemical shifts in Me$_2$SO-d_6

Peptide Residue[b]	Boc-APGVGV-OMe[a]		HCO-APGVGV-OMe[a]		HCO-(APGVGV)$_2$-OMe[a]		HCO-(APGVGV)$_3$-OMe[a]	
	Slope (Hz/10°C)	0° Intercept	Slope (Hz/10°C)	0° Intercept	Slope (Hz/10°C)	0° Intercept	Slope (Hz/10°C)	0° Intercept
Ala^1	-17.6[c]	7.10[c]	-12.0	8.45	-12.8	8.48	-12.5	8.48
Ala^7					-13.6	8.37	-13.6	8.36
Ala^{13}							-13.6	8.36
Gly^3	-8.5	8.25	-8.6	8.27	-8.5	8.31		
Gly^9					10.1	8.29		
Gly^{15}								
Val^4	-9.2	7.75	-9.0	7.75	-9.7	7.79	-8.8	7.77
Val^{10}					-9.7	7.79	-8.8	7.77
Val^{16}							-8.8	7.77
Gly^5	-11.3	8.38	-11.5	8.37	-11.9	8.42		
Gly^{11}					-10.3	8.36		
Gly^{17}								
Val^6	-13.4[d]	8.18[d]	-13.0[d]	8.19[d]	-11.7	7.86	-12.3	7.87
Val^{12}					-13.2[d]	8.19[d]	-12.3	7.87
Val^{18}							-12.5[d]	8.18[d]

[a] APGVGV is an abbreviation of Ala-Pro-Gly-Val-Gly-Val.
[b] Gly^3 and Gly^5 assignments are tentative.
[c] This residue is displaced because of the Boc derivative.
[d] This residue is the carboxyl terminus.

Delineation of the glycine NH resonances is not assisted by a terminal position in Ala-Pro-Gly-Val-Gly-Val, but tentative assignments can be achieved by comparison with positions of glycine NH resonances in HCO-Val-Pro-Gly-Val-Gly-OMe, Boc-Val-Ala-Pro-Gly-Val-Gly-OMe, Boc-Ala-Pro-Gly-Val-Gly-Val-OMe, and HCO-Ala-Pro-Gly-Val-Gly-Val-OMe. In the pentapeptide, the NH triplet at highest field is assigned to Gly^3; the Gly^5 NH triplet is at a considerably lower field position. The Gly^3 NH triplet of the pentapeptide at 19°C in Me_2SO-d_6 corresponds closely to the upfield Gly NH triplet in HCO-Ala-Pro-Gly-Val-Gly-Val-OMe. This resonance in the hexapeptide, with a slope of –8.6 for a 10° increment in temperature and a 0°C intercept of 8.27 ppm in Me_2SO-d_6, is tentatively assigned to Gly^3 NH. The upfield glycine triplet in Boc-Val-Ala-Pro-Gly-Val-Gly-OMe has a slope of –9.0/10°C and an intercept of 8.29 ppm, which correlates well with the upfield, tentatively Gly^3 NH, triplet in HCO-Ala-Pro-Gly-Val-Gly-Val-OMe and with that of Boc-Ala-Pro-Gly-Val-Gly-Val-OMe with a slope of –8.5/10°C and an intercept of 8.25 ppm. These correlations could be complicated by specific conformational effects and therefore must be considered tentative until synthesis of a Gly^3-d_2 hexapeptide. This work is in progress.

As seen in the α-CH region of Fig. 2 and also by the presence of a satellite formyl proton peak in Figs. 2 and 3, about 20% of the molecules are *cis* (24). Fig. 3 gives the peptide NH region of the dodecapeptide HCO-(Ala-Pro-Gly-Val-Gly-Val)$_2$-OMe in Me_2SO-d_6 at 20°C with assignments of the Val 4, 6, 10, and 12, and the Ala 1 and 6 resonances.

The temperature dependences of peptide proton chemical shifts for the four peptides Boc-Ala-Pro-Gly-Val-Gly-Val-OMe, HCO-Ala-Pro-Gly-Val-Gly-Val-OMe, HCO-(Ala-Pro-Gly-Val-Gly-Val)$_2$-OMe, and HCO-(Ala-Pro-Gly-Val-Gly-Val)$_3$-OMe are given in Table 1. Throughout the series, the Val^4 NH resonance and its counterpart in the higher oligomers Val^{10} and Val^{16} exhibit a decreased temperature coefficient. So too does the glycine NH that is tentatively assigned Gly^3. Also the 0° intercepts for the valine 4, 6, and 10 resonances are found upfield near 7.77 ppm.

The temperature dependences of peptide proton chemical shifts in methanol for Boc-Ala-Pro-Gly-Val-Gly-Val-OMe, HCO-Ala-Pro-Gly-Val-Gly-Val-OMe, and HCO-(Ala-Pro-Gly-Val-Gly)$_2$-OMe are given in Table 2. The most evident features of this table are the low temperature coefficients for the Val^4 and Val^{10} peptide protons and their higher-field positions, 8.01 ppm at 0°C.

The Val 4, 10, and 16 peptide NH resonances have been delineated in Me_2SO-d_6 and in methanol by their lower temperature coefficients of chemical shift and by their higher-field position. Further delineation of the Val^4 NH resonance can be seen in Fig. 4, which illustrates the dependence of chemical shift on the mole percent of methanol and trifluoroethanol (TFE). The Val^4 NH resonance shifts upfield by about 25 Hz on going from methanol to TFE, whereas the Ala^1 NH and Val^6 NH resonances shift upfield more than 150 Hz and the Gly^5 NH and Gly^3 NH resonances shift upfield by about 130 Hz and 80 Hz, respectively.

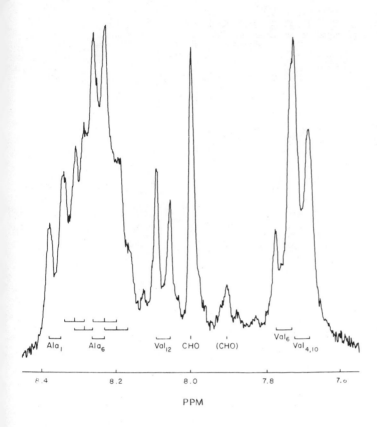

Figure 3. Proton magnetic resonance spectrum at 220 MHz of the dodecapeptide HCO-(Ala-Pro-Gly-Val-Gly-Val)$_2$-OMe in the peptide NH region at 20° (in Me$_2$SO-d_6). The Val and Ala assignments are given.

DISCUSSION

The Val[4] NH has been delineated by its low temperature coefficient of chemical shift in Me$_2$SO-d_6 (Table 1) and in methanol (Table 2) and by its small chemical shift difference on going from a methanol to a TFE solvent system (see Fig. 4). These findings indicate that the Val[4] NH is shielded from the solvent. The Val[4] NH also is the resonance with the highest-field position in the CHO derivatives (Tables 1 and 2). The high-field position in combination with the chemical shift dependences on temperature and solvent mixture is characteristic of a β turn. This situation is analogous to that of the pentapeptide Val[4] (17) [Fig. 1 (a)] and is consistent with the β turn shown in Fig. 5.

As seen in Fig. 2, there is a *cis-trans* isomerization of the Val-Pro peptide bond wherein about 80% of the molecules are *trans* and 20% are *cis*. In the *cis*

Table 2. Temperature dependence of peptide proton chemical shift in methanol

Peptide Residue	Boc-APGVGV-OMe[a]		HCO-APGVGV-OMe[a]		HCO-(APGVGV)$_2$-OMe[a]	
	Slope (Hz/10°C)	0° Intercept	Slope (Hz/10°C)	0° Intercept	Slope (Hz/10°C)	0° Intercept
Ala^1	−18.7[b]	7.08[b]	−16.5	8.64	−14.1	8.61
Ala^7					−16.5	8.48
Gly^3	−13.0	8.59	−14.7	8.56	−13.6	8.53
Gly^9					−13.6	8.53
Val^4	−8.8	8.01	−8.1	8.01	−9.0	8.01
Val^{10}					−9.0	8.01
Gly^5	−14.7	8.68	−16.7	8.79	−13.3	8.72
Gly^{11}					−13.3	8.72
Val^6					−13.9	7.97
Val^{12}	−14.5[c]	8.21[c]	−16.3[c]	8.29[c]	−13.6[c]	8.20[c]

[a]APGVGV is an abbreviation for Ala-Pro-Gly-Val-Gly-Val.

[b]This residue is displaced because of the Boc derivative.

[c]This residue is the carboxyl terminus.

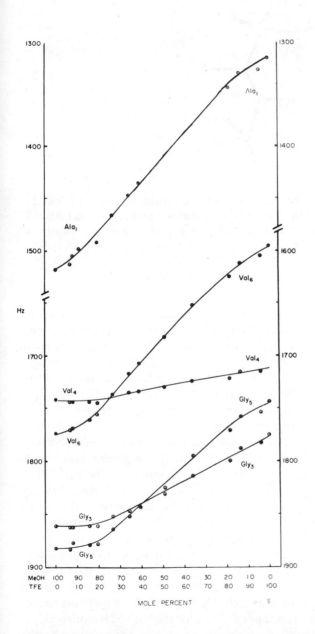

Figure 4. Methanol-trifluoroethanol (MeOH-TFE) solvent-mixture dependence of the peptide NH resonances of Boc-Ala-Pro-Gly-Val-Gly-Val-OMe. The Val[4] NH resonance is delineated by its small change in chemical shift. The glycine assignments are tentative.

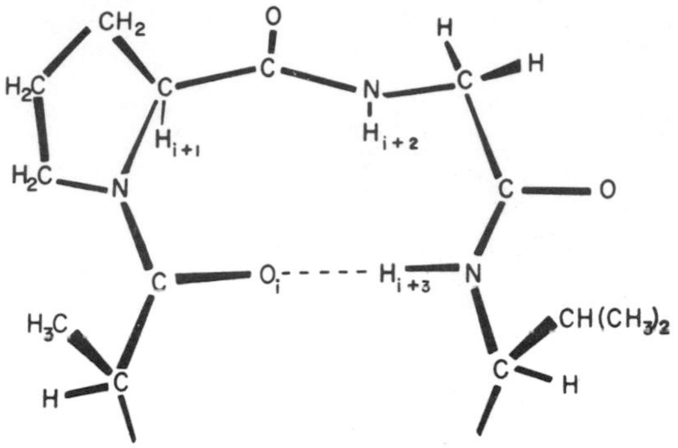

Figure 5. Proposed β turn for the hexapeptide of elastin. As seen in Tables 1 and 2, solvent-shielding of the Val^4 NH resonance is seen to repeat in higher oligomers, so that this conformational feature is expected to recur regularly as the hexapeptide repeats. As with the pentapeptide and tetrapeptide of elastin, Pro^2 and Gly^3 are at the corners of the β turn in positions $i + 1$ and $i + 2$. In the hexapeptide, Ala^1 is in position i and Val^4 is in position $i + 3$.

conformation, the β turn shown in Fig. 5 is not possible. Accordingly, it is of interest to know what conformational features are implied in proton magnetic resonance studies of the *cis* isomer. In the *cis* isomer, the Val^4 NH could not be solvent-shielded by the hydrogen-bonding of the β turn in Fig. 5. The low percentage of *cis* isomer in the Ala-Pro-Gly-Val-Gly-Val hexapeptide and its oligomers does not lend itself to characterization. In the Pro-Gly-Val-Gly-Val-Ala permutation, however, there is a conformational equilibrium between two nearly equally populated states (see Fig. 6).

At 20°C in Me_2SO-d_6 (Fig. 6), two conformational states, A and B, are observed with probabilities of 0.45 and 0.55, respectively. A glycine triplet tentatively assigned as Gly^3 is split into two well-resolved states as is the Val^4 NH doublet. When the temperature is raised to 49°C the peaks coalesce and the second glycine triplet becomes better resolved. A plot of the peptide NH chemical shifts as a function of temperature is given in Fig. 7. During temperature-induced coalescence, the temperature dependence of chemical shift will be anomalously small for the high-field component and anomalously large for the low-field component, for example, -4.3/10°C for Val^{4A} and -15.3/10°C for Val^{4B} and -7.3/10°C and -10.3/10°C for Gly^B and Gly^A, respectively. For the glycine residue, the slope above coalescence (up to 100°C) is closely the mean of the two resolved slopes. This is not the case for the Val^4 NH resonances. The study in Me_2SO-d_6 is limited, unfortunately, at lower temperatures because of the freezing point of the solvent, which makes it impossible to obtain accurately the slopes once the resolution into two states at lower temperature is complete. If we take the lower two temperatures, the coefficients for Val^{4A}

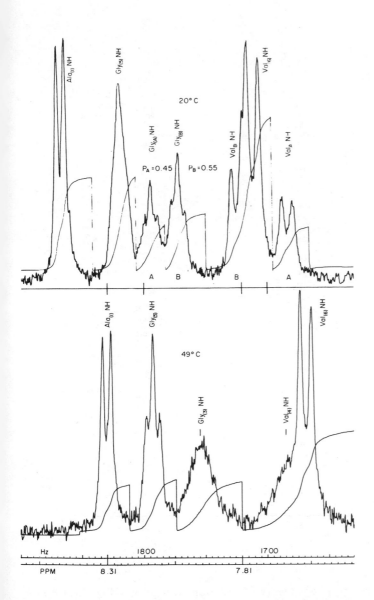

Figure 6.   Peptide NH region of Boc-Pro-Gly-Val-Gly-Val-Ala-OMe in Me$_2$SO-d_6. In the upper figure, the Val[4] NH resonance at 20°C is split, giving two resolvable conformations. A glycine resonance also is split into conformers A and B with probabilities P_A = 0.45 and P_B = 0.55. In the lower figure, the split resonances are seen to coalesce at 49°C.

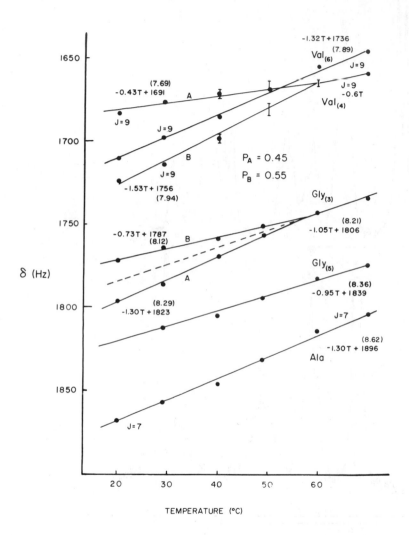

Figure 7. Temperature dependence of the resonances given in Fig. 6 for Boc-Pro-Gly-Val-Gly-Val-Ala-OMe in Me_2SO-d_6. The slopes in Hz/°C and 0°C intercepts in Hz and in ppm are given in brackets. Also included are the coupling constants for the doublets. The Val[6] NH resonance maintains a value of $J = 9$ Hz over the temperature range and those of Val[4] NH also are 9 Hz before and after coalescence. The coupling constant for Ala[1] remains approximately 7 Hz over the temperature range. The glycine assignments are tentative.

NH and Val^{4B} NH are approximated as –7.8/10°C and –11.4/10°C, respectively. The slope for Val^{4A} NH is lower than seen in Table 1, implying that the β turn is present. The slope of –11.4/10°C for Val^{4B} indicates that the shielding does not occur in the *cis* isomer. This situation is to be expected for a β turn utilizing the carbonyl in peptide linkage to the Pro nitrogen. The interesting effects on the glycine residue imply that there may be an intramolecular hydrogen bond involving this residue in conformation B. Detailed discussion of this conformation will be possible after the glycine assignments are more certain. It is of interest, though, to note that the temperature coefficient for this peptide NH (tentatively given as Gly^3 in Table 2) is low.

By observing the Boc CH$_3$ resonances, which are singlets, as a function of temperature (see Fig. 8) it is possible to approximate the thermodynamic properties of the transition state between A and B. Using tetramethylsilane as an internal reference for homogeneity-limited line width and assuming the states to be equally populated (25), the following values are obtained: $\Delta F^{\ddagger}$ = 16 kcal/mole, $\Delta H^{\ddagger}$ = 15.5 kcal/mole, and $\Delta S^{\ddagger}$ = 1.9 e.u. For HCO-Pro-Gly-Val-Gly-Val-Ala-OMe the formyl singlet can be used. In this case P_A = 0.48 and P_B = 0.52 and coalescence occurs just above 100°C: $\Delta F^{\ddagger}$ = 19 kcal/mole, $\Delta H^{\ddagger}$ = 18.5 kcal/mole and $\Delta S^{\ddagger}$ = 1.9 e.u. If equally populated states are not assumed and a single resolved peak is used, similar values are obtained.

The above results further support the β turn indicated in Fig. 5 as a significant conformational feature. The more difficult question is the probability of this conformation or the percentage of molecules that are in this conformation in solution at a given time and temperature. The problem could be approached by the general mole-fraction expression

$$
x_i = \frac{a_{obs} - \sum_{j \neq i} \left(1 + \sum_{j \neq i} x_k\right) a_j}{a_i - \sum_{j \neq i} a_j}
\tag{1}
$$

which reduces to

$$
x_i = \frac{a_{obs} - a_j}{a_i - a_j}
\tag{2}
$$

when only two states are considered. Taking the two states as shielded from solvent, s, and solvent exposed, e, Eq. (2) is written

$$
x_s = \frac{a_{obs} - a_e}{a_s - a_e}
\tag{3}
$$

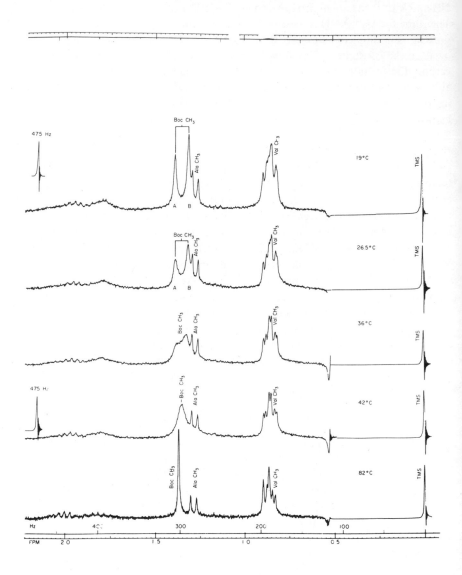

Figure 8. Temperature study in Me$_2$SO-d_6 of the Boc, Ala, Val CH$_3$ protons. The line widths of TMS are used to correct the Boc CH$_3$ line widths in the calculation of the rate constant for conversion between states A and B. Coalescence is seen to occur near 42°C. At 19°C $\nu_A - \nu_B = 17$ Hz.

The problem, as with efforts to calculate percent of α helix, β-pleated sheet and other conformations, is to obtain appropriate reference states and means of delineation. If we take gramicidin S in methanol as a reference state, temperature coefficients are obtainable: $a_s{}^T = -6.7$ and $a_e{}^T = 16$ (5, 6). Taking the value of $a_{obs}^T = -9$ (see Table 2) for Val^4 NH, we calculated $\chi_s = 0.75$; in other words, 75% of the time the β turn is formed in solution. Needless to say that such an approach is fraught with all of the difficulties attending similar calculations of α-helix, β-pleated-sheet, and random-coil percentages. Using temperature coefficients is even more tenuous because

$$a_{obs}^T = \sum_i a_i{}^T P_i \qquad (4)$$

where i are all the states in equilibrium and

$$P_i = \frac{e^{-\Delta F_i/RT}}{\sum_i e^{-\Delta F_i/RT}} \qquad (5)$$

Thus, two rapidly interchanging states, seen as a single peak in the proton magnetic resonance experiment, but with different free energies and different chemical shifts when resolved, could give fortuitously high or low temperature coefficients.

The second method of delineating solvent-shielded versus solvent-exposed peptide NHs could also be used in terms of Eq. (1). Again, gramicidin S could be the model and solvent mixtures of methanol-TFE the method. The observable a is defined as the change in chemical shift in going from methanol to TFE. For gramicidin S we have $a_s{}^{SM} = 44$ Hz and $a_e{}^{SM} = -240$ Hz (9), and from Fig. 4, for the Val^4 NH resonance we have $a_{obs} = -26$ Hz. These values give $\chi_s{}^{SM} = 0.74$; that is, 74% of the time the β turn is formed. This approach has the special problem of different average conformations in the two solvents. Table 3 gives the mole fractions of solvent-shielded peptide NH protons using the two methods and applying them to the β turns of the elastin peptides.

After the preceding discussion, an adequate disclaimer is required to prevent literal interpretation. At most, the calculations could be only a means of qualitatively distinguishing what is meant by the presence of a β turn that is distinguished by the above proton magnetic resonance methods. The results should not be interpreted in terms of rigid structures. And the calculated mole fractions are only to give a qualitative sense the frequency of occurrence.

Table 3. Calculated mole fractions of solvent-shielded protons

Elastin β Turn	Calculated Mole Fraction[a]	
	Temperature-Coefficient Method[b]	Solvent-Mixture Method
Hexapeptide[c] Val^4 NH	0.75	0.74
Pentapeptide[d] Val^4 NH	0.66	0.76
Tetrapeptide[e] Gly^4 NH	0.97	0.63

[a]See text for discussion. These values should be considered only in the most qualitative sense.

[b]For methanol solvent system.

[c]Ala-Pro-Gly-Val-Gly-Val.

[d]Val-Pro-Gly-Val-Gly.

[e]Val-Pro-Gly-Gly.

ACKNOWLEDGMENTS

This work was supported by the National Institutes of Health Grant No. He-11310. The authors gratefully acknowledge the technical assistance of B.J. Harrison, P. King, L.W. Mitchell, and D. Yarborough and the advice on peptide synthesis of B. Johnson.

REFERENCES

1. J.A. Foster, E. Bruenger, W.R. Gray, and L.B. Sandberg, *J. Biol. Chem.*, **248**, 2876 (1973).
2. W.R. Gray, L.B. Sandberg, and J.A. Foster, *Nature* (London), **246**, 461 (1973).
3. D.W. Urry, *Proc. Nat. Acad. Sci. U.S.*, **68**, 810 (1971).
4. D.W. Urry, *Proc. Nat. Acad. Sci. U.S.*, **69**, 1610 (1972).
5. M. Ohnishi and D.W. Urry, *Biochem. Biophys. Res. Commun.*, **36**, 194 (1969).
6. D.W. Urry and M. Ohnishi, in *Spectroscopic Approaches to Biomolecular Conformation*, D.W. Urry, ed., *American Medical Association Press*, Chicago, 1970, p. 263.
7. M. Ohnishi and D.W. Urry, *Science*, **168**, 1091 (1970).
8. D.W. Urry and N.G. Kumar, *Biochemistry*, **13**, 1829 (1974).
9. T.P. Pitner and D.W. Urry, *J. Am. Chem. Soc.*, **95**, 1399 (1972).
10. J.D. Glickson, D.F. Mayers, J.M. Settine, and D.W. Urry, *Biochemistry*, **11**, 477 (1972).
11. D.W. Urry, J.D. Glickson, D.F. Mayers, and J. Haider, *Biochemistry*, **11**, 487 (1972).

12. D.W. Urry, M. Ohnishi, and R. Walter, *Proc. Nat. Acad. Sci. U.S.*, **66**, 111 (1970).

13. D.W. Urry and R. Walter, *Proc. Nat. Acad. Sci. U.S.*, **68**, 956 (1971).

14. T.P. Pitner and D.W. Urry, *Biochemistry*, **11**, 4132 (1972).

15. N.G. Kumar and D.W. Urry, *Biochemistry*, **12**, 3811 (1973).

16. N.G. Kumar and D.W. Urry, *Biochemistry*, **12**, 4392 (1973).

17. D.W. Urry, W.D. Cunningham, and T. Ohnishi, *Biochemistry*, **13**, 609 (1974).

18. D.W. Urry, L.W. Mitchell, and T. Ohnishi, *Proc. Nat. Acad. Sci. U.S.*, in press.

19. D.W. Urry and T. Ohnishi, *Biopolymers*, in press.

20. D.W. Urry, L.W. Mitchell, and T. Ohnishi, *Biochem. Biophys. Res. Commun.*, in press.

21. D.W. Urry, *Research/Development*, **25** (June), 18 (1974).

22. C.C. McDonald and W.D. Phillips, *J. Am. Chem. Soc.*, **91**, 1513 (1969).

23. C.M. Deber, F.A. Bovey, J.P. Carver, and E.R. Blout, *J. Am. Chem. Soc.*, **92**, 6191 (1970).

24. For the method of calculation, see, for example F.A. Bovey, *Nuclear Magnetic Resonance Spectroscopy*, Academic Press, New York, 1969, Chapter VII.

PART III. CYCLIC AND LINEAR OLIGOPEPTIDES

A COMPARISON OF CYCLIC PEPTIDE CONFORMATIONS DEDUCED BY NMR AND X-RAY

F.A. BOVEY, Bell Laboratories, Murray Hill, N.J. 07974

SYNOPSIS: This paper is a review and a comparison of published structures (i.e., conformations) of cyclic polypeptides determined in solution by NMR and in the crystal by x-ray diffraction. It appears that there are more instances of agreement than of disagreement between conformations deduced by these two methods. Generally, there is good agreement when strong stabilizing elements such as proline units, bulky side chains, and complexed metal ions are present.

INTRODUCTION

The purpose of this paper is to describe some cyclic polypeptides and one depsipeptide that have been examined by both high-resolution NMR and x-ray diffraction and to compare the conformations deduced by each method. This comparison is not primarily intended to raise the question of which method gives the correct result, for there are many cases in which the two conformations substantially agree and others in which they do not. It seems most probable at present that the latter represent real differences between the conformations in the crystal and in solution.

The NMR parameters that have been employed in conformational studies of polypeptides include

1. Measurement of scalar couplings, particularly of α-CH and NH protons, termed $J_{N\alpha}$, to yield the torsional angle ϕ (see below)
2. Effects of temperature, fluoroalcohols (1–4), or nitroxide radicals (2) on NH resonances as a means of distinguishing solvent-exposed from internally shielded or hydrogen-bonded peptide protons
3. NH exchange rates for the same purpose
4. Other proton and carbon-13 chemical shifts, particularly as a function of solvent, metal complexation, etc.
5. Carbon-13 spin lattice relaxation.

Conformational energy calculations are frequently a valuable and necessary support for NMR measurements and vice versa, particularly for more complex molecules. It is not our purpose to review this matter in detail here, because it has been done elsewhere (5–17), but rather to consider the results of such studies with a minimum of supporting description.

DIKETOPIPERAZINES

Information on the simplest cyclic peptides, the diketopiperazines, is now quite extensive. There are NMR data on at least 20 diketopiperazines (11) and

248

a substantial body of x-ray data as well. Unfortunately, there are some difficulties, beginning with the prototype molecule, diketopiperazine itself, or *cyclo*-(Gly-Gly). X-ray diffraction shows that in the crystal the ring is planar (18, 19). We may suppose that in solution it maintains this conformation or, as molecular models suggest, inverts rapidly between boat conformers:

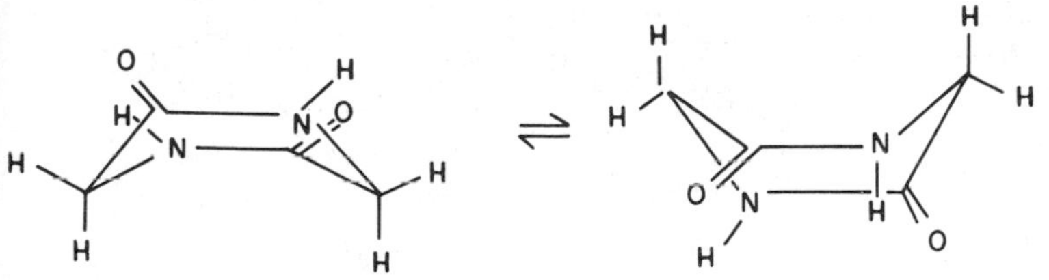

If planar, $J_{N\alpha}$ should be about 2.0 Hz, corresponding to dihedral angles of about $60°$ for both glycine protons, based on the Karplus-type relationship proposed by Bystrov *et al.* (20, 21), which is similar to those proposed by several other groups [see (11), (12) and (17) for reviews] :[1]

$$J_{N\alpha} = 8.9 \cos^2 \phi' - 0.9 \cos \phi' + 0.9 \sin^2 \phi' \tag{1}$$

If inverting, we average $J_{N\alpha}$ over dihedrals of about $0-20°$ and $90-110°$ and its value should be 3.5–4.0 Hz. A value of about 2.0 Hz in Me_2SO has been reported by Kopple and Marr (24); this appears to be consistent with a planar conformation. In trifluoroacetic acid, however, the coupling is immeasurably small, probably less than 0.5 Hz, which is not consistent with either alternative.

For monosubstituted diketopiperazines, that is, *cyclo*(Gly-X), it is expected that, in addition to the planar conformation (a) shown below, there will be two possible boat conformations: (b) with the side chain in the axial position and (c) with the side chain in the equatorial, or "bowsprit," position:

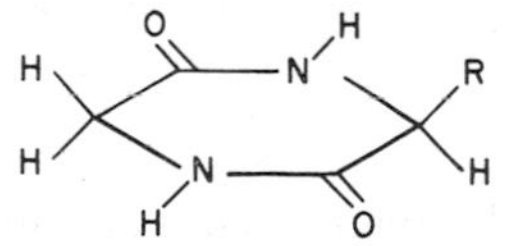

(a)

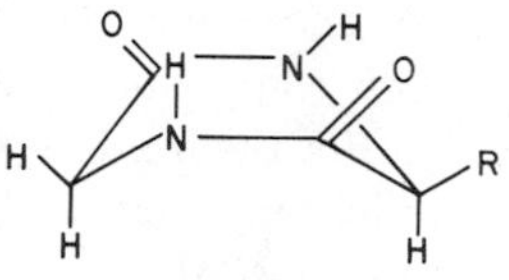

(c) 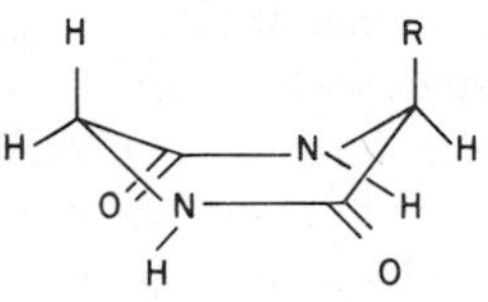   (b)

[1]Here ϕ' is the *dihedral* angle; the form of this function and the relationship of θ in the 1966 and 1970 conventions (22, 23) are summarized in Fig. 1.

For LL (or DD) diketopiperazines, that is, the *cis* isomers, a diequatorial (e) or a planar conformation may be expected to have preference over the boat conformation, which, in general, will be sterically unfavorable.

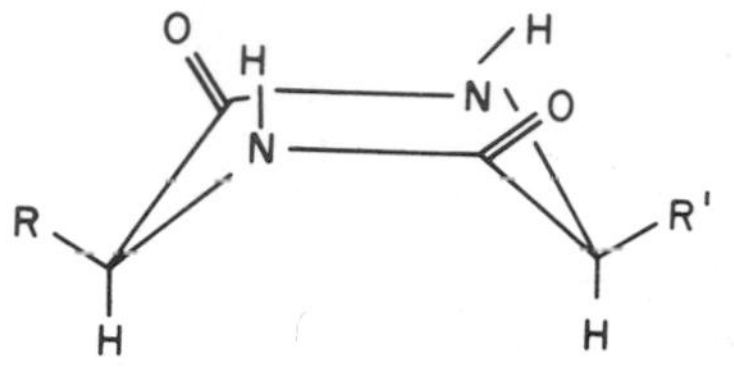

(d)

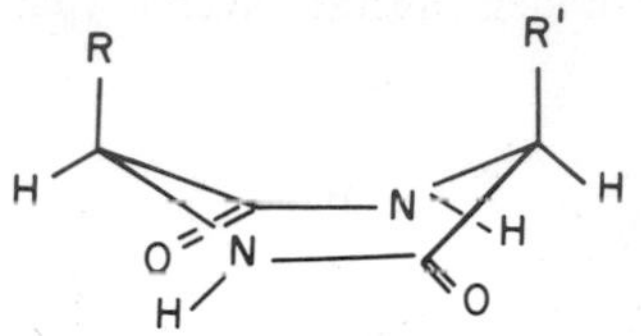

(e)

For the DL (or *trans*) compounds, either an inverting boat

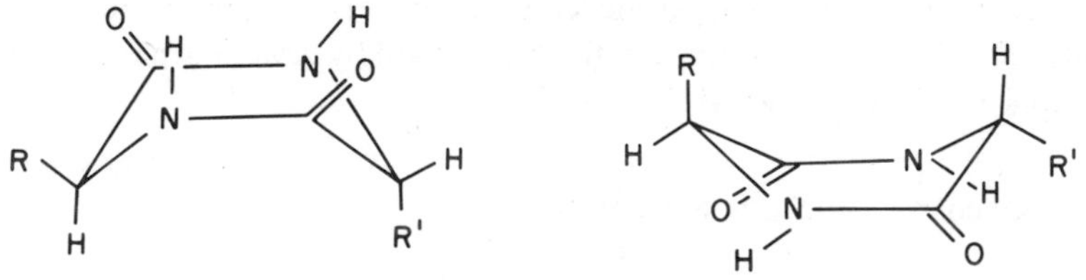

(f)

or a planar conformation seems plausible from models:

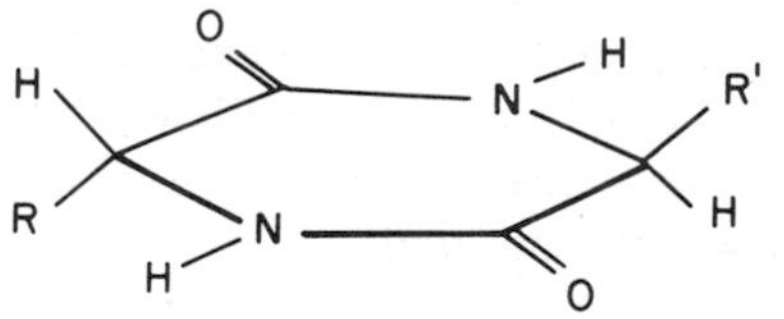

(g)

Benedetti *et al.* (25) and Sletten (26) have reported that the crystalline state *cyclo*(Ala-Ala), that is, L-*cis*-3,6-dimethyl-2,5-piperazinedione, has the equatorial boat conformation (e) with planar amide groups inclined at about $30°$ to each other, and the DL or *trans* isomer has a planar ring, corresponding to (g). Both compounds exhibit very small values of $J_{N\alpha}$, probably not exceeding 1 Hz (27). These data are consistent with the x-ray findings for the *cis* compound because the NH—C_αH dihedral is close to $90°$ here (see Fig. 1). For the *trans* compound, $J_{N\alpha}$ should be about 2.5–3.0 Hz if planar, but for inverting boat conformers the averaged coupling should be about 4.0 Hz. Thus the NMR results presently do not appear to provide a convincing picture of the behavior of these compounds in solution. Likewise, for *cyclo*(Gly-Ala), $J_{N\alpha}$ values for

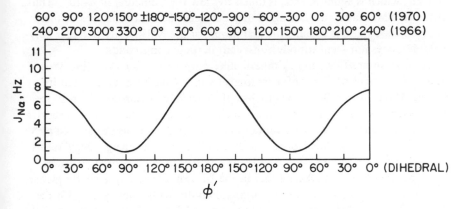

ϕ

60° 90° 120°150° ±180°–150°–120°–90° –60° –30° 0° 30° 60° (1970)
240° 270°300° 330° 0° 30° 60° 90° 120°150° 180°210° 240° (1966)

ϕ'

Figure 1. The dependence of the vicinal coupling $J_{N\alpha}$ on the H–N–C$_\alpha$–H dihedral angle expressed as a function of the torsional angle ϕ [in the 1966 and 1970 conventions (22, 23)].

both the glycine and the alanine residues are zero (in Me$_2$SO) within experimental error (27), which is not consistent with any of the conformers shown.

The Gly-X diketopiperazines, in which X is an aromatic amino acid, behave in a particularly interesting fashion. The glycine protons of *cyclo*(Gly-Tyr), or monobenzyldiketopiperazine, show a differentiation in chemical shift of the order of one part per million (24, 28, 29), in contrast to a negligible difference when X is nonaromatic (24, 28). This feature has been found in a number of similar compounds (28) and strongly suggests a folded conformation that corresponds to (b) above, with the aromatic ring positioned over the glycine methylene group, as shown below, presumably as a result of attractive π–π interaction between the aromatic ring and the carbonyl groups.

For the tyrosine residue, $J_{\alpha\beta}$ is in agreement with this conformation, and Webb and Lin (30) have found that x-ray results fully confirm it for the crystal. The $J_{N\alpha}$ values for the glycine residue are not consistent, however. The approximately zero coupling of the most shielded glycine proton, H$_h$, is reasonable because the dihedral to NH is nearly 90°, but the coupling of the low field proton, H_ϱ, which is about 3.0 Hz, is much too low for a dihedral of 0–20°. Thus, although NMR and x-ray data agree in general for the conformations of these compounds, $J_{N\alpha}$, one of our principal props for the NMR study of larger polypeptides, does not seem entirely consistent in these compounds.

Another interesting pair of isomeric diketopiperazines are *cyclo*(Pro-Pro) and *cyclo*(Pro-DPro), the conformations of which have been worked out by Young, Madison, and Blout (31) by use of the vicinal proton couplings in the proline ring, aided by the europium complex Eu(fod)$_3$ to sort out the ring *syn* and *anti* protons. A set of conformational parameters was also generated by energy-minimization calculations. Both the NMR and the energy calculations agree that the LL isomer is a boat with the β carbons in equatorial positions, corresponding to (e) above, and that in the DL isomer the ring is nearly planar. The proline ring conformations are similar in both, with a negative χ_1 (in the sense indicated below), corresponding to a puckering of the β carbon out of the plane of the other four atoms in the *syn* (or "endo") direction:

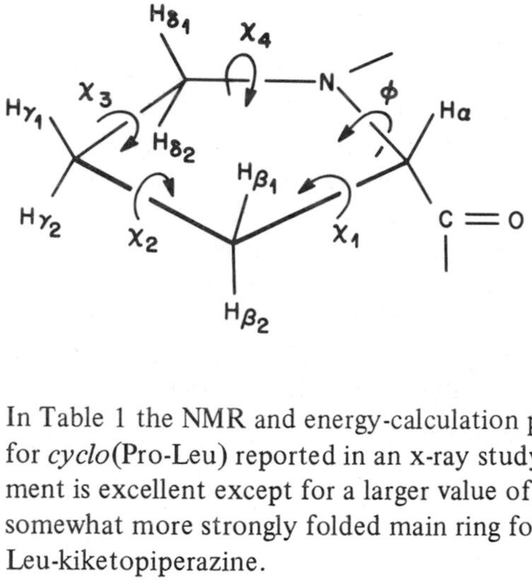

In Table 1 the NMR and energy-calculation parameters are compared to those for *cyclo*(Pro-Leu) reported in an x-ray study by I.L. Karle (32). The agreement is excellent except for a larger value of ϕ in the latter, corresponding to a somewhat more strongly folded main ring for the LL compound than for Pro-Leu-kiketopiperazine.

CYCLO(PRO)$_3$ AND CYCLO(PRO-PRO-HYP)

Two particularly interesting molecules are the tripeptides *cyclo*(tri-L-proline) and *cyclo*(Pro-Pro-Hyp). *Cyclo*(tri-L-proline) was first synthesized by Rothe (33). Its NMR spectrum and those of its derivatives have been reported by Deber, Torchia, and Blout (34), who used the vicinal proton couplings to determine the proline ring conformations, as with the proline diketopiperazines. Very recently, x-ray structures have been described by Kartha *et al.* (35). Unlike most of the molecules that we shall discuss, these are highly rigid—comparable to Pro-Pro diketopiperazines—and therefore distortions that may occur through intermolecular hydrogen bonding and other interactions peculiar to the crystal

Table 1.   Bond torsion angles in proline diketopiperazines [from Ref. (31)]

Angles	cyclo-(Pro-Leu) Crystal Structure	cyclo(Pro-Pro)		cyclo(Pro-DPro)	
		Calc'd	Expt'l	Calc'd	Expt'l
x_1	−32	−33	−30	−37	−40
x_2	36	34	42	36	29
x_3	−25	−23	−21	−22	−12
x_4	4	2		−1	
ϕ	−42	−16		−6	
ψ	34	26		5	
ω	6	−10		−14	

should have a relatively small effect. In other words, the x-ray and NMR structures should agree well. An additional feature is that the NMR structure was determined three years before the x-ray structure.

The net result of the NMR study is a molecule with three equivalent rings, each with the nitrogen atom fairly strongly puckered *exo* ("anti") to the carbonyl group. Three such units can be assembled only into a ring with a *cis* conformation ($\omega \cong 0°$) at the peptide bonds, which must be at least approximately planar. The resulting structure is shown in stereo in Fig. 2. The conclusions for *cyclo*(Pro-Pro-Hyp) and for its acetate and benzoate esters were essentially the same.

The x-ray results are in good agreement with the NMR conclusions. A major difference is that the Hyp unit has a torsional angle ω (for C_α–C'–N–C_α) as high as 18°. [This finding is in agreement with the predictions of Venkatachalam (36), which are not supported by either the x-ray or the NMR data for the Pro units.] The deviation of the Hyp unit from the NMR conformation may be due,

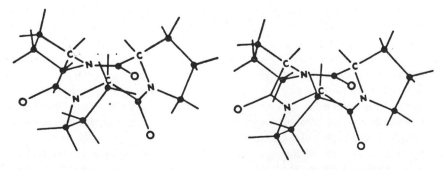

Figure 2. A stereoview of *cyclo*(tri-L-proline). [Reprinted with permission from Deber *et al., J. Am. Chem. Soc.,* **93**, 4893 (1971). Copyright by the American Chemical Society.]

at least in part, to the fact that in the crystal the OH group is linked by a strong hydrogen bond (2.7 Å) to a carbonyl group of a neighboring molecule.

SARCOSINE PEPTIDES

Dale and Titlestad (37) have reported the synthesis and NMR spectra of cyclic oligopeptides of sarcosine, for example, N-methylglycine,

$$\left[\begin{array}{c} \overset{\displaystyle CH_3}{\underset{\displaystyle |}{}} \\ N-CH_2-\overset{\displaystyle O}{\underset{\displaystyle \|}{C}} \end{array} \right]_n$$

where n = 2, 3, 4, 5, 6 and 8. The spectra do not give any detailed information concerning torsion angles because the only J coupling present is that of the geminal methylene protons. They are highly informative, however, of the symmetry of the molecules and also indicate very slow ring inversion, possibly due to strong transannular interactions.

For the tripeptide there is no x-ray data. The NMR spectrum of the tripeptide, however, indicates threefold symmetry and a highly nonplanar ring. These observations make it virtually certain that the conformation is similar to that of cyclo-triproline.

The tetrapeptide spectra are consistent only with the *cis-trans-cis-trans* structure and is in full agreement with the x-ray structure determined by Groth (38).

The pentapeptide spectrum shows five methyl resonances and five methylene quartets. This gives no detailed information, but does show that there is only one predominant conformer out of six possible ones in solution and that this one cannot be symmetric; that is, it is neither all *cis* nor all *trans*. The x-ray structure confirms this finding and shows that, in fact, the chain conformation is *cis-cis-cis-trans-trans* (39).

Cyclo-octasarcosyl shows four equal methyl resonances, resolvable in $CD_3OD:C_6D_6$. In their original 1969 publication, Dale and Titlestad stated that of the 38 possible conformational isomers only a *cis-cis-trans-trans-cis-cis-trans-trans* sequence could be consistent with this observation. Actually, two other isomers, *cis-cis-cis-trans-cis-cis-cis-trans* and *trans-trans-trans-cis-trans-trans-trans-cis*, are equally consistent, as stated in their 1973 paper (40) on the x-ray determination of the crystal structure. In fact, their original proposal was correct. The x-ray data show a notably open structure with four molecules of water of hydration. The infrared spectra of the crystal and chloroform solutions were identical, and so the solution structure is very probably the same. It is a notably open structure with all ψ angles close to the planar zigzag value ($-180°$). The two portions of the structure with *trans* peptide bonds have a form-II polyproline conformation, and the *cis* portions correspond to form I. There are no transannular interactions to appeal to for stabilization, and so the stability of this conformation must be attributed to the intrinsic preference of the chain itself.

CYCLIC HEXAPEPTIDES

The class of cyclic polypeptides that has received the most attention is the cyclic hexapeptides. The emphasis is attributable to the relative ease with which they can be synthesized and to the fact that many bind cations and/or have anti-biotic activity. Hruby (11) has listed 31 compounds for which solution confor-mations have been suggested, principally on the basis of NMR. The conforma-tion that has been most frequently proposed, and in some cases convincingly demonstrated, is the antiparallel β conformation, usually with β turns at the ends and two cross-ring hydrogen bonds, as suggested originally by Schwyzer (41-43). The internally hydrogen-bonded protons are generally more shielded, exhibiting small or zero temperature coefficients and reduced exchange rates.

Despite extensive solution studies, opportunities to compare NMR and x-ray structures are limited for this group of compounds. A well-documented example is that of *cyclo*(Gly-Gly-DAla-DAla-Gly-Gly):

$$
\begin{array}{ccc}
6 & 1 & 2 \\
\text{DAla-Gly-Gly} \\
| & \rightleftharpoons & | \\
\text{DAla-Gly-Gly} \\
5 & 4 & 3
\end{array}
$$

The proton spectrum is relatively complex (44) because of the lack of symmetry; there are separate spectra for each of the six residues. The absence of proline or bulky side chains suggests the possibility of a number of low-energy conforma-tions between which equilibration will be rapid. It is observed that although all of the glycine residues exchange NH protons rapidly, one of them has a tempera-ture coefficient about one-third as great as the others, suggesting one internal hydrogen bond. The x-ray crystal conformation (45) (Fig. 3) shows the expected two intramolecular hydrogen bonds with all *trans* peptide bonds, the two D-alanine residues forming one of the turns. If the two methyl side chains are disregarded, the conformation has an approximate center of symmetry. This is not a typical antiparallel β structure (the ϕ values of Gly_1 and Gly_4 deviate from the appropriate values), and in fact, with regard to intramolecular forces alone it is a high-energy structure. In the crystal, however, this is more than counter-balanced by the presence of nine external hydrogen bonds to three H_2O mole-cules of crystallization (per peptide molecule) and one intermolecular $NH\cdots O=C$ hydrogen bond, in addition to the two Gly_1-Gly_4 intramolecular hydrogen bonds.

In contrast, energy calculations indicate that in solution there are 25 low-energy conformations consistent with the observed values of $J_{N\alpha}$, all of which have energies about 15 ± 1 kcal less than that of the crystal conformation.

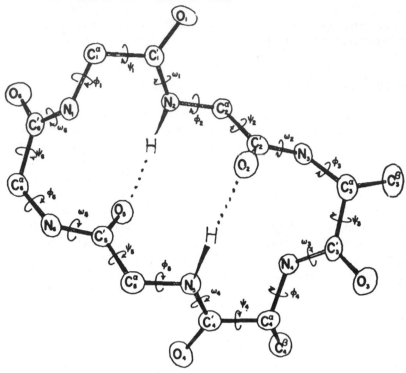

Figure 3. The x-ray conformation of *cyclo*(Gly-Gly-DAla-DAla-Gly-Gly). [Reprinted with permission from Karle *et al.*, *J. Am. Chem. Soc.*, **92**, 3755 (1970). Copyright by the American Chemical Society.]

Eight of these are calculated to have one $(NH)_{Gly_3} \cdots (O=C)_{Ala_2}$ hydrogen bond that is different from both of those in the crystal. This molecule is thus fairly flexible in solution, and none of the solution conformations corresponds to the crystal conformation.

GRAMICIDIN S

Gramicidin S is celebrated in the annals of polypeptide NMR as the first molecule to have its solution conformation determined by the NMR criteria we have already discussed (46). There have been other NMR studies (47–49) generally supportive of the conclusion that this decapeptide, whose sequence is

$$\begin{array}{ccc} \text{Val-Orn-Leu-DPhe-Pro} \\ | & \rightleftharpoons & | \\ \text{Pro-DPhe-Leu-Orn-Val} \end{array}$$

has an extended antiparallel β conformation with four cross-ring hydrogen bonds forming a rather rigid ring. This conformation had been proposed for gramicidin S years earlier by Schwyzer (50). The DPhe-Pro peptide bonds, like all the others, are *trans*, as shown particularly clearly by the proline carbon-13 spectrum (51). This general type of structure with C_2 symmetry was also suggested by the early x-ray study of Hodgkin and Oughton (52) and has been confirmed by a more recent (but still not completely detailed) x-ray study by De Santis *et al*. (53).

ANTAMANIDE

Antamanide occurs in the poisonous mushroom *amanita phalloides* and acts as an antidote to its poisons, amanitin and phalloidin, by preventing their accumulation in the liver. It is of particular interest because it forms quite strong complexes with Li^+, Na^+, K^+, and Ca^{2+}. The strongest complex is formed with Na^+, and it is probably in this form that it is biologically active (54).

Antamanide is a decapeptide without symmetry and with an unusually high proportion of proline and phenylalanine:

$$
\begin{array}{ccccc}
8 & 9 & 10 & 1 & 2 \\
\text{Pro-Phe-Phe-Val-Pro} \\
| & & \rightleftharpoons & & | \\
\text{Pro-Phe-Phe-Ala-Pro} \\
7 & 6 & 5 & 4 & 3
\end{array}
$$

It has a number of possible conformations in solution, depending upon the nature of the solvent. In weak hydrogen-bond-accepting solvents such as acetic acid and acetonitrile, the most probable structure is now believed to be an antiparallel β structure with four cross-ring hydrogen bonds, as shown in Fig. 4 (55). This conformation is consistent with the proton data and with the partial carbon-13 spectrum shown in Fig. 5. Here we see (shaded) the resonances of the β and γ carbons of the proline residues. (The four peaks at the left are the four β carbons of the phenylalanines and those at the right are the valine and alanine methyls.) The spectrum shows two peaks in each region of the spectrum that now can be recognized (56–61) as corresponding to *cis* and *trans* X-Pro peptide bonds, that is, peptide bonds to proline units: *γ-cis*, *γ-trans*, *β-trans*, and *β-cis*—a total of eight. Clearly, then, there are two *cis* and two *trans* X-pro bonds, but the spectrum does not reveal which is which. Note that the spectrum of the sodium complex is very similar. The antiparallel β-structure conformation requires that the Val_1-Pro_2 and Phe_6-Pro_7 bonds be *trans* and that the Pro_2-Pro_3 and Pro_7-Pro_8 bonds be *cis*.

When the sodium complex is formed, there are major conformational changes that are clearly evident in the NMR data (55), but there is no change in the conformations about these peptide bonds. This is shown not only by the similarity of the carbon-13 spectra, but also by the fact that as NaSCN is added to an antamanide solution, there are shifts and broadenings of many peaks, but no doubling of any of them (55, 62). Doubling is to be expected if *cis* $\rightleftharpoons$ *trans* peptide bond isomerizations occur, because such processes require an activation energy of about 20 kcal and are therefore slow on the NMR scale. The proposed structure

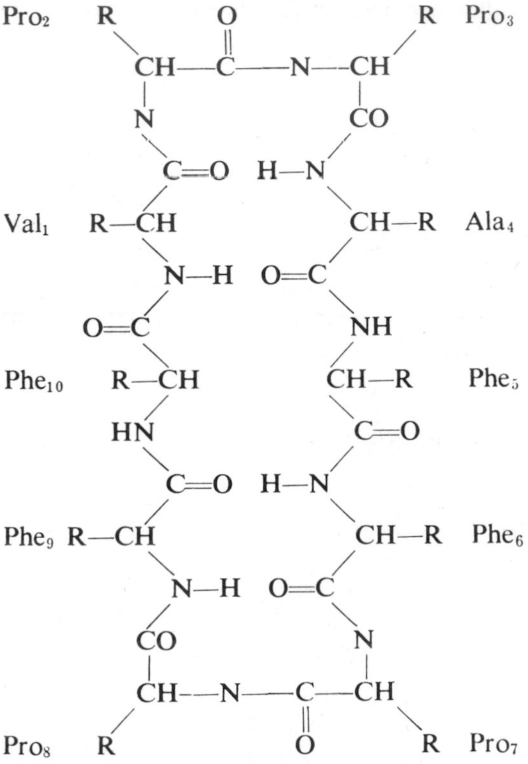

```
Pro₂    R        O              R  Pro₃
          \       ‖            /
          CH——C——N——CH
          |             |
          N            CO
           \           /
           C=O   H—N
          /           \
Val₁  R—CH          CH—R   Ala₄
          \          /
          N—H   O=C
         /           \
       O=C           NH
      /               \
Phe₁₀  R—CH        CH—R    Phe₅
        HN            \
         \             C=O
         C=O   H—N
        /            \
Phe₉ R—CH          CH—R   Phe₆
         \           /
         N—H   O=C
        /            \
       CO            N
      /               \
     CH——N——C——CH
    /            ‖       \
Pro₈  R          O       R  Pro₇
```

Figure 4. The antiparallel β conformation of antamanide. [Reprinted with permission from Patel, *Biochemistry*, **12**, 667 (1973). Copyright by the American Chemical Society.]

of the sodium complex, based on NMR data, is a saddle-shaped structure with a twofold pseudo-axis and with the carbonyl groups turned inward to line the cavity in which the sodium ion is held. There are two hydrogen bonds: Phe_6NH to $Pro_3C{=}O$ and Val_1NH to $Pro_8C{=}O$. The carbonyl carbon spectrum shows that four carbonyl groups form ligands to the metal (62, 63). (An alternative proposal consistent with the NMR data in which the Val_1-Pro_2 and Phe_6-Pro_7 peptides are *cis* and the Pro_2-Pro_3 and Pro_7-Pro_8 bonds are *trans* (61) can now be excluded for both the complexed and uncomplexed forms.)

Fig. 6 shows the x-ray structure of the lithium complex of antamanide reported by Karle *et al.* (64). It is believed that the sodium complex does not differ significantly. The close similarity to the NMR conformation, including the hydrogen bonds, is clear. There are some angular differences, particularly

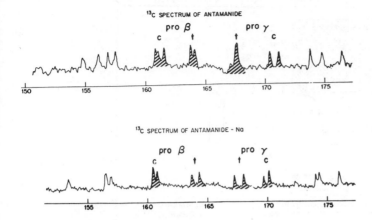

Figure 5. The 25 MHz proton-decoupled carbon-13 spectrum of the proline residues (shaded) of antamanide.

in the values of ψ for Ala_4, Phe_5, Phe_9, and Phe_{10}, but these tend to compensate, giving a very similar overall appearance. In Fig. 7 the two structures are compared on a Ramachandran (φ, ψ) plot (65). The x-ray structure is of relatively high energy, as can be seen by the positions of Ala_4, Phe_5, Phe_9, and Phe_{10}. In the NMR structure, these move to lower-energy positions. This movement probably corresponds to a relaxation in solution upon elimination of stabilizing forces present only in the crystal.

The coordination of the metal in the crystal is unusual in that the four carbonyl ligands are arranged pyramidally to the metal, with a molecule of acetonitrile (from which the material was crystallized) forming a fifth ligand at the apex of the pyramid.

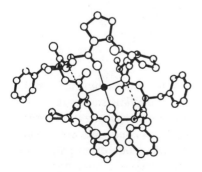

Figure 6. One view of the x-ray structure of the lithium complex of antamanide [from Ref. (64)].

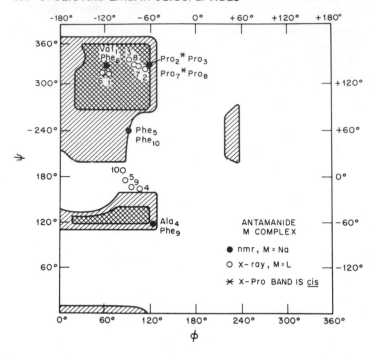

Figure 7. Ramachandran plot of the x-ray (64) and NMR (62, 64) structures of the metal complex of antamanide.

VALINOMYCIN

The well-known, much-studied cyclic depsipeptide antibiotic valinomycin has the structure

It is capable of forming a potassium complex, and both the complexed and uncomplexed forms have been the subject of a number of NMR studies (66–70). Like antamanide, valinomycin undergoes conformational isomerization with change of temperature and solvent, but always exhibits a spectrum consistent with threefold symmetry on the NMR time scale. (A pseudorotating structure that may be unsymmetric at any one instant is not necessarily ruled out, however; we will discuss this point shortly.) In a predominantly hydrocarbon environment, the "pore," or "bracelet," structure (Fig. 8) is dominant; in this conformation all the valine NH protons are internally hydrogen-bonded. In

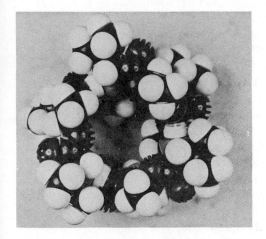

Figure 8. The pore, or bracelet, conformation of valinomycin.

ether solvents at low temperatures, propellerlike conformations are formed in which the DVal hydrogen bonds are retained, but the LVal NH protons are turned outward to bond with the solvent, with the result that the pore closes up. Still other conformations may exist under other conditions (69).

The NMR structure of the potassium complex (66–68, 70) resembles the pore, or bracelet, structure, but differs in that all six of the ester carbonyls, which are turned outward in the bracelet, turn *inward* to form ligands to K^+. This is indicated by the carbon-13 spectrum (70, 71) (Fig. 9), in which two of the four carbonyl resonances, each representing three residues, move downfield on complexation.

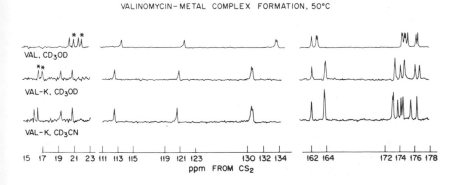

Figure 9. The 25 MHz proton-decoupled carbon-13 spectrum of valinomycin and its K^+ complex in methanol and acetonitrile at 50°C. [Reprinted with permission from Patel, *Biochemistry*, 12, 486 (1973). Copyright by the American Chemical Society.]

The crystal structures of the uncomplexed form determined by Duax (72) and the K^+ complex determined by Pinkerton (73) are shown as (a) and (b), respectively, in Fig. 10. The structure of the complex is threefold symmetric and is in good agreement with the NMR structure deduced by all published NMR studies. In contrast, the x-ray structure of the uncomplexed form possesses a pseudocenter of symmetry, but does not possess threefold symmetry, and therefore, it differs from any of the solution structures. Eight of the twelve residues are in conformations closely resembling those of the complex, but the other four (one each of LVal, DVal, LLac, and DHyIv) are different. It might be suggested that, as mentioned earlier, this structure is one of several pseudorotamers, of which we see the average in solution, and that the instantaneous solution structure likewise is not symmetric. Such a proposal requires averaged valine $J_{N\alpha}$ values of about 5.0 Hz instead of the observed values of 6.1–7.6 Hz. In addition, the Russian group (66) has measured the dipole moment of uncomplexed valinomycin in carbon tetrachloride and found a value of 3.5 D; in benzene, a value of 3.4 D has been found by Patel and Tonelli (69). In all probability, both values refer to the bracelet structure. These values agree with the calculated value (69) of 3.9 D for this structure, but the calculated value for the Duax crystal structure is 6.9 D. Thus, it appears that the solution conformation is probably truly threefold symmetric, or nearly so, and really differs from that of the crystal.

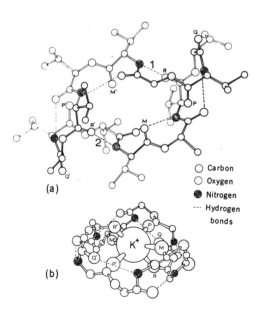

◯	Carbon
◯	Oxygen
●	Nitrogen
---	Hydrogen bonds

(a)

(b)

Figure 10. X-ray conformations of valinomycin (top) and its K^+ complex (bottom). [Reprinted with permission from Duax *et al., Science,* **176, 911** (1972). Copyright 1972 by the American Association for the Advancement of Science.]

It is noteworthy that a recent Raman spectroscopic study (74) strongly indicates that the crystal conformation is not unique, but that the structure, and particularly the number of internal hydrogen bonds, depends on the solvent from which this polypeptide is crystallized.

CONCLUSION

This survey is admittedly a selective one, for reasons of space and time. Several examples of considerable interest, including ferrichrome and alumichrome (75, 76), have been omitted from consideration. Again, although there is no complete x-ray structure for oxytocin as yet, there is a structure for the cysteinyl-prolyl-leucyl-glycine side chain (77), which agrees with the β-turn structure proposed (78) for this portion of the molecule. In summary, it appears at present that there are more instances of agreement than of disagreement between x-ray and NMR conformations of cyclic polypeptides. There is generally good agreement in the presence of strong stabilizing factors such as proline units, bulky side chains, and complexed metal ions.

REFERENCES

1. T.P. Pitner and D. Urry, *J. Am. Chem. Soc.,* **94**, 1399 (1972).
2. K.D. Kopple and T.J. Schamper, in *Chemistry and Biology of Peptides,* J. Meienhofer, ed., Ann Arbor, Mich.: Ann Arbor Science Publishers, 1972, p. 75.
3. K.D. Kopple and T.J. Schamper, *J. Am. Chem. Soc.,* **94**, 3644 (1972).
4. D.A. Torchia, *Biochemistry,* 11, 1462 (1972).
5. J.J.M. Rowe, J. Hinton, and K.L. Rowe, *Chem. Rev.,* **70**, 1 (1970).
6. G.C.K. Roberts and C. Jardetzky, *Adv. Protein Chem.,* **24**, 447 (1970).
7. J. Engel and G. Schwarz, *Angew. Chem., Int. Ed.,* **9**, 389 (1970).
8. B. Sheard and E.M. Bradbury, *Progr. Biophys. Mol. Biol.,* **20**, 187 (1970).
9. A. Allerhand and E.A. Trull, *Ann. Rev. Phys. Chem.,* **21**, 317 (1970).
10. D.W. Urry and M. Ohnishi, in *Spectroscopic Approaches to Biomolecular Conformation,* D.W. Urry, ed., Chicago: American Medical Association, 1970.
11. V.J. Hruby, in *Chemistry and Biochemistry of Amino Acids, Peptides and Proteins,* Vol. 3, B. Weinstein, ed., New York: Marcel Dekker, 1974.
12. F.A. Bovey, A.I. Brewster, D.J. Patel, A.E. Tonelli, and D.A. Torchia, *Acc. Chem. Res.,* **5**, 193 (1972).
13. W.A. Thomas, *Ann. Reports on NMR,* **6** (1974).
14. F.A. Bovey, *High Resolution NMR of Macromolecules,* New York: Academic Press, 1972, Chap. 13.
15. C.H. Hassall and W.A. Thomas, *Chem. in Britain,* **7**, 145 (1971).
16. M. Goodman, A.S. Verdini, Nam S. Choi and Y. Masuda, *Topics in Stereochemistry,* **5**, 69 (1970).
17. F.A. Bovey, *Macromol. Rev.,* **6** (1974).
18. R.B. Corey, *J. Am. Chem. Soc.,* **60**, 1598 (1938).
19. B. Degeilh and R.E. Marsh, *Acta Crystallogr.,* **12**, 1007 (1959).
20. V.F. Bystrov, S.L. Portnova, V.I. Tsetlin, V.T. Ivanov, and Yu.A. Ovchinnikov, *Tetrahedron,* **25**, 493 (1969).

21. T.A. Balashova and Yu.A. Ovchinnikov, *Tetrahedron,* **29**, 873 (1973).

22. J.T. Edsall, P.J. Flory, J.C. Kendrew, A.M. Liquori, G. Nemethy, and G.N. Ramachandran, *Biopolymers,* **4**, 121 (1966); *J. Biol. Chem.,* **241**, 1004 (1966); *J. Mol. Biol.,* **15**, 399 (1966).

23. J.C. Kendrew, W. Klyne, S. Lifson, T. Miyazawa, G. Nemethy, D.C. Phillips, G.N. Ramachandran, and H.A. Scheraga, *Biochemistry,* **9**, 3471 (1970); *J. Biol. Chem.,* **245**, 489 (1970); *J. Mol. Biol.,* **52**, 1 (1970).

24. K.D. Kopple and D.H. Marr, *J. Am. Chem. Soc.,* **89**, 6193 (1967).

25. E. Benedetti, P. Corradini, M. Goodman, and C. Pedone, *Proc. Nat. Acad. Sci. U.S.,* **62**, 650 (1969).

26. E. Sletten, *J. Am. Chem. Soc.,* **92**, 172 (1970).

27. Unpublished results in the author's laboratory.

28. K.D. Kopple and M. Ohnishi, *J. Am. Chem. Soc.,* **91**, 962 (1969).

29. F.A. Bovey, *Nuclear Magnetic Resonance Spectroscopy,* New York: Academic Press, 1969, p. 67.

30. L.E. Webb and C.-F. Lin. *J. Am. Chem. Soc.,* **93**, 3818 (1971).

31. P.E. Young, V. Madison, and E.R. Blout, *J. Am. Chem. Soc.,* **95**, 6142 (1973).

32. I.L. Karle, *J. Am. Chem. Soc.,* **94**, 81 (1972).

33. M. Rothe, K.D. Steffen, and I. Rothe, *Angew. Chem., Int. Ed.* (English), **4**, 356 (1965).

34. C.M. Deber, D.A. Torchia, and E.R. Blout, *J. Am. Chem. Soc.,* **93**, 4893 (1971).

35. G. Kartha, G. Ambady, and P.V. Shankar, *Nature,* **247**, 204 (1974).

36. C.M. Venkatachalam, *Biochim. Biophys. Acta,* **168**, 397 (1968).

37. J. Dale and K. Titlestad, *Chem. Comm.,* **1969**, p. 657.

38. P. Groth, *Acta Chem. Scand.,* **24**, 780 (1970).

39. K. Titlestad, P. Groth, and J. Dale, *J. Chem. Soc., Chem. Comm.,* **1973**, p. 646.

40. K. Titlestad, P. Groth, J. Dale, and M.Y. Ali, *J. Chem. Soc., Chem. Comm.,* **1973**, p. 346.

41. R. Schwyzer, P. Sieber, and B. Gorup, *Chimica,* **12**, 90 (1958).

42. R. Schwyzer, *Rec. Chem. Progr.,* **20**, 147 (1959).

43. R. Schwyzer, J.D. Carrion, B. Gorup, H. Nolting, and A. Tun-Kyi, *Helv. Chim. Acta,* **47**, 441 (1964).

44. A.E. Tonelli and A.I. Brewster, *J. Am. Chem. Soc.,* **94**, 2851 (1972).

45. I.L. Karle, J.W. Gibson, and J. Karle, *J. Am. Chem. Soc.,* **92**, 3755 (1970).

46. A. Stern, W.A. Gibbons, and L.C. Craig, *Proc. Nat. Acad. Sci. U.S.,* **61**, 735 (1968).

47. R. Schwyzer and U. Ludescher, *Biochemistry,* **7**, 2514, 2519 (1968).

48. M. Ohnishi and D.W. Urry, *Biochem. Biophys. Res. Comm.,* **34**, 803 (1969).

49. Yu.A. Ovchinnikov, V.T. Ivanov, V.F. Bystrov, A.I. Miroshnikov, E.N. Shepel, N.D. Abdullaev, E.S. Efremov, and L.B. Senyavina, *Biochem. Biophys. Res. Comm.,* **39**, 217 (1970).

50. R. Schwyzer, in *Amino Acids and Peptides with Antimetabolic Activity,* G.E.W. Wolstenholme, ed., London: Churchill, 1958, p. 171.

51. W.A. Gibbons, J.A. Sogn, A. Stern, L.C. Craig, and L.F. Johnson, *Nature* (London), **227**, 841 (1970).

52. D.C. Hodgkin and B.M. Oughton, *Biochem. J.,* **65**, 752 (1957).

53. P. De Santis, in *Proceedings of the International Symposium on the Conformation of Biological Molecules and Polymers,* Jerusalem, 1973, p. 493.

54. Th. Wieland, in *Chemistry and Biology of Peptides,* J. Meienhofer, ed., Ann Arbor, Mich.: Ann Arbor Science Publishers, 1972, p. 377 *et seq.;* see also *Angew. Chem.,* **80**, 209 (1968).

55. D.J. Patel, *Biochemistry,* **12**, 667 (1973).

56. F.A. Bovey, in *Chemistry and Biology of Peptides,* J. Meienhofer, ed., Ann Arbor, Mich.: Ann Arbor Science Publishers, 1972, p. 3.

57. I.C.P. Smith, R. Deslauriers, and R. Walter, *ibid.,* p. 29.

58. W.A. Thomas and M.K. Williams, *J. Chem. Soc., Chem. Comm.,* **1972**, p. 788.

59. K. Wuthrich, A. Tun-Kyi, and R. Schwyzer, *FEBS Lett.,* **25**, 104 (1972).

60. D.E. Dorman and F.A. Bovey, *J. Org. Chem.,* **38**, 2379 (1973).

61. W. Voelter and O. Oster, *Chemiker Zeitung,* **96**, 586 (1972).

62. D.J. Patel, *Biochemistry,* **12**, 677 (1973).

63. V.F. Bystrov, V.T. Ivanov, S.A. Kozmin, I.I. Mikhaleva, K.Kh. Khalilulina, Yu.A. Ovchinnikov, E. Fedin, and P.V. Petrovski, *FEBS Lett.,* **21**, 34 (1972).

64. I.L. Karle, J. Karle, Th. Wieland, W. Burgermeister, H. Faulstich, and B. Witkop, *Proc. Nat. Acad. Sci. U.S.,* **70**, 1836 (1973).

65. D.J. Patel and A.E. Tonelli, *Biochemistry,* **13**, 788 (1974).

66. V.T. Ivanov, I.A. Laine, N.D. Abdullaev, L.B. Senyavina, E.M. Popov, Yu.A Ovchinnikov, and M.M. Shemyakin, *Biochem. Biophys. Res. Comm.,* **34**, 803 (1969).

67. M. Ohnishi and D.W. Urry, *Biochem. Biophys. Res. Comm.,* **36**, 194 (1969).

68. D.W. Urry and M. Ohnishi, in *Spectroscopic Approaches to Biomolecular Conformation,* D.W. Urry, ed., Chicago: American Medical Association, 1970, p. 263.

69. D.J. Patel and A.E. Tonelli, *Biochemistry,* **12**, 486 (1973).

70. D.J. Patel, *ibid.,* p. 496.

71. M. Ohnishi, M.-C. Fedarko, J.D. Baldeschwieler, and L.F. Johnson, *Biochem. Biophys. Res. Comm.,* **46**, 312 (1972).

72. W.L. Duax, H. Hauptman, C.M. Weeks, and D.A. Norton, *Science,* **176**, 911 (1972).

73. M. Pinkerton, L.D. Steinrauf, and P. Dawkins, *Biochem. Biophys. Res. Comm.,* **35**, 512 (1969).

74. K.J. Rothschild, I.M. Asher, E. Anastassakis, and H.E. Stanley, *Science,* **182**, 384 (1973).

75. M. Llinas, M.P. Klein and J.B. Neilands, *J. Mol. Biol.,* **52**, 399 (1970) [NMR].

76. A. Zalkin, J.D. Forrester, and D.H. Templeton, *J. Am. Chem. Soc.,* **88**, 1810 (1966) [x-ray].

77. A.D. Rudko, F.M. Lovell, and B. Low, *Nature New Biol.,* **232**, 18 (1971).

78. D.W. Urry and R. Walter, *Proc. Nat. Acad. Sci. U.S.,* **68**, 965 (1971).

CYCLIC PEPTIDES

ELKAN R. BLOUT, CHARLES M. DEBER, and LILA G. PEASE, Department of Biological Chemistry, Harvard Medical School, Boston, Mass. 02115

SYNOPSIS: The conformations of some synthetic and naturally-occurring cyclic peptides are briefly reviewed and recent results from our laboratory on two series of proline-containing cyclic peptides are presented. The conformations of hexapeptides in the series, $cyclo(\text{X-Pro-Y})_2$, which serve as models for β turns in proteins, are discussed. The effects of various side chains on the conformations are indicated. The other series of peptides has the general formula $cyclo(\text{Pro-Gly})_n$, where $n = 2, 3, 4. \ldots$ $Cyclo(\text{Pro-Gly})_3$ and $cyclo(\text{Pro-Gly})_4$ show symmetric conformations with γ turns, as well as other conformations depending upon solvent. These compounds bind mono- and divalent cation salts; binding constants for some of these cation complexes are reported.

INTRODUCTION

Recent interest in cyclic peptides derives from the pioneering work of Vincent du Vigneaud on the isolation, biological properties, and synthesis of the hormones oxytocin and vasopressin (1). Methods for the synthesis of cyclic peptides have been established and hundreds of such compounds have been made and tested for biological action; more recently, cyclic peptides have been the subject of conformational determinations both in solution and in the solid state (2). Conformational determinations of cyclic peptides are of interest to investigators working with new techniques of conformational analysis because these compounds can be obtained in several series of increasing complexity. In addition, and perhaps equally important, cyclic peptides have been recognized as good models for certain structural features of polypeptide chains of proteins. Before turning to recent results from our laboratory, it is appropriate to review some of the important contributions from other workers in this field.

CONFORMATIONS OF CYCLIC PEPTIDES

The most extensively studied naturally occurring cyclic peptide is gramicidin S, $cyclo(\text{Val-Orn-Leu-DPhe-Pro})_2$. Proposed solution conformations for gramicidin S have resulted from the work of Schwyzer *et al.* (3), Stern *et al.* (4), and from Ovchinnikov *et al.* (5). The suggested structures (all derived from NMR data) differ in some respects, but they establish that this cyclic peptide exists in a C_2-symmetric conformation with four intramolecular hydrogen bonds in an antiparallel β-sheet structure with all peptide bonds *trans*. Perhaps the most significant result of these studies was the finding that the DPhe-Pro units occur in corner positions around which the cyclic peptide chain reverses direction. This general structural feature, including the preceding and following residues, is known as the β turn (or β bend, 3_{10} bend, hairpin bend, or reverse turn). Intrigued by the observation that the D residue and the structurally rigid proline residue were present in the β turn of the peptide, investigators addressed themselves to the more general question: What residues are likely to form a β turn and reverse the direction of a peptide or protein chain?

Calculations of sterically permissible arrangements of four amino acids linked to form a 1-4 hydrogen bond, and hence a β turn, were first performed by Venkatachalam (6). He revealed that there are two basic ways, shown in Fig. 1, of accomplishing a reversal of chain direction without affecting hydrogen bonding:[1] Type I in which $\phi_2 = -60°$, $\psi_2 = -30°$; $\phi_3 = -90°$, $\psi_3 = 0°$ and Type II in which $\phi_2 = -60°$, $\psi_2 = 120°$; $\phi_3 = 80°$, $\psi_3 = 0°$. The mirror images of both, Type I' and Type II', are fully allowed. Recent refinements (8) of Venkatachalam's original calculations have not altered his general conclusions.

Based on the hypothesis that a proline residue is an important conformational determinant, particularly with respect to its corner locations, the cyclic hexapeptides that have been synthesized to date may be grouped logically into two general categories: those containing proline and those without proline. The research groups of Ovchinnikov and Kopple have been especially interested in the latter category.

[1] For an explanation of conventions used in dihedral angle nomenclature, see IUPAC-IUB Commission on Biochemical Nomenclature, *Biochemistry,* **9,** 3471 (1970).

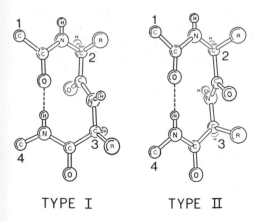

TYPE I   TYPE II

Figure 1. Diagrammatic representation of the two general types of β turns. [Adapted from R.E. Dickerson *et al.* (7).]

Ovchinnikov and his collaborators have synthesized a series of 21 cyclic hexa-peptides (9), which include all possible combinations of Gly and LAla residues, as well as all possible diastereomeric cyclo-hexaalanyls. The NMR spectra of these compounds in dimethylsulfoxide solution invariably displayed two groups of NH protons: four (downfield) protons and two (upfield) protons, of which the latter two had reduced temperature dependences. For example, in a cyclic hexapeptide such as $cyclo(\text{Gly}_4\text{-Ala}_2)$, three pairs of two intramolecular hydrogen bonds are possible, as illustrated in Fig. 2. The NMR spectra, which begin to broaden and coalesce as the temperature is lowered, indicate that a rapidly interconverting mixture of some or all of the possible conformers represents the true situation. The conclusion that one conformer is not clearly preferred over the others is attributed to the small energy differences expected between the A, B, and C forms.

Kopple obtained similar results in his investigations of mono-, di-, and tri-substituted glycine-containing cyclic hexapeptides, such as $cyclo(\text{Gly}_5\text{-Tyr})$ (10). The NMR data were consistent with the presence of two "solvent-shielded" NH protons versus four exposed ones, but it was not possible to describe a unique conformation in solution for this cyclic peptide other than to note that the large tyrosyl side chain is likely to occur at one of the two possible corner positions; in fact, Kopple notes in this study that the preference for a corner might not be exhibited by residues, such as alanine, which have smaller side chains than tyrosine.

The ambiguities arising from rapidly (on the NMR time scale) interconverting conformers are largely resolved in the proline-containing cyclic hexapeptides. The simplest compound of this type is $cyclo(\text{Gly-Pro-Gly})_2$, for which two C_2-symmetric intramolecularly hydrogen-bonded conformations are possible (Fig. 3). Schwyzer et al. (11), who first synthesized and examined this molecule, demonstrated that only one of the alternative structures A or B occurred. He was, however, unable to determine which conformer was observed. Only recently, with the aid of specifically deuterated and ^{13}C-enriched preparations, the conformation of $cyclo(\text{Gly-Pro-Gly})_2$ in aqueous solution has been found (12, 13) to be

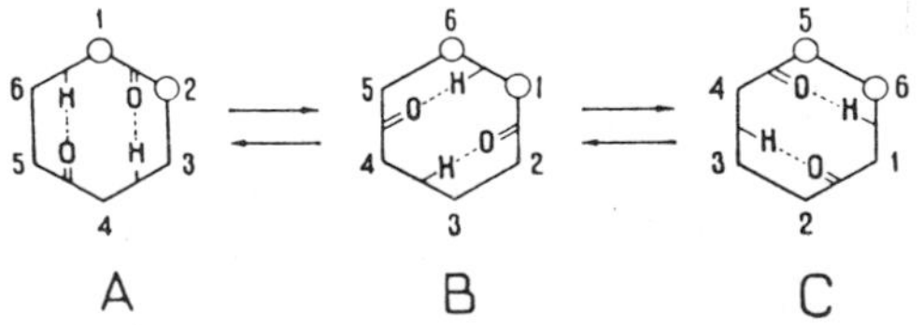

Figure 2. Possible intramolecularly hydrogen-bonded conformations of $cyclo(\text{Gly}_4\text{-Ala}_2)$. [Reprinted with permission from S.L. Portnova et al., Tetrahedron Lett., **33**, 3085 (1971). Copyright by Pergamon Press.]

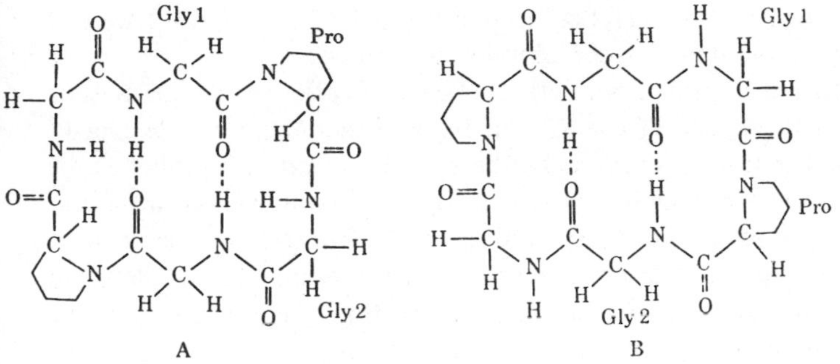

Figure 3. Two C_2-symmetric intramolecularly hydrogen-bonded conformations of *cyclo*(Gly-Pro-Gly)$_2$. [Reprinted with permission from Pease *et al., J. Am. Chem. Soc.*, **95**, 258 (1973). Copyright by the American Chemical Society.]

the structure with the glycine preceding the proline intramolecularly hydrogen-bonded, as in Fig. 3A. The NMR data may be interpreted in terms of a single conformation with the following (ϕ,ψ) angles:

	Gly$_1$	Pro	Gly$_2$
ϕ	~0°	-60°	90°
ψ	~0°	120°	0°

The data are equally consistent with weighted averages of similar conformational states related by rotations of perhaps ± 30° around the C_α bonds of the four glycyl residues. The proposed conformer falls in the (ϕ,ψ) region of a Type-II β turn. However, inspection of the dihedral angles given by Venkatachalam reveals that the same (Pro-Gly) sequence could be accommodated in a Type-I turn. The NMR data definitely exclude a Type-II' turn, which would require Pro in position 3, but do not unequivocally confirm whether Type-I or Type-II turns occur in this peptide.

Subsequent to the initial *cyclo*(Gly-Pro-Gly)$_2$ studies, NMR measurements have been widely employed to determine the position of proline as a function of the primary sequence of the cyclic peptide. The first proline-containing cyclic hexapeptides to be studied in detail, namely, *cyclo*(Gly-Pro-Ser)$_2$ and *cyclo*(Ser-Pro-Gly)$_2$, were found to take up solution conformations with the residue preceding Pro involved in intramolecular hydrogen bonding (14, 15). This information places Pro in position 2 of a β turn in both peptides. It should be noted here that the introduction of proline into a peptide chain allows a further conformational variable, namely, that X-Pro peptide bonds, where X is any amino acid, may occur in energetically competitive *cis* and *trans* forms. Indeed, additional conformers of both cyclic hexapeptides containing one *cis* Gly-Pro bond [*cyclo*(Gly-Pro-Ser)$_2$] or two *cis* Ser-Pro bonds [*cyclo*(Ser-Pro-Gly)$_2$] were found to be in equilibrium with the all-*trans* conformers.

Kopple prepared a series of synthetic cyclic hexapeptides of the general formula *cyclo*(DPhe-Pro-X)$_2$ (where X = LAla, LOrn, or LHis), that assumed the gramicidin S conformation, namely, the Type II'-β turn with proline in the position 3 (16). No evidence for *cis* peptide bonds was obtained in this series. However, the retroisomers of these peptides, *cyclo*(XPro-DPhe)$_2$, were also studied (17) and, upon conformational analysis, gave results that paralleled the *cyclo*(Ser-Pro-Gly)$_2$ findings, including the observation of conformers containing two *cis* X-Pro bonds. Recent efforts of our laboratory to determine more precisely the priority of factors influencing the relationship between primary sequence and solution conformation of proline-containing cyclic hexapeptides are described below.

Dale and Titlestad have synthesized and studied a series of sarcosine-containing cyclic tetrapeptides (18, 19). These peptides are rigid, conformationally restricted molecules that, like *cyclo*(tri-L-prolyl), provide excellent models for use in relating aspects of NMR spectra to specific conformational features. Their early observations on *cyclo*(tetra-Sar) proved to be quite general, namely, that its solution [and crystal (20)] conformation is centrosymmetric with two kinds of magnetically (and physically) nonequivalent Sar residues built into a *cis-trans-cis-trans* peptide-bond backbone (as in peptide bonds to Pro, the *cis* or *trans* configurations of X-Sar bonds are of comparable energy, because they both are N-substituted imino acids). When additional cyclic tetrapeptides were synthesized and examined by NMR spectroscopy, compounds such as *cyclo*(Sar$_3$-Ala), *cyclo*(Sar$_3$-Gly), and *cyclo*(Sar-Gly-Sar-Gly) were all found to possess the *cyclo*(Sar)$_4$ *cis-trans-cis-trans* conformation. In fact, the stability of this class of structures is so high that *cyclo*(Gly$_2$-Sar$_2$) and *cyclo*(Ala$_2$-Sar$_2$) also take this backbone structure, despite the fact that in the latter two cases one *cis*-NH-type amide group (i.e., Sar-Gly and Sar-Ala) must be present.

Antamanide, a cyclic decapeptide [*cyclo*(Phe-Phe-Val-Pro-Pro-Ala-Phe-Phe-Pro-Pro) (all-L)] with antitoxin activity, has been studied extensively with a view toward explaining function by analyzing conformation. Because the primary sequence of the peptide is pseudo-C_2-symmetric, but not actually C_2-symmetric, proton magnetic resonance spectra are rather complex, and precise conformational details were not easily obtained (21). However, with the advent of readily available ^{13}C spectra, used in conjunction with conformational energy calculations, Patel (22) and Tonelli (23) were able to show that antamanide is a naturally occurring material that contains *cis*-peptide bonds in its backbone conformation. Patel found that two *cis* X-Pro bonds are present in aqueous solution—a deduction aided by the observation of chemical shifts of Pro C_β and Pro C_γ carbons in diagnostic positions—but was unable ultimately to distinguish between a conformation with *cis* bonds at Val-Pro and Phe-Pro and one with *cis* bonds at the two Pro-Pro sequences. A further finding was that the addition of sodium cation, with which antamanide forms a strong complex (24), does not change the backbone conformation with respect to peptide-bond geometry, although some variation in dihedral angles occurs in the formation of the ion-binding cavity of the cyclic peptide (22). Karle *et al.* (25) have performed x-ray–crystallographic analysis on Li-antamanide and Na-(Phe[4]Val[6])-antamanide (a C_2-symmetric derivative) and found that in both instances the two *cis* Pro-Pro conformer was present. The finding of a conformation in the solid state may not imply that the same conformation exists in solution, but in this instance, Patel and Tonelli have analyzed their findings in comparison with the x-ray results (26) and concluded that the conformation in the crystal is very close to that observed for the complex in solution.

Another cyclic peptide whose conformation has been reexamined recently is the neurohypophyseal hormone oxytocin [*cyclo*(Cys-Tyr-Ile-Gln-Asn-Cys)-Pro-Leu-Gly-NH$_2$] (27). In addition to the parent compound, a series of oligomeric precursors of oxytocin was studied by ^{13}C NMR in order to form a solid empirical basis for resonance assignment. Although ^{13}C chemical shifts cannot yet be generally interpreted in terms of specific conformational features, substantial variations in these shifts were observed over the series of compounds, particularly for α and β carbons. It was proposed that such variation could not be due to primary structural changes, but must be attributed to alterations of the conformation of the main chain and, probably, the side chains as well.

In order to understand better the conformations of cystine-containing peptides such as the hypophyseal hormones, considerable effort has been directed toward a detailed description of the disulfide bond. Ludescher and Schwyzer (28) have applied the knowledge gained from the conformational analysis of gramicidin S to study the chirality of the cystine disulfide group. They synthesized a derivative of gramicidin S in which the two Orn residues were replaced with two Cys residues, and the disulfide bridge was then closed over the top of the cyclic decapeptide ring. The rigidity and stereochemical requirements of the model were such that the disulfide group should have a dihedral angle of $\pm 120°$, so it was possible to deduce from NMR and CD spectra that the disulfide bridge had so-called P-helical chirality, and that the signs of Cotton effect bands observed were consistent with predictions arising from quadrant rule theory.

Studies on solution conformations of cyclic peptides, especially by identifying the presence and types of β-turn structures, have facilitated proton NMR studies on the penta-, deca-, and pentadecapeptides of the basic sequence, Val-Pro-Gly-Val-Gly, one of the repeating units found on sequence analysis of the protein tropoelastin (29). By using criteria to establish the presence of solvent-shielded protons, particularly their relative lack of movement with increasing temperature, Urry proposed that a Type-II β turn with Val-Pro-Gly-Val residues is a characteristic, recurrent structural feature in the solution conformations of these linear peptides.

The work in our laboratory has involved synthesis, experimental conformational determinations, and computed conformational investigations of cyclic peptides. At the outset of these studies, we decided to emphasize the use of proline-containing peptides because this imino acid residue previously had been shown to prefer the non-α-helical portions of peptide chains of proteins (30) for which the cyclic systems may serve as appropriate models and also because proline confers conformational constraints on any peptide, due to the fact that the N–C_α bond is part of a five-membered-ring system. We shall describe some recent results—and the implication of the results—that have been obtained with two classes of cyclic peptides: cyclic hexapeptides of the general formula *cyclo*(X-Pro-Y)$_2$ and the series *cyclo*(Pro-Gly)$_n$, where n = 2, 3, 4... .

CYCLO(X-PRO-Y)$_2$

In the cyclic hexapeptide series, X and Y are glycine, alanine, serine, phenylalanine, and valine residues. Many of the possible combinations of X and Y have been synthesized using both D- and L-amino acids. These cyclic hexapeptides serve as useful models for β turns; such turns have recently been recognized to comprise a large portion (27–33%) of the polypeptide chains of globular proteins (31, 32). In general, the favored all-*trans* peptide bond conformer found for the cyclic hexapeptides examined is C_2-symmetric with the residue preceding the proline hydrogen-bonded (Fig. 4). [The notable exception has been Kopple's series of (non-Gly-containing) peptides *cyclo*(DPhe-Pro-X)$_2$ in which the NHs of the residue following Pro are solvent-shielded and the Pros are *cis'*.] Evidence for these conformations comes primarily from studies of the temperature dependence of the chemical shifts of the NH protons. Shifts of the resonances of NH protons with increasing temperature are generally small for those protons that are intramolecularly hydrogen-bonded or buried and much larger for those NH protons capable of forming hydrogen bonds to the solvent. In Fig. 5 the glycine and alanine NH peak positions for *cyclo*(Gly-Pro-Ala)$_2$ are plotted

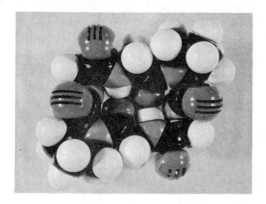

Figure 4. Photograph of a CPK model showing the proposed solution conformation of *cyclo*(Gly-Pro-Gly)$_2$. (Compare Figure 3A.)

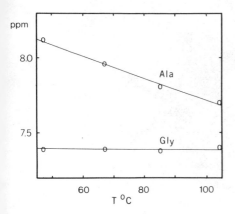

Figure 5. Temperature dependence of peptide-NH resonances of *cyclo*(Gly-Pro-Ala)$_2$. Chemical shifts δ scale, downfield from internal TMS [from (32)].

versus temperature. These data strongly suggest that the glycyl peptide protons are solvent-shielded, most likely through participation in internal hydrogen bonds, and those of the alanine residues are externally hydrogen-bonded. The fact that only a single spectrum is obtained for each pair of like residues indicates that the conformation has C_2 symmetry.

It is clear, however, that other conformers may exist even in these simple model compounds. For example, if the prolines are preceded by L residues, it is found that a conformer with two *cis* X-proline peptide bonds exists. This conformer is stabilized in solvents such as dimethylsulfoxide, which is a strong hydrogen-bond acceptor. The two *cis* conformers (Fig. 6) contain no intramolecular hydrogen bonds. Model studies on the all-*trans* conformers reveal steric problems between the side chain of the L residue and the proline δ methylene group. Consistent with this, the amount of the conformer with *cis* X-Pro peptide bonds, which appears to relieve the unfavorable contacts (compare Figs. 4 and 6) is observed to increase as the side-chain size increases (33) (Table 1).

Investigations of *cyclo*(Gly-Pro-X)$_2$ peptides, where X is an L residue, indicate that these compounds exist in a C_2-symmetric conformation that is apparently stabilized by two intramolecular Gly-Gly hydrogen bonds. In addition, several cyclic hexapeptides of this type display an asymmetric conformation that has been shown to contain one *cis* X-Pro bond. When valine is the X residue, preliminary results indicate that in dimethylsulfoxide solution the asymmetric conformer is favored over the symmetric. In contrast, in the phenylalanine cyclic hexapeptide, only an all-*trans* conformer is observed. These results suggest that the predominant conformation is determined not only by the size of the X residue, but also by subtleties of its side chain. Table 2 summarizes preliminary data on the amounts of *cis* conformers for these cyclic hexapeptides as a function of varying side chains. These data were obtained from ^{13}C NMR analysis,

Figure 6. Photograph of a CPK model of the proposed two *cis* X-Pro bond-containing conformer. Model shown is *cyclo*(Val-Pro-Gly)$_2$.

Table 1. Distribution[a] of all-*trans* and two *cis* conformers for *cyclo*(X-Pro-Gly)$_2$ hexapeptides

Peptide	Solvent	All-*trans* (%)[b]	Two *cis* (%)[b]
cyclo(Gly-Pro-Gly)$_2$	D_2O	~100	~0
	Me_2SO-d_6	~100	~0
cyclo(Ala-Pro-Gly)$_2$	D_2O	88	12
cyclo(Ser-Pro-Gly)$_2$[c]	D_2O	75	25
	Me_2SO-d_6	20	80
cyclo(Phe-Pro-Gly)$_2$	Me_2SO-d_6	20	80
cyclo(Val-Pro-Gly)$_2$	D_2O	13	87
	Me_2SO-d_6	<20	>80

[a]Measured by ^{13}C NMR.

[b]Uncertainty = ±5%.

[c]Measured by ^1H NMR.

Table 2. Distribution[a] of all-*trans* and one *cis* conformers for *cyclo*(Gly-Pro-X)$_2$ hexapeptides

Peptide	Solvent	All-*trans* (%)[b]	One *cis* (%)[b]
cyclo(Gly-Pro-Gly)$_2$	D$_2$O	~100	~0
cyclo(Gly-Pro-Ala)$_2$	D$_2$O	76	24
cyclo(Gly-Pro-Ser)$_2$	D$_2$O	77	23
cyclo(Gly-Pro-Phe)$_2$[c]	Me$_2$SO-d_6	~100	~0
cyclo(Gly-Pro-Val)$_2$	D$_2$O	58	42
	Me$_2$SO-d_6	23	77

[a]Measured by ^{13}C NMR.
[b]Uncertainty = ±10%.
[c]Synthesized by P.E. Young.

utilizing the Pro C$_\beta$ and C$_\gamma$ resonances as indicators of the state of isomerization around the X-Pro bond.

The fact that multiple conformations are found in this series of cyclic hexapeptides, depending on solvent, side-chain size, and character of the side chain, suggests that the proline residue at or near β turns in proteins may have at least two structural roles: It confers conformational rigidity because of the restriction to rotation about the N–C$_\alpha$ bond and, in addition, allows for the existence of two distinct isomers (*trans* or *cis*) with respect to the preceding peptide bond residue. Hexapeptides without L residues (other than the prolines) show *only* the all-*trans* conformer, with the residue preceding Pro intramolecularly hydrogen-bonded. We have not yet completely analyzed the results, but it is necessary to examine the importance of 1–4 hydrogen bonds as a stabilizing feature for β-turn conformations. Calculations (34) indicate that the formation of 1–4 hydrogen bonds may not be a significant stabilizing influence for such turns in the cyclic-hexapeptide system. Infrared-spectroscopic studies in solution and in the solid state now underway possibly will yield information concerning the strength of hydrogen-bonded interactions of these cyclic hexapeptides. Within the next year, collation and comparison of all data on the many synthetic cyclic hexapeptides now available will help to answer several fundamental questions, such as: (1) Why do some cyclic hexapeptides assume asymmetric conformations while others take up only symmetric conformations? (2) What features disrupt the β turn? (3) What residues can be tolerated in the various positions of the turn? (4) Are hydrogen bonds important in these β turns? (5) How do solvents influence the stability of β turns?

CYCLO(PRO-GLY)$_n$ PEPTIDES

As indicated above, we have been examining another series of proline-containing cyclic peptides, namely, those of the general formula *cyclo*(Pro-Gly)$_n$ (n = 2, 3, 4...). All of these peptides have high sequential symmetry, and those with n = 3, 4... may form central cavities suitable for the binding of ions or molecules.

Recent work on *cyclo*(Pro-Gly)$_2$ (34) [and studies by Dale and Titlestad on *cyclo*(Sar-Ala)$_2$ (35)] indicates the presence of conformational interconversions that are slow on the NMR time scale; these peptides can adopt *cis-trans-cis-trans* peptide bond backbones, but configurationally, they do not have the usual "LLDD" (where Gly or Sar is "D") sequence. The observed equilibria in *cyclo*(Pro-Gly)$_2$ have been interpreted in terms of a hindered ψ angle rotation between the *cis'* and *trans'* rotamers of the Pro C$_\alpha$–C=0 bond.

Cyclo(Pro-Gly)$_3$ and *cyclo*(Pro-Gly)$_4$ are the compounds of this series that have been examined in the greatest detail. Both *cyclo*(Pro-Gly)$_3$ and *cyclo*(Pro-Gly)$_4$ are incapable of assuming conformations in which there are 1–4 transannular hydrogen bonds. Both of these cyclic peptides, however, may form a type of polypeptide chain bend, or turn, that has been found, albeit less frequently than β turns, in proteins (37). This type of turn, termed the γ *turn* (38), requires three residues to reverse chain direction—compared to four residues in the β turn. The hydrogen bonding in γ turns occurs between the carbonyl oxygen of residue 1 and the NH group of residue 3, as shown in Fig. 7 (34, 39). It has been proposed that one of the conformers of *cyclo*(Pro-Gly)$_3$ has three such γ turns.

Cyclo(Pro-Gly)$_3$. The original NMR studies (40) identified three conformational states for *cyclo*(Pro-Gly)$_3$. Two of these are C$_3$-symmetric, as evidenced by the magnetic equivalence of three Pro-Gly units. The third conformer, however, is asymmetric and gives separate resonances for each Pro-Gly unit. In a subsequent theoretical study (41), the intramolecular potential energy (the sum of the van der Waals and dipolar interactions) was computed for all possible *cyclo*(Pro-Gly)$_3$ conformers. This investigation considered dihedral angle variations and, in addition, the possible isomerization of the Gly-Pro peptide bonds. The complete exploration of conformational space via computer revealed a class of C$_3$-symmetric conformers that had not been previously considered, namely, a class having all peptide bonds *trans* and three intramolecular 1–3 hydrogen bonds.

In a recent investigation (39), CD and both ^{1}H and ^{13}C NMR spectroscopy were employed, so that it is now possible to identify unambiguously all *cyclo*-(Pro-Gly)$_3$ conformers. One conformer, designated S, occurs in relatively inert solvents (such as chloroform and dioxane). Conformer S is C$_3$-symmetric, has all *trans*-peptide bonds, and is stabilized by 1–3 intramolecular hydrogen bonds (Fig. 8). Upon addition of cations to solutions of conformer S, it is found that a new C$_3$-symmetric conformation is formed, designated S*. For the formation of

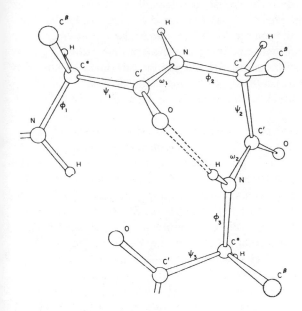

Figure 7. Diagrammatic representation of a γ turn. [Reprinted with permission from G. Némethy and M.P. Printz, *Macromolecules*, **5**, 818 (1972). Copyright by the American Chemical Society.]

Figure 8. Photograph of a CPK model of the S conformer of *cyclo*(Pro-Gly)$_3$. [Reprinted with permission from Madison *et al.*, *J. Am. Chem. Soc.*, in press. Copyright by the American Chemical Society.]

conformer S* from conformer S, the Pro-Gly peptide unit must be rotated approximately 80° about the proline C_α–C' bonds and concomitantly about the glycine N–C_α bonds. This rotation breaks the intramolecular 1–3 hydrogen bonds, exposes the NHs to solvents, and brings the three proline carbonyl groups closer together on one face of the molecule. The glycine-carbonyl oxygens lie in a similar relation to each other on the opposite face of the molecule. These arrays of carbonyl oxygens are in positions suitable to bind some solvents or cations.

In addition to the symmetric conformer (S), which exists in apolar solvents, and the cation complex (conformer S*), both of which are C_3-symmetric, an asymmetric conformer (A) occurs in polar solvents, such as water and dimethyl-sulfoxide. This conformer appears to have one *cis* Gly-Pro peptide bond and one *cis'* proline ψ angle. The intramolecular potential energy computed for conformer A is several kcal/mole above that of conformers S and S*. Computations (41) have shown that interactions of *cyclo*(Pro-Gly)$_3$ with bulk solvent will not shift the equilibrium to populate conformer A. Thus, it may be concluded that the large intramolecular potential energy difference between conformer A and the C_3-symmetric conformers must be compensated by strong specific peptide-solvent interactions. Water would be particularly effective in stabilizing conformer A, because models indicate that four water molecules can form eight nearly linear hydrogen bonds to bridge the six carbonyls in two groups of three.

Although the binding of cations to *cyclo*(Pro-Gly)$_3$ in aqueous solution is weaker than in less polar solvents, it is still stronger than that expected from its component amide groups acting singly. From the data of Von Hippel *et al.* (42), sodium and calcium perchlorates in water solution bind to an individual amide group of polyacrylamides with binding constants of 0.10 and 0.21 M^{-1}, respectively. The binding constants of these two salts with *cyclo*(Pro-Gly)$_3$ are 3.7 and 110 times the polyacrylamide values for six peptide groups.

The binding behavior of *cyclo*(Pro-Gly)$_3$ parallels that of the naturally occurring cyclic peptide, antamanide (24). Antamanide shows higher selectivity among cations of a given charge, although both antamanide and *cyclo*(Pro-Gly)$_3$ show preference for Na^+ and Ca^{2+} over larger cations. An additional property of *cyclo*(Pro-Gly)$_3$ is that its cation complexes are completely extracted from organic phases by water, suggesting that this cyclic peptide may show unusual effects in lipid bilayer mediated transport systems.

Cyclo(Pro-Gly)$_4$. Preliminary analysis of NMR data suggests that in chloroform *cyclo*(Pro-Gly)$_4$ has a conformation similar to that designated "S" for *cyclo*(Pro-Gly)$_3$ in that all eight peptide bonds are *trans*, and the four Gly-NH peptide protons are involved in γ turns. The addition of salts apparently causes the peptide units joining the Pro and Gly residues to rotate inward, forming the binding cavity for cations. In *cyclo*(Pro-Gly)$_4$ this cavity may be larger than that in *cyclo*(Pro-Gly)$_3$. There is now evidence that the size of the binding site varies with the size of the cation.

We have been able to isolate crystalline complexes of *cyclo*(Pro-Gly)$_4$ with all the alkali metal thiocyanates and with calcium perchlorate. Salts, which are practically insoluble in chloroform, are solubilized in this solvent by the presence of *cyclo*(Pro-Gly)$_4$. This suggests that the apparent solubility of the salt is related to the formation of a peptide complex.

Preliminary ion-binding constants for *cyclo*(Pro-Gly)$_4$ cation complexes in water and acetonitrile (43) have been obtained. Assuming a 1:1 stoichiometry, the binding constants (in M^{-1}) in acetonitrile range from approximately 10^3 for Na^+ to 10^5–10^6 for Mg^{2+}, Ca^{2+}, and Ba^{2+}.

Many interesting experiments can be performed to take advantage of the larger size of the binding cavity of *cyclo*(Pro-Gly)$_4$. One of us (C.M.D.) has been exploring the binding of amino acid salts to *cyclo*(Pro-Gly)$_4$. Carbon-13 NMR data obtained from chiral amino acid ester salts, such as D- and L-phenylalanine methyl ester hydrochloride, show separate resonances for certain carbons of the L salts and the D salts in the presence of *cyclo*(Pro-Gly)$_4$, suggesting an enantiomeric differentiation by the host peptide. Evidence for similar diastereomeric complexes has been obtained with several other pairs of enantiomeric amino acids, and these data suggest that still more highly stereoselective cyclic-peptide host molecules can be synthesized.

ACKNOWLEDGMENTS

This work has been supported, in part, by U.S. Public Health Service Grants AM-07300 and AM-10794. One of us (L.G.P.) held a National Science Foundation Predoctoral Fellowship. We are grateful to our colleague, Vincent S. Madison, for helpful discussions and critical readings of this manuscript.

REFERENCES

1. For a review of previous work, see: Vincent Du Vigneaud, *Johns Hopkins Med. J.,* **124**, 53 (1969).
2. F.A. Bovey, in *Peptides, Polypeptides, and Proteins,* Proceedings of the Rehovot Symposium 1974, Wiley(Interscience), New York, 1974.
3. (a) R. Schwyzer and U. Ludescher, *Biochemistry,* **7**, 2514 (1968); (b) R. Schwyzer and U. Ludescher, *Helv. Chim. Acta,* **52**, 2033 (1969).
4. A. Stern, W.A. Gibbons, and L.C. Craig, *Proc. Nat. Acad. Sci. U.S.,* **61**, 734 (1968).
5. Yu.A. Ovchinnikov, V.T. Ivanov, V.F. Bystrov, A.I. Miroshnikov, E.N. Shepel, N.D. Abdullaev, E.S. Efremov, and L.B. Senyavina, *Biochem. Biophys. Res. Comm.,* **39**, 217 (1970).
6. C.M. Venkatachalam, *Biopolymers,* **6**, 1425 (1968).
7. R.E. Dickerson, T. Takano, D. Eisenberg, O.B. Kallai, L. Samson, A. Cooper, and E. Margoliash, *J. Biol. Chem.,* **246**, 1511 (1971).
8. R. Chandrasekaran, A.V. Lakshminarayanan, U.V. Pandya, and G.N. Ramachandran, *Biochim. Biophys. Acta,* **303**, 14 (1973).

9. S.L. Portnova, V.V. Shilin, T.A. Balashova, J. Biernat, V.F. Bystrov, V.T. Ivanov, and Yu.A. Ovchinnikov, *Tetrahedron Lett.*, **33**, 3085 (1971).

10. (a) K.D. Kopple, M. Ohnishi, and A. Go, *J. Am. Chem. Soc.*, **91**, 4264 (1969); (b) K.D. Kopple, M. Ohnishi, and A. Go, *Biochemistry*, **8**, 4087 (1969); (c) K.D. Kopple, A. Go, R.H. Logan, Jr., and J. Savdra, *J. Am. Chem. Soc.*, **94**, 973 (1972).

11. (a) R. Schwyzer, P. Sieber, and B. Gorup, *Chimia (Switz.)*, **12**, 90 (1958); (b) R. Schwyzer, *Rec. Chem. Progr.*, **20**, 147 (1959); (c) R. Schwyzer, J.P. Carrion, B. Gorup, H. Nolting, and A. Tun-Kyi, *Helv. Chim. Acta*, **47**, 441 (1964).

12. R. Schwyzer, Ch. Grathwohl, J.-P. Meraldi, A. Tun-Kyi, R. Vogel, and K. Wüthrich, *Helv. Chim. Acta*, **55**, 2545 (1972).

13. L.G. Pease, C.M. Deber, and E.R. Blout, *J. Am. Chem. Soc.*, **95**, 258 (1973).

14. D.A. Torchia, A. Di Corato, S.C.K. Wong, C.M. Deber, and E.R. Blout, *J. Am. Chem. Soc.*, **94**, 609 (1972).

15. D.A. Torchia, S.C.K. Wong, C.M. Deber, and E.R. Blout, *J. Am. Chem. Soc.*, **94**, 616 (1972).

16. K.D. Kopple, A. Go, T.J. Schamper, and C.S. Wilcox, *J. Am. Chem. Soc.*, **95**, 6090 (1973).

17. K.D. Kopple, T.J. Schamper, and A. Go, *J. Am. Chem. Soc.*, in press.

18. J. Dale and K. Titlestad, *Chem. Comm.*, 656 (1969).

19. J. Dale and K. Titlestad, *Chem. Comm.*, 1403 (1970).

20. P. Groth, *Acta Chem. Scand.*, **24**, 780 (1970).

21. A.E. Tonelli, D.J. Patel, M. Goodman, F. Naider, H. Faulstich, and Th. Wieland, *Biochemistry*, **10**, 3211 (1971).

22. D.J. Patel, *Biochemistry*, **12**, 667 (1973).

23. A.E. Tonelli, *Biochemistry*, **12**, 689 (1973).

24. Th. Wieland, H. Faulstich, W. Burgermeister, W. Otting, W. Möhle, M.M. Shemyakin, Yu.A. Ovchinnikov, V.T. Ivanov, and G.G. Malenkov, *FEBS Lett.*, **9**, 89 (1970).

25. I.L. Karle, J. Karle, Th. Wieland, W. Burgermeister, H. Faulstich, and B. Witkop, *Proc. Nat. Acad. Sci. U.S.*, **70**, 1836 (1973).

26. D.J. Patel and A.E. Tonelli, *Biochemistry*, **13**, 788 (1974).

27. A.I. Richard Brewster, V.J. Hruby, A.F. Spatola, and F.A. Bovey, *Biochemistry*, **12**, 1643 (1973).

28. U. Ludescher and R. Schwyzer, *Helv. Chim. Acta*, **54**, 1637 (1971).

29. D.W. Urry, W.D. Cunningham, and T. Ohnishi, *Biochemistry*, **13**, 609 (1974).

30. A.G. Szent-Györgyi and C. Cohen, *Science*, **126**, 697 (1957).

31. J.L. Crawford, W.N. Lipscomb, and C.G. Schellman, *Proc. Nat. Acad. Sci. U.S.*, **70**, 538 (1973).

32. P.Y. Chou and G.D. Fasman, *Biochemistry*, **13**, 222 (1974).

33. L.G. Pease, Ph.D. Thesis, Harvard University, 1974.

34. V.S. Madison, in *Peptides, Polypeptides and Proteins*, Proceedings of the Rehovot Symposium 1974, Wiley(Interscience), New York, 1974.

35. C.M. Deber, E.T. Fossel, and E.R. Blout, *J. Am. Chem. Soc.*, in press.

36. J. Dale and K. Titlestad, *Chem. Comm.*, 255 (1972).

37. B.W. Matthews, *Macromolecules,* **5**, 818 (1972).
38. G. Némethy and M.P. Printz, *Macromolecules,* **5**, 755 (1972).
39. V. Madison, M. Atreyi, C.M. Deber, and E.R. Blout, *J. Am. Chem. Soc.,* in press.
40. C.M. Deber, D.A. Torchia, S.C.K. Wong, and E.R. Blout, *Proc. Nat. Acad. Sci. U.S.,* **69**, 1825 (1972).
41. V. Madison, *Biopolymers,* **12**, 1837 (1973).
42. P.H. von Hippel, V. Peticolas, L. Schack, and L. Karlson, *Biochemistry,* **12**, 1256 (1973).
43. P.A. Kosen and V.S. Madison, unpublished results.

CYCLIC PEPTIDES CONTAINING DL SEQUENCES

K.D. KOPPLE, T.J. SCHAMPER, and **A. GO**, Department of Chemistry, Illinois Institute of Technology, Chicago, Ill. 60616

SYNOPSIS: The backbone conformation of a series of cyclic hexapeptides of the sequence type (Gly-D-X-X-Gly-D-X-X) was investigated by proton magnetic resonance and model-building. All of the peptides studied appear to have the same backbone; variation among the side chains of Leu, Val, Orn, and Phe does not produce changes in backbone-determining interactions. The general region of conformation space occupied by these peptides is most likely a β structure in which the glycine residues are most nearly extended and chain reversal occurs primarily in the DL sequence. There are no good transannular hydrogen bonds, probably because rotational freedom of the D-residue side chain requires that the Gly-D-X peptide bond be more nearly perpendicular to than in the mean plane of the hexapeptide ring. Considering this result and the available data on proline-containing cyclic hexapeptides, it is concluded that a sequence D-X-Pro-D-X, where the D residues may be glycine, is the sequence most likely to form a stable β turn in an unconstrained system.

INTRODUCTION

Cyclic hexapeptides of C_2 symmetry provide some of the simplest constrained peptide backbones in which the amino acid residues may adopt all values of ϕ and ψ. As is now well established, they frequently adopt backbone conformations, originally suggested by Schwyzer (1), that can be described as two β turns connected by two extended residues. There may or may not be good hydrogen bonds across the β turns. Using the peptides that adopt this β conformation, it is possible to study factors that influence backbone conformation, at least at the level that distinguishes among glycine, proline, and other substituted residues, by considering frame shifts in the relation between sequence and backbone.

When proline occurs in a cyclic hexapeptide with the general backbone described, it cannot occur in the extended position, numbered 1. [We use the numbering adopted by Venkatachalam in his discussion of such systems (2).] The experimental data for proline-containing cyclic hexapeptides, obtained by the groups of Blout and of Schwyzer, and in our own laboratory, are collected in Table 1, which lists the predominant arrangements of residues within the β-turn framework, in backbones with *trans* X-Pro peptide bonds, as deduced from NMR studies, chiefly of dimethylsulfoxide and aqueous solutions. These observations are in accord with the following generalizations:

1. The β turn X_2-Pro_3 is excluded by interference between $C_X{}^\beta$ and C_{Pro}^δ (10,11). No such restriction exists for D-X_2-Pro_3.

Table 1. Stable arrangements in proline-containing cyclic-hexapeptide backbones containing β turns joined by extended residues

| Position | | | Another | Internal | |
1	2	3	Structure?[a]	N-H, δ[b]	Reference
Gly	Pro	Gly	No	7.6	3, 4
D-X	Pro	Gly			5
X	Pro	Gly	Yes	8.0	6
X	Pro	D-X	Yes	8.0–8.2	7
Gly	Pro	X	Yes	7.6	8
X	D-X	Pro	No	7.0–7.4	9

[a] In some cases, structures containing *cis* X-Pro peptide bonds are also present, to an extent dependent on solvent. The structure containing β turns and *trans* X-Pro bonds in these cases is presumably less stable than in the peptides in which only the one structure occurs.
[b] Chemical shift in Me$_2$SO solution, internal TMS reference, 20–30°. Where a range is given, more than one compound is included.

2. For Pro$_2$-X$_3$ turns, LD [Venkatachalam Type II (2)] are favored over LL turns (Type I). In the Type-II turns, there is more allowable variation at small cost in energy in the conformational angles of both residues (12), so that entropy is likely to be an important component of this preference.

3. Proline prefers position 2 ($\psi = 120°$) to position 3 [ψ nearer $-60°$ (9)]. Here again, greater torsional freedom about $\psi = 120°$ (11, 13) may be important.

4. Glycine readily takes position 3, probably by default, and possibly because it can more readily adapt itself to values of ψ near 0° by opening the C'–C$_\alpha$–N bond. Glycine with $\psi = 0° \pm 30°$ occurs frequently in proteins.

5. At the level at which the above preferences operate, distinctions among side chains of nonproline residues is not important (7, 9) and accommodation of any nonproline residue to the extended conformation (position 1) is not a barrier.

These generalizations support predictions of the relevant conformational calculations and are in agreement with the experimental observations in proteins (14) that proline is the most likely residue to occur in position 2 of a β turn and glycine is highly likely to appear in position 3.

For cyclic hexapeptides without the constraint of proline residues, the data are more limited. Peptides of the type *cyclo*(Gly-X-Gly)$_2$ are stable in that region of conformational space in which the substituted residue is in position 2, as might have been expected if the principal operative factor was that the substituted residue itself be in the most favorable region of its dipeptide energy map

(15, 16). The observations of Ovchinnikov *et al.* (16) indicate that *cyclo*(DAla-Ala-Ala)$_2$ has a strongly preferred structure[1] in which one pair of peptide protons is solvent-shielded, that is, directed to the inside of a β turn, but they do not specify the positions of the residues within the β-turn frame. Similarly, their observations on *cyclo*(Gly-Ala-Ala)$_2$ indicate that one pair of alanine residues is more solvent-shielded than glycine or other alanine residues, but this again does not precisely fix the more favored sequence.

We have lately prepared and studied by proton magnetic resonance a group of peptides of the sequence *cyclo*(Gly-D-X-X)$_2$. These are listed in Table 2. If β turns are formed by these peptides, it is not certain from what has gone before which of the possibilities, Gly$_2$-D-X$_3$, D-X$_2$-X$_3$, or X$_2$-Gly$_3$, will be favored.

EXPERIMENTAL METHOD

The cyclic peptides were prepared by azide coupling in dimethyl formamide, as in previous work (7, 9). The cyclization yields of isolated pure crystalline products are given in Table 2. The open-chain hexapeptide precursors were prepared by standard methods, *N*-hydroxysuccinimide ester and nonaqueous azide couplings, although analytically pure intermediates were not obtained in all stages. The final products were chromatographically homogeneous and were characterized by elemental analysis, mass spectra, and their high-resolution proton magnetic resonance spectra.

Proton magnetic resonance spectra were obtained partly by using a Bruker HX-270 instrument at the University of Chicago (17) and partly by using the 250 MHz spectrometer of the NMR Facility for Biomedical Research at Carnegie-Mellon University (18).

---

[1] In a peptide exchanging rapidly on the NMR time scale between conformations of comparable energy, the NMR observables are averaged. Therefore we take as a guiding, but not overriding, principle that a spectrum that shows a wider than average range of those parameters that change with conformation (e.g., peptide proton chemical shifts, coupling constants, and temperature coefficients) is more likely to correspond to a single stable backbone than one in which these parameters all have similar values.

Table 2. Cyclic peptides *cyclo*(Gly-D-X-X)$_2$ examined in this study

		Cyclization Yield (%)
I.	*cyclo*(Gly-DVal-Leu)$_2$	56
II.	*cyclo*(Gly-DLeu-Leu)$_2$	>31
III.	*cyclo*(Gly-DVal-Leu-Gly-DOrn-Orn)	53
IV.	*cyclo*(Gly-DPhe-Phe-Gly-DLeu-Leu)	>30
V.	*cyclo*(Gly-DPhe-Phe-Gly-DOrn-Orn)	65

RESULTS

The two peptides, I and II in the present series, with C_2 sequence symmetry exhibit proton magnetic resonance lines in which like residues have identical lines (Fig. 1). Spectral assignments are unambiguous for *cyclo*(Gly-DVal-Leu)$_2$ and for the DVal and Leu of the unsymmetrical peptide *cyclo*(Gly-DVal-Leu-Gly-DOrn-Orn). In all of the peptides except I, a distinction was made by

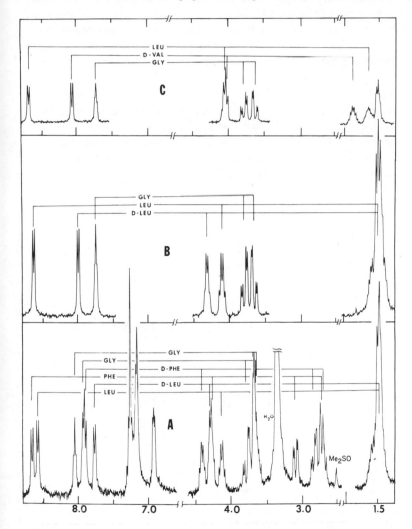

Figure 1.   270 MHz proton resonance spectra (Me$_3$Si reference) of (A) *cyclo*(-Gly-DPhe-Phe-Gly-DLeu-Leu-), (B) *cyclo*(-Gly-D-Leu-Leu-)$_2$, and (C) *cyclo*(-Gly-DVal-Leu-)$_2$ all in Me$_2$SO-d_6, at 20°C, concentration about 45 mg/ml. Assignments made by decoupling are shown. Methyl resonances at 1.0 ppm are not shown.

analogy between D and L residues of the same kind, for example, DPhe and Phe, because the unsymmetrical peptides show peptide proton resonances closely paired in chemical shift and coupling constant. Fig. 1 shows these distinctions. The pairing of peptide proton resonances strongly suggests that the backbone is dependent on the sequence Gly-D-X-X, and not on the nature of X, whether it be Val, Leu, Phe, or Orn. In dimethylsulfoxide (Me$_2$SO), methanol, and water, the D residue peptide protons of all the peptides are in the range 7.7–8.2 ppm, and the L residue peptide protons are in the range 8.6–9.0 ppm. The chemical shift of the peptide protons of the glycine residues is more strongly dependent on solvent. These observations are illustrated graphically in Fig. 2 for a subset of peptides from which the magnetic-shielding effects of the aromatic side chains are absent.

A criterion for a stable β turn in these peptides, apart from consistent coupling constants, is the presence of a pair of peptide protons shielded from the solvent. Evidence from the usual experiments follows.

Temperature coefficients of the peptide proton resonances are given in Table 3. In dimethylsulfoxide and methanol, the glycine residues have the low temperature coefficients expected for internal protons. The L-residue protons

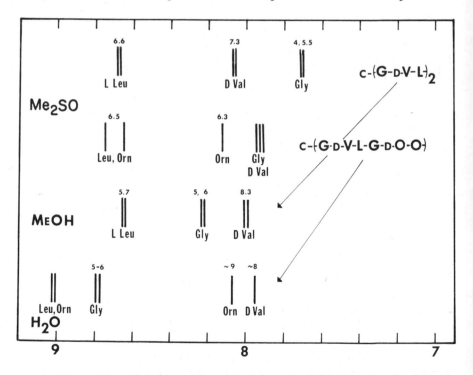

Figure 2. Peptide proton resonances of *cyclo*(Gly-DVal-Leu)$_2$ and *cyclo*(Gly-DVal-Leu-Gly-DOrn-Orn) indicating pairing of lines. Coupling constants are given above the lines.

Table 3. Temperature coefficients of peptide proton resonances of $cyclo$(Gly-D-X-X)$_2$

Peptide	Solvent	Range	Coefficient (ppm upfield/deg)		
			Gly	D-X	L-X
1	Me$_2$SO	35–100	0.001	0.006	0.007
1	HFP	5–35	0.003	0.006	0.007
1	TFE	−25–+65	0.005	0.004	0.009
2	Me$_2$SO	35–100	0.001	0.006	0.008
3	MeOH/Me$_2$SO	−10–+20	0.003	0.0045	0.008
3	MeOH	−40–+20	0.002[a]	−[b]	0.010 (Leu) 0.008 (Phe)

[a] One glycine; the other is obscured by overlapping resonances.
[b] Overlap of resonances over much of the range.

uniformly have the large coefficients of solvent-exposed amide protons. Consistent with this is the experiment of Fig. 3, which shows that the addition of a nitroxide radical to a dimethylsulfoxide solution of I (or II) broadens the resonance of the Leu peptide protons most and affects the resonance of the Gly protons least. This is the result also in hexafluoro-2-propanol (HFP). The differential effect is not large in either solvent, however, and judging by previous experience (9, 19), none of the protons is greatly sequestered. The glycine peptide proton resonances also shift upfield least when HFP is substituted for Me$_2$SO (19, 20); they do shift upfield, however, and in general, it appears that the resonances of fully sequestered protons shift slightly downfield under this change (7, 9, 19, 20). [In this connection, the variability of the chemical shift of the glycine proton upon going from Me$_2$SO to methanol or water (Fig. 2) also argues against a completely shielded glycine proton.]

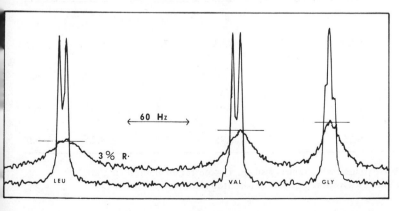

Figure 3. 270 MHz spectrum of peptide proton region showing the differential line broadening effect of 3% 3-oxyl-2,2,4,4-tetramethyloxazolidine in Me$_2$SO.

There are difficulties in interpreting all of the NH resonance data in terms of a stable β-turn structure with glycine in position 1 (extended), the D residue at 2, and the L residue at 3. An exchange study of $cyclo$(Gly-DLeu-Leu)$_2$ in Me$_2$SO containing methanol indicates that all three kinds of peptide protons exchange in minutes, at the same rate. This behavior is not unusual for cyclic hexapeptides. In HFP, however, for both I and II, the L-residue proton has a half-life of about 10 hours, the D-residue proton has a half-life of about 1 hour, and the glycine proton has a half-life of about 15 minutes. By the other methods (temperature dependence, exposure to neutral radical, and solvent shift) the L-residue proton is highly exposed, and as just indicated, some case can be made for a somewhat buried Gly proton. (If it were required, a rationalization of this exchange experiment could be made on the basis of base catalysis of exchange by nearby carbonyl groups). Furthermore, if the glycine residue is extended, with an internally directed peptide proton, the D residue is in position 2 of the β turn. Previous indications with other likely β turns [gramicidin S (20, 21), evolidine (19, 22), oxytocin, and related substances (23–25), and $cyclo$(Gly-X-Gly)$_2$ (15, 19)] are that the peptide proton in position 2 is highly exposed to solvent, judged by temperature coefficient and the effect of shifting to a nonbasic solvent. In the present case it is the L residue that appears most exposed, and the peptide proton of the D residue appears, by these criteria, less exposed. In some solvents (Table 3) the temperature coefficients of the D residue and the glycine residue approach each other.

A study of the peptide proton resonances of II in the four butyl alcohols (Table 4) shows that the peptide proton resonance of the DLeu residue is almost independent of most of the solvent alkyl group. A possible explanation for this

Table 4. Peptide proton resonances of $cyclo$(Gly-DLeu-Leu)$_2$

Solvent	Gly		DLeu		Leu	
	δ[a]	J[b]	δ[a]	J	δ[a]	J
n-BuOH	8.22	5.8 4.8	8.02	7.8	8.43	6.0
i-BuOH	8.25	5.7 4.2	7.99	7.8	8.41	6.0
s-BuOH	8.12	6.0 4.5	8.03	7.4	8.37	6.3
t-BuOH	8.05	–	7.95	7.0	8.24	6.0

[a] At 20°C, relative to internal tetramethylsilane, concentration about 50 mg/ml.
[b] Obtained from NH resonances plus part of α-proton multiplet, $\pm$ <0.5 Hz.

phenomenon is that the residue is either completely shielded from the solvent or freely exposed; the NH resonance of *N*-methylacetamide is similarly unaffected (9). The other two kinds of peptide proton show some sensitivity to the shape of the alcohol solvent, indicating some hindrance to solution.

The HC_α–NH coupling-constant data argue against a highly stable backbone conformation for these peptides. The coupling constants show a distinct trend upon going to the less hindered hydrogen-bond-accepting solvents (water and methanol) from Me_2SO and HFP, in that coupling for the D residue tends to increase from about 7 Hz (140° dihedral angle) to near 9 Hz (160°) and coupling for the L residue tends to decrease from 7 Hz toward 5 Hz (130°). The glycine coupling constants, which are consistent with averaging about $\phi = 180°$ or ±60°, tend to increase on going to water. (For $\phi = 180°$, that is, oscillations about a 120° dihedral angle, this phenomenon indicates greater flexibility in water.)

If there is a possibility that the conformational population varies with solvent, it is probably necessary to consider only minor environmental perturbations in the search for solvent exposure of peptide protons, and to exclude the results of comparisons of spectra obtained in different solvents.

For Me_2SO, then, the results obtained may be considered consistent with a β-type cyclic-hexapeptide backbone, in which the glycine peptide protons are not completely shielded from solvent by the backbone and probably not strongly transannularly hydrogen-bonded. A backbone conformation with C_2 symmetry, having average H–C–N–H dihedral angles near 140° ($J \approx 7$ Hz) for the D and L residues, and near 120° ($J \approx 4.5$) for the glycine, is

Gly		D-X		L-X	
ϕ	ψ	ϕ	ψ	ϕ	ψ
180°	160°	80°	-110°	-80°	-40°
(180°)	(160°)	(100°)	(-100°)	(-70°)	(-50°)

(A)

A backbone oscillating about this conformation does not possess good transannular $NH_{Gly}\cdots OC_{Gly}$ hydrogen bonds, the average N$\cdots$O distance being about 4 Å. All residues are in stable regions of their dipeptide energy maps, however, and side-chain rotation of the substituted residues is not appreciably inhibited in CPK models. If a conformation close to this is retained in solvents other than Me_2SO, it is not difficult to see that the chemical shift and temperature dependence of the glycine peptide proton is variable because this proton is not strongly shielded from the solvent. It is not clear from models, however, why the D-residue peptide proton should be any less exposed to solvent than that of the L residue (see Fig. 2 and the non-Me_2SO data in Table 3).

A modification (B) of conformation (A) is not excluded by the coupling constants:

Gly		D-X		L-X	
ϕ	ψ	ϕ	ψ	ϕ	ψ
$180°$	$120°$	$160°$	$-120°$	$-80°$	$-50°$
$(180°)$	$(120°)$	$(130°)$	$(-110°)$	$(-70°)$	$(60°)$

(B)

This conformation has nothing approaching a transannular hydrogen bond; the Gly-D-X peptide bond is almost perpendicular to the average peptide ring plane, and the D-residue peptide proton, although not shielded from the solvent, is at least in a position, relative to the peptide ring, that is different from the NHs of position-2 residues of other suggested β turns. The environment of the L residue and glycine protons is not affected by the rotation of the Gly-D-X peptide bond that converts (A) into (B).

For conformation (A) or (B) the change in coupling constant upon going from Me$_2$SO to aqueous solutions suggests the modified conformations shown in parentheses. The effect on (A) of this change in angles is to open the peptide structure, thereby increasing the accessibility of the glycine carbonyls to the solvent.

The other backbone conformation types consistent with the coupling-constant data are less consistent with all facts. Conformations of the type

D-X		L-X		Gly	
ϕ	ψ	ϕ	ψ	ϕ	ψ
$160°$	$-170°$	$-80°$	$120°$	$70°$	$60°$
	$(or -90°)$	$(or -160°)$			

(C)

have the turn X$_2$-Gly$_3$, which has already been encountered in cyclo(Gly-Leu-Gly)$_2$ and related peptides. The glycine residue in position 3 in these peptides has a high temperature coefficient in Me$_2$SO and in methanol (15) and undergoes a large upfield chemical shift upon going to HFP (19), in contrast to what was observed for the present set of peptides.

The remaining class of conformations, which has the L residue extended and the glycine, in effect, in position 2, should be associated with solvent shielding of the L-residue peptide protons and relative exposure of the glycine peptide protons. This expectation also is contradicted by observation.

DISCUSSION

We conclude, although not with certainty, that the preferred general form of the peptide backbone of cyclic hexapeptides cyclo(Gly-X-D-X) is one in which the chain reversal occurs in the sequence X-D-X, but that an especially stable, transannularly hydrogen-bonded β turn is not formed, probably because of the rotational freedom of the side chain of the residue in position 2. Fig. 4 compares on a (ϕ,ψ) map the regions in which all three α-β rotamers of a residue with a butyl side chain are readily allowed, as estimated by Ponnuswamy and Sasisekharan (26), with the region in which good hydrogen bonds are formed in an LD β turn, as estimated by Chandrasekaran *et al.* (12). The mismatch for residue 2 is apparent. This problem would be less serious for alanine at position 2.[1] On the other hand, when proline is in position 2, the rotational freedom of the side chain is replaced by a side-chain constraint to $\phi = -60°$.

   In sum, we anticipate, considering the evidence from cyclic hexapeptides, that to construct stable β turns in oligopeptides without the constraint of ring formation, sequences should be employed of the type $D-X_1-Pro_2-D-X_3$, where the D-X residues in positions 1 and 3 may be replaced by glycine.

[1] The fact that J is approximately 5.7 Hz for Ala in *cyclo*(Gly-Ala-Gly)$_2$ (16) and 6.0 Hz in *cyclo*(Gly-Leu-Gly)$_2$ and *cyclo*(Gly-Tyr-Gly) (15), which corresponds to $\phi = -80°$ rather than $-60°$, suggests that there is restricted mobility of the alanine residue as well.

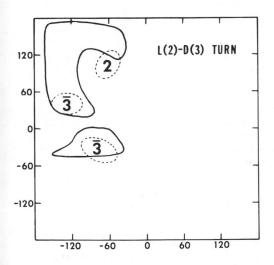

Figure 4. (ϕ,ψ) plane showing (heavy contours) regions within 4 kcal of the calculated energy minimum for all three α-β rotamers ($\chi_1 = \pm60°$, $180°$) of a butyl side chain, as reported by Ponnuswamy and Sasisekharan (26), and the regions in which the conformations of the residues 2 and 3 of a β turn with the sequence X-D-X occur, according to Venkatachalam (2) and Chandrasekaran *et al.* (12).

ACKNOWLEDGMENTS

This work was supported by grants GM-14069 and GM-47357 from the National Institutes of Health of the U.S. Public Health Service. Allan Oliver, a National Science Foundation Undergraduate Research Participant, prepared the cyclo(Gly-DLeu-Leu)$_2$.

REFERENCES

1. R. Schwyzer, *Rec. Chem. Progr.*, **20**, 147 (1959).
2. C.M. Venkatachalam, *Biopolymers*, **6**, 1425 (1968).
3. L.G. Pease, C.M. Deber, and E.R. Blout, *J. Am. Chem. Soc.*, **95**, 258 (1973).
4. R. Schwyzer, Ch. Grathwohl, J.P. Meraldi, A. Tun-Kyi, R. Vogel, and K. Wüthrich, *Helv. Chim. Acta.*, **55**, 2545 (1972).
5. E.R. Blout, C.M. Deber, and L.G. Pease, private communication.
6. D.A. Torchia, S.C.K. Wong, C.M. Deber, and E.R. Blout, *J. Am. Chem. Soc.*, **94**, 616 (1972).
7. K.D. Kopple, T.J. Schamper, and A. Go, *J. Am. Chem. Soc.*, **96**, in press.
8. D.A. Torchia, A. Di Corato, S.C.K. Wong, C.M. Deber, and E.R. Blout, *J. Am. Chem. Soc.*, **94**, 609 (1972).
9. K.D. Kopple, A. Go, T.J. Schamper, and C.S. Wilcox, *J. Am. Chem. Soc.*, **95**, 6090 (1973).
10. A. Damiani, P. DeSantis, and A. Pizzi, *Nature*, **226**, 542 (1970).
11. P.R. Schimmel and P.J. Flory, *J. Mol. Biol.*, **34**, 105 (1968).
12. R. Chandrasekaran, A.V. Lakshminarayanan, U.V. Pandya, and G.N. Ramachandran, *Biochem. Biophys. Acta.*, **303**, 14 (1973).
13. V. Madison and J. Schellman, *Biopolymers*, **9**, 511 (1970).
14. P.Y. Chou and G.D. Fasman, *Biochemistry*, **13**, 222 (1974).
15. (a) K.D. Kopple, M. Ohnishi and A. Go, *Biochemistry*, **8**, 4087 (1969).
    (b) K.D. Kopple, A. Go, R.H. Logan, Jr., and J. Savrda, *J. Am. Chem. Soc.*, **94**, 973 (1972).
16. (a) S.L. Portnova, V.V. Shilin, T.A. Balashova, J. Biernat, V.F. Bystrov, V.T. Ivanov, and Yu. A. Ovchinnikov, *Tetrahedron Lett.*, 3085 (1971).
    (b) V.T. Ivanov, S.L. Portnova, T.A. Balashova, V.F. Bystrov, V.V. Shilin, J. Biernat, and Yu. A. Ovchinnikov, *Khim. Prir. Soedin*, 339 (1971).
    (c) S.L. Portnova, T.A. Balashova, V.F. Bystrov, V.V. Shilin, J. Biernat, V.T. Ivanov, and Yu. A. Ovchinnikov, *ibid.*, 323 (1971).
17. Supported by Grant GP-33116 from the National Science Foundation.
18. Supported by Grant RR-00292 from the National Institutes of Health.
19. K.D. Kopple and T.J. Schamper in *Chemistry and Biology of Peptides*, J. Meienhofer, ed., Ann Arbor Science Publishers, Ann Arbor, Mich., 1972, p. 75.
20. T.P. Pitner and D.W. Urry, *J. Am. Chem. Soc.*, **94**, 1399 (1972).
21. M. Ohnishi and D.W. Urry, *Biochem. Biophys. Res. Commun.*, **36**, 194 (1969).
22. K.D. Kopple, *Biopolymers*, **10**, 1139 (1971).
23. D.W. Urry, M. Ohnishi and R. Walter, *Proc. Nat. Acad. Sci. U.S.*, **66**, 111 (1970).
24. A.I. Richard Brewster, V.J. Hruby, J.A. Glasel, and A.E. Tonelli, *Biochemistry*, **12**, 5294 (1973).
25. R. Walter and J.D. Glickson, *Proc. Nat. Acad. Sci. U.S.*, **70**, 1199 (1973).
26. P.K. Ponnuswamy and V. Sasisekharan, *Biopolymers*, **10**, 565 (1971).

NMR STUDIES OF CYCLIC DIPEPTIDES CONTAINING HISTIDINE AND TRYPTOPHAN RESIDUES

M. ANDORN and **M. SHEINBLATT**, Department of Chemistry, Tel Aviv University, Tel Aviv, Israel

SYNOPSIS: The NMR spectra of some cyclic dipeptides containing histidine and tryptophan residues were measured in D_2O and Me_2SO solutions as functions of pH and temperature. An upfield shift of the imidazole C_4H proton of *cyclo*-L-histidyl-L-tryptophan (LL) was observed, in contrast to what was observed for the LD isomer and other, related compounds. This result is explained by the preferred conformation of the LL isomer in which the indole ring faces the diketopiperazine and the imidazole ring is turned away from it. Calculated values of the indole ring current explain the measured upfield shifts.

INTRODUCTION

The interaction between peptide bonds and aromatic rings was demonstrated by NMR studies of linear and cyclic dipeptides containing one aromatic amino acid residue (1–7). These studies show that the preferred conformation is one in which the aromatic ring faces the peptide backbone (folded form) (3, 5, 6). The stabilization of this conformation is believed to arise from the dipole–induced-dipole interaction between the peptide bond and the aromatic ring. Following a report on the existence of an interaction between indole and imidazole rings (8, 9), we studied cyclic dipeptides containing two aromatic amino acid residues. In this paper we present NMR measurements on *cyclo*-L-histidyl-L-tryptophan (LL), *cyclo*-L-histidyl-D-tryptophan (LD), and *cyclo*-L-histidyl-L-phenylalanine (LLPhe).

EXPERIMENTAL METHOD

Cyclic dipeptides were purchased from Miles-Yeda Ltd., Rehovot, Israel. The NMR spectra were recorded on a Varian HA-100 spectrometer, with a time-averaging computer (Varian C-1024). (Some spectra were recorded on a Varian HR-220 spectrometer at Columbia University.) Chemical shift values were determined relative to internal references: *t*-butyl alcohol and dioxane in D_2O and Me_2SO solutions, respectively. The pH of the D_2O solutions was adjusted by small amounts of DCl or NaOD using a Radiometer type PHM 26 pH meter. The calculations of the molecular conformations were carried out on a C.D.C. 6600 computer in the Tel Aviv computation center.

DISCUSSION OF RESULTS

Typical NMR spectra of the aromatic protons of *cyclo*(His-Trp) (LL) and *cyclo*(His-DTrp) (LD) at relatively low pH values are given in Fig. 1. The most pronounced difference between the spectra of these compounds is the approximately 1 ppm upfield shift of the histidyl-C_4H (His-C_4H) absorption line of *cyclo*(His-Trp) compared to that of its disastereoisomer. The behavior of the His-C_4H proton in LL was compared to its behavior in LD and *cyclo*(His-Phe) (LLPhe) (which were used as reference compounds) by detailed measurements of the pH dependence of the chemical shift of these lines. The differences in the shifts of these protons in LD and LLPhe are small over the measured pH region (1–10), but in LL the line shifts upfield over the entire range of measurement. Also, the His-C_4H line of *cyclo*(His-Trp) is almost pH independent in the region where ionization of the imidazole ring occurs, in contrast to its behavior in the reference compounds.

These results are understandable if we assume the existence of a preferred conformation in LL. In order to study the nature of this conformation, we measured the temperature dependence of the chemical shift of the His-C_4H proton in all three systems. The measurements were carried out in D_2O and Me_2SO solutions. The results (given in Fig. 2) show clearly a pronounced temperature effect in LL, in contrast to a very small temperature dependence in the reference compounds. The behavior of the LL compound is typical for systems in which

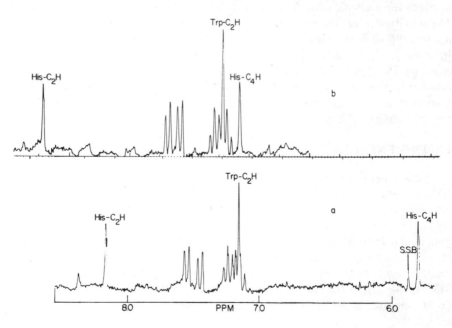

Figure 1. 220 MHz spectra of the aromatic protons in D_2O solutions of (a) *cyclo*(His-Trp) and (b) *cyclo*(His-DTrp) at pH = 2 (internal reference: TSP).

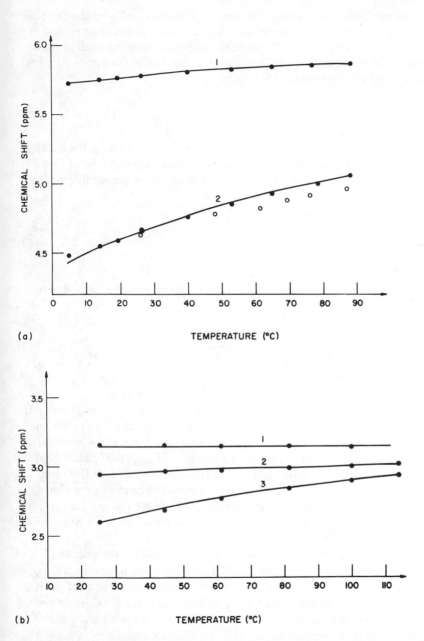

Figure 2. (a) The temperature dependence of the chemical shift of the His-C$_4$H proton in D$_2$O solutions of (1) *cyclo*(His-DTrp) and (2) *cyclo*(His-Trp). Dark circles show results at low pH values; open circles show results at high pH values. (b) The temperature dependence of the chemical shift of the His-C$_4$H proton in Me$_2$SO solutions of (1) *cyclo*(His-DTrp), (2) *cyclo*(His-Phe), and (3) *cyclo*(His-Trp).

there is an equilibrium between a preferred conformation and several other possible conformations. Raising the temperature tends to equalize the population of all possible conformations, whereas in the reference compounds all possible conformations are already equally populated. Accordingly, the measured chemical shift of the His-C_4H proton (δ) is given by

$$\delta = p\delta_B + (1-p)\,\delta_F \tag{1}$$

where δ_B is the chemical shift of the preferred conformation (B), p the fraction of molecules in this conformation, and δ_F the average chemical shift of all other possible conformations (F). Following Kopple et al. (5, 6), the equilibrium constant can be written in the form:

$$K = \frac{p}{1-p} = \frac{\delta - \delta_F}{\delta_B - \delta} = \frac{\Delta\delta}{\Delta\delta_{max} - \Delta\delta} \tag{2}$$

where $\Delta\delta = \delta - \delta_F$ and $\Delta\delta_{max} = \delta_B - \delta_F$. The relation between $\Delta\delta$ and temperature is given by

$$\log\left(\frac{\Delta\delta}{\Delta\delta_{max} - \Delta\delta}\right) = -\left(\frac{\Delta H^\circ}{R}\right)\frac{1}{T} + \frac{\Delta S^\circ}{R} \tag{3}$$

Accordingly, a plot of $\log K$ versus $1/T$ should yield a straight line.

It is not possible to determine experimentally the values of δ_B and δ_F in our system. On the basis of our previous discussion we assume that δ_F is given by the chemical shift values measured in the reference compounds. In order to find the value of $\Delta\delta_{max}$, we had to fit three parameters, $\Delta\delta_{max}$, ΔH°, and ΔS°, into Eq. (3). This analysis was performed for the His-C_4H line of LL in D_2O and Me_2SO, using LD and LLPhe as reference compounds. The results are plotted in Fig. 3. The thermodynamic parameters calculated in Me_2SO solution are similar for each reference compound. Thus, our choice of reference compounds can be considered reliable.

At this stage we can propose a model that will account for the calculated $\Delta\delta_{max}$ values. In our model we assume that the only contribution to $\Delta\delta_{max}$ arises from the effect of the indole ring current on the chemical shift of the His-C_4H proton. This effect depends only on the location of the His-C_4H proton relative to the indole ring. In order to find this location we must know the geometry of the diketopiperazine ring and four dihedral angles: $\chi_1, \chi_2, \chi_1', \chi_2'$. Inspection of the chemical shift values and the $J_{NH-\alpha-CH}$ coupling constants, which are listed in Table 1, leads to the conclusion that the upfield shift of the His-C_4H proton of LL relative to the corresponding line in LD and LLPhe indicates that the preferred conformation exists also in Me_2SO solution. It was found by Ramachandran et al. (10) that in a diketopiperazine, equal values of

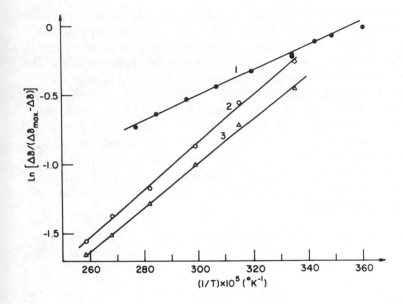

Figure 3. Plots of calculated log K versus $1/T$. (1) D_2O solution; reference compound LD. (2) Me_2SO solution; reference compound LD. (3) Me_2SO solution; reference compound LLPhe.

the $J_{NH-\alpha-CH}$ coupling constants of approximately 2 Hz are characteristic of a planar ring. The measured coupling constants in LL and LLPhe show clearly the planarity of the diketopiperazine ring of these peptides in Me_2SO. Thus, in the following calculations we restrict ourselves only to the case of a planar diketopiperazine ring. In this case, any conformation is defined by the dihedral angles χ_1, χ_2, χ'_1, and χ'_2. The angles that characterize the preferred conformation were found in three steps. First, because crystallographic data for the measured

Table 1. Chemical shift values and coupling constants in Me_2SO

Compound	Chemical Shift Values (ppm)[a]					$J_{NH-\alpha-CH}$ (Hz)	
	His-C_4H	His-C_2H	NH(ℓ)[b]	NH(h)[b]	$\Delta\delta_{NH}$	NH(ℓ)[b]	NH(h)[b]
Cyclo(His-Trp)	2.53	3.93	4.42	3.97	0.45	2.0	2.0
Cyclo(His-DTrp)	3.15	3.95	4.40	4.28	0.12	2.0	<0.5
Cyclo(His-Phe)	3.05	3.95	4.48	4.19	0.29	~2	~2

[a]Measured relative to the line of 1–4 dioxane that appears in the lowest field.
[b]ℓ and h are low- and high-field NH protons of the diketopiperazine ring, respectively.

peptides are not available in the literature, we calculated the geometry of the molecules from crystallographic data for histidine (11), tryptophan (12), and diketopiperazine (13). In the second step, we eliminated conformations that are not allowed by the criterion of excluded volume (14). In these calculations the rotation steps were 9° about each rotation axis. (This step reduced the number of possible conformations from 2.5×10^6 to about 1.1×10^4.) Finally, using the chemical shift calculations of Giessner-Prettre and Pullman (15, 16), we were able to define a region (below and above the plane of the indole ring) in which protons are subjected to an upfield shift equal to our calculated $\Delta\delta_{max}$, as a result of the indole ring current. Calculations of the conformations of LL in which the His-C_4H proton occupies this region show that about 250 such conformations are possible. The range of the dihedral angles that characterize the preferred conformations are:

$$\chi_1' \ (H_\alpha',C_\alpha',C_\beta',C_\gamma') \ = \ 173°(L) - 189°(L)$$

$$\chi_2' \ (C_\alpha',C_\beta',C_\gamma',C_2) \ = \ 57°(R) - \ 93°(R)$$

$$\chi_1 \ (H_\alpha,C_\alpha,C_\beta,C_\gamma) \ = \ 52°(L) - \ 79°(L)$$

$$\chi_2 \ (C_\alpha,C_\beta,C_\gamma,N_2) \ = \ 30°(R) - 120°(R)$$

A zero dihedral angle is defined by the synplanar conformation (17), and L and R stand for left-hand and right-hand rotations, respectively.

Fig. 4 shows one typical conformation in which the plane of the indole ring lies above the plane of the diketopiperazine ring and the imidazole ring lies away from it. This feature is typical for all the calculated preferred conformations.

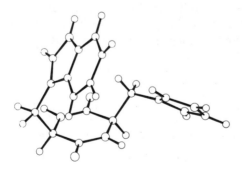

Figure 4. A characteristic folded conformation of *cyclo*(His-Trp). The dihedral angles defined in the text are $\chi_1 = 61°(L)$, $\chi_2 = 77°(R)$, $\chi_1' = 177°(L)$, and $\chi_2' = 84°(R)$.

ACKNOWLEDGMENTS

We wish to thank professor R. Bersohn for measurements of 220 MHz spectra and C. Giessner for detailed calculations of the indole ring-current effect.

REFERENCES

1. P. Gupta-Bhaya, Ph.D. Thesis, Columbia University, 1972.
2. F.A. Bovey and G.V.D. Tiers, *J. Am. Chem. Soc.,* **81,** 2870 (1959).
3. C.M. Deber and H. Joshua, *Biopolymers,* **11,** 2493 (1972).
4. H. Joshua and C.M. Deber, in *Chemistry and Biology of Peptides,* J. Meienhofer, ed., Ann Arbor Science Publishers, Ann Arbor, Mich., 1972.
5. K.D. Kopple and D.H. Marr, *J. Am. Chem. Soc.,* **89,** 6913 (1967).
6. K.D. Kopple and M. Ohnishi, *J. Am. Chem. Soc.,* **91,** 962 (1969).
7. G. Gawne, G.W. Kenner, N.H. Rogers, R.C. Sheppard, and K. Titlestad, in *Peptides 1968,* E. Bricas, ed., North-Holland, Amsterdam, 1968, p. 28.
8. M. Shinitzki and M. Fridkin, *Eur. J. Biochem.,* **9,** 176 (1969).
9. M. Shinitzki, Ph.D. Thesis, Weitzman Institute of Science, 1968.
10. G.N. Ramachandran, R. Chandrasekaran, and K.D. Kopple, *Biopolymers,* **10,** 2113 (1971).
11. J. Donahue and A. Caron, *Acta Cryst.,* **17,** 1178 (1964).
12. T. Takaji *et al., Bull. Chem. Soc. Japan,* **39,** 2369 (1966).
13. E. Sletten, *J. Am. Chem. Soc.,* **92,** 172 (1970).
14. G.N. Ramachandran, ed., *Conformation of Biopolymers,* Vol. 1, Academic Press, New York, 1967, p. 83.
15. C. Giessner-Prettre and B. Pullman, *J. Theor. Biol.,* **39,** 287 (1971).
16. C. Giessner-Prettre, private communication.
17. *Biochemistry,* **9,** 3471 (1970).

CIS, TRANS, AND NONPLANAR PEPTIDE BONDS IN OLIGOPEPTIDES: ^{13}C NMR STUDIES

KURT WÜTHRICH, CHRISTOPH GRATHWOHL, and **ROBERT SCHWYZER,**
Institut für Molekularbiologie und Biophysik, Eidgenössische Technische
Hochschule, 8049 Zürich, Switzerland

SYNOPSIS: This paper presents data on peptide bond conformations obtained from ^{13}C NMR investigations of *cyclo*-tetraglycyl and a selection of linear oligopeptides each containing a prolyl residue. In *cyclo*-tetraglycyl, the four peptide groups are found to be magnetically equivalent and different from a standard *trans* or *cis* peptide group. It is suggested that the observed NMR features correspond to a nonplanar form of the peptide groups. The previously described manifestation of *cis* and *trans* X-Pro bonds in the proline ring carbon chemical shifts was employed for further studies of the structural factors that govern the equilibrium between *cis-* and *trans*-proline in selected linear oligopeptides. This is part of a systematic investigation of the potential of the X-Pro sequence as a natural probe for monitoring preferred molecular conformations in nonglobular linear polypeptides.

INTRODUCTION

The concept of the standard planar *trans* peptide group (1, 2) has been a great asset in investigations of polypeptide conformations over the past twenty years. Today, however, in light of the details of peptide conformation that are observable by modern experimental techniques, it appears no longer to be universally adequate even in the absence of severe constraints, such as those imposed in small cyclic peptides (3). The occurrence of *cis* peptide groups has been demonstrated in numerous linear and cyclic polypeptides containing proline and other N-substituted amino acid residues (4–30), and experimental and theoretical studies of model compounds indicates that nonplanar peptide groups might be quite common in peptides and proteins (31–33).

Although some information about *cis* peptide groups resulted from x-ray studies (4–6), most of the data came from nuclear magnetic resonance (NMR) studies. The use of proton NMR spectroscopy led to the discovery of a number of typical peptides with *cis* amide bonds, including linear oligopeptides (7, 10, 12, 13), polysarcosine (8), and a variety of cyclic penta-, hexa-, and octapeptides (9, 14–17, 27). Proton NMR techniques were applied also to the study of thermodynamic and kinetic aspects of the interconversion between *cis* and *trans* N-substituted amide groups. The results obtained from a variety of peptides and model amide compounds, including polyproline (11), N-substituted formamides (34, 35), *N*-acyl-proline (35), and methyl-*N*-acetyl sarcosinate (8, 37), indicate that the *trans* form is more stable ($\Delta G°$ of the order of 0.1–2.0 kcal mole^{-1} and $\Delta G^{\ddagger} \cong 20$ kcal mole^{-1}). These data are supported by recent theoretical studies (24, 38).

Technical difficulties limited proton NMR investigations of peptide-bond conformations to relatively few examples, but the advent of ^{13}C NMR has greatly expanded the scope of these studies. The ready detection of *cis-trans* isomerism about the X-Pro bond (18–21, 23, 25, 26, 30) appears to be an important contribution of ^{13}C NMR to the field of conformational studies of polypeptides. This paper reports on some ^{13}C NMR studies of peptide-group conformations in oligopeptides.

MATERIALS AND METHODS

The synthesis and characterization of *cyclo*-tetraglycyl has been described previously (39). *Cyclo*(Phe-Pro), *cyclo*(DPhe-Pro), H-Phe-Pro-OH, H-DPhe-Pro-OH, H-Thr-Phe-Pro-OH, H-Thr-Phe-Pro-Gen-Thr-Ala-Ile-Gly-OH, and a variety of partial sequences from different peptide hormones mentioned in the text were obtained from W. Rittel and M. Brugger, Ciba-Geigy AG, Basel. The peptides of H-Ala-Pro-OH, H-Ala-Ala-Pro-Ala-OH, H-Ala-Ala-Pro-Ala-Ala-OH, H-Ala-Ala-Ala-Pro-Ala-OH, H-Ala-Ala-Ala-Pro-Ala-Ala-OH, $C_6H_5-CH_2-O-CO-$Ala-Pro-NH$_2$, and $F_3C-CO-Gly-Gly-Pro-Ala-OCH_3$ were purchased from Bachem AG, Liestal, Switzerland. All ^{13}C NMR spectra were recorded on a Varian XL-100 spectrometer using the Fourier transform method.

CYCLO-TETRAGLYCYL

The chemical shifts of the carbonyl carbon resonances in *cyclo*-tetraglycyl were compared with those in glycyl residues in diketopiperazines, which are taken to be representative of the standard planar *cis* peptide group of glycine (2), and in linear oligopeptides, which are employed to represent the standard planar *trans* peptide group of glycine (2). Details of the NMR spectra will be presented elsewhere (40).

Fig. 1 shows that in trifluoroacetic acid (TFA), dimethylsulfoxide (Me$_2$SO), and D$_2$O, the carbonyl carbon resonances of the *cis* peptide bonds of glycine in diketopiperazine are consistently 1–2 ppm further upfield than the corresponding resonance of the *trans* peptide bonds in linear oligopeptides. In the ^{13}C NMR spectrum of *cyclo*-tetraglycyl, a single line at 175.5 ppm in TFA and at 169.8 ppm in Me$_2$SO corresponds to the four carbonyl carbon atoms. The magnetic equivalence of the four peptide groups is also observed in the proton NMR spectrum (40), in apparent contrast with earlier observations of this molecule (41).

Because the four peptide groups are equivalent (Fig. 1), the possibility of a molecular conformation in which *cyclo*-tetraglycyl contains a combination of *cis* and *trans* amide bonds can be excluded (41). Even if there were rapid *cis-trans* isomerization around the ring, the resulting average resonance would be

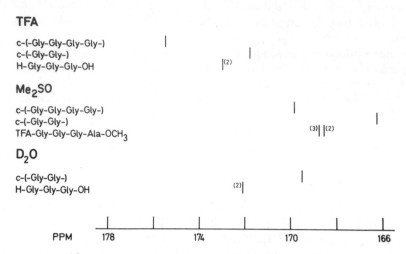

Figure 1. ^{13}C NMR chemical shifts in ppm from TMS of the carbonyl carbon atoms of *cyclo*-tetraglycyl, *cyclo*-diglycyl, and the nonterminal glycyl residues in H-Gly-Gly-Gly-OH and TFA-Gly-Gly-Gly-Ala-OCH$_3$ in three deuterated solvents.

expected to lie in the spectral region between the lines corresponding to the *cis* and *trans* forms and not at the extreme downfield position observed in Fig. 1. The carbonyl-carbon chemical shift in *cyclo*-tetraglycyl also is distinct from that of either compound chosen to represent the standard *trans* or standard *cis* peptide unit [Fig. 1 and (40)].

Previous investigations of the ring geometry had led to the prediction that *cyclo*-tetrapeptides could not accommodate four standard *trans* peptide groups (3). The observed downfield shift of the carbonyl carbons in *cyclo*-tetraglycyl leads us to suggest that the distortion of the peptide bonds from the planar standard dimension consists not only in possible variations of the angles τ^2, but includes also out-of-plane bending of the N—H and N—C$_\alpha$ bonds, which would give rise to the previously described pyramidal structure at the nitrogen atom (31-33). It is conceivable that such a pyramidal distortion of the peptide group could result in the observed deshielding of the carbonyl carbon atom (40).

THE EQUILIBRIUM BETWEEN *CIS* AND *TRANS* X-PRO AMIDE GROUPS IN LINEAR POLYPEPTIDES

With the distinct manifestation of the *cis-trans* isomers of the X-Pro bond in the proline ring carbon resonances (18-21), the simultaneous presence of *cis* and *trans* proline has been demonstrated in a considerable number of peptides (21-23, 25, 26, 28-30, 42). Some new examples are shown in the Figs. 2 and 3. In *cyclo*(Phe-Pro) and *cyclo*(DPhe-Pro), only the resonances of *cis*-proline are

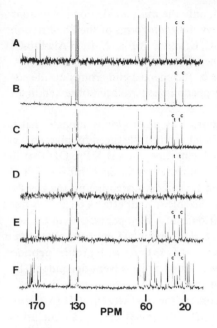

Figure 2. Proton noise-decoupled ^{13}C NMR spectra at 25.14 MHz of six proline-containing peptides. Approximately 0.1 *M* solutions in D_2O (pH = 5.0) were studied at 26°C. (A) *cyclo*(Phe-Pro), (B) *cyclo*(DPhe-Pro), (C) H-Phe-Pro-OH, (D) H-D-Phe-Pro-OH, (E) H-Thr-Phe-Pro-OH, (F) H-Thr-Phe-Pro-Gln-Thr-Ala-Ile-Gly-OH. The lines marked "c" and "t" correspond to the β- and γ-carbon atoms of proline in the *cis* and *trans* forms of the Phe-Pro amide bond, respectively.

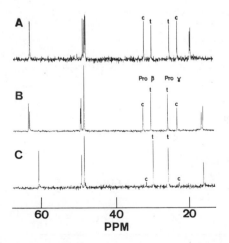

Figure 3. Proton noise-decoupled ^{13}C NMR spectra at 25.14 MHz of 0.1 *M* aqueous solution of H-Ala-Pro-OH at (A) pH 12.2, (B) pH 5.4, (C) pH 1.4.

seen (Fig. 2A and B). In H-DPhe-Pro-OH, only the *trans* amide bond is present (Fig. 2D). The equilibrium between the *cis* and *trans* forms of the Phe-Pro bond varies with the overall length of peptides C, E, and F in Fig. 2. In H-Ala-Pro-OH, as shown in Fig. 3, the percentage of *cis* peptide conformations increases at higher pH. This emphasizes the influence of the charged end groups on the *cis-trans* equilibrium in this peptide. Similar phenomena have been observed in a variety of peptides (42).

Overall, our observations and those reported by other groups indicate that the *cis-trans* equilibrium of the X-Pro amide bond can be influenced by many different factors. Thus, it was found (23) in different dipeptides H-X-Pro-OH that the negative charge on the carboxyl group of proline favors the *cis* conformation (Fig. 3). The same conclusion was reached for sarcosyl-sarcosine on the basis of proton NMR and theoretical studies (24). Sizable solvent effects on the *cis-trans* equilibrium have also been observed in various linear oligopeptides with protected uncharged end groups [Table 1 and (25, 26, 42)], in poly(L-proline) (11), and in cyclic peptides containing prolyl residues (14–17, 43). As with proline peptides, the *cis-trans* equilibrium in polysarcosyl was found to be solvent dependent (8). In Fig. 2 it is quite obvious that the X-Pro conformation is influenced by the primary peptide structure. Previously a comparison of H-Gly-Pro-OH (61% *trans*), H-Ala-Pro-OH (57% *trans*), and H-Val-Pro-OH (59% *trans*) led to the conclusion that the amino acid preceding proline might have a negligible effect on the *cis-trans* equilibrium (19). The data on H-Phe-Pro-OH and H-DPhe-Pro-OH (Fig. 2) indicate that this cannot be generalized to include all the common amino acid

Table 1. Occurrence of *cis*-proline in a selection of linear oligopeptides

Peptide[a]	cis Form (%) of the X-Pro Group				
	In Aqueous Solution (P_2O)			In Other Solvents	
	pH 1.5	pH 5.5	pH 10.5		
H-A-P-OH	10	40	55	CD_3OD:	60
H-A-A-P-A-OH	5	10	10		
H-A-A-P-A-A-OH	5	10	5		
H-A-A-A-P-A-OH	10	10	5		
H-A-A-A-P-A-A-OH	5	10	5		
Z-A-P-NH$_2$				Me_2SO:	15
				$CDCl_3$:	5
F_3C-CO-G-G-P-A-OCH$_3$		20		Me_2SO:	40
				$CDCl_3$:	5

[a] A = L-alanine, P = L-proline, G = glycine, Z = $C_6H_5-CH_2-O-CO-$.

residues. The influence of the configuration of the amino acid preceding proline is particularly striking (26). Additional experiments show further that the proline *cis-trans* equilibrium can be affected also by changes of the primary structure outside the amino acid residues immediately preceding and following proline (e.g., by variation of chain length) (42).

Among the above-mentioned factors that appear to influence the proline *cis-trans* equilibrium in linear oligopeptides, the effects of the amino acid preceding proline and of the charge localized on the carboxyl group of proline can probably be rationalized to a large extent from considerations of the structure in the immediate environment of the observed X-Pro peptide group (42). On the other hand, many of the variations observed after changing the solvent medium or the primary structure outside the immediate environment of proline actually might arise in part or entirely from changes of the overall molecular conformation (44). If the effects of the local covalent structure and the postulated conformational factors (44) determining the *cis-trans* equilibrium could be separated, ^{13}C NMR of X-Pro bonds in principle might serve as a natural probe of peptide conformations that are not readily accessible to observation by other techniques. This might be of particular interest for studies of certain linear polypeptides that appear from their proton NMR spectra (45–47) to exist predominantly in an extended random-coil form. In these molecules, the occurrence of energetically favored extended molecular conformations might be traceable through their influence on the X-Pro bonds (44). It would seem that in the absence of a sizable stabilization via the overall molecular conformation the *trans* conformation of N-substituted amides is favored over the *cis* conformation by no more than 2 kcal mole^{-1}, so all-*cis* X-Pro bonds might be expected to be found in flexible medium-sized polypeptide chains (44). Preferred conformations including all-*cis* X-Pro bonds have previously been reported for poly(L-proline) (4, 11) and certain globular proteins (5, 6).

Readily available naturally occurring peptides invariably contain many different amino acids and, hence, a multitude of different structural features that might affect the *cis-trans* equilibrium of X-Pro. For more systematic studies of the factors governing this parameter in linear oligopeptides, therefore, we have had recourse to a series of simple synthetic peptides. These model compounds have the structures H-(Ala)$_m$-Pro-(Ala)$_n$-OH. The variations of structure so far include $n = 1, 2$, or $3, m = 0, 1$, or 2, and different charges localized on the terminal amino and carboxyl groups. The results of an incomplete study of this series of molecules in aqueous solution are summarized in Table 1. Details of the NMR spectra will be presented elsewhere (44).

A tentative interpretation of the data in Table 1 may illustrate how the concept of distinguishable structural and conformational effects on the *cis-trans* equilibrium of X-Pro (44) might be applied. If we start with the hypothesis that there are no energetically favored molecular conformations in aqueous solutions of the peptides H-(Ala)$_m$-Pro-(Ala)$_n$-OH, the data of the Table 1 indicate that a *cis:trans* ratio of approximately 1:10 is characteristic of the structural entity

$$\overset{O}{\underset{\parallel}{-C}} - Ala - Pro - \overset{H}{\underset{\mid}{N}} -$$

Neither the introduction of additional alanyl residues preceding or following this sequence nor the variations of the charged end groups in these longer peptides appear to noticeably affect the conformation of the Ala-Pro bond. Replacement by glycine of the alanine preceding proline was previously found to have a negligible effect on the *cis:trans* ratio (19). Hence, it might be further concluded that the increased amount of *cis*-proline in $F_3C-CO-Gly-Gly-Pro-Ala-OCH_3$ (Table 1) arises as a consequence of the preferred formation of molecular conformations requiring *cis*-proline (44).

ACKNOWLEDGMENTS

We would like to thank W. Rittel and M. Brugger for the gift of various peptides used in this study. Financial support by the Schweizerischer Nationalfonds is gratefully acknowledged.

REFERENCES

1. R.B. Corey and L. Pauling, *Proc. Roy. Soc.* (London), **B141**, 10 (1953).
2. G.N. Ramachandran and V. Sasisekharan, *Adv. Protein Chem.*, **23**, 283 (1968).
3. N. Go and H.A. Scheraga, *Macromolecules*, **3**, 178 (1970).
3A. C. Ramakrishnan and K.P. Sarathy, *Biochim. Biophys. Acta*, **168**, 402 (1968).
4. W. Traub and U. Shmueli, in *Aspects of Protein Structure*, G.N. Ramachandran, ed., Academic Press, New York, 1963, p. 81.
5. H.W. Wyckoff, K.D. Hardman, N.M. Allewell, T. Inagami, L.N. Johnson, and F.M. Richards, *J. Biol. Chem.*, **242**, 3984 (1967).
6. C.S. Wright, R.A. Alden, and J. Kraut, *Nature* (London), **221**, 235 (1969).
7. M. Goodman and M. Fried, *J. Am. Chem. Soc.*, **89**, 1264 (1967).
8. F.A. Bovey, J.J. Ryan, and F.P. Hood, *Macromolecules*, **1**, 305 (1968).
9. J. Dale and K. Titlestad, *Chem. Comm.*, 656 (1969).
10. C.M. Deber, F.A. Bovey, J.P. Carver, and E.R. Blout, *J. Am. Chem. Soc.*, **92**, 6191 (1970).
11. D.A. Torchia and F.A. Bovey, *Macromolecules*, **4**, 246 (1971).
12. C.M. Deber, F.A. Bovey, J.P. Carver, and E.R. Blout, in *Peptides 1969*, E. Scoffone, ed., North-Holland, Amsterdam, 1971.
13. V.J. Hruby, A.I. Brewster, and J.A. Glasel, *Proc. Nat. Acad. Sci. U.S.*, **68**, 450 (1971).
14. D.A. Torchia, A. Di Corato, S.C.K. Wong, C.M. Deber, and E.R. Blout, *J. Am. Chem. Soc.*, **94**, 609 (1972).
15. D.A. Torchia, S.C.K. Wong, C.M. Deber, and E.R. Blout, *J. Am. Chem. Soc.*, **94**, 616 (1972).
16. C.M. Deber, D.A. Torchia, S.C.K. Wong, and E.R. Blout, *Proc. Nat. Acad. Sci. U.S.*, **69**, 1825 (1972).

17. J.P. Meraldi, R. Schwyzer, A. Tun-Kyi, and K. Wüthrich, *Helv. Chim. Acta,* **55**, 1962 (1972).
18. K. Wüthrich, A. Tun-Kyi, and R. Schwyzer, *FEBS Lett.,* **25**, 104 (1972).
19. W.A. Thomas and M.K. Williams, *Chem. Comm.,* 994 (1972).
20. F.A. Bovey, in *Chemistry and Biology of Peptides,* J. Meienhofer, ed., Ann Arbor Science Publishers, Ann Arbor, Mich., 1972, p. 3.
21. I.C.P. Smith, R. Deslauriers, and R. Walter, in *Chemistry and Biology of Peptides,* J. Meienhofer, ed., Ann Arbor Science Publishers, Ann Arbor, Mich., 1972, p. 29.
22. D.E. Dorman and F.A. Bovey, *J. Org. Chem.,* **38**, 1719 (1973).
23. D.E. Dorman and F.A. Bovey, *J. Org. Chem.,* **38**, 2379 (1973).
24. J.C. Howard, F.A. Momany, R.H. Andreatta, and H.A. Scheraga, *Macromolecules,* **6**, 535 (1973).
25. R. Deslauriers, R. Walter, and I.C.P. Smith, *Biochem. Biophys. Res. Comm.,* **53**, 244 (1973).
26. R. Deslauriers, C. Garrigou-Lagrange, A. Bellocq, and I.C.P. Smith, *FEBS Lett.,* **31**, 59 (1973).
27. J.P. Meraldi, H. Moeschler, R. Schwyzer, A. Tun-Kyi, and K. Wüthrich, *J. Phys.,* **34**, C8–41 (1973).
28. D.J. Patel, *Biochemistry,* **12**, 667 (1973).
29. D.J. Patel, *Biochemistry,* **12**, 677 (1973).
30. W. Voelter, O. Oster, and K. Zech, *Angew. Chem.,* **86**, 46 (1974).
31. F.K. Winkler and J.D. Dunitz, *J. Mol. Biol.,* **59**, 169 (1971).
32. G.N. Ramachandran, A.V. Lakshminarayanan, and A.S. Kolaskar, *Biochim. Biophys. Acta,* **303**, 8 (1973).
33. G.N. Ramachandran and A.S. Kolaskar, *Biochim. Biophys. Acta,* **303**, 385 (1973).
34. L.A. La Planche and M.T. Rogers, *J. Am. Chem. Soc.,* **86**, 337 (1964).
35. R.C. Neuman, V. Jonas, K. Anderson, and R. Barry, *Biochem. Biophys. Res. Comm.,* **44**, 1156 (1971).
36. H.L. Maia, K.G. Orrell, and H.N. Rydon, *Chem. Comm.,* 1209 (1971).
37. A.L. Love, T.D. Alger, and R.K. Olson, *J. Phys. Chem.,* **76**, 853 (1972).
38. A.E. Tonelli, *J. Am. Chem. Soc.,* **93**, 7153 (1971).
39. R. Schwyzer, B. Iselin, W. Rittel, and P. Sieber, *Helv. Chim. Acta,* **39**, 872 (1956).
40. Ch. Grathwohl, A. Bundi, R. Schwyzer, and K. Wüthrich, *Helv. Chim. Acta,* to be submitted.
41. J. Dale and K. Titlestad, *Chem. Comm.,* 255 (1972).
42. Ch. Grathwohl and K. Wüthrich, to be published.
43. J.P. Meraldi, Ph.D. Thesis, Eidgenössische Technische Hochschule Zürich, 1974.
44. K. Wüthrich and Ch. Grathwohl, *FEBS Lett.,* submitted.
45. D.J. Patel, *Macromolecules,* **3**, 448 (1970).
46. D.J. Patel, *Macromolecules,* **4**, 251 (1971).
47. A. Masson, Ph.D. Thesis, Eidgenössische Technische Hochschule Zürich, 1974.

CONFORMATIONAL ANALYSIS OF MONODISPERSE ALIPHATIC LINEAR HOMO-OLIGOPEPTIDES

M. GOODMAN, Department of Chemistry, University of California (San Diego), La Jolla, Calif. 92037, C. TONIOLO, Institute of Organic Chemistry, University of Padua, Padua 35100, Italy, and F. NAIDER, Department of Pure and Applied Sciences, Richmond College (CUNY), Staten Island, N.Y. 10301

SYNOPSIS: Research on the conformations of alanyl, methionyl, leucyl, valyl, and isoleucyl homo-oligopeptides is presented. Conformations deduced are based upon circular dichroism, nuclear magnetic resonance, ultraviolet, and infrared spectroscopy. We have established the critical size for the onset of helical, β-associated and folded forms, if applicable, for each of the peptide series. A detailed examination of the nuclear magnetic resonance spectra in the alanine series is included and a structural model proposed to explain the "double peaks" for the pentamer and longer peptides in chloroform in the presence of small amounts of trifluoroacetic acid.

INTRODUCTION[*]

Since our initial work in the early 1960s (1), the conformational analysis of linear homo-oligopeptides has become the subject of increasing interest. We have investigated the conformation of glutamate and aspartate homo-oligomers using UV, ORD, CD, IR, and NMR spectroscopies (2-7). Early studies showed that the γ-alkyl-L-glutamate oligomers adopt helical conformations at chain lengths as short as the heptamer. In contrast, under similar conditions β-methyl-L-aspartate oligomers show no helicity until the undecamer. Polypeptides composed of the former residues assume right-handed helical structures; those composed of the latter residues form left-handed helices.

Conformational investigations of oligopeptides provide information on end effects and on the effects of short- and medium-range interactions on conformations. Furthermore, stereochemical analyses of oligopeptides allow more precise conclusions to be drawn concerning the conformations assumed by synthetic polypeptides and proteins. These analyses contribute to the solution of basic questions in spectroscopy. Because experimental approaches can be considered to

---

[*]The following abbreviations have been used throughout the text: UV, ultraviolet; ORD, optical rotatory dispersion; CD, circular dichroism; IR, infrared; NMR, nuclear magnetic resonance; TFE, trifluorethanol; HFIP, hexafluoroisopropanol; TFA, trifluoroacetic acid; HFA, hexafluoroacetone sesquihydrate; BOC, *tert*-butyloxycarbonyl; Z, benzyloxycarbonyl; MEEA, 2-methoxy-[2-ethoxy-(2-ethoxy)] acetyl; OMe, methyl ester; OEt, ethyl ester; Mo, morpholide; Ala, alanine; Val, valine; Leu, leucine; Met, methionine; Ile, isoleucine. All amino acid residues have the L configuration.

complement conformational energy calculations, the combination of these techniques provides a comprehensive picture of the structural state of peptides and proteins. Oligopeptide studies can be used also to test the validity of models proposed for the mechanism of chain growth and crystallization during the course of polymerization of activated aminoacyl monomeric units under synthetic and biosynthetic conditions.

Most globular proteins have short segments in helical conformations, β structure, and turns. Cyclic oligopeptides provide information on turns and β conformations. Side-chain geometries are often more easily assigned in cyclic oligopeptides because the ring system can severely limit available conformational space. Cyclic peptides, however, are less appropriate models for proteins than linear peptides because proteins contain end residues, sequence directions along the chain and α-helical segments. An additional limitation of the use of cyclic peptides as models for proteins can occur if there is strain in the ring systems, because such effects lead to distortions of bond angles and even to nonplanar peptide bonds or to a combination of both. We have commenced a systematic study of homologous series of linear oligopeptides. In this, the first stage, we have dealt primarily with homo-oligopeptides because their conformations are the most amenable to investigation by spectroscopic techniques and semiempirical calculations. As we accumulate information, we will extend our work to include co- and sequence oligopeptides.

In this paper, we will summarize the approaches used in studies of linear homo-oligopeptides and discuss our recent findings on the conformations of specific aliphatic oligopeptides in solution (8–12). We have employed UV, IR, CD, and NMR spectroscopy to deduce conformational characteristics for these peptides. In particular, CD provides information about the average state of the secondary structure existing in solution based upon chromophore analysis. The superconducting NMR spectrometers that are currently available often allow us to probe the local environments of individual NH and α-CH protons from which we can deduce more subtle conformational characteristics of oligopeptides.

Preliminary clues to which secondary structure linear aliphatic homo-oligopeptide series might assume in solution were obtained using UV absorption in the $\pi \rightarrow \pi^*$ peptide transition region. This technique often gives indications of the presence of unordered, α-helical, or β conformations (13), but it is difficult to interpret because the $\pi \rightarrow \pi^*$ transitions substantially overlap each other. Thus, in making our conformational assignments, we have relied heavily on CD data. In the sections below, we will discuss our results for Ala, Met, Leu, Val, and Ile homo-oligomers. It should be noted that an advantage of studies of homo-oligomers is that smaller peptides generally exist as essentially unordered molecules in solution. Changes in the nature of the CD curves as the chain length increases therefore can be directly correlated with the onset of ordered secondary structures (14).

Alanine oligomers. A considerable amount of work has been completed on the analysis of L-alanine homo-oligomers in solvents such as TFE, HFIP, HFA, and mixtures of TFE with sulfuric acid or water (11, 12). The results of detailed

CD analysis of five different series, namely, $Z(Ala)_n OEt$, $Z(Ala)_n Mo$, $BOC(Ala)_n$-OMe, $MEEA(Ala)_n OEt$, and $MEEA(Ala)_n Mo$, indicate that at chain lengths of seven and larger, Ala oligomers can exist in unordered, β- and α-helical forms depending on the solvent, temperature, and oligomer concentration. Fig. 1 shows that in 99% TFE–1% sulfuric acid, spectral patterns for the higher oligomers in the $Z(Ala)_n Mo$ series clearly resemble those for α-helical polypeptides. An analogous study in pure TFE revealed that $BOC(Ala)_7 OMe$, $Z(Ala)_7 OEt$, $Z(Ala)_8 Mo$, and $MEEA(Ala)_9 Mo$ assume β structures in this solvent. Moreover, lowering the concentration of several of the β-forming L-alanine oligomers eventually resulted in CD patterns that reflect a mixture of disordered and helical conformations. Studies on oligopeptide conformation must be carried out at several different concentrations because oligomers containing aliphatic side chains tend to aggregate even at concentrations approaching $5 \times 10^{-4} M$ (0.5 mg/ml). Only when this association is prevented by dilution or the addition of small amounts of strong acid can their true tendency to assume folded species be determined. For all of the oligopeptides we examined, HFIP proved to be a structure-breaking solvent, and in fact, CD studies showed that up to the nonamer the L-alanine oligomers are essentially disordered in this fluoroalcohol.

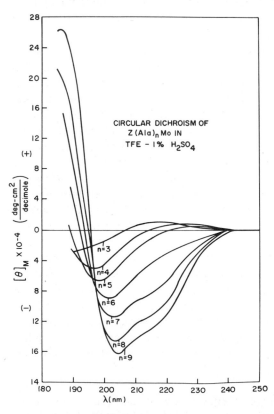

Figure 1. CD spectra of $Z(Ala)_n Mo$ in 99% TFE–1% H_2SO_4 at 25°C.

The NMR technique is also extremely useful in elucidating oligopeptide structures (4, 5, 7). The earliest systematic study of the NMR of a homo-oligopeptide series was reported by our group on the γ-ethyl-glutamate peptides in TFE and Me_2SO (4, 5). We were able to assign each of the NH protons in the lower oligomers and to detect the onset of helicity. In a subsequent paper (7), we reported evidence for the existence of an intramolecular hydrogen bond in the tetrapeptide derived from γ-ethyl-glutamate dissolved in chloroform-TFA mixtures. We have continued to use this technique and believe that superconducting spectrometers, coupled with computer averaging methods, will allow us to compare NMR and CD results at identical concentrations.

By use of NMR spectroscopy on oligopeptides, we were able to examine the nature of the so-called "double-peak" phenomenon observed for a number of polypeptides in chloroform-TFA and dimethylsulfoxide-methanol systems. Ferretti (15) initially reported this phenomenon and attributed it to coexisting helical and coil forms. Ullman (16) proposed that the double peaks arise from polydispersity of the synthetic polypeptides. Experimental evidence to support this contention has been reported by Bradbury et al. (17) and by Wada (18). In previous papers on polyalanine (19, 20) we uncovered a variant of the polydispersity explanation for the double peak. We assigned the upfield peak from the α-CH resonances to rapidly interchanging helical and coil structures, the downfield peak to oligopeptides with chain lengths too short to form helices. In order to further clarify the reasons for the double peak, we undertook a study of the NMR spectra of monodisperse $MEEA(Ala)_n Mo$ (n = 3–9) oligopeptides (21) using a 220 MHz spectrometer and time averaging to improve signal-to-noise ratio.

The NMR behavior of the α-CH protons of the L-alanine trimer (Fig. 2) and tetramer in various mixtures of $CDCl_3$-TFA is similar. The peak at 4.8–5.0 ppm is assigned to the α-CH resonances of the C-terminal residue protected by the morpholino group and the peak near 4.5 ppm is assigned to the combined α-CH resonances of the N-terminal and internal residues. Under all conditions employed, the upfield peak does not separate into its component parts. The ratio of the intensity of the lowfield to the upfield peaks is 1:2 and 1:3 for the trimer and tetramer, respectively (Table 1). The pentamer and longer oligomers exhibit the same C-terminal downfield peak at 4.8–5.0 ppm and the same double-peak characteristic for all the other residues (4.3–4.6 ppm). As with the low-molecular-weight polydisperse polyalanines ($\overline{DP}$ = 37), which we have discussed in previous papers (19–21), the double peaks for these monodisperse pure oligopeptides can be disrupted by addition of TFA to the chloroform solution. Table 1 shows the amount of TFA necessary to bring about coalescence of the double peak as a function of the size of the oligopeptides. The location of the α-CH of the C-terminal residue downfield is due to the fact that it is linked to a morpholino group (tertiary amide), which causes greater solvation and deshielding. The remaining residues are linked by peptide bonds. Preliminary results on $CH_3CO(Ala)_5NHCH_3$ provide support for the above explanation. A single peak is observed for all the α-CH protons in $CDCl_3$ containing 2% TFA.

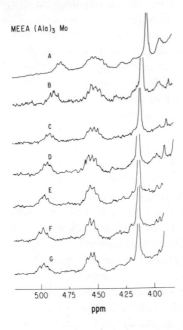

Figure 2. Partial 220 MHz NMR spectra for the α-CH region of MEEA(Ala)$_3$Mo as a function of TFA concentration in CDCl$_3$. The TFA concentrations were: A, 0%; B, 0.2%; C, 0.4%; D, 0.8%; E, 1.2%; F, 2.0%; G, 3.0% [sample concentration = 0.3% w/v (7×10^{-3} M); 32 scans].

Table 1.   Relative intensities of the downfield peak and the upfield split peaks for MEEA(Ala)$_n$OMe at 99.6% CDCl$_3$–0.4% TFA and coalescence points as a function of TFA concentration at 25°C

MEEA(Ala)$_n$Mo	Downfield Peak Intensity[a]	Upfield Peak Intensities[b]	TFA in CDCl$_3$ for Coalescence (%)
3	1	2	–
4	1	3	–
5	1	2:2	0.9
6	1	2:3	1.2
7	1	2:4	3.2
8	1	2:5	3.6
9	1	2:6	6.8

[a]Assigned to the α-CH of the C-terminal residue.

[b]Assigned to the α-CHs of the N-terminal and internal residues.

Table 1 also compares the relative intensities of the peaks for the entire oligomer series. It is clear that the upfield envelope for the pentamer (~4.5 ppm) splits into two equally intense resonances that interchange slowly on an NMR time scale (Fig. 3). The intensities of the lower field component of the double peak remain constant at two protons for the pentamer through the nonamer, but the upfield peak of the pair continuously increases in intensity from two protons for the pentamer to six for the nonamer (Fig. 4).

Changes in temperature can cause the double peaks for the pentamer to coalesce. As the temperature is raised, the upfield α-CH peak (4.25 ppm) moves toward the downfield peaks (4.50 ppm) and actually coalesces completely with it above 40°C. The α-CH peaks of the longer oligopeptides show no such coalescence over the same temperature range. At a TFA concentration of 0.5% coalescence of the double peak for the pentamer also can be brought about by reducing the sample concentration to 0.1% (w/v). Partial coalescence is observed for the hexamer under similar conditions. The longer oligomers do not exhibit a concentration dependence in their NMR spectra. We interpret these results to indicate that the aggregates in solution are very stable for the oligopeptides with chain lengths greater than six.

It is evident that these double peaks do not arise from polydispersity or from an equilibrium between helical and "random" forms. Furthermore, the double-peak effect for these alanine homo-oligomers does not depend on the nature of the protecting groups because we have noted similar effects when other protecting groups, such as benzyloxycarbonyl and ethyl ester, are used.

We believe that the double peaks for these alanine oligopeptides arise from folded forms that commence at the pentamer stage. These folded forms involve intramolecular hydrogen bonds stabilized by aggregation. We envisage the model as having exposed and solvated C-terminal and N-terminal regions that are constant in size throughout the entire oligopeptide series. The internal sections of

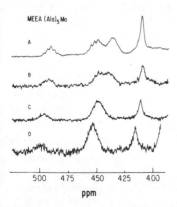

Figure 3. Partial 220 MHz NMR spectra for the α-CH region of MEEA(Ala)$_5$Mo as a function of TFA concentration in CDCl$_3$. The TFA concentrations were: A, 0.2%; B, 0.4%; C, 0.8%; D, 1.0% [sample concentration = 0.3% w/v (5 × 10^{-3} M); 32 scans].

MEEA (Ala)$_9$ Mo

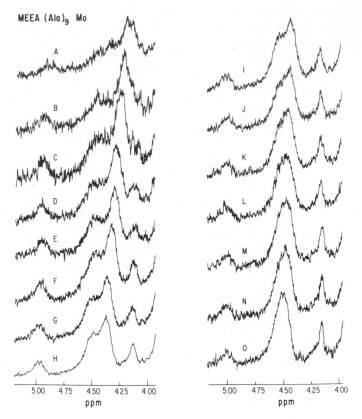

Figure 4. Partial 220 MHz NMR spectra for the α-CH region of MEEA(Ala)$_9$Mo as a function of TFA concentration in CDCl$_3$. The TFA concentrations were: A, 0.4%; B, 0.8%; C, 1.2%; D, 1.6%; E, 2.0%; F, 2.4%; G, 2.8%; H, 3.2%; I, 3.6%; J, 4.0%; K, 4.4%; L, 4.8%; M, 5.2%; N, 5.6%; O, 6.8% [sample concentration = 0.3% w/v (3 × 10^{-3} M); 32 scans].

the longer Ala oligomers are involved both in intramolecular hydrogen bonding and aggregation. The α-CH protons from internal residues therefore should be less solvated and should exhibit their resonances upfield. Consequently, the constant two protons of the downfield components of the double peaks are assigned to the α-CH protons of the ultimate and penultimate N-terminal residues of the oligomers, which are more exposed to solvent. According to the proposed model, slow interchange on the NMR time scale among the classes of residues is reasonable and double peaks should result. This explanation is consistent with the NMR results. The existence of such hydrogen bonds in the oligopeptide series was recently confirmed by our IR study of MEEA(Ala)$_n$Mo (n = 5–9) in TFA-CDCl$_3$ solvent mixtures. This double peak phenomenon differs from those observed for polydisperse polypeptides. In our case, specific aggregations of folded forms in solution lead to the effect and the broad molecular-weight distribution leads to the double peaks for the polypeptides.

Methionine and leucine oligomers. Recently we commenced the synthesis and conformational analysis of the BOC(Met)$_n$ OMe oligopeptides (12). Fig. 5 presents the CD spectra in TFE. We attribute the long-wavelength band ($\sim$220 nm) to the $n{\to}\pi^*$ transition of the peptide chromophore and the short-wavelength band (197–205 nm) to the $\pi{\to}\pi^*$ transition. The CD curves for the smaller oligomers (n = 4,5,6) are similar in shape and peak position. An abrupt change for the heptamer is indicated by a red shift of the $\pi{\to}\pi^*$ transition ($\sim$4 nm) and the development of significant negative ellipticity at 222.5 nm. These changes continue in the nonamer, where the $\pi{\to}\pi^*$ band maximum occurs at 205.5 nm and the total molar intensity of the $n{\to}\pi^*$ band increases to 59,000. We believe that these data suggest that the L-methionine oligopeptides begin forming secondary structures at the heptamer in TFE, whereas the trimer through hexamer appear to exist essentially in a disordered state. The CD spectra of the heptamer and nonamer are remarkably similar to those reported by us for the γ-ethyl-glutamate series in TFE (4, 5). Thus, the secondary structures for these oligopeptides in this fluoroalcohol may be similar. In HFIP and HFA, however, the CD patterns show no significant changes from dimer to heptamer. This result is not surprising because oligopeptides in HFIP and HFA are known to require much longer chain lengths for the onset of secondary structures.

Unfortunately, these oligomers are insoluble in water. Nevertheless, we did examine their CD properties in TFE-water mixtures. The addition of 80% water (v/v) to a solution of the L-methionine heptamer in TFE induces a dramatic variation in the ellipticity values that is reminiscent of the onset of a β structure. Dilution or increasing the temperature (T_m = 44°C for the heptamer) disrupts this β-type conformation.

We have also synthesized and investigated the BOC(Leu)$_n$OMe series (12a). Very similar CD results have been obtained in TFE, HFIP, HFA, and TFE-water

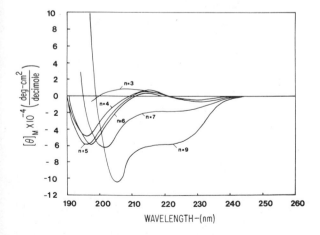

Figure 5. CD spectra of BOC(Met)$_n$OMe in TFE at 25°C. Oligomer concentration = 1 mg/ml (3 × 10^{-3}–7 × 10^{-4} M).

mixtures. The homopolymers derived from L-methionine and L-leucine are known to form very stable right-handed α helices in nondenaturing organic solvents (22). Recent statistical analysis of the ordered and disordered regions of globular proteins showed that L-methionine and L-leucine residues can conform to either helical or β-sheet regions in these biopolymers (23). Thus, it is not unreasonable that oligomers composed solely of L-methionine or L-leucine residues might form helical secondary structures above a certain chain length in appropriate solvents. Our results indicate that the heptamers of L-methionine and L-leucine do exist partially in helical structures in TFE and that the critical size for helix formation in this solvent occurs at the heptamer. Also, our studies have shown that the addition of water to solutions of these oligomers in TFE forces the higher oligomers to assume associated structures. Undoubtedly, the fact that water is a poor solvent for these hydrophobic peptides causes them to aggregate. Because many oligopeptides have been found to assume β structures in the solid state, it is quite reasonable that the aggregated species assume this conformation in organic solvent–water mixtures.

Valine and isoleucine oligomers. The homologous series $BOC(Val)_nOMe$ and $BOC(Ile)_nOMe$ have been prepared and examined by chiro-optical techniques (8, 9, 12a). The CD spectra of the oligovalines in TFE are shown in Fig. 6. As the chain length increases from two to six, the Cotton effects in both the 210–220 nm and 190–200 nm spectral regions increase gradually in intensity. The

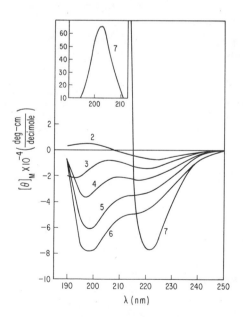

Figure 6. CD spectra of $BOC(Val)_nOMe$ in TFE at 25°C. Oligomer concentration = 3×10^{-4} M.

spectral band associated with the $\pi \to \pi^*$ transition (195 nm) of the peptide chromophore also exhibits a small red shift (~5 nm) from the trimer to the hexamer. The CD pattern of the heptamer is completely different from those of the lower members of the series. It exhibits a negative Cotton effect centered at 222 nm and a positive one at 202 nm with a crossover at 214.5 nm. The overall shape and position of the crossover and Cotton effects are typical of CD spectra of polypeptides in β conformations (14). Our CD study has shown also that the ordered conformation assumed by the valine heptamer in TFE can be destroyed by increasing the temperature, or adding HFIP. In contrast, it was found that addition of water to a solution of BOC(Val)$_7$OMe in TFE does not change the nature of the spectral pattern of the heptamer.

Quite comparable CD results have been obtained for the L-isoleucine series in the same solvents. In TFE, both the heptamer and octamer were found to exist in β conformations. Although BOC(Ile)$_7$OMe exhibits a β-disordered transition as the oligomer concentration is lowered, dilution studies on the corresponding octamer did not indicate any change in conformation at concentrations as low as 0.02 mg/ml. These results suggest that the octamer forms an extremely stable β conformation. Although our evidence is not sufficient to justify a definite conclusion, the existing data do imply that the octamer may exist in an intramolecularly hydrogen-bonded β form. We feel that the additional stability of the octamer cannot be explained by the effect of one added residue on an intermolecular β structure. In contrast, at these short chain lengths, the extra residue might be expected to lead to an intramolecular species that is appreciably more stable than that formed by the heptamer.

We have commenced an investigation of the NMR spectra of the isoleucine oligopeptides using CDCl$_3$-TFA mixtures (24). Fig. 7 shows the partial NMR spectra of the BOC-Ile octamer at 0.35% concentration with differing amounts of trifluoroacetic acid in the solvent. We believe that under the conditions of measurement, the BOC end group is cleaved and that we are dealing with the isoleucine octamer methyl ester trifluoroacetate salt. As with the alanine oligopeptides, there appears to be a downfield peak in each of the spectra that is most likely attributable to the α-CH of a C-terminal residue. A broad peak, which we assign to the α-CH adjacent to the ammonium ion end group, appears at highest frequency. The remaining α-CH resonances can be seen at intermediate frequencies. All the peaks shift downfield with increasing TFA concentration.

At present we are carrying out a detailed NMR analysis of the methionine and isoleucine oligopeptides similar to that described above for the alanine series. These results will be presented in a forthcoming publication.

CONCLUSION

The results presented in this paper represent an ongoing study of the synthesis and conformations of acyclic oligopeptides. Pure oligopeptides allow us to employ the various spectroscopic techniques described in this paper to deduce conformations as a function of the chain length and the nature of the peptide side chains. Extensions of our work in this area include studies on oligopeptides with specific sequences.

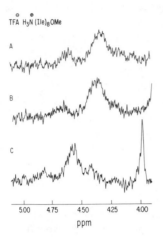

TFA $H_3\overset{\oplus}{N}$ (Ile)$_8$OMe

A

B

C

| 5.00 | 4.75 | 4.50 | 4.25 | 4.00 |

ppm

Figure 7. Partial 220 MHz NMR spectra for the α-CH region of t-BOC(Ile)$_8$OMe as a function of TFA concentration in CDCl$_3$. The TFA concentrations were: A, 2%; B, 13%; C, 100% [sample concentration 0.35% w/v (4 × 10^{-3} M); 16 scans].

These studies provide important information that may not be deducible from cyclic peptides. Cyclic peptides generally lead to simpler spectroscopic patterns than linear peptides because they have no N-terminal or C-terminal residues. Cyclic peptides may not be useful as model substances for linear natural products or protein structures, however. The cyclic character of a given peptide may induce conformational restrictions and preferences that are absent in the acyclic analogs. Consequently, we view our major effort in the conformational studies of linear oligopeptides as a basic program in the elucidation of peptide and protein conformations.

REFERENCES

1. M. Goodman, E.E. Schmitt, I. Listowsky, F. Boardman, I.G. Rosen, and M.A. Stake, in *Polyamino Acids, Polypeptides and Proteins*, M.A. Stahmann, ed., University of Wisconsin Press, Madison, Wis., 1962, p. 195 and references therein.
2. M. Goodman, I. Listowsky, Y. Masuda, and F. Boardman, *Biopolymers*, **1**, 33 (1963).
3. M. Goodman and I. Rosen, *Biopolymers*, **2**, 537 (1964).
4. M. Goodman, A.S. Verdini, C. Toniolo, W.D. Phillips, and F.A. Bovey, *Proc. Nat. Acad. Sci. U.S.*, **64**, 444 (1969).
5. M. Goodman, C. Toniolo, and A.S. Verdini, in *Peptides 1969*, E. Scoffone, ed., North-Holland, Amsterdam, 1971, p. 207.
6. M. Goodman, Y. Masuda, and A.S. Verdini, *Biopolymers*, **10**, 1031 (1971).
7. P. Temussi and M. Goodman, *Proc. Nat. Acad. Sci. U.S.*, **68**, 1767 (1971).
8. C. Toniolo, *Biopolymers*, **10**, 1707 (1971).
9. M. Goodman, F. Naider, and C. Toniolo, *Biopolymers*, **10**, 1719 (1971).

10. M. Goodman, R. Rupp, and F. Naider, *Bioorg. Chem.,* **1**, 294 (1971).
11. M. Goodman, F. Naider, and R. Rupp, *Bioorg. Chem.,* **1**, 310 (1971).
12. (a) C. Toniolo and G.M. Bonora, to be submitted; (b) J.M. Becker and F. Naider, to be submitted.
13. W.B. Gratzer, in *Poly-α-Amino Acids,* Vol. 1, G.D. Fasman, ed., Marcel Dekker, New York, 1967, p. 177 and references therein.
14. S. Beychok, in *Poly-α-Amino Acids,* Vol. 1, G.D. Fasman, ed., Marcel Dekker, New York, 1967, p. 293 and references therein.
15. J.A. Ferretti, *Chem. Commun.,* 1030 (1967).
16. R. Ullman, *Biopolymers,* **9**, 471 (1970).
17. E.M. Bradbury, P.D. Cary, C. Crane-Robinson, and P.G. Hartman, *Pure Appl. Chem.,* **36**, 53 (1973) and references therein.
18. K. Nagayama and A. Wada, *Biopolymers,* **12**, 2443 (1973).
19. M. Goodman, F. Toda, and N. Ueyama, *Proc. Nat. Acad. Sci. U.S.,* **70**, 331 (1973).
20. M. Goodman and N. Ueyama, *Biopolymers,* **12**, 2639 (1973).
21. M. Goodman, N. Ueyama, and F. Toda, *Biopolymers,* in press.
22. G.D. Fasman, in *Poly-α-Amino Acids,* Vol. 1, G.D. Fasman, ed., Marcel Dekker, New York, 1967, p. 499 and references therein.
23. P.Y. Chou and G.D. Fasman, *Biochemistry,* **13**, 211 (1974).
24. M. Goodman, C. Toniolo, N. Ueyama, unpublished results.

OPTICAL ACTIVITY OF POLYPEPTIDES IN THE INFRARED: SYMMETRY CONSIDERATIONS

JOHN A. SCHELLMAN, Department of Chemistry, University of Oregon, Eugene, Ore. 97403

SYNOPSIS: Cyclic boundary conditions are applied to examine the infrared circular dichroism of helices and pleated-sheet structures for polypeptides. It is shown that with a dipole coupling mechanism, the anisotropy factor for an exciton is proportional to frequency, but independent (on the average) of the electronic or vibrational nature of the absorption band. Simple geometrical criteria for rotatory strength are derived. The results are in semiquantitative agreement with detailed computer calculations on the same systems. Several bands that should have observable circular dichroism (CD) are discussed.

Much of the work on proteins and polypeptides concerns the relationship of structure with chemical, physical, and biological properties. As a result, there has been a steady pressure to extract the greatest possible amount of structural information from any appropriate technique. Barring those important cases where x-ray–diffraction methods are applicable on crystals or fibers, the most successful procedures have been spectroscopic (hypochromicity, ORD-CD, IR, NMR, Raman spectroscopy, etc.), although important information has also been obtained from hydrogen-exchange and hydrodynamic techniques. With regard to the conformation of the peptide backbone in polypeptides, measurements of circular dichroism and infrared spectra have been important because both, in different ways, give direct information about the coupling of peptides with one another and this information, in turn, is intimately related to backbone geometry.

In principle, there is no reason why the technique of circular dichroism, which is a specialized type of difference spectrum, cannot be applied to the infrared bands of polypeptides. It is certainly conceivable that the additional information obtained in this way would enhance our interpretation in the same way that circular dichroism in electronic bands has supplemented electronic absorption spectroscopy. Until recently, technical difficulties have prevented such measurements. The detection of infrared radiation in a typical infrared spectrometer is too slow and insensitive to measure the small signals expected to be associated with the circular dichroism of vibrational transitions. Technical advances of the past five years have obviated these difficulties. A light modulation device now available works in the infrared and permits the use of sophisticated procedures for processing signals (1). In addition, highly sensitive and fast infrared detectors have been developed (2). Not all the technical problems have been solved, but they are in the hands of renowned research groups (3–5) and it appears that success is merely a matter of time. In fact, the first clear-cut measurement of vibrational circular dichroism has appeared just recently (3).

On the theoretical side, the problem of the circular dichroism of vibrational bands strongly resembles the problem of electronic excitation. In general, if the system can be regarded as small compared to the wavelength of light, then for a quantum jump from state A to state B, the integrated circular dichroism and the rotatory strength of the band are related by the formula

$$\int \frac{\Delta\epsilon}{\nu} \, d\bar{\nu} = \left(\frac{32\pi^3 N}{3(2303) \, hc}\right) R_{AB} \tag{1}$$

where $R_{AB} = \text{Im}(\mu_{AB} m_{BA})$, N is Avogadro's number, $\bar{\nu}$ the frequency in cm^{-1}, and μ and m are the magnetic-dipole-moment operators for *all* particles in the system, that is,

$$\mu = \mu_{el} + \mu_{nuc} = -e\sum_{i} \mathbf{r}_i + e\sum_{j} Z_j \mathbf{r}_j$$

$$\mathbf{m} = \mathbf{m}_{el} + \mathbf{m}_{nuc} = -\frac{\beta}{\hbar}\sum_{i} \mathbf{r}_i \times \mathbf{p}_i + \frac{\beta_N}{\hbar}\sum_{j} Z_j M_j \mathbf{r}_j \times \mathbf{p}_j \tag{2}$$

In the above, e is the unit positive charge, Z_j is the number of charges on nucleus j, β is the Bohr magneton, β_N is the nuclear magneton, M_j is the mass number of nucleus j, and $\mathbf{r}$ and $\mathbf{p}$ represent the position and momentum vectors of electrons and nuclei. In an electronic band, nuclear moments do not contribute because of the Franck–Condon principle. The nuclear *levels* do contribute to give fine structure to the electronic band. By contrast, electronic moments contribute to vibrational bands. The physical interpretation of this fact is that the electronic distribution of the molecule follows the comparatively slow variation in nuclear position, leading to electronic, as well as nuclear, current densities. A complete theory of the electronic following of nuclear motions is a formidable problem that has been solved only by the application of high-powered computer techniques for specialized cases. Theoreticians in the field have evaded this difficulty (1) by assuming an idealized model of vibrating charges, (2) by ascribing to each atom a fixed partial charge that travels with it during molecular vibrations, or (3) by utilizing experimentally determined transition moments. Consequently, the breakdown into nuclear and electronic motions is unnecessary.

The last procedure has already been applied to electronic transitions (6, 7) and may be applied without change to vibrational transitions. Exciton splittings, polarization, hypochromic effects, and circular dichroism are generated automatically by this procedure. This method of calculation takes full advantage of all information available about the chromophore and takes proper care of such problems as accidental degeneracies, end effects, one-electron optical activity, and so on. The result is a representation of real systems that is limited only by the accuracy of some of the matrix elements and by the fact that a limited number of excited states are considered. The disadvantage of the method is that although the equations are general relations of linear algebra, their solution, except for a simple dimer, requires a computer. As a result, it is sometimes difficult to keep an intuitive grasp on the way in which various physical factors enter into the solution.

A procedure that leads to simple, analytical results for coupled identical systems (8) makes use of the most elementary assumption of solid-state spectroscopy, namely, cyclic boundary conditions for the repeating units. Since the appearance of the paper by Moffitt *et al.* (9), it has been known that cyclic boundary conditions, although they give a satisfactory account of the absorption spectra of an ordered array, must be used with great caution in considering optical activity. The nature of this theoretical difficulty has been clarified only in recent years (10-12). In this paper we shall apply this simple analytical approach to helical and β structures, keeping in mind that the formulas so obtained can have only a qualitative application for real systems. What is gained from this approach is a clear picture of how the symmetry elements of an ordered structure and the symmetry properties of transition moments within the structure determine the qualitative features of absorption and circular dichroism. These results then aid in interpreting the more quantitative results obtained by realistic but complex methods of calculation.

IDEALIZED EXCITONS

For background we present here the simplest results of helical exciton theory. In accordance with the use of cyclic boundary conditions, the exciton wave functions are given by an expression of the form

$$\psi_K = \Sigma a_{jK} \phi_j \tag{3}$$

where

$$a_{jK} = \frac{1}{\sqrt{N}} e^{ijK 2\pi/N} \tag{4}$$

Here ϕ_j is the wave function for a state in which group j is excited and all other groups are in their ground state; N is the number of groups in the helix, and K is the exciton index, which takes on values from 0 to $\pm N/2$ and is a measure of the wavelength of the exciton. To a good approximation, the only bands that are of spectroscopic importance are those in which each unit cell of the helix is in an identical state of excitation. We therefore designate N as the number of residues in a unit cell. If the helix has an integral number of residues per turn, n, then $N = n$. The basic helical parameters for a number of important helices are given in Table 1. The angular advance per residue is given by $\epsilon = 2\pi/n$, and following Ramachandran *et al.* (13), n and ϵ will be defined as negative for left-handed helices, t is the rise per residue, and P is the pitch, or rise, per turn; $P = nt$.

Table 1. Some helical parameters

	K (allowed)	ϵ	P (Å)
α helix	0, ±5	100°	5.36
DNA (B)	0, ±1	36°	34.6
DNA (A)	0, ±1	32.8°	28.6
‖ β sheet	0, 1	180°	6.50
⇈ β main chain	0, 1	180°	6.89
⇅ β interchain	0, 1	180°	9.46

Of the N unit cell excitons, only three are reached by electric-dipole-allowed transitions:

$$K = 0 \qquad a_{j,0} = \frac{1}{\sqrt{N}} \quad \text{for all } j$$

$$K = \pm\frac{N}{n} \qquad a_{j,\pm N/n} = \frac{1}{\sqrt{N}} e^{\pm ij\epsilon} \tag{5}$$

These unit cell excitons correspond to an exciton in which there is no phase change in going from one group to the next (in this case, polarization in the direction of the helix axis results) and excitons in which the phase advance per residue is equal in magnitude to the angular rotation per residue of the helix. The latter mode is degenerate and polarized perpendicular to the helix axis. In the special case of $N = 2$, there is only one perpendicular transition and the degenerate pair becomes a single transition. The exciton modes for which $K = \pm N/n$ obviously can be combined—and frequently they are— to give coefficients of the form $\sin j\epsilon$ and $\cos j\epsilon$. We are especially concerned with the α helix for which $N = 18$, with a parallel β structure for which $N = 2$, and with the antiparallel β structure that has two perpendicular screw axes, each with $N = 2$.

The energy of an excitation is obtained by evaluating

$$\int \psi_K^* \sum_{ij} V_{ij} \, \psi_K^* \, d\tau$$

where the summation is over the pairwise interactions of groups i and j and V is the potential energy of interaction. The frequencies of the allowed excitons are

$$\tilde{\nu}_\| = \tilde{\nu}_0 + \frac{2}{h \sum_j V_{0j}} \qquad \text{for } K = 0$$

$$\tilde{\nu}_\perp = \tilde{\nu}_0 + \frac{2}{h \sum_j V_{0j} \cos j\epsilon} \qquad \text{for } K = \pm\frac{N}{n} \tag{6}$$

where V_{0j} is the matrix element for the pairwise interaction of a central group labeled 0 and a neighbor j units along the chain (8).

The theory just outlined is known to give a good representation of the absorption and dichroism of crystals and large molecular aggregates. Its application to circular dichroism, on the other hand, causes some problems. To apply the above selection rules to an optically active system requires the use of Rosenfeld's equation as well as the cyclic boundary conditions. The two assumptions are contradictory. If a molecule is long enough to justify the use of cyclic boundary conditions, it is too long to use Rosenfeld's equation; if it is short enough to use Rosenfeld's equation, end effects tend to dominate and the selection rules break down. Nevertheless, the wave functions described by Eqs. (3) and (4) can be useful. They provide simple analytical formulas that permit an analysis of the way in which distinct structural features enter into the optical properties of molecular aggregates. In addition they can be used as a basis for discussing optical properties. The correct set of eigenfunctions for a finite system, the localized excitations and the cyclic excitons are all related to one another by unitary transformations.

THE MIYAZAWA–BLOUT THEORY OF IR ABSORPTION AND DICHROISM

Miyazawa and Blout (14) have applied the results of the preceding section to the infrared absorption and dichroism of polypeptides in the α-helical conformation and in the parallel and antiparallel β-sheet conformation. They assumed that only nearest-neighbor interactions were important in the determination of energy via Eq. (6) and distinguished two kinds of nearest neighbors: those that are covalently linked and those that are hydrogen-bonded to one another. Because covalent linkages occur in the α helix for $j = \pm1$ and hydrogen-bonded linkages for $j = \pm3$, they wrote the energy equation for the α helix as

$$\bar{\nu}_{\parallel} = \bar{\nu}_0 + D_1 + D_3 \qquad\qquad K = 0$$

$$\bar{\nu}_{\perp} = \bar{\nu}_0 + D_1 \cos\theta + D_3 \cos 3\theta \qquad K = \pm\frac{N}{n} = \pm5 \tag{7}$$

where D_1 and D_3 represent covalent- and hydrogen-bond interactions, respectively. Thus, $D_1 = 2/hV_{01}$ and $D_3 = 2/hV_{03}$. These two frequencies are associated with parallel and perpendicular polarization, respectively. The formulas thus combine predictions of infrared dichroism and frequency. Rather than try to predict the coupling energy of vibrations, Miyazawa and Blout utilized experimental data on a number of peptides and peptide analogs. They applied the same procedure to β sheets.

A modified form of the theory, since proposed by Krimm (15), considers interactions other than 01 and 03. In particular, in the case of the α helix, the 02 interaction plays an important role. It turns out that the measurement of infrared circular dichroism will probably provide a way to test the relative merits of the two theories because detailed calculations (16) show that the two models should have quite different optically active behavior.

In this paper and in the complete theory (16), we build directly on the Miyazawa–Blout–Krimm analysis. Thus, we assume their energies and polarizations at the outset. The only additional requirement is the matrix elements necessary for optical activity. These will be considered in the next section.

THE ROTATORY STRENGTH TENSOR

We will find it convenient to consider the rotatory strength tensor rather than the average rotatory strength discussed in Eq. (1) (17). The rotatory strength tensor is so defined that if $\mathbf{L}$ is a unit vector in the direction of propagation of the radiation, the circular dichroism of an absorption band is given by the formula

$$\Delta\epsilon = K\phi(\tilde{\nu})\mathbf{L}\underline{\mathbf{R}}\mathbf{L} \tag{8}$$

where $K = 32\pi^3 N\nu_{a0}/3hc(2303)$ and $\phi(\tilde{\nu})$ is a normalized band shape factor. $\underline{\mathbf{R}}$ is given by

$$\underline{\mathbf{R}} = \frac{-3e}{2mc} \text{Im} [\mu_{0a}(p\mathbf{r})_{a0}] \tag{9}$$

It will suffice in general to know the diagonal elements only of this tensor, which are given by

$$R_{xx} = \text{Im} \frac{3e}{2mc} [(\mu_z)_{0a}(p_y x)_{a0} - (\mu_y)_{0a}(p_z x)_{a0}] \tag{10}$$

We obtain R_{yy} and R_{zz} from (16) by cyclic permutation of the indices. In general, the z direction will be the direction of the helix axis; x and y will be equivalent axes in the α-helical case. In the sheet structures, x will be defined to lie in the polypeptide plane with y perpendicular to the plane. In the above, the subscript 0a indicates the matrix element for a jump from the ground state to excited state a; μ is the electric dipole moment, p is the momentum operator and $\mathbf{r}$ (with components x, y, z) is the position vector. The orientations of the electric dipole transition moments are known from infrared dichroic studies on model compounds (18) and are shown in Fig. 1. We were unable to find integrated absorption spectra of the amide-I and -II bands in the literature. Charles Swenson measured the integrated intensity of these two bands (19) on N-methylacetamide, finding $\mu = .29$ and $.21$ Debyes, respectively.

Quantities such as

$$\frac{3e}{2mc} \text{Im} [(p_{y_j} x_j)_{a0}]$$

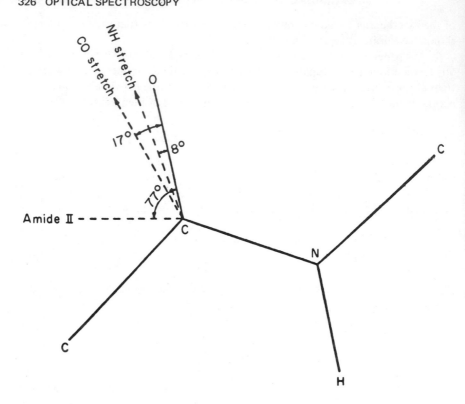

Figure 1. Infrared transition moments after Sandeman (18).

can be related to the electric dipole matrix elements by an important and useful approximation. The matrix element for the whole unit cell may be expressed in terms of the matrix element for the groups giving $\Sigma_j a_{jK}(p_{yj}x_j)_{a0}$. Now x_j is the x coordinate of charged particles in this group relative to an arbitrary origin somewhere in the molecule and it can be written as $x = X_j + \rho$ where X_j is a fixed vector within the jth group. Because ρ contributes to the higher-order local multipoles of the jth group, it can be neglected in a first approximation.[1] If we ignore the local multipoles, then for each group we can write

$$\frac{3e}{2mc} \mathrm{Im}\,[(p_y x)_{a0}] \cong 3\pi\tilde{v}_{a0}X(\mu_y)_{a0} \tag{11}$$

where we have utilized the standard transformation $(p_y)_{a0} = 2\pi\,\mathrm{Im}\,c\bar{v}_{a0}(\mu_y)_{a0}/e$. Substituting (11) into (10), we have

---

[1] It may turn out that in infrared transitions there are analogs to $n\pi^*$ transitions that will invalidate this assumption. If so, these transitions will have to be added to the theory just as in the electronic case.

$$R_{xx} = 3\pi\tilde{\nu}_{a0} \left[\{\mu_z\} \{X\mu_y\} - \{\mu_y\} \{X\mu_z\} \right] \tag{12}$$

where the terms in braces are a shorthand notation for

$$\{\mu_z\} = \Sigma a_{Kj} (\mu_{zj})_{0K}$$

$$\{X\mu_y\} = \Sigma a_{Kj}^* X_j (\mu_{yj})_{K0}$$

and expressions for R_{yy} and R_{zz} as well as other matrix elements may be obtained by cyclic permutation of cartesian indices.

Thus we see that the ingredients necessary for the calculation of CD at this approximation are the coefficients a_{Kj}, obtainable from the cyclic exciton assumption, the orientations of transition moments, and the "positions" of transition moments. Transition-moment orientations are known for the peptide group from dichroic studies (Fig. 1) and can easily be transformed into the unit cell for any given structure. The best positions for the transition dipoles are those that minimize the multipole contribution coming from the local coordinate ρ. Because these are unknown, the positions must be guessed at and will be taken as in the center of the most active atoms of the transition, that is, halfway between C and O for the amide-I transition and halfway between N and H for the amide-II transition. This assumption can cause problems with the amide-II transition of β structures, but this is discussed elsewhere (6).

The average rotatory strength is given by one-third of the trace of Eq. (12), that is, by

$$R_{a0} = \pi\bar{\nu}_{a0} \{\mu\} \times \{RX\mu\} \tag{13}$$

where the braces again indicate the exciton-weighted summation over all members of the unit cell. This relation is clearly related to the Kirkwood formula for the rotatory strength generated by coupled oscillators.

THE CIRCULAR DICHROISM OF HELICES

We have recently discussed the rotatory strength of oriented helices for the cyclic exciton limit (as well as for real cases (17) and so need only cite the results. The sums of Eq. (12) can be evaluated in closed form for an arbitrary helix. Taking the helix axis as the z axis we have $\mu_z = \mu_\parallel$. The moment perpendicular to the helix axis is $\mu_\perp = (\mu_x^2 + \mu_y^2)^{1/2}$. The tangential component of the transition moment occurs in the formulas and is given by $\mu_t = \mu \cos \delta$, where δ is the angle between the xy projections of the transition moment and one tangent to the helix at the center of the chromophore. The characteristic distances for the helical problem are r_0, the radial distance of the transition moment from the helix axis, and the translation per turn, P, which is defined as positive. The product $\mu_\parallel\mu_\perp$ is positive for transition moments that go around the helix in a right-handed sense and negative for left-handed. There is no necessary correlation between the sense of the transition moment and the sense of the helix.

The results of the calculation are given in Table 2. Only three transitions are allowed. These are labeled 0, S, and C to designate the $K = 0$ exciton and the sine and cosine form of the $K = \pm N/n$ excitons, respectively. The A and B terms are simple functions of the helical parameters.

The A terms come from the original Moffitt theory and are given by

$$A_\| = \frac{3\pi}{4} \bar{\nu}_\| \mu_\perp \mu_\| r_0 \cos \delta$$

$$A_\perp = - \frac{3\pi}{4} \bar{\nu}_\perp \mu_\perp \mu_\| r_0 \cos \delta$$

(14)

The subscripts on A indicate the direction of electric polarization relative to the helix axis. The terms $A_\|$ and $A_\perp$ differ from one another in magnitude only because of the exciton split that separates $\bar{\nu}_\|$ from $\bar{\nu}_\perp$. We note that this contribution does not depend on P or ϵ, but requires a transition moment that simultaneously is separated from the helix axis and has both a tangential and a parallel component.

The experimental quantity of interest is $\Delta\epsilon/\epsilon$ because this is the concentration-independent quantity that determines the feasibility of a measurement. Apart from correction terms for refractive index, this quantity is given by

$$\frac{\Delta\epsilon}{\epsilon} = \frac{4R_{0a}\phi(\bar{\nu})}{D_{0a}\rho(\bar{\nu})}$$

(15)

where ϕ and ρ are normalized shape factors for circular dichroism and absorption, respectively, and $D_{0a} = (\mu_{0a})(2)$ is the dipole strength. We will use $4R/D_{tot}$ as a first approximation to $\Delta\epsilon/\epsilon$, recognizing that this is an average value that is exceeded in some regions of the spectrum and not attained in others. An obvious refinement would be to calculate $4R_\|/D_\|$, $4R_\perp/D_\perp$, and so on, but

Table 2. Helical rotatory strengths

State	0	S	C	S + C
Polarization	z, ‖	y, ⊥	x, ⊥	⊥
$R = R_{zz}$	2A	−A	−A	−2A
$R = (R_{xx} + R_{yy})/2$	0	B	−B	0
R_{av}	4/3A	−(2A −B)/3	−(2A +B)/3	−4/3A

for survey purposes division by the total dipole strength will be adequate. One result is that the calculated anisotropies will be lower limits. Putting $\mu_\parallel = \mu \cos \theta$ and $\mu_\perp = \mu \sin \theta$ and dividing by μ^2 we obtain

$$\frac{4R_A}{D} \cong 4\pi\tilde{\nu} \cos \theta \sin \theta \; r_0 \cos \delta \quad \text{(parallel polarized band)} \quad (16)$$

where θ is the angle the transition moment makes with the helix axis. The perpendicularly polarized band is of opposite sign. This formula states that the value of $\Delta\epsilon/\epsilon$ is determined by the position and orientation of the transition moment as well as the frequency, regardless of the nature of the transition. Because electronic and nuclear motions have the same kind of structural constraints, we come to the important conclusion that *on the average, the ratio of the magnitude of infrared and ultraviolet anisotropy factors is controlled only by the ratio of IR to UV frequencies.* For the most prominent UV and IR chromophores of peptides, this ratio should be about 1/15 to 1/50. Given a helical array with transition-moment geometry, factors like the relative masses of electrons and nuclei do not enter into the picture.

The amide-I and amide-II transitions are given in Table 3. The contributions of the A terms to $\Delta\epsilon/\epsilon$ are in the next-to-last column. The value for the amide-I band is well within the realm of possible measurement, but that for the amide-II band is probably too small to measure. The amide-II transition moment is essentially perpendicular to the helix axis so that μ is almost negligibly small. In this respect, the amide-II band of the α helix resembles the strong electronic transitions of the B form of DNA, which also have perpendicular transition moments. The N–H stretch at 3300 cm^{-1} is essentially parallel to the helix axis and should also produce a very small rotatory strength for the converse reason.

In a real helix with end effects, the idealized picture represented by the A terms is altered. The degeneracies are split and the A terms are projected over those excitons that most closely resemble the $K = 0$ and $K = 5$ excitons. This feature leads to a spreading of the bands, particularly the perpendicular band.

Table 3. Optical parameters for the amide-I and amide-II transitions of the α helix[a]

	$\tilde{\nu}_\parallel(\text{cm}^{-1})$	$\tilde{\nu}_\perp(\text{cm}^{-1})$	$\mu_\perp$	$\mu_\parallel$	δ	$\Delta\epsilon/\epsilon(A)$	$\Delta\epsilon/\epsilon(B)$
Amide I	1650	1652	0.17	–0.24	25°	0.92×10^{-4}	-0.17×10^{-4}
Amide II	1516	1546	0.21	0.004	17°	3.7×10^{-6}	-0.46×10^{-4}

[a] The frequencies are the assignments of Miyazawa and Blout (19). Transition moment directions are those of Sandeman (18). Helical geometry is that of Arnott and Dover (23).

In addition, the B terms, which sum to zero with the degenerate C and S modes of Table 2, are spread out into positive and negative wings which are superimposed on the perpendicular A contribution. The general situation is depicted in Fig. 2 and detailed calculations are given in Ref. (6). The quantity B is given by

$$B = \frac{3\pi}{4} \bar{\nu}_\perp \mu_\perp^2 P \cot \epsilon \tag{17}$$

It depends solely on the transition moment perpendicular to the helix axis. As pointed out by Tinoco (20), terms of this form are responsible for the characteristic CD bands of the B form of DNA. In analogy to Eq. (16), the contribution of the B terms to $\Delta\epsilon/\epsilon$ can be estimated by dividing by μ^2

$$\frac{\Delta\epsilon}{\epsilon} \cong \pi\bar{\nu} \sin^2\theta \, P \cot \epsilon$$

The calculated values for the amide-I and -II bands are given in the last column of Table 3. The contributions are appreciable, especially for the amide-II band.

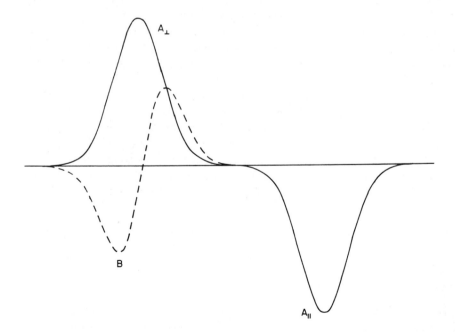

Figure 2. A typical finite exciton system. The A terms are spread over two envelopes for parallel and perpendicular polarization, which may overlap more strongly than shown here. Superimposed on the A terms is a couplet arising from B terms depicted by the dashed curves.

In summary, we concluded that the relative magnitudes of helical circular dichroism measured by $\Delta\epsilon/\epsilon$ are of the order of the ratio of IR to UV frequencies. The amide-I band should display circular dichroism with both A and B contributions (Moffitt and helical contributions); in this respect, it resembles the $\pi\pi^*$ band of the α helix. The amide-II band should resemble B-DNA with a couplet centered about the perpendicular band. The NH stretch should show very low optical activity because it does not have the perpendicular component required for either kind of circular dichroism. The sign of the B couplet cannot be determined by these qualitative considerations because it depends on the way in which the energies of the active modes are distributed. Detailed calculations given in Ref. (16) indicate that division by the total dipole strength leads to a significant underestimate of $\Delta\epsilon/\epsilon$. The predictions for the amide-I band are of the order of 10^{-3}, which is a relatively large anisotropy factor even by electronic standards.

THE PARALLEL β STRUCTURE

The parallel β structure is one of two theoretical structures for sheets of polypeptides originally proposed by Pauling and Corey (21). The structure has been found in distorted form in small segments of globular proteins, but evidence for its presence in β fibers has diminished in recent years (22). Two unit cells of the structure are shown on the left in Fig. 3. There are two peptides per unit cell related by a twofold screw axis. It is not appropriate to utilize the helix formulas of the preceding section for a twofold screw axis because the perpendicular

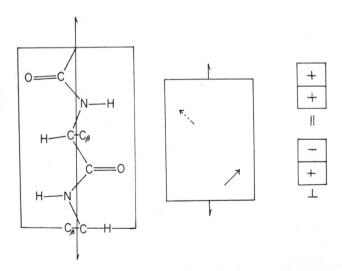

Figure 3. Parallel β structure. Coordinates are from (26). In the center figure, the solid arrows are pointing up out of the plane, dashed arrow is pointing down. The structure has been transformed to represent L-amino acids.

degeneracy does not exist and a number of the summed terms are incorrect for this special case. The wave functions are given by Eq. (5), however, for the allowed transitions with $\epsilon = \pi$. We will label the two allowed excited states $\parallel$ and $\perp$ and represent their eigenfunctions with the vectors $a_\parallel = (1/\sqrt{2})(1,1)$ and $a_\perp = (1/\sqrt{2})(1,-1)$. From Eq. (6), the frequencies of the two bands are

$$\nu_I = \nu_0 + D_1 + D_1'$$

$$\nu_{II} = \bar{\nu}_0 - D_1 + D_1'$$

where D_1 is the interaction parameter along the polypeptide chain and D_1' is a frequency shift associated with interchain hydrogen bonding. We are now dealing with two-dimensional excitons with simple translational symmetry in the x direction and a twofold axis in the z direction. The juxtaposition of the two transition moments per unit cell is shown in the center of Fig. 3 and the phases of the wave functions are schematically represented on the right.

The result of the screw symmetry operation on position vectors and pertinent matrix elements is readily obtained. These data together with the two wave functions facilitate calculation of the formulas for the rotatory strength. The results are shown in Table 4. The coordinates of the transition moment relative to an origin on the screw axis are defined as (α,β,γ). The perpendicular terms R_{xx} and R_{yy} are the equivalent of the A terms in the previous section on helices. As before, the rotatory strength is simultaneously proportional to distance from the screw axis and the parallel and perpendicular components of μ. The B terms do not arise because a twofold screw axis involving perpendicular moments has neither a right nor a left sense.

Table 4. Rotatory strength of the parallel β structure[a]

Transition	$\parallel$	$\perp$
Polarization	z	x,y
R_{xx}	$3\pi\tilde{\nu}\alpha\mu_y\mu_z$	$-3\pi\tilde{\nu}\alpha\mu_y\mu_z$
R_{yy}	$-3\pi\tilde{\nu}\beta\mu_x\mu_z$	$3\pi\tilde{\nu}\beta\mu_x\mu_z$
R_{zz}	0	0
R_{av}	$\pi\tilde{\nu}\mu_z(\alpha\mu_y - \beta\mu_x)$	$-\pi\tilde{\nu}\mu_z(\alpha\mu_y - \beta\mu_x)$

[a] Unit cell results have been divided by 2 to put rotatory strengths on a residue basis.

Estimating $\Delta\epsilon/\epsilon$ for the parallel band b and dividing R by μ^2, we obtain

$$\frac{\Delta\epsilon}{\epsilon(\parallel)} = 4\pi\bar{\nu}\cos\theta\sin\theta\,(\alpha\sin\phi - \beta\cos\phi)$$

where θ is the angle the transition moment makes with the twofold axis and ϕ is the azimuthal angle in the xy plane. Numerical results for this β structure are given in Table 5. The predicted anisotropy for amide I is very small because of the almost-nonexistent value of θ. The amide-II band should be in measurable range. It turns out that the reasonable range of positions of the latter transition moment is close to the screw axis so that calculated results are quite sensitive to this parameter. The problem is discussed in Ref. (16).

THE ANTIPARALLEL β STRUCTURE

Our procedure for calculating rotatory strengths of ordered peptide arrays is now sufficiently well established that some of the details can be left out of the development. The steps are: (1) use cyclic boundary conditions to establish wave functions for allowed states; (2) use the symmetry element(s) to generate transition moments and their positions for the array of groups; and (3) use the combined results for the summations of Eq. (12) and its permutations.

The structure of the antiparallel β sheet is shown on the left in Fig. 4. There are two twofold screw axes in the sheet, one along the chain axis (z), as in the parallel structure, and another perpendicular to the chain direction along the x axis. There is also a twofold rotational axis in the y direction, perpendicular to the polypeptide sheet. Four transition moments per unit cell that are related by the symmetry operations are shown in the center of the figure. Applying cyclic boundary conditions to both screw axes leads to four wave functions that change phase by 0 or π on the application of either of the two screw axes. These functions are labeled $0, x, y$, and z and are depicted on the right in Fig. 4. It turns out to be convenient to locate the origin at the intersection of the screw axes in

Table 5. Optical parameters of the parallel β structure[a]

	$\bar{\nu}_\parallel(cm^{-1})$	$\bar{\nu}_\perp(cm^{-1})$	$\alpha(Å)$	$\beta(Å)$	μ_x	μ_y	μ_z	$\Delta\epsilon/\epsilon(\parallel)$
Amide I	1645	1630	0.98	−0.22	0.29	−0.04	−0.06	−3.6 × 10⁻⁶
Amide II	1530	1550	−0.76	0.22	0.07	−0.10	−0.18	−4.6 × 10⁻⁵

[a]Coordinates are from Pauling and Corey (21). Other parameters from Table 3.

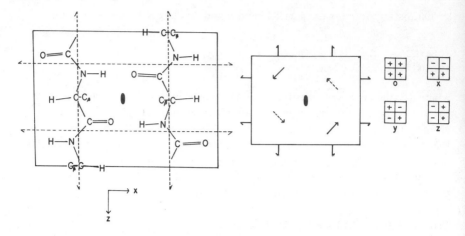

Figure 4. Antiparallel β structure. Coordinates are from (27). In the center figure, the solid arrows are pointing up out of the plane, dashed arrow is pointing down.

the (x,z) plane. The position of a transition moment vector relative to this origin will be symbolized by (α,β,γ). The energies of the levels are given by

$$\nu_0 = \bar{\nu}_0 + D_1 + D_1'$$

$$\nu_x = \bar{\nu}_0 - D_1 + D_1'$$

$$\nu_y = \bar{\nu}_0 - D_1 - D_1'$$

$$\nu_z = \bar{\nu}_0 + D_1 - D_1'$$

These results are obtained by applying Eq. (6) to both screw axes. The terms D_1 and D_1' are as defined for the parallel β chain. The resulting formulas for the rotatory strength are presented in Table 6.[2] The three contributions to R_{av} all have the form of an electric dipole along the polarization direction and an angular momentum around this direction. Polar coordinates do not offer any advantage for this kind of geometry so the anisotropy factor will be expressed in terms of a unit vector (e_x, e_y, e_z) in the direction of the transition moment. This leads to

[2] State 0 has not been listed because it is electrically not allowed. The phase arrangement of transition dipoles of this transition is depicted in the center of Fig. 4. As can be seen, there is potentially a large magnetic dipole for this transition, so that with the breakdown of idealized selection rules, it can be expected to contribute.

$$\frac{\Delta\epsilon}{\epsilon(x)} \cong 4\pi\bar{\nu}\, e_x\, (\beta e_z - \gamma e_y)$$

$$\frac{\Delta\epsilon}{\epsilon(y)} \cong 4\pi\bar{\nu}\, e_y\, (\gamma e_x - \alpha e_z)$$

$$\frac{\Delta\epsilon}{\epsilon(z)} \cong 4\pi\bar{\nu}\, e_z\, (\alpha e_y - \beta e_x)$$

Results are calculated for the antiparallel β structure in Table 7. All three bands for amide I are very small and probably at or beyond the lower limit of measurability (3), because for all three axes either the electric or magnetic moment is very small. The y and z polarized bands of the amide-II transition may be measurable.

It should be possible to get larger signals and more information from oriented CD studies. Observation down a fiber axis will measure R_{zz}; measurement perpendicular to the sheets will give R_{yy}. Because the exciton splittings are relatively small, there is considerable cancellation of overlapping bands of opposing rotatory strength. Studies of oriented circular dichroism can bring out subbands in the same fashion as linear dichroism studies.

DISCUSSION

In the foregoing we extended the most elementary applications of exciton theory to the matrix elements required for calculating the rotatory strength of helical and two-dimensional structures of polypeptides. Though the intent was to determine the geometrical and optical factors that lead to appreciable circular dichroism in vibrational bands, the formulas obtained are general and become specialized to infrared bands only when vibrational frequencies and transition moments are substituted into the formulas. Although the results can be considered only qualitative, they have heuristic value. The major geometrical factors involved in circular dichroism are brought out in simple terms. Moreover, even though there are no experimental data with which to compare the results, they

Table 7. Results for the antiparallel β-chain coordinates[a]

	Amide I	Amide II
μ_x, μ_y, μ_z (D)	0.28, -0.02, -0.10	-0.06, -0.08, 0.19
α, β, γ (Å)	-0.97, -0.03, 0.51	0.75, 0.04, -0.42
$\nu_x, \Delta\epsilon/\epsilon(x)$	1632, 9.0 × 10^{-6}	1540, 6.5 × 10^{-6}
$\nu_y, \Delta\epsilon/\epsilon(y)$	1668, 2.3 × 10^{-6}	1550, 3.9 × 10^{-5}
$\nu_z, \Delta\epsilon/\epsilon(z)$	1685, -6.9 × 10^{-6}	1530, -4.5 × 10^{-5}

[a]Structural coordinates are from Arnott et al. (24). Other parameters from Table 3.

do reproduce the shape of CD curves obtained from a full exciton calculation for a large but finite set of residues (16). The details of this comparison will be given with the results of the complete calculation. It turns out that in all the cases discussed there are regions of the exciton bands where absorption is low and $\Delta\epsilon$ is relatively high. Division by the total dipole strength strongly underestimates the anisotropy factor in such spectral regions. The difference is a factor of about 10, so even the β structures should be well within the range of experiment.

That IR and UV transitions should have experimental anisotropy factors in the ratio of their frequencies is a natural result of the dipole coupling mechanism applied to single excitons. There is no reason, however, to expect this relation to hold for the coupling of single absorption bands of different frequencies of the coupling of two different exciton bands. This is because vibrational bands can couple not only by way of the external fields of their transition dipoles but also mechanically, through the approximate Hooke's law forces of bonds and bond angles. It is only when coupling modes are determined by symmetry that electronic and vibrational effects are directly comparable.

There is a potential defect in the dipole coupling mechanism. Orbital electronic magnetic moments depend on current density, not on charge density. The usual effect of electronic currents is such that magnetic moments calculated from vibrational electric dipoles tend to be underestimated. This effect has been discussed earlier (25) and has been confirmed by Holzwarth *et al.* (3, 26) in comparison of the theoretical and experimental vibrational circular dichroism of (R)-$(-)$-neopentyl-1-d-chloride. The cited results pertain to the fixed-charge model. If the same effect arises when experimental transition dipoles are the basis for calculations, the formulas derived above could be underestimates of vibrational circular dichroism. This finding would favor experimental work, but would complicate the theoretical picture considerably.

There is no difficulty applying the methods to other structures. One could, in fact, apply them systematically to all of the optically active crystallographic space groups. Also, the problem could have been attacked utilizing the full apparatus of space group theory. Because the main purpose of the paper was to test the feasibility of experimental work, however, a direct physical approach was preferable.

ACKNOWLEDGMENTS

The author was greatly aided in this study by extensive discussions with Y. Snir and R. Frankel, who have performed full quantitative calculations on these systems. This research was supported by grants from the National Institutes of Health (GM 20195, GM 15423) and the National Science Foundation (GB 41459).

REFERENCES

1. J. Kemp, *J. Opt. Soc. Am.,* **59**, 950 (1969).
2. J. Jamieson, R. McFec, G. Plass, R. Grube, and R. Richards, *Infrared Physics and Engineering,* McGraw-Hill, New York, 1963.
3. G. Holzwarth, E. Hsu, H. Mosher, T. Faulkner, and A. Moscowitz, *J. Am. Chem. Soc.,* **96**, 252, (1974).
4. P.J. Stephens, personal communication.
5. R. Dudley, S. Mason, and R. Peacock, *J. Chem. Soc. Chem. Commun.,* 1084 (1972).
6. P. Bayley, E. Nielsen, and J. Schellman, *J. Phys. Chem.,* **73**, 228 (1969).
7. V. Madison and J. Schellman, *Biopolymers,* **11**, 1041 (1972).
8. W. Moffitt, *J. Chem. Phys.,* **25**, 467 (1956).
9. W. Moffitt, D. Fitts, and K. Kirkwood, *Proc. Nat. Acad. of Sci. U.S.,* **43**, 723 (1957).
10. T. Ando, *Prog. Theor. Phys.* (Kyoto), **40**, 471 (1968).
11. F. Loxsom, *J. Chem. Phys.,* **51**, 4899 (1969).
12. C. Deutsche, *J. Chem. Phys.,* **52**, 3703 (1970).
13. G. Ramachandran, C. Ramakrishnan, and V. Sasisekharan, *J. Mol. Biol.,* **7**, 95 (1963).
14. T. Miyazawa, *J. Chem. Phys.,* **32**, 1647 (1960); T. Miyazawa and E. Blout, *J. Am. Chem. Soc.,* **83**, 712 (1961).
15. S. Krimm, *J. Mol. Biol.,* **4**, 528 (1962).
16. J. Snir, R. Frankel, and J. Schellman, submitted to *Biopolymers.*
17. J. Snir and J. Schellman, *J. Phys. Chem.,* **77**, 1653 (1973).
18. I. Sandeman, *Proc. Roy. Soc.* (London), **A232**, 105 (1955).
19. C. Swenson, unpublished results.
20. I. Tinoco, *J. Am. Chem. Soc.,* **86**, 297 (1964).
21. L. Pauling and R. Corey, *Proc. Nat. Acad. Sci. U.S.,* **39**, 253 (1953).
22. V. Ananthanarayanan, *J. Sci. Ind. Res.,* **31**, 593 (1972).
23. S. Arnott and S. Dover, *J. Mol. Biol.,* **30**, 209 (1967).
24. S. Arnott, S. Dover, and A. Elliott, *J. Mol. Biol.,* **30**, 201 (1967).
25. J. Schellman, *J. Chem. Phys.,* **58**, 2882 (1973).
26. T. Faulkner, A. Moscowitz, G. Holzwarth, E. Hsu, and H. Mosher, *J. Am. Chem. Soc.,* **96**, 251 (1974).

STUDIES OF THEORETICAL CIRCULAR DICHROISM OF POLYPEPTIDES: CONTRIBUTIONS OF β TURNS

ROBERT W. WOODY, Department of Chemistry, Arizona State University, Tempe, Ariz. 85281

SYNOPSIS: Theoretical studies of oligopeptides in various β-turn conformations have been carried out. The β turn has recently been recognized as an important element of globular protein structure as well as being fundamental in the conformation of cyclic peptides. We have assessed the contributions of such β turns to the circular dichroism of proteins and polypeptides by calculating the $n\pi^*$ and $\pi\pi^*$ rotational strengths for tripeptides in various β-turn conformations. Circular dichroism (CD) curves were calculated from the theoretical rotational strengths. For most variants of the β turn, the predicted CD curves resemble the β-II type spectrum described by Fasman and Potter (27). This spectrum is like that of an ordinary β structure [e.g., poly(L-lysine)] except that it has extrema red-shifted by 5–10 nm. This shift suggests that polypeptides with the β-II CD pattern have a large fraction of their residues in β-turn regions. The type of β turn required by the DPhe-Pro sequence of gramicidin S is unique in its strong tendency to exhibit an α-helixlike CD spectrum. This feature is consistent with the overall CD spectrum of gramicidin S.

INTRODUCTION

Circular dichroism (CD) and optical rotatory dispersion (ORD) have been widely used for the study of protein conformation. A number of schemes have been suggested (1–3) for extracting the fraction of residues in helical and β-pleated-sheet regions from the observed CD or ORD curve of a globular protein. A major obstacle to the success of such schemes has been the contribution of peptide groups that are neither in helical nor in β-pleated-sheet regions. In this paper, I will describe calculations of the circular dichroism of oligopeptides in various types of β-turn conformations.

Until recently, it has been conventional to consider the α helix and β-pleated sheets as the only types of regular structural elements in globular proteins. All residues not participating in one of these types of organized structures were considered to be part of "disordered" or "random" regions. In the last few years, however, a third type of organized structure has been recognized as playing an important structural role in globular proteins. These structures have been variously termed β turns, β bends, reverse turns, or 3_{10} bends. Two examples of these are shown in Fig. 1. In these structures, the peptide chain folds back on itself and may or may not form a hydrogen bond between the carbonyl group of residue i and the peptide hydrogen or residue $i + 3$. The two examples shown have such a hydrogen bond. The difference between these β turns is that in Type I, the middle peptide residue is oriented with the carbonyl oxygen downward, whereas in Type II, the carbonyl group points upward.

338

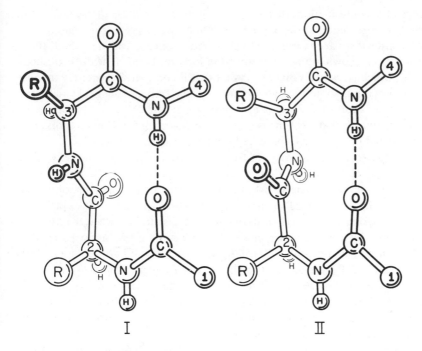

Figure 1. Two basic types of β-turn conformation described by Venkatachalam. [Reprinted with permission from R.E. Dickerson *et al.*, *J. Biol. Chem.*, **246**, 1511 (1971). Copyright by the American Society of Biological Chemists.]

Structures of this type were apparently first postulated independently by Schmidt *et al.* (4) and by Schwyzer *et al.* (5). Both groups were investigating the cyclic decapeptide gramicidin S, the former by x-ray diffraction, the latter in synthetic studies. A structure in which four residues form an antiparallel β structure with three residues at each end completing a 180° turn of the polypeptide chain and also forming a hydrogen bond is consistent with the observed C_2 symmetry in the crystal (4) and provides a reasonable explanation for the ease of cyclization in synthesis (5).

Subsequently, Shields and coworkers (6, 7) observed that the concentration dependence of the infrared spectrum of certain linear, blocked tetrapeptides in nonaqueous media indicated the existence of an intramolecular hydrogen bond. The ability to assume this hydrogen bond depended on the nature of the side chains and their sequence. These factors could be rationalized by the hypothesis of a β-turn conformation. X-ray-diffraction studies of *Chrysopa* silk by Geddes *et al.* (8) led them to propose β-turn structures in which the chain folds back upon itself repeatedly to form a so-called cross-β structure (9). Urry and Ohnishi (10) provided NMR evidence for β turns in a variety of cyclic oligopeptides.

Venkatachalam (11) was the first to carry out detailed conformational analyses of possible β-turn structures. Using a hard-sphere approximation, he surveyed the conformations that are sterically feasible for L-amino acids and that allow a hydrogen bond of the $4 \to 1$ type. He found several broad regions of conformational space consistent with these requirements, centered about the conformations shown in Fig. 1. In addition to types I and II, he distinguished type III, which is really a variant of type I that is considered separately because the conformations about the two α-carbons in the turn are identical and therefore the turn is part of a helical conformation—specifically, it comprises one turn of a 3_{10} helix. Venkatachalam also was the first to point out that β turns are present in globular proteins, specifically in lysozyme and myoglobin.

Dickerson et al. (12) pointed out that cytochrome c has six bends like those described by Venkatachalam, three of type I and three of type II. Lewis et al. (13) compiled a more extensive list of β turns occurring in lysozyme, ribonuclease S and α-chymotrypsin. They used more relaxed criteria for the existence of turns, considering such structures to occur if the distance between the ith α-carbon and the $(i + 3)$rd α-carbon is less than 7Å and if residues $(i + 1)$ and $(i + 2)$ are not in an α-helical conformation. Thus, they did not require hydrogen-bond formation. They analyzed also the relative frequency of occurrence of various amino acid residues in β turns and found that aspartic acid, glycine, serine, threonine, tyrosine, and proline are especially abundant in these regions. Finally, they suggested that the formation of β turns may provide an important directing influence in the folding of polypeptide chains.

Crawford et al. (14) analyzed the structures of seven globular proteins and found that about one-third of the residues are involved in turn regions, with a comparable number in helical regions and only about half this number in β sheets. They also analyzed the composition of the turn regions and arrived at conclusions similar to those of Lewis et al. (13) as to the preferences of certain amino acids.

Lewis et al. (15) extended their previous work, considering eight different globular proteins. They introduced a notation that we shall use here. In addition to types I, II, and III described by Venkatachalam (11), Lewis et al. defined type IV in which two or more of the ϕ and ψ angles defining the turn differ by more than $40°$ from those given by Venkatachalam. They also defined three other types of turns, called V, VI, and VII, which are much less common in the proteins examined and which I shall not discuss here.

From the recent work in this area, it is clear that β turns are ubiquitous and important elements of globular protein structure, as well as important in small cyclic and linear peptides, homopolypeptides, and some forms of silk. Because CD is very sensitive to peptide conformation and is widely used as an experimental method for studying both peptides and proteins, it was of interest to investigate theoretically the circular dichroism contributions of peptides in the β-turn conformation. Is there a characteristic CD pattern for β turns? Are the different types of turns distinguishable from one another by CD? Are the contributions of the turn regions in proteins significant for the overall CD and hence do they affect helix-content determinations? These are some of the questions that a theoretical treatment of the CD of β turns might answer.

METHODS

The methods that were used for calculating the rotational strengths and CD curves for various oligopeptides have been described in previous papers (16, 17). The $n\pi^*$ and $\pi\pi^*$ transitions of the peptide group were considered. Transition energies, monopole charges and positions, transition-moment directions and magnitudes were essentially identical to those used previously in calculations on α helices (16) and β-pleated sheets (17).

Our approach is based upon the work of Kirkwood (18), Moffitt (19), and Tinoco (20) and closely resembles that of Schellman and coworkers (21) and of Pysh (22). We consider the interaction of transition charge densities (approximated by monopoles) in different groups of the oligomer, as well as the mixing of excited states within a group under the influence of the permanent field of the rest of the molecule. The former interactions give rise to coupled-oscillator terms, whereas the latter are analogous to one-electron effects. The excited states of the oligomer are taken to be linear combinations of monomer excited states. Electric- and magnetic-dipole transition moments are calculated from the mixing coefficients.

The rotational strength of the transition from the oligomer ground state to its excited state K, R_K, is then calculated:

$$R_K = \text{Im} \left\{ \mu_{0K} \, m_{K0} \right\}$$

where μ_{0K} and m_{K0} are, respectively, electric- and magnetic-dipole transition moments.

In a calculation of this type, the energies of the various oligomer levels are obtained as the eigenvalues from the matrix diagonalization procedure used to solve the secular determinantal equation. For the electrically allowed transitions, these energies may differ significantly from the energy of the isolated group because of interchromophoric coupling, which is analogous to exciton splitting. For an amide trimer, we obtain six oligomer excited states, corresponding to the two transitions ($n\pi^*$ and $\pi\pi^*$) in each monomer. Three of these are predominantly $\pi\pi^*$ in character and may be spread in energy from about 185–205 nm. The other three are predominantly $n\pi^*$ and show only small splittings because of the electrically forbidden character of the $n\pi^*$ transition. In calculating CD curves from the theoretical rotational strengths, the $\pi\pi^*$ bands are located at the positions predicted by theory, whereas the $n\pi^*$ levels are summed and located at the zero-order wavelength, 220 nm.

Gaussian bands are assumed in calculating CD spectra, with an arbitrary uniform half-width at e^{-1} of the maximum of 12.5 nm.

The geometries of the systems under consideration were generated by standard matrix transformation methods outlined by Ooi et al. (23). The peptide-group geometry was also taken from Ooi et al. (23). The ϕ and ψ angles quoted in this work are consistent with the IUPAC recommendations (24).

RESULTS

The geometry of the β turns was adapted from the work of Venkatachalam (11). Each of the 15 conformations he reported was considered. Because of the breadth of the allowed regions and the corresponding flexibility in β-turn conformations, each of the four dihedral angles (ϕ_2, ψ_2, ϕ_3, ψ_3) required to specify a tripeptide in a β-turn conformation was allowed to vary by $\pm 10°$ about the values given in Table 1 of Venkatachalam's paper (11). Thus, there are 3^4, or 81, variants for each basic conformation. The rotational strengths and CD curves were calculated for each of these variants.

First, let us consider the results for the three fundamental types, I-III. Fig. 2 shows the calculated CD curves for the conformations designated by Venkatachalam (11) as types I, II, and III. The vertical lines indicate the rotational strengths and wavelengths of the various $\pi\pi^*$ CD bands. There is little

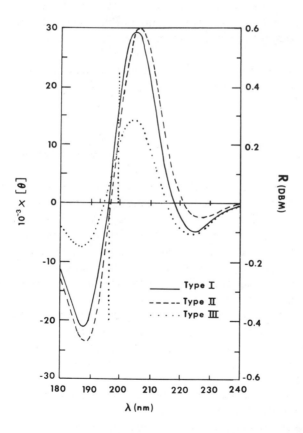

Figure 2. Calculated CD curves for Venkatachalam's β turns (11). Type I (——), type II (– –), and type III (· · · ·). These are all class-B spectra. The vertical dotted lines indicate the $\pi\pi^*$ rotational strengths for the type-III conformation.

difference between the CD predicted for type-I and type-II trimers. Both exhibit a minimum at about 225 nm, a maximum at ~205 nm and a minimum just below 190 nm. The type-III (3_{10}-helix) trimer has extrema at essentially the same wavelengths as do the other two types, but the amplitude of the shorter wavelength extrema is significantly smaller.

The similarity of these three types suggests that all of the many variants we have considered will have the same CD pattern. This is indeed the dominant CD pattern, but other types are also predicted. In Fig. 3, we show the CD spectrum of a type-III variant in which ϕ_2 and ψ_3 have been changed by $10°$. We now observe a negative CD band at about 210 nm and a positive band at 192 nm. The shape of this CD curve (but not the amplitude) is reminiscent of that of an α helix, although there is no discrete $n\pi^*$ minimum because the $n\pi^*$ band and the negative long-wavelength $\pi\pi^*$ band overlap and give rise to only one minimum at ~210 nm. Also shown in Fig. 3 are two other characteristic CD patterns. One, given by a variant of Venkatachalam's type 10 [a type-IV turn in the notation of Lewis *et al.* (15)], has a minimum at 225 nm, a maximum at about 210, a minimum at 197 and a maximum at 182. The other, given by a type-I turn with a ϕ_2 variation of $30°$ from Venkatachalam's value, looks rather like the curves shown in Fig. 2. However, the long-wavelength minimum occurs at about 220 nm and the maximum is below 200 nm. Thus, both extrema are significantly blue-shifted relative to the spectra previously discussed. This spectrum resembles that of a standard β structure [e.g., poly(L-lysine)] (1), which has a positive band at about 195 nm and a negative band at 217 nm.

The spectra that were obtained could be classified, with only a few ambiguities, into eight categories—the four basic classes displayed previously, together with their mirror images. Table 1 shows the criteria of classification.

The results are difficult to represent in compact form. Fig. 4 is an attempt to convey the results, as completely as is practical in terms of our classification scheme, for type-I and -II β turns. The large-scale grid represents the variation in ϕ_2 and ψ_2, the smaller-scale grid shows the variation of each value of ϕ_2 and ψ_2 with ϕ_3 and ψ_3. It is not possible to present detailed maps like these for all types considered, but they are available from the author upon request.

Table 2 contains a summary of the results on variants of the β turn. For each of the 15 types considered, the 81 calculated CD spectra were classified, and the relative frequency of each class is given in the table. It is clear that B is by far the most frequently occurring pattern. For type-I turns, virtually all of the variants give this spectrum. For type-II turns, about 80% of the variants give pattern B, but about 20% give an inverted α-helix pattern (class C). About two-thirds of the type-III turns give B patterns, about 15% give class A and another 20% give class C.

We can also discern three distinct regions in which patterns other than B are relatively abundant. For types 3–7 in Venkatachalam's notation (11), there is

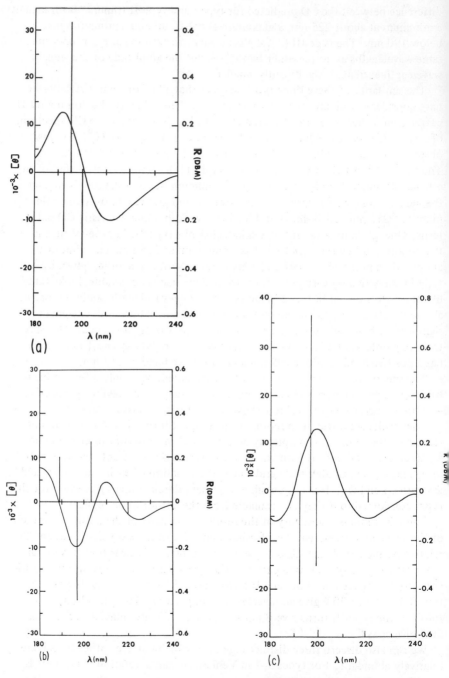

Figure 3. Calculated CD curves for variants of Venkatachalam's β turns (11): (a) A type-3 β-turn (–50,–30; –60,–20). This is a class-C spectrum. (b) A type-10 β-turn (–70,–10; 80,20) (class-D spectrum). (c) A type-4 β-turn (–60,–30; –60,0) (class-A spectrum). The vertical lines indicate the wavelengths and magnitudes of the component CD bands.

Table 1. Classification scheme for calculated CD spectra

Class	Extrema		Polypeptide Conformation[a]
	Sign	Wavelength Range	
A	–	210–220	β-I
	+	195–200	
	–	<190	
B	–	>220	β-II
	+	200–210	
	–	<190	
C	–	200–210[b]	α_R
	+	180–195	
D	–	>225	Charged poly(Glu) or poly(Lys)
	+	210–220	
	–	190–200	
	+	<190	

[a] The polypeptide system or conformation that gives a similar CD spectrum.
[b] Class-C spectra are associated with a negative $n\pi^*$ rotational strength, although in general there is no discrete minimum in the $n\pi^*$ region.

Table 2. Predicted CD characteristics of β turns

Venkatachalam Structure[a]	Lewis et al. Structure[b]	Occurrence of CD Types[c] (%)							
		A	B	C	D	A'	B'	C'	D'
1	(I)		90		2		1	5	1
2	I	11	88				1		
3	III	15	65	19	1				
4	(I)	24	38	24			13	1	
5	(III)	16	56	5	4	3	10	1	5
6	(I)	28	62	4			5		1
7	(III)	19	35	34			12		
8	IV		83	4	12				
9	IV		90	4	6				
10	IV		67	10	23				
11	IV		84	7	9				
12	IV		85	7	7				
13	IV		53					47	
14	II		81					19	
15	IV		83					17	

[a] C.M. Venkatachalam, *Biopolymers*, **6**, 1425 (1968).
[b] P.N. Lewis, F.A. Momany, and H.A. Scheraga, *Biochim. Biophys. Acta*, **303**, 211 (1973). Numbers in parentheses indicate proximity to a structural type.
[c] A', B', and so on are mirror images of A, B, ...

B B B B A B B B B (-70,-20)	B B B B B B B B B (-60,-20)	B A A A A A B' A A (-50,-20)
B B B B B B B B B (-70,-30)	B B B B B B B B B (-60,-30)	B B B B B B B B B (-50,-30)
B B B B B B B B B (-70,-40)	B B B B B B B B B (-60,-40)	B B B B B B B B B (-50,-40)

(-100,10)	(-90,10)	(-80,10)
(-100,0)	(-90,0)	(-80,0)
(-100,-10)	(-90,-10)	(-80,-10)

Type I B-turn

B B B B C'B B B B (-70,130)	B C'C' B B B B B B (-60,130)	C'C'C' C'C'C' B C'C' (-50,130)
B B B B B B B B B (-70,120)	B B B B B B B B B (-60,120)	C'C'C' B B C' B B B (-50,120)
B B B B B B B B B (-70,110)	B B B B B B B B B (-60,110)	B B B B B B B B B (-50,110)

(80,10)	(90,10)	(100,10)
(80,0)	(90,0)	(100,0)
(80,-10)	(90,-10)	(100,-10)

Type II B-turn

Figure 4. Representation of calculated CD spectra for variants of type-I and type-II β turns. The large-scale grid represents variation in (ϕ_2, ψ_2). The smaller-scale grid represents variation in (ϕ_3, ψ_3) as denoted below grid. Each point represents a β-turn conformation and the letter denotes the nature of the calculated CD spectrum.

an appreciable probability of the A and C patterns. For types 9–12, about 10–20% of the variants give C or D patterns. For types 13–15, an appreciable number (as many as one-half in the case of type 13) give an inverted α-helix pattern.

To determine whether the patterns predicted might be the result of considering only tripeptide fragments, calculations have been performed for pentapeptide fragments from lysozyme (25) and subtilisin BPN (26), which have been found by Lewis *et al.* (15) to contain β-turn segments. The dominance of the class B pattern is apparent for pentapeptides although it is perhaps not quite so strong. In 14 of the 40 cases considered, the calculated pattern for the pentapeptide did not coincide with either the most probable or the next most probable pattern predicted for the tripeptide. It must be noted, however, that in a number of cases these protein fragments deviate rather widely from the Venkatachalam (11) β turns and the predictions were based upon the closest structure.

DISCUSSION AND CONCLUSIONS

The calculations on the tripeptides and on pentapeptide fragments from known protein structures indicate that β turns generally make CD contributions that are of the type we have denoted as class B. This pattern is characterized by a negative $n\pi^*$ band with a minimum above 220 nm and a maximum in the 200–210 nm region. This pattern resembles that of poly(L-lysine) in the β-pleated sheet conformation (1), but the extrema are shifted to the red by 5–10 nm. It is interesting to note that Fasman and coworkers (27, 28) have obtained similar CD spectra for films and solutions of certain homopolypeptides, particularly those of serine, threonine, and cysteine esters and ethers. They have suggested that there are two types of β structures, which they refer to as β-I and β-II. Poly(L-lysine) and poly(L-valine) are representative of β-I polypeptides, which give a negative band at ~217 nm and a positive band at ~195 nm. Poly(S-carboxymethyl)-L-cysteine, has a negative band at ~225 nm and a positive band at 198 nm in solution and at even longer wavelengths in solid films.

The difference between the β-I and β-II CD patterns observed by Fasman and coworkers is exactly like that between our class A and class B spectra. The present results suggest that the β-II polypeptides differ from the β-I polypeptides in that they have a larger fraction of their residues in β-turn regions. Both types are probably largely intramolecular β structures, but the frequency with which the chain reverses direction and hence the length of the regular antiparallel β segments differs. This proposed interpretation is consistent with the nature of the amino acids forming β-II polypeptides. Serine, threonine, and cysteine are concentrated in β turns in globular proteins (13, 14). It is consistent also with the infrared data of Fasman and Potter (27), who found that most of the β-II polypeptides give infrared spectra of the cross-β type, which indicates short transverse β segments and frequently occurring β turns.

In a sense, the finding that the various β-turn conformations tend to give a single type of CD spectrum is disappointing. Initially we hoped that different types of β turns would be distinguishable on the basis of CD spectra. This would make CD a much more useful technique for characterizing the conformations of cyclic polypeptides.

One result that emerged from our calculations is that conformations that can accommodate the amino acid sequence (D-X)-(Pro) have a high probability of exhibiting an α-helixlike CD pattern (class C spectra). The calculations of Chandrasekharan et al. (29) on dipeptides with a D,L sequence show that β bends containing this sequence fall in the region of Venkatachalam's type 13', where the prime indicates a mirror image of type 13. From Table 2, it can be seen that type 13 has a 50% probability of giving a class C' spectrum. Type 13 therefore will have a high probability of yielding a type-C or an α-helixlike CD pattern.

This result is consistent with the remarkable resemblance of the CD spectrum of gramicidin S to that of an α-helical polypeptide. Laiken et al. (30) showed that hydrogenated gramicidin S (in which contributions of aromatic side chains are eliminated) has two distinct negative bands in the long-wavelength region and that the amplitude is comparable to that of an α helix. The NMR studies of Stern et al. (31) have led to a model that supports the early suggestions of Schmidt et al. (4) and Schwyzer et al. (5). In this structure, there are two β turns with an intervening stretch of antiparallel β structure. The turns are formed by the DPhe-Pro sequence. Thus our results provide a qualitative explanation of the gramicidin-S CD pattern and indicate that at least one type of β turn can be recognized in cyclic oligopeptides.

Finally, we may consider what effect β turns may have on the overall CD spectrum of a globular protein. The prevalent class-B spectra of β turns has a negative $n\pi^*$ band with an amplitude per residue that is about one-fifth that of an α-helical segment. It also has a positive band in the region of the second minimum of the α-helix spectrum. In this case, the amplitude is about one-half to one-fourth of the α-helix spectrum, but in the opposite direction. It might be anticipated that a globular protein with substantial fractions of α helix and β turns (e.g., lysozyme and subtilisin) (13, 14) should exhibit a relatively stronger negative band at 222 nm and a weaker 208–210 nm band. The CD spectrum of subtilisin (32) does show such an effect, but that of lysozyme (4) does not. Further work is necessary to assess the importance of these contributions for globular protein CD.

ACKNOWLEDGMENTS

I want to thank James Shields of Eli Lilly and Co., who first aroused my interest in the β-turn conformation. The development of the computer programs used here was assisted by K.-P. Li, and Gary and Julie Wen and John Charles helped with the data analysis. I am also grateful to Harold Scheraga for several helpful discussions and for sending me a preprint of a paper in advance of publication. This work was supported in part by USPHS GM-17850.

REFERENCES

1. N. Greenfield and G.D. Fasman, *Biochemistry*, **8**, 4108 (1969).
2. V.P. Saxena and D.B. Wetlaufer, *Proc. Nat. Acad. Sci. U.S.*, **68**, 969 (1971).
3. Y.-H. Chen, J.T. Yang, and H.M. Martinez, *Biochemistry*, **11**, 4120 (1972).
4. G.M.J. Schmidt, D.C. Hodgkin, and B.M. Oughton, *Biochem. J.*, **65**, 744 (1957).
5. R. Schwyzer, P. Sieber, and B. Gorup, *Chimia*, **12**, 53 (1958).
6. J.E. Shields and S.T. McDowell, *J. Am. Chem. Soc.*, **89**, 2499 (1967).
7. J.E. Shields, S.T. McDowell, J. Pavlos, and G.R. Gray, *J. Am. Chem. Soc.*, **90**, 3549 (1968).
8. A.J. Geddes, K.D. Parker, E.D.T. Atkins, and E. Brighton, *J. Mol. Biol.*, **32**, 343 (1968).
9. W.T. Astbury, *Proc. Royal Soc.* (London), **B134**, 303 (1947).
10. D.W. Urry and M. Ohnishi in *Spectroscopic Approaches to Biomolecular Conformation*, D.W. Urry, ed., American Medical Association, Chicago, 1970, p. 263.
11. C.M. Venkatachalam, *Biopolymers*, **6**, 1425 (1968).
12. R.E. Dickerson, T. Takano, D. Eisenberg, O. Kallai, L. Samson, A. Cooper, and E. Margoliash, *J. Biol. Chem.*, **246**, 1511 (1971).
13. P.N. Lewis, F.A. Momany, and H.A. Scheraga, *Proc. Nat. Acad. Sci. U.S.*, **68**, 2293 (1971).
14. J.L. Crawford, W.N. Lipscomb, and C.G. Schellman, *Proc. Nat. Acad. Sci. U.S.*, **70**, 538 (1973).
15. P.N. Lewis, F.A. Momany, and H.A. Scheraga, *Biochim. Biophys. Acta*, **303**, 211 (1973).
16. R.W. Woody, *J. Chem. Phys.*, **49**, 4797 (1968).
17. R.W. Woody, *Biopolymers*, **8**, 669 (1969).
18. J.G. Kirkwood, *J. Chem. Phys.*, **5**, 479 (1937).
19. W. Moffitt, *J. Chem. Phys.*, **25**, 476 (1956).
20. I. Tinoco, Jr., *Adv. Chem. Phys.*, **4**, 113 (1962).
21. P.M. Bayley, E.B. Nielsen, and J.A. Schellman, *J. Phys. Chem.*, **73**, 228 (1969).
22. E.S. Pysh, *J. Chem. Phys.*, **52**, 4723 (1970).
23. T. Ooi, R.A. Scott, G. Vanderkooi, and H.A. Scheraga, *J. Chem. Phys.*, **46**, 4410 (1967).
24. IUPAC-IUB Commission on Biochemical Nomenclature, *Biochemistry*, **9**, 3471 (1970).
25. P.K. Warme and H.A. Scheraga, *Biochemistry*, **13**, 757 (1974).
26. R.A. Alden, J.J. Birktoft, J. Kraut, J.D. Robertus, and C.S. Wright, *Biochem. Biophys. Res. Commun.*, **45**, 337 (1971).
27. G.D. Fasman and J. Potter, *Biochem. Biophys. Res. Commun.*, **27**, 209 (1967).

28.   L. Stevens, R. Townend, S.M. Timasheff, G.D. Fasman, and J. Potter, *Biochemistry*, **7**, 3717 (1968).
29.   R. Chandrasekharan, A.V. Lakshminarayanan, U.V. Pandya, and G.N. Ramachandran, *Biochim. Biophys. Acta,* **303**, 14 (1973).
30.   S. Laiken, M. Printz, and L.C. Craig, *J. Biol. Chem.,* **244**, 209 (1967).
31.   A. Stern, W.A. Gibbons, and L.C. Craig, *Proc. Nat. Acad. Sci. U.S.,* **61**, 734 (1968).
32.   T.T. Herskovits and H.H. Fuchs, *Biochim. Biophys. Acta,* **263**, 468 (1972).

APPLICATION OF CIRCULAR POLARIZATION OF LUMINESCENCE TO THE STUDY OF PEPTIDES, POLYPEPTIDES, AND PROTEINS

I.Z. STEINBERG, J. SCHLESSINGER, and **A. GAFNI,** Department of Chemical Physics, The Weizmann Institute of Science, Rehovot, Israel

SYNOPSIS: The light emitted by asymmetric luminescent molecules may be partially circularly polarized. Circular polarization of luminescence (CPL) is related to the molecular conformation in the excited state in the same way that circular dichroism (CD) is related to the molecular conformation in the ground state. Thus, CPL may serve as a tool for the study of molecules in their excited state or for the detection of changes in molecular conformation that take place upon electronic excitation. In complex systems, CPL is simpler to interpret than CD because it is limited to only a few transitions in luminescent chromophores (usually to a single electronic transition in organic chromophores). Furthermore, the rotatory power of forbidden transitions may be readily studied by CPL, whereas CD is inapplicable.

Circular polarization of luminescence has been applied to the study of peptides, polypeptides, and proteins. Cyclic dipeptides containing tyrosine or tryptophan undergo a pronounced change in molecular conformation upon electronic excitation. This change is arrested in highly viscous media. It is shown by CPL that both monomeric and dimeric acridine dyes acquire optical activity when bound to helical poly(glutamic acid), the different contributions being well separated. Apparently, 9-amino acridine forms excimers when bound to this polypeptide. Aromatic side chains in small linear peptides and in fully reduced and denatured proteins in 6 M guanidine hydrochloride produce no detectable CPL, whereas some circular polarization was observed in the fluorescence of all native proteins studied. Thus, the tertiary structure of the proteins is responsible for the CPL of the aromatic side chains. The spectral behavior of the CPL of human serum albumin clearly reflects the heterogeneity of the population of molecules of this protein. In studies of a few proteins where CD was inapplicable, CPL was found to be a convenient substitute. Studies of the Tb^{3+} complexes of transferrin and conalbumin revealed that the metal binding sites of these two proteins and the two sites per transferrin molecule have similar structure.

INTRODUCTION

Materials containing asymmetric molecules interact differently with right-handed and left-handed circularly polarized light. This property has given rise to special branches of spectroscopy, namely, optical rotatory dispersion (ORD) and circular dichroism (CD), which are manifested, respectively, by the rotation of the direction of polarization of plane-polarized light and by a difference in absorption of circularly polarized light of opposite sense of rotation. Recently attention has been given to the presence of some degree of circular polarization in the light emitted by asymmetric molecules (1–8). The phenomenon of circular polarization of luminescence (CPL) of asymmetric molecules was shown to be related to the conformation of the molecules in the electronic excited state, when light emission occurs, in the same way that CD is related to the molecular conformation in the electronic ground state (1–8). The special characteristics of CPL make this spectroscopic method a useful tool for tackling a variety of problems of biochemical interest (3–6, 8).

We shall review the theoretical background of CPL and describe some experimental aspects of this spectroscopic technique. The potentialities of CPL will be evaluated briefly, and some recent application of this method to the study of peptides and proteins will be presented.

THEORETICAL BACKGROUND

The difference in absorption of right-handed and left-handed circularly polarized light by asymmetric molecules was studied quantum mechanically by Condon *et al.* (9) using time-dependent perturbation theory. The phase variation of the alternating electric field of the light beam over the dimensions of the absorbing molecule had to be taken into account in the treatment of this problem. A linear approximation for the variation of the field was used in the above theory, obviously justified for molecules that are small relative to the wavelength of the excitation light. For a collection of absorbing molecules of random orientation, the contribution of the transition quadrupole moment vanished, and the following expression was obtained for a transition from state A to state B,

$$g_{AB} = \frac{\epsilon_\ell - \epsilon_r}{\epsilon} = \frac{P_\ell - P_r}{(P_\ell + P_r)/2} = 4 \frac{\text{Im}(A|\mathbf{p}|B) \cdot (B|\mathbf{m}|A)}{|(A|\mathbf{p}|B)|^2} \tag{1}$$

where g_{AB} is the absorption anisotropy factor (10), ϵ_ℓ and ϵ_r are the absorption coefficients for left-handed and right-handed circularly polarized light, respectively, ϵ is the average absorption coefficient $[\epsilon = (\epsilon_\ell + \epsilon_r)/2]$, P_ℓ and P_r are the probabilities for excitation by unit intensities of right-handed and left-handed circularly polarized light, respectively, and $\mathbf{p}$ and $\mathbf{m}$ are the electric- and magnetic-dipole-moment operators, respectively. The symbol Im denotes taking the imaginary part of the expression that follows.

The above considerations and results are directly applicable to induced emission of circularly polarized light of one sense or the other. The spontaneous emission of circularly polarized light can be evaluated with Einstein's classical treatment, balancing absorption and emission of light in a cavity at thermodynamic equilibrium, subject to the assumption that the absorption and emission processes of circularly polarized light of either sense should be balanced independently (8). The extent of circular polarization of the spontaneously emitted light therefore should obey the expression

$$g_{em} = \frac{\Delta f}{f/2} = \frac{P_{\ell} - P_r}{(P_{\ell} + P_r)/2} = 4 \frac{\text{Im}(A'|\mathbf{p}|B') \cdot (B'|\mathbf{m}|A')}{|(A'|\mathbf{p}|B')|^2} \qquad (2)$$

where g_{em} is the emission anisotropy factor and Δf and f are the intensities of of the circularly polarized component and of the total emitted light, respectively (Δf is defined as positive for left-handed net polarization). The symbols A' and B' refer to the quantum states involved in the emission process.

From Eqs. (1) and (2) it is evident that for absorption and emission transitions that involve the same quantum states, the values of the corresponding anisotropy factors, g_{AB} and g_{em}, should be the same. Great care should be exercised, however, in drawing conclusions from the above statement. Thus, the fluorescence emitted from substances in condensed media, as a rule, involves a transition from the first singlet excited state to the ground state, $S_1 \rightarrow S_0$, whereas absorption at the long-wavelength band of the absorption spectrum is attributable to the opposite process, $S_0 \rightarrow S_1$. However, the anisotropy factors characterizing these transitions will not necessarily assume the same value. The reason for this is rather simple: the electronic wave functions for both the states S_0 and S_1 depend explicitly on the coordinates of the nuclei that make up the molecule. By the Franck–Condon principle, the molecular conformation in the ground state is pertinent to absorption processes, whereas the molecular conformation in the excited state is relevant to emission processes. If the molecule undergoes a change in its conformation or experiences a change in asymmetric perturbation by its environment, upon electronic excitation, g_{AB} and g_{em} generally will not assume the same numerical value. The factors g_{AB} and g_{em} therefore do not carry redundant information; they are related to the conformation and environmental conditions of the molecule in different electronic states.

It is instructive to restate the above considerations in more precise terms (8). Within the framework of the Born–Oppenheimer approximation, the molecular wave function, ψ, may be written as the product of the electronic part, $\phi(q,Q)$, and the nuclear part, $\theta(Q)$; that is

$$\psi = \phi(q,Q)\theta(Q) \qquad (3)$$

where q and Q are the coordinates of the electrons and the nuclei, respectively. Because of the orthogonality of the electronic part of the wave function,

$$(A|\mathbf{p}|B) = \int \theta_A^*(Q) \theta_B(Q) \int \phi_A^*(q,Q) \mathbf{p}\phi_B(q,Q) \, dq \, dQ \qquad (4)$$

The expression for $(B|\mathbf{m}|A)$ is analogous. Let us focus attention on the ground state and excited state vibrational wave functions, $\theta_A(Q)$ and $\theta_B(Q)$, respectively. In absorption processes, nearly all absorbing molecules are in the lowest vibrational level of the electronic ground state, and $\theta_A(Q)$ is finite in the vicinity of Q_0^{gr}, that is, the equilibrium nuclear coordinates that are characteristic of the ground state. Thus, using the conventional approximation procedure, which is applicable to allowed transitions, one obtains for absorption processes,

$$(A|\mathbf{p}|B) \cong \int \theta_A^*(Q)\theta_B(Q)\,dQ \int \phi_A^*(q,Q_0^{gr})\mathbf{p}\phi_B(q,Q_0^{gr})\,dq \tag{5}$$

with an analogous expression for $(B|\mathbf{m}|A)$. In contrast, in emission processes, $\theta_B(Q)$ is finite only in the vicinity of the equilibrium nuclear coordinates for the excited-state conformation, denoted Q_0^{ex}. In the approximate expression for emission analogous to Eq. (5), Q_0^{ex} should be substituted for Q_0^{gr}.

Substituting Eq. (5) in Eq. (1) yields (11)

$$g_{AB} = 4\frac{\mathrm{Im}\,[\int \phi_A^*(q,Q_0^{gr})\mathbf{p}\phi_B(q,Q_0^{gr})\,dq \cdot \int \phi_B^*(q,Q_0^{gr})\mathbf{m}\phi_A(q,Q_0^{gr})\,dq]}{[\int \phi_A^*(q,Q_0^{gr})\mathbf{p}\phi_B(q,Q_0^{gr})\,dq]^2} \tag{6}$$

Appropriate substitution in Eq. (2), replacing Q_0^{gr} with Q_0^{ex} throughout yields a similar expression for g_{em}:

$$g_{em} = 4\frac{\mathrm{Im}\,[\int \phi_{A'}^*(q,Q_0^{ex})\mathbf{p}\phi_{B'}(q,Q_0^{ex})\,dq \cdot \int \phi_{B'}^*(q,Q_0^{ex})\mathbf{m}\phi_{A'}(q,Q_0^{ex})\,dq]}{[\int \phi_{A'}^*(q,Q_0^{ex})\mathbf{p}\phi_{B'}(q,Q_0^{ex})\,dq]^2}$$

$$\tag{7}$$

The above results illustrate the earlier statements that g_{AB} is characteristic of the electronic wave functions of the molecule in the ground state conformation, whereas g_{em} is characteristic of the electronic wave functions of the molecule in the excited state conformation. Thus, the values of g_{AB} and g_{em} may be different even if the *electronic* states A and B are the same as A' and B', respectively.

It is pertinent to note that in the expressions for g_{AB} and g_{em} described in Eqs. (6) and (7), respectively, the vibrational wave functions, $\theta(Q)$, cancel. The important theorem stated by Moscowitz [for CD spectra (11)] that the absorption anisotropy factor for a single allowed electronic transition should be constant is based on the above result. The same applies, of course, to the emission anisotropy factor (8). It may be recalled in this connection that in contrast to absorption spectra, the fluorescence spectrum of a pure compound in condensed media generally involves a single electronic transition, that is, $S_1 \to S_0$. Thus, the emission anisotropy factor is normally expected to assume an approximately constant numerical value throughout the fluorescence spectrum for a single compound in a well-defined environment (8). The significance of this conclusion will be discussed below.

EVALUATION OF CPL AS A SPECTROSCOPIC TOOL

As discussed above, the circular polarization of luminescence is one of the manifestations of molecular asymmetry; the information it yields, however, is different, in principle, from that obtained by CD and ORD. Briefly, some of the potentialities of CPL are as follows:

1. Circular polarization of luminescence is related to the asymmetry of the luminescent molecule in the excited state molecular conformation. The pertinent excited state is that in which emission takes place. Thus, CPL is especially suited as a spectroscopic tool when the excited state is of primary interest, as is the case with chlorophyll (12).
2. A comparative study of CD and CPL may disclose changes in molecular conformation that take place upon electronic excitation. Such information may help, for example, to determine the significance of data obtained by fluorescent probes to problems related to the ground state (3).
3. Because CD and CPL of a chromophore under specific conditions are studied at different spectral ranges and may assume different numerical values, circumstances may arise in which it is experimentally more feasible to use CPL than CD (5, 13) as a diagnostic tool for conformational problems.
4. The CPL studied for chromophores in condensed media usually is associated with a single electronic transition. Complications due to overlap of electronic bands are thus avoided, and the emission anisotropy factor is expected to be approximately constant across the emission spectrum. Deviations from such behavior are readily discernible and may be informative (4, 13).
5. The study of CPL is limited to fluorescent chromophores. This limitation may be advantageous, however, in complex systems containing many chromophores, such as proteins. All chromophores may contribute to the CD spectrum, and it may be difficult, if not impossible, to assign the contribution of the various chromophores to the observed spectrum. More specificity is obtained in the study of the CPL. For example, in the case of proteins, only the contribution of tryptophan residues, and to some extent that of tyrosines, is observed in CPL.

Furthermore, the spectral resolution between tyrosine and tryptophan residues and among different tryptophan residues in a protein molecule, is appreciably better in their emission spectrum than in their absorption spectrum.

6. In the study of the optical rotatory power associated with forbidden transitions, CPL has a decisive advantage over CD. Such transitions are usually too weak to be detected by light absorption. In contrast, the luminescence associated with such transitions is observed often. Examples of emission processes involving such transitions are phosphorescence and luminescence caused by internal transitions of the rare-earth elements (5).

The study of CPL offers some additional possibilities. These potentialities are of interest, however, because of their physical rather than biochemical aspects and will be discussed elsewhere (14).

APPLICATION TO PEPTIDES

Of the amino acids that proteins contain, tryptophan, tyrosine, and phenylalanine are fluorescent. The study of the circular polarization of peptides that contain these amino acids has yielded information regarding changes in molecular conformation that occur upon electronic excitation of the aromatic side chains. The results obtained with small peptides also have contributed to the understanding of the spectral behavior of more complex systems, namely, proteins.

The spectra of the absorption and emission anisotropy factors, g_{AB} and g_{em} respectively, for the cyclic peptide *cyclo*(Gly-Tyr) in dioxane solution are presented in Fig. 1. [The CD of this cyclic peptide was reported previously (15).] The absorption and fluorescence spectra also are included. The corresponding spectral data for *cyclo*(Val-Trp) are shown in Fig. 2. In both cases, the absorption anisotropy factor is not constant in the spectral range presented, probably because of the occurrence of more than one electronic transition in this range [a possibility that has been extensively discussed, at least in the case of tryptophan (16–19)]. It should be noted that for Tyr and Trp, as well as for linear peptides containing these amino acids (such as N-acetyl-L-tryptophanamide, L-Trp-(Gly)$_3$-COOH, carbobenzoxy-(Gly)$_3$-L-Trp(Gly)$_3$-O-benzyl ester, and carbobenzoxy-(Gly)$_4$-L-Trp(Gly)$_3$-O-benzyl ester), g_{AB} is much smaller than for the cyclic peptides studied. It is thus plausible to assume that the interaction of the aromatic rings with the diketopiperazine ring is responsible for the relatively high optical activity observed for the cyclic peptides (15). Evidence of a conformation that permits such an interaction has been put forward by nuclear magnetic resonance studies (20, 21) and x-ray diffraction (22) and is supported by theoretical investigations (23, 24). Because a single aromatic side chain per molecule is sufficient to produce the observed CD, exciton interactions proposed to play a role in the CD of cyclic dipeptides containing two aromatic side chains (25) obviously are not essential to obtain the observed CD. The striking feature

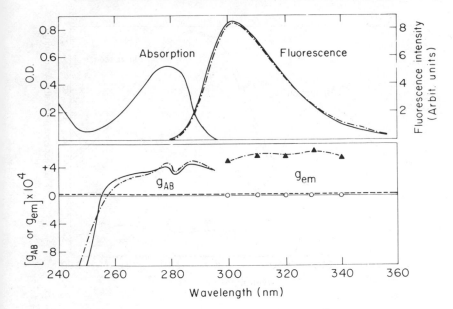

Figure 1. Spectroscopic data for *cyclo*(Gly-Tyr) at ~22°C. Absorption in 3.8×10^4 *M* dioxane is shown; a similar spectrum was obtained in a highly viscous solvent composed of polyoxypropylene and dioxane (POP-D). Fluorescence in 3.10^{-5} *M* dioxane (—); in POP-D (·—·—·). Absorption anisotropy factor, g_{AB}, in 3.8×10^{-4} *M* dioxane (—), in POP-D (·—·—·). Emission anisotropy factor, g_{em}, in 3.8×10^4 *M* (○); in POP-D (▲). For comparison, g_{AB} and g_{em} of Tyr in dioxane or POP-D are shown (– –).

of Figs. 1 and 2 is that the emission anisotropy factor vanishes within experimental error (estimated to be $\pm 3 \times 10^{-5}$). Thus, it is evident that the conformation of the cyclic peptides described is different when the aromatic chromophore is excited than when it is in the ground state. Of course, the distribution of the electronic charge of the chromophores is different in the two states, resulting in a difference in interaction with the rest of the molecule and hence in different conformations in the two states. Similar behavior was observed for a variety of other peptides (26).

Strong support for the above conclusions may be drawn from the findings that in very highly viscous solution (e.g., in a mixture of polyoxypropylene and dioxane), g_{em} does not vanish, but assumes a magnitude comparable to that of g_{AB} (see Figs. 1 and 2). Obviously, under these conditions conformational change within the lifetime of the excited state is arrested by the resistance of the medium.

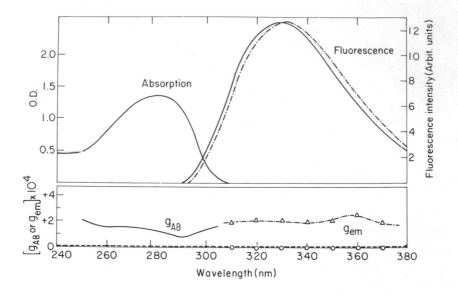

Figure 2. Spectroscopic data for *cyclo*(Trp-Val) at ~22°C. Absorption in 2.5 × 10^{-4} *M* dioxane is shown. A similar spectrum was obtained in a highly viscous solvent composed of polyoxypropylene and dioxane (POP-D). Fluorescence in 1.5 × 10^{-5} *M* dioxane (——); in POP-D (·—·—·). Absorption anisotropy factor, g_{AB}, in 2.1 × 10^{-4} *M* dioxane (——). Emission anisotropy factor, g_{em}, in 5 × 10^{-4} *M* dioxane (○); in POP-D (△). For comparison, g_{AB} and g_{em} of *N*-acetyl-L-tryptophanamide in dioxane and in POP-D are shown (– –).

COMPLEXES OF POLY(GLUTAMIC ACID) WITH ACRIDINE DYES

Blout and Stryer found that acridine dyes, which by themselves are not asymmetric, may acquire optical activity upon binding to poly(L-glutamic acid) or poly(D-glutamic acid) and thus contribute to the CD or ORD spectra (27, 28). The CD spectra of the bound dyes at their absorption bands are rather complex. There is good evidence that the long-wavelength absorption band of the acridine nucleus is composed of two electronic transitions (29–31). Furthermore, dimers or other aggregates of the dyes are known to form under a variety of circumstances; splitting of each transition is then possible because of exciton interactions, thereby increasing the number of electronic levels involved in the spectrum (31). Because the absorption spectra of dye monomers and dimers (or higher aggregates) show marked overlap, a more complex CD spectrum is expected when monomers and higher aggregates coexist. The circumstances are by far simpler in the case of emission. The fluorescence spectra of the monomers and dimers are better resolved than those of absorption, and the spectrum of each species involves a single electronic transition. Thus, the CPL spectra may be more easily interpreted than the corresponding CD spectra.

The spectra of the absorption and emission anisotropy factors of the complexes of acridine orange with poly(L-glutamic acid) and poly(D-glutamic acid) at pH 4.3 are presented in Fig. 3. At each wavelength the magnitude of g_{AB} and g_{em} for the two enantiomeric polypeptides is opposite in sign and comparable in absolute value. It should be noted that at pH values higher than 6.0, at which these polypeptides assume a random conformation, both g_{AB} (27, 28) and g_{em} vanish, indicating that measurable optical activity in the excited state, as well as in the ground state, is induced in the dye molecules only when the polymers are in the helical form. As seen from Fig. 4, the CD spectrum of the band extending between 400 and 550 nm is rather complex, the absorption anisotropy factor being highly variable, even changing several times, across the band. The peak of the absorption spectrum of the complexes described (monomer:dye ratio = 40:1) is at 470 nm (see Fig. 3), which shows that the adsorbed acridine orange

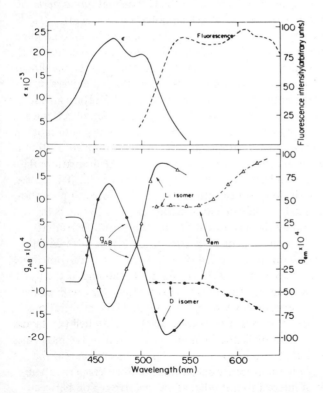

Figure 3. Spectroscopic data for the complexes between acridine orange and poly(L-glutamic acid) or poly(D-glutamic acid) at ~22°C. The monomer:dye ratio is 40. The solvent is aqueous phosphate buffer, pH 4.3.

molecules exist in dimeric form (32) to an appreciable extent. These circumstances, in conjunction with the fact that the acridine nucleus exhibits two electronic transitions at the spectral range under discussion, readily accounts for the
complexity of the CD. In contrast, g_{em} exhibits simple behavior, although it is
definitely not constant across all of the emission band. A transition occurs at
the spectral range above 560 nm, indicating that more than one species is responsible for the luminescence. Because the emission above 600 nm was attributed
to AO dimers, it was concluded that the optical activity measured above 560 nm
contains contributions from dye dimers or higher aggregates. Due to solubility
limitations, complexes of AO with poly(glutamic acid) at various degrees of saturation could not be studied; however, the above conclusions were strongly supported by such studies with the corresponding complexes with poly(A) (4). It is
of interest that g_{em} of the dimer dye molecules is much greater than their g_{AB}
at any point of the long-wavelength absorption band. The cause is probably a
pronounced cancellation of circular dichroism of different transition in the absorption spectrum, but it also may reflect a change in conformation upon electronic excitation. An important fact revealed by the CPL is that *monomeric* AO
may exhibit optical activity when bound to helical poly(glutamic acid).

The spectral data for complexes of poly(L-glutamic acid) and 9-amino acridine
(9-AA) are presented in Fig. 4. This dye tends not to aggregate in solution (34).
Its absorption spectrum in the complex is similar to that reported for the dye in
dilute solution, except for a slight shift to longer wavelengths and a decrease in
the molar extinction coefficient (34). The spectrum of the absorption anisotropy factor in this case is indeed simpler than that of the complex of AO with
poly(glutamic acid) and yet g_{AB} is not constant across the last absorption band.
The occurrence of two electronic transitions that contribute to the absorption in
this spectral range probably account for this phenomenon. The dramatic variation of g_{em} across the emission band, however, is quite unexpected. The possibility that the emission in the long-wavelength range is attributable to phosphorescence was rendered unlikely by lifetime measurements (the lifetime being
shorter than a few μsec) and by the insensitivity of the emission spectrum to
changes in oxygen concentration. The obvious conclusion is that different species that have different emission spectra and different optical activity contribute
to the observed fluorescence spectrum. Different modes of binding of the dye
to the polypeptide could account for this result. However, analogy with similar
studies performed on complexes of 9-AA with poly(A), in which preparations
with different ratios between the constituents were made and studied (4), favors
the interpretation that the emission is due to monomer and dimer dye molecules,
the latter being formed after electronic excitation. Thus, the emission in the
long-wavelength region of the fluorescence spectrum should be properly attributed to exciplexes. It is of interest to note that at the overlap region between
the absorption and the emission spectra, where the contribution of the monomer
to the fluorescence predominates, g_{AB} and g_{em} assume nearly the same value,
indicating a similar mode of binding of the dye to the polymer in the ground and

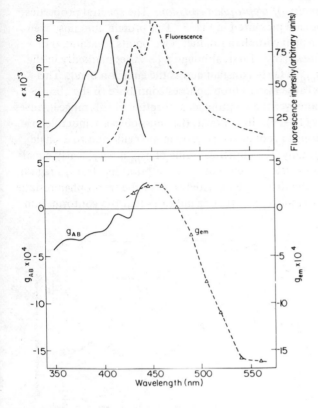

Figure 4. Spectroscopic data for the complex between 9-amino acridine and poly(L-glutamic acid) at ~22°C. The monomer:dye ratio is 40. The solvent was an aqueous acetate buffer, pH 4.6.

excited states. Also in the present case, both g_{AB} and g_{em} become too small to be detected at pH values at which the polymer is in the random-coil conformation.

APPLICATION TO PROTEINS

The chromophores that are amenable to study by CPL in the field of proteins are not limited to the tryptophan and the tyrosine residues. Many of the coenzymes are fluorescent or can be made to luminesce by relatively mild chemical modifications. Similarly, fluorescent ligands of proteins are encountered frequently. These ligands include enzyme inhibitors and haptens that bind to antibodies, as well as the ligands in protein-dye complexes and in protein complexes with luminescent rare-earth ions. As a matter of fact, proteins are so varied that they can offer many illustrations of the different aspects, potentialities, and limitations of CPL as a spectroscopic tool. A few examples will be presented below.

Detection of heterogeneity of tryptophan emission. The spectral properties of staphylococcal nuclease are presented in Fig. 5. This protein contains seven tyrosine residues and a single tryptophan residue. Two points regarding the anisotropy factors are noteworthy. First, although g_{AB} varies markedly in the spectral region studied, g_{em} is fairly constant across the emission band. This result is not surprising, because many chromophores contribute to the absorption band studied, and moreover the tryptophan absorption in this range involves two electronic transitions (16–19). In contrast, the emission is attributable primarily to the single tryptophan residue in the protein molecule and to a single electronic transition. Secondly, the numerical value of g_{em} is very nearly equal to that of g_{AB} at the long-wavelength edge of the absorption spectrum, a region in which the transition to the first singlet excited state of the tryptophan residue predominates. As discussed above, this finding indicates that the conformation

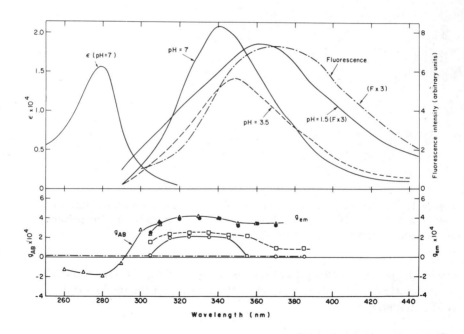

Figure 5. Spectroscopic data for staphylococcal nuclease at ~22°C. Absorption is shown for 6.15×10^{-4} M protein concentration in aqueous 0.1 M tris buffer, pH 7.0. Fluorescence in 0.1 M tris buffer, pH 7.0; in aqueous solution, pH 3.5; in aqueous solution, pH 1.5; and in 6 M guanidine hydrochloride (·—·—·). Absorption anisotropy factor, g_{AB}, in 0.1 M tris buffer, pH 7.0 (——); in 6 M guanidine hydrochloride (·—·—·); protein concentration 6.15×10^{-4} M. Emission anisotropy factor, g_{em}, in 0.1 M tris buffer, pH 7.0 (△); in 0.1 M tris buffer, pH 7.0, and the presence of calcium ions (5×10^{-4} M) (●); in aqueous solution, pH 3.5 (□); in aqueous solution, pH 1.5 (○); in 6 M guanidine hydrochloride (·—·—·); protein concentration 2×10^{-4} M.

of the indole side chain and its interaction with the environment is very similar in the ground state and in its first singlet excited state. Calcium ions, which are known to bind to this protein and are necessary for its enzymic activity (35) have no measurable effect on the CPL spectrum; thus the environment of the tryptophan residue is not affected noticeably by the binding of the calcium ions. The effect of pH on g_{em} also is shown in Fig. 5. At pH 3.5, at which the structure of the protein is believed to be disrupted by a variety of criteria (36), g_{em} decreases throughout the spectrum relative to its value at pH 7.0, the decrease being more pronounced at the long-wavelength range of the spectrum. This feature indicates that the population of protein molecules at this pH is not homogeneous: one fraction emits at longer wavelengths and has little optical activity whereas another fraction emits at shorter wavelengths and still possesses appreciable rotatory power. When the pH is lowered to 1.5, g_{em} continues to change, indicating that a "fully" denatured state was not reached at pH 3.5. It should be noted that only in 6 M guanidine hydrochloride solution does g_{em} vanish completely (see Fig. 5); thus some degree of structure seems to prevail in this protein even at pH 1.5 in aqueous solution.

Azurine is another protein that has a single tryptophan. As with nuclease, its emission anisotropy factor is nearly constant across the emission band and is of comparable value to its absorption anisotropy factor at the long-wavelength edge of the absorption spectrum (37). In marked contrast to the above, g_{em} of human serum albumin varies markedly across its emission spectrum (see Fig. 6). It varies also in the range of the tryptophan emission in spite of the fact that it contains a single tryptophan residue. This behavior probably reflects a heterogeneity in the population of the molecules of this protein, different molecules having different emission spectra and different asymmetric environments for their tryptophan residue. Such heterogeneity has been indicated before by a variety of other methods (38–40).

If a protein molecule contains more than one tryptophan residue, the various indole side chains generally have different environments and thus have somewhat different emission spectra and different optical activity. The value of g_{em} therefore is not expected to be constant across the emission spectrum of the protein. An example of this phenomenon is chicken pepsinogen (see Fig. 7), a protein that contains six tryptophan residues (41). It is of interest that in D_2O, the value of g_{em} changes in the long-wavelength region of the emission spectrum, where the tryptophan residues exposed to the polar solvent contribute more heavily to the protein fluorescence. It may be recalled that the quantum yield of the indole chromophore is enhanced in D_2O, relative to H_2O (42). This finding gives more weight to the above interpretation of the nonconstancy of g_{em} in terms of heterogeneity of environment of the various tryptophan residues in the protein molecule.

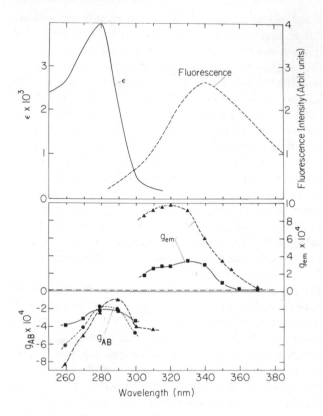

Figure 6. Spectroscopic data for human serum albumin at ~22°C. Absorption in 0.1 M phosphate buffer, pH 6.8, for a protein concentration of $3.4 \times 10^{-4}\,M$. Fluorescence in 0.1 M phosphate buffer, pH 6.8, for a protein concentration of $3 \times 10^{-5}\,M$. Absorption anisotropy factor, g_{AB}, in 0.1 M phosphate buffer, pH 6.8 (▲); in aqueous solution, pH 3.7 (●); in 6 M guanidine hydrochloride (■); reduced with dithioerythritol, in 6 M guanidine hydrochloride (·—·—·); protein concentration $-3.4 \times 10^{-4}\,M$. Emission anisotropy factor, g_{em}, in 0.1 M phosphate buffer, pH 6.8 (▲); in 6 M guanidine hydrochloride (■); reduced by dithioerythritol, in 6 M guanidine hydrochloride (·—·—·); protein concentration $-3.4 \times 10^{-4}\,M$.

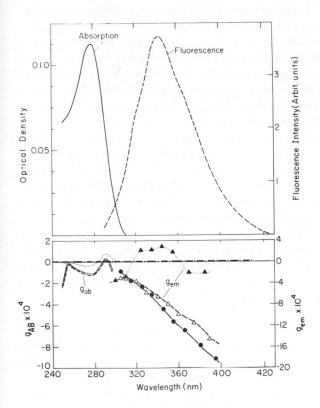

Figure 7. Spectroscopic data for chicken pepsinogen at ~22°C. Absorption for 2.2 × 10⁻⁶ M concentration in 0.1 M phosphate buffer, pH 7.0. Fluorescence for 3 × 10⁻⁶ M concentration in 0.1 M phosphate buffer, pH 7.0. Absorption anisotropy factor, g_{AB}, for a protein concentration of -3 × 10⁻⁵ M, in 0.1 M phosphate buffer, pH 7.0, in H₂O (— —); in 0.1 M phosphate buffer, pH 7.0, in D₂O (——); in 8 M guanidine hydrochloride (· · · ·); reduced with dithioerythritol in 8 M guanidine hydrochloride (·—·—·). Emission anisotropy factor, g_{em}, for a protein concentration of -1.2 × 10⁻⁴ M, in 0.1 M phosphate buffer, pH 7.0, in H₂O (△); in 0.1 M phosphate buffer, in D₂O (●); in 8 M guanidine hydrochloride (▲); reduced with dithioerythritol in 8 M guanidine hydrochloride (·—·—·).

Protein denaturation. In all the cases studied, native proteins exhibit optical activity in their tryptophan fluorescence, as measured by their CPL. A question of interest is whether this optical activity is a result of asymmetry induced by the secondary and tertiary structure of the protein or whether the primary structure alone may induce asymmetry in the indole chromophores. As shown above, short linear peptides of tyrosine and tryptophan do not have detectable circular polarization in their fluorescence. Figs. 5, 6, and 7 demonstrate that the proteins staphylococcal nuclease, human serum albumin, and chicken pepsinogen also completely lose their CPL when denatured under drastic conditions of high concentrations of guanidinium hydrochloride and reduction of internal disulfide bonds (where present) by dithioerythritol. Other proteins tested behave in much the same way (27). Thus, it may be concluded that in all cases studied the tryptophan residues lose their optical activity in the excited state when the native structure of the proteins is destroyed. Interestingly, the tryptophan residues in human serum albumin and chicken pepsinogen show pronounced CPL even in high concentrations of guanidine hydrochloride, if the internal disulfide bonds are not disrupted (see Figs. 6 and 7). The internal bridges (18 in human serum albumin and 17 in chicken pepsinogen) help the molecules retain some structure even under strongly denaturing conditions.

CPL as a substitute for CD. In a variety of circumstances CPL may be a more useful analytic or diagnostic tool than CD in the study of polypeptides and proteins. For example, the CD of a chromophore may be too small to be measured conveniently, whereas its CPL may be large. Furthermore, the absorption bands of a chromophore of interest, that is, a ligand, may overlap the bands of other chromophores in the system studied, such as those of the macromolecules that bind the ligand. Cases of special interest in this connection are forbidden transitions, which may be very difficult to observe by light absorption, but are often readily observed by emission processes; thus CPL is the only choice for the measurement of their optical rotatory power.

When derivatives of the 1-dimethylamino-5-naphthalenesulfonyl chromophore (DNS) bind to anti-DNS antibodies, no detectable CD is induced in their long-wavelength absorption band. Bands at shorter wavelengths could not be studied because of overlap with the protein absorption spectrum. The fluorescence of the bound DNS, however, was found to exhibit some circular polarization, reflecting the asymmetric environment of the antibody binding site. The value of g_{em} varied widely across the emission band and depended upon the wavelength of the excitation light. These results were interpreted in terms of the heterogeneity of the population of the antibody molecules (13). The marked difference between g_{AB} and g_{em} shows that a change in mode of binding of the hapten in the antibody binding site takes place upon electronic excitation of the ligand. Changes in mode of interaction between proteins and bound small molecules upon electronic excitation were also observed in the case of anthraniloyl chymotrypsin and in the complex of chymotrypsin with 2-*p*-toluidinylnaphthalene-6-sulfonate (3). These results emphasize that information derived from the luminescence of fluorescent probes strictly applies to the excited state of the probe and that caution should be exercised in the interpretation of the results in terms of the properties of the complexes in the ground state.

The possibility of measuring the optical rotatory power of forbidden transitions by CPL has a useful application in the study of the metal binding sites of human blood transferrin and chicken egg conalbumin. These proteins, whose function is apparently transport and storage of iron, have two binding sites for ferric ions per protein molecule, but they can also bind trivalent terbium ions (43). That the latter ions are luminescent is indicated by four structured emission bands in the visible range of the spectrum. These bands are the product of a multitude of electronic transitions, all of which are internal transitions in the $4f$ shell and, therefore, are essentially forbidden. Thus, these transitions cannot be observed by light absorption at reasonable concentrations of the protein-metal complexes. Because of the multitude of the electronic transitions involved, the emission anisotropy factor varies dramatically across the emission spectrum. It is of interest that g_{em} assumes exceptionally high values in certain regions of the spectrum. Thus, the spectrum of g_{em} may serve as a sensitive "fingerprint" of the environment of the bound metal ions. With this technique it was shown that the metal binding sites of transferrin and conalbumin are essentially identical in structure. The same spectrum of g_{em} also was obtained upon partial saturation of the sites with Tb^{+3} and in tertiary complexes containing protein:Tb^{3+}:Fe^{3+} or protein:Tb^{3+}:Ho^{3+}, and we concluded, therefore, that the two sites on each protein molecule are most probably identical under physiological conditions (5).

CONCLUSION

The circular polarization of luminescence is a measure of the optical rotatory power of molecules in the excited state, which is responsible for emission, in the same way that circular dichroism is a measure of the optical rotatory power of molecules in their ground state. In common with other experimental methods, CPL has characteristic potentialities and limitations as a spectroscopic tool. It is applicable only to luminescent molecules, and then only to a limited number of electronic transitions (generally, to a single transition). In complex systems, however, these limitations may turn out to be advantageous, supplying more specific information about the emitting species. Moreover, CPL sometimes may be applied when CD fails.

REFERENCES

1. A. Gafni and I.Z. Steinberg, *Photochem. Photobiol.*, **15**, 93 (1972).
2. I.Z. Steinberg and A. Gafni, *Rev. Sci. Instr.*, **43**, 409 (1972).
3. J. Schlessinger and I.Z. Steinberg, *Proc. Nat. Acad. Sci. U.S.*, **67**, 769 (1972).
4. A. Gafni, J. Schlessinger, and I.Z. Steinberg, *Israel J. Chem.*, **11**, 423 (1973).

5.   A. Gafni and I.Z. Steinberg, *Biochemistry*, **13**, 800 (1974).
6.   S. Veinberg, S. Shaltiel, and I.Z. Steinberg, *Israel J. Chem.*, in press.
7.   C.A. Emeis and L.J. Oosterhoff, *J. Chem. Phys.*, **54**, 4809 (1971).
8.   I.Z. Steinberg, in *Concepts in Biochemical Fluorescence*, R. Chen and H. Edelhoch, eds., Marcel Dekker, New York, 1974.
9.   E.U. Condon, W. Altar, and H. Eyring, *J. Chem. Phys.*, **5**, 753 (1937).
10. W. Kuhn, *Ann. Rev. Phys. Chem.*, **9**, 417 (1958).
11. A. Moscowitz, in *Modern Quantum Chemistry*, Part III, O. Sinanoglu, ed., Academic Press, New York, 1965, p. 31.
12. A. Gafni, H. Hardt, and I.Z. Steinberg, submitted for publication.
13. J. Schlessinger, I. Pecht, and I.Z. Steinberg, *J. Mol. Biol.*, submitted for publication.
14. I.Z. Steinberg and B. Ehrenberg, in preparation.
15. H. Edelhoch, R.E. Lippoldt, and M. Wilchek, *J. Biol. Chem.*, **243**, 4799 (1968).
16. G. Weber, *Biochem. J.*, **75**, 335 (1960).
17. E. Yeargers, *Biophys. J.*, **8**, 1505 (1968).
18. E.H. Strickland and C. Billups, *Biopolymers*, **12**, 1989 (1973).
19. L.J. Andrews and L.S. Forster, *Biochemistry*, **11**, 1875 (1972).
20. K.D. Kopple and D.M. Marr, *J. Am. Chem. Soc.*, **89**, 6193 (1967).
21. K.D. Kopple and M. Ohnishi, *J. Am. Chem. Soc.*, **91**, 962 (1969).
22. L.E. Webb and Chi-Fan Lin, *J. Am. Chem. Soc.*, **93**, 3818 (1971).
23. J. Caillet, B. Pullman, and B. Maigret, *Biopolymers*, **10**, 221 (1971).
24. R. Chandrasekaran, A.V. Lakshminarayanan, P. Mohanakrishnan, and G.N. Ramachandran, *Biopolymers*, **12**, 1421 (1973).
25. E.H. Strickland, M. Wilchek, J. Horwitz, and C. Billups, *J. Biol. Chem.*, **245**, 4168 (1970).
26. J. Schlessinger and I.Z. Steinberg, in preparation.
27. E.R. Blout and L. Stryer, *Proc. Nat. Acad. Sci. U.S.*, **45**, 1591 (1959).
28. L. Stryer and E.R. Blout, *J. Am. Chem. Soc.*, **83**, 1411 (1961).
29. A. Wittwer and V. Zanker, *Z. Physik. Chem., Neue Folge*, **22**, 417 (1959).
30. A.C. Albrecht, *J. Mol. Spectr.*, **6**, 84 (1961).
31. K. Yamaoka and R.A. Resnik, *J. Phys. Chem.*, **70**, 4051 (1966).
32. M.E. Lamm and D.M. Neville, Jr., *J. Phys. Chem.*, **69**, 3872 (1965).
33. A. Blake and A.R. Peacocke, *Biopolymers*, **6**, 1225 (1968).
34. K. Yamaoka and R.A. Resnik, *Biopolymers*, **8**, 289 (1969).
35. P. Cuatrecasas, S. Fuchs, and C.B. Anfinsen, *J. Biol. Chem.*, **242**, 1541 (1967).
36. H.F. Epstein, A.N. Schechter, and J.S. Cohen, *Proc. Nat. Acad. Sci. U.S.*, **68**, 2042 (1971).
37. J. Schlessinger, I. Pecht, and I.Z. Steinberg, unpublished results.
38. J.F. Foster, M. Soyami, H.A. Peterson, and W.J. Leonard, *J. Biol. Chem.*, **240**, 2495 (1965).
39. W.B. de Lauder and P. Wahl, *Biochem. Biophys. Res. Commun.*, **42**, 398 (1971).
40. G. Hazan, Ph.D. Thesis, The Weizmann Institute of Science, Rehovot, Israel (1973).

41.   G.R. Tristram and R.H. Smith, *Adv. Protein Chem.*, **18**, 227 (1963).
42.   L. Stryer, *J. Am. Chem. Soc.*, **88**, 5708 (1966).
43.   C.K. Luk, *Biochemistry*, **10**, 2838 (1971).

CONFORMATIONAL CHANGES IN POLYPEPTIDES AND PROTEINS FROM LASER RAMAN SCATTERING

TAIN-JEN YU and WARNER L. PETICOLAS, Department of Chemistry, University of Oregon, Eugene, Ore. 97403

SYNOPSIS: Laser Raman spectra of poly(L-lysine) (PLL) have been taken of single crystals of the α-helical form, platelike particles of the β form, and films at different humidities. The spectra of these films and crystals are compared with previous measurements in solution. Evidence is presented that the crystal forces alter both the backbone structure and the structure of the side groups so that the α-helical structures of PLL in the solid phases and in solution are not exactly the same. Furthermore, the structure of the solid phases are very sensitive to the relative humidity. The $\alpha \rightleftharpoons \beta$ transition in PLL films has been followed using Raman spectroscopy.

In contrast, the Raman spectra of a protein, α-chymotrypsin have been taken in solution, crystals, and films as a function of humidity. In this case, there is no change in the Raman spectrum; thus the conformation of this protein is remarkably stable with regard to gentle changes in the environment.

Thus certain polypeptides appear to be more sensitive structurally to their environment than certain proteins. Consequently, synthetic homopolymers should be used judiciously as models for protein behavior.

INTRODUCTION

Proteins may be considered to be linear polymers of amino acids with a polypeptide backbone and occasional disulfide cross-linking. Consequently, synthetic linear poly(amino acids) are often used as model compounds for the more complex proteins. We have made a comparative study of the structure of both a poly(amino acid), poly(L-lysine) (PLL), and a common protein, α-chymotrypsin, in different physical environments (crystalline, solutions, and cast films under varying humidity conditions).

In general, x-ray diffraction can be used to determine structure only in the crystalline state. Consequently, the effect of crystalline forces on the structure of the protein or poly(amino acid) has been unknown. Recently, Raman spectroscopy has been shown to be a method of obtaining information about the structures of polypeptides and proteins both in solution and in the solid state (1-11). From an examination of the Raman spectrum, however, we can discern only changes in the backbone (peptide) structure of proteins because of the large number of conformations of the amide (peptide) groups in a protein. In contrast, polypeptides generally have rather simple structures—the α-helical form, the β structure, and disordered forms. Thus it seemed worthwhile to study the effect of crystallinity, humidity, and water content on these relatively simple peptide structures and to compare these results with data on proteins.

A rather thorough study has already been made of the various conformations of poly(L-lysine) by laser Raman scattering (10). We have extended this work by comparing the Raman spectra of films, crystals, and solutions to determine what changes in structure can be brought about by fairly mild changes in the environment. These changes are contrasted with those undergone by a protein, α-chymotrypsin.

EXPERIMENTAL METHOD

Poly(L-lysine) hydrobromide (PLL·HBr) was purchased from Pilot Chemicals Inc. Single crystals of PLL·HPO$_4$ were grown by the methods of Padden, Keith, and Giannoni (12). Approximately 0.3 ml of 0.2 M ammonium monohydrogen phosphate was added to 1.0 ml of 0.2% PLL·HBr and a precipitate of PLL·HPO$_4$ was formed. Typical crystals are shown in Fig. 1. In these crystals, electron-diffraction patterns confirm the α-helical structure.

The crystals and their mother liquor were sealed in a glass capillary tube and centrifuged to one end. The capillary was placed in a thermostatted block and the Raman spectra taken with the 5145 cm^{-1} line of an argon laser at 450 milliwatts (mW). The monochromator slits of a Spex-1400 double-grating monochromator were set at 350 μ.

Figure 1. Single crystals of α-poly(L-lysine) · HPO$_4$ dried from a suspension in 80% water, 20% 2-ethoxyethanol (electron micrograph).

The single crystals of α-helical PLL in the mother liquor dissolve at about 45°C. On further heating, the solution becomes cloudy and another precipitate appears at about 75°C. The stable high-temperature precipitate has been shown to consist of small platelike particles of indefinite outline that give both x-ray and electron-diffraction (12) reflections characteristic of the extended β conformation. Thus we have been able to obtain a crystalline $\alpha \rightleftharpoons \beta$ transition to compare with the solution $\alpha \rightleftharpoons \beta$ transition studied earlier.

To study the effect of humidity on films or fibers of PLL·HCl, films were cast from their concentrated aqueous solution. These were placed in a small humidity chamber over different saturated salt solutions and left for one to two days to reach equilibrium. The spectra then were taken with the laser power turned down to 200–350 mW.

Worthington Chemical Co. was our source of α-chymotrypsin. A few grains of the powder were placed at one end of a very small glass capillary and a humidifying agent was placed at the other end. The humidity was varied from 0% with a drying agent to 95%.

Crystals of α-chymotrypsin about 1 mm on an edge were made by standard techniques. A saturated solution of enzyme-grade ammonium sulfate was diluted to 95% strength with water (Solution A). Spectroscopic-grade dioxane was added to the amount 4% to a 50:50 mixture of 0.1 M sodium citrate and 0.1 M citric acid (Solution B). The protein was dissolved 4% by weight in solution B and the solution filtered through 3 μ millipore filter paper. One-half milliliter of the protein solution was placed in each of five test tubes. Then 0.48 ml of solution A was added to the first, 0.485 ml to the second, and so on to 0.50 ml to the fifth test tube. The test tubes were thermostatted at 20°C, and after one week good crystals were found in several of the tubes.

RESULTS

We now can compare the Raman spectra of the α and β form of PLL in single crystals, in films, and in solution or suspension. Similarly, we can compare the Raman spectra of α-chymotrypsin powder (as a function of humidity), α-chymotrypsin in solution and α-chymotrypsin in crystalline form. In this way, we might gain some idea of the effect of crystalline forces and water content on the structure of polypeptides and proteins.

Before discussing these results, we must review briefly the Raman spectra of polypeptides as they pertain to the backbone structure. The pioneering work of Lord and Yu (2, 5, 6) on the Raman spectra of proteins has shown that practically all of the Raman bands of proteins arise from the vibrations of the side groups of the constituent amino acids and that they are relatively insensitive to conformation. Exceptions to this are the frequencies of the Amide-I and Amide-III bands that arise from the backbone of the polypeptide chain, the frequencies of the disulfide bond vibration, and the intensities of the $C-C_\alpha$ or $C_\alpha-N$ bonds that appear to be sensitive to conformation.

Table 1 lists the Amide-I and Amide-III Raman frequencies of various polypeptides and simple proteins in which the conformation of the polypeptide chain is well established. For comparison, Table 2 lists the Raman frequencies

Table 1. Amide-I and Amide-III Raman frequencies (cm^{-1}) in polypeptides and proteins

| Substance | α Helix | | β Structure | | Random Coil or Structure Ionized | |
	Amide I	Amide III	Amide I	Amide III	Amide I	Amide III
Polyglycine (4)	1654[a]	1261[a]	1674 (S)	1234		
Poly(L-alanine) (10)	1658 (W)	1261 (W)				
Poly(L-glutamic acid) (6)					1665	1248
Glucagon (7)	1658	1266	1672	1232		1248
Poly(L-lysine) (10)		Weak		1240		1243
H$_2$O	1645 (W)		1670 (S)		~1660 (VW)	
D$_2$O	1632 (W)		1658 (S)		1660 (S)	

[a]Polyglycine II is a threefold helix.

Table 2.   Amide-II Raman frequencies (cm^{-1}) from *cis* amides

Substance	Amide II
Diketopiperazine (13)	1456 (powder), 1385 (aqueous)
Hypoxanthine (14)	1464
1-Methyluracil (15)	1417 (aqueous)

in several *cis* amides of the Amide-II group. The Amide-II vibrations are Raman active only in *cis* amides and consequently will be of no interest in Raman spectroscopy of proteins where almost all of the amides are *trans*. We note also that there is no exact correlation between the frequency of the Amide-III band and the conformation of the backbone, especially in the β structure, in which the band varies between 1230 cm^{-1} and 1240 cm^{-1}. Furthermore, in the α structure of PLL and poly(L-alanine) the Amide-III band is extremely weak. As has been demonstrated for PLL in our previous paper (10), however, the $\alpha \rightleftharpoons \beta$ and other transitions can be followed easily by Raman spectra if the Amide-III frequencies are first determined independently for each type of conformation.

A much more reliable structure-frequency relationship exists between the Amide-I frequency and the structure of the various polypeptide models. Clearly, the Raman frequency of the Amide-I group is always about 1674 cm^{-1} for the β structure and 1650 cm^{-1} for the helical structures; the random coil is midway between them at 1660 cm^{-1}.

With these frequencies in mind we can discuss the new spectra of the single crystals of $PLL \cdot HPO_4$ in the α form in Fig. 2, bottom. The striking first observation is that in single crystals of $PLL \cdot HPO_4$, the Amide-III band is moderately strong, whereas in the previous measurements in solution the Amide-III band could not be seen at all. Notice that in this region we have two bands—one at 1246 cm^{-1} and another at 1210 cm^{-1}. There seems to be little doubt that these single crystals are composed of polymers in the α-helical form, as indicated by the Amide-I band. Consequently, we can see indications from the Amide-III region that either the crystal forces in the single crystals are strong enough to change the structure slightly over that of an α helix of PLL in solution (10) or we are seeing the turns at the surfaces of the crystals.

Another unusual feature of this spectrum is the band at 1325 cm^{-1}. In our solution spectrum of α-helical PLL, this band had the shape of a doublet at 1311 cm^{-1} (S) and 1341 cm^{-1} (M). This band undoubtedly arises from the lysine side chains. Thus we have evidence that the side-chain geometry of helical PLL in the single crystals differs from its geometry in solution (10).

The bands in the range 900–1100 cm^{-1} also differ from those of the previous solution spectra. However, $(NH_4)_2HPO_4$ itself shows three bands in this range at about 899 cm^{-1}, 988 cm^{-1}, and 1065 cm^{-1}. These bands might be described as P–O(H) stretch and symmetric and asymmetric PO_3^{2-} stretch. Consequently, it is difficult to sort out the polymer bands from the solvent bands.

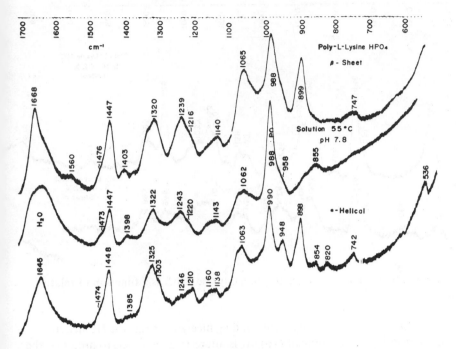

Figure 2. Laser Raman spectra of PLL·HPO$_4$ in single crystals, solution, and β sheet.

When the crystals are heated in their mother liquor, they go into solution. In the spectrum in the center of Fig. 2, we see by the 1243 cm^{-1} band of Amide-III and the absence of a sharp Amide-I band that the spectrum is the same as the one we previously (10) found for the random, ionized form of PLL·HBr at neutral pH. The β-sheet conformation that results from further heating gives the spectrum shown at the top of Fig. 2. Here again we see differences in the region of 1320 cm^{-1}, indicating a different side-chain conformation in the β form prepared in this manner than in the β form previously prepared. Note also that the Amide-III band at 1239 cm^{-1} is broader and weaker than in the previous spectrum. The Amide-I band, however, shifts from 1645 cm^{-1} to 1668 cm^{-1}, which, within experimental error, is exactly the same frequency that was observed previously. This result demonstrates the reliability of this band as a guide to the overall structure.

For the same transitions (α-helical single crystals to ionized random solution to precipitated β sheet) carried out in D$_2$O the Amide-I' band shifts to lower frequencies, but the transitions can be determined from the shift of the Amide-I band from 1635 cm^{-1} (α) to 1660 cm^{-1} (ionized) to 1656 cm^{-1} (β). Indeed, in this case, there is no problem with the water band, and the Amide-I' band can be seen even in the ionized form, in which it is very weak.

Now let us examine the spectrum of the PLL·HCl films shown in Fig. 3. First, it is apparent from the position of the Amide-I bands that there is a change from the α form (1647 cm^{-1}) at 92% relative humidity, to the β form

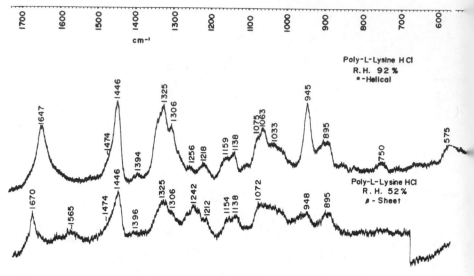

Figure 3. Laser Raman spectra of poly(L-lysine)·HCl as a function of relative humidity.

(1670 cm^{-1}) at 52% humidity. This finding nicely confirms the IR measurements of Blout and Lenormant (16). It is interesting, however, to note that the 1325 cm^{-1} band in these films is very similar to that found in PLL·HPO$_4$. Thus the lysine side-chain conformations seem to be alike in the films and crystals and different from that found previously in solution. Similarly, in the PLL·HCl films we see for the α helix a stronger Amide-III band, which appears to be a doublet at 1256 cm^{-1}, 1218 cm^{-1} that is similar to the doublet in the PLL·HPO$_4$ α-helical single crystals, but again different from the unobservably weak Amide-III band of the α helix in solution.

One other band of interest in the film at 92% relative humidity is the 945 cm^{-1} band characteristic of the α helix. The exact assignment of this band is unknown, but it falls in the region of the C–C stretch (10).

Now we shall contrast the rather large structural changes observed in model polypeptides such as PLL with the almost total lack of change in the Raman spectra of proteins. Fig. 4 shows the Raman spectra of α-chymotrypsin single crystals and in solution at 30% concentration (pH 5.0) with no buffer added. Within experimental error, there is absolutely no change in the frequencies of either the Amide-I band (1668–69 cm^{-1}) or the Amide-III band (1245,1260 cm^{-1} doublet). Similar studies by Yu et al. (11) have also shown very little change in the structure of α-chymotrypsin between crystals and solution. (The band at 982 cm^{-1} is due to the presence of SO$_4{}^{2-}$ ion used in the preparation of α-chymotrypsin (6) and in the preparation of single crystals.)

CONCLUSIONS

The Raman frequencies of α-helical PLL have been shown to differ slightly according to whether the α-helical PLL is in the solid (crystalline) state or in

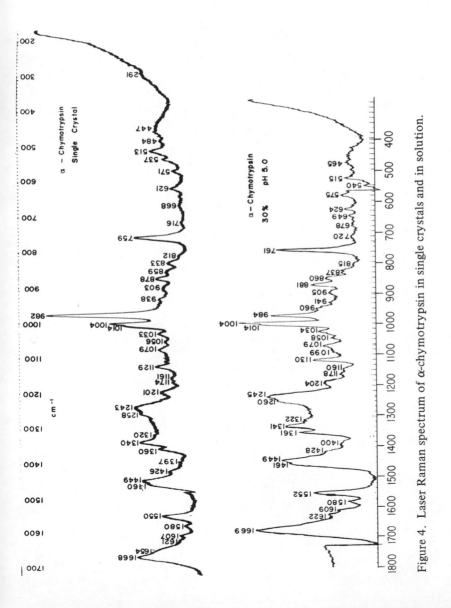

Figure 4. Laser Raman spectrum of α-chymotrypsin in single crystals and in solution.

solution. This observation has been interpreted to be evidence for slight structural changes in the α helix in solution and in the crystal or in the presence of chain folding. Similar changes in the β structure depend on environment. Films and fibers of PLL·HCl have shown the structure to be very sensitive to humidity, in confirmation of the work of Blout and Lenormant (16). In contrast, there is no evidence for a change in either the backbone or side-chain structure of α-chymotrypsin in the transition from dry powder to humid powder to single crystal to aqueous solution.

ACKNOWLEDGMENTS

The authors' work reported in this paper was supported in part by a grant from the National Institutes of Health. W.L. Peticolas was a Guggenheim Fellow during 1973–74, on sabbatical from the Institute Max von Laue – Paul Langevin, B.P. 156, 38100 Grenoble, France.

REFERENCES

1.   J.L. Koenig, *J. Poly. Sci.*, Part D, **6**, 59 (1972).
2.   R.C. Lord, *Pure Appl. Chem. Suppl.*, **7**, 179 (1971).
3.   B. Fanconi, B. Tomlinson, L.A. Nafie, E.W. Small, and W.L. Peticolas, *J. Chem. Phys.*, **51**, 3993 (1969).
4.   E.W. Small, B. Fanconi, and W.L. Peticolas, *J. Chem. Phys.*, **52**, 4369 (1970).
5.   R.C. Lord and Nai-Teng Yu, *J. Mol. Biol.*, **50**, 509 (1970).
6.   R.C. Lord and Nai-Teng Yu, *J. Mol. Biol.*, **51**, 203 (1970).
7.   Nai-Teng Yu and C.S. Liu, *J. Am. Chem. Soc.*, **94**, 5127 (1972).
8.   Nai-Teng Yu and B.H. Jo, *J. Am. Chem. Soc.*, **95**, 5033 (1973).
9.   Nai-Teng Yu and B.H. Jo, *Arch. Biochem. Biophys.*, **156**, 469 (1972).
10. Tain-Jen Yu, J.L. Lippert, and W.L. Peticolas, *Biopolymers*, **12**, 2161 (1973).
11. Nai-Teng Yu, B.H. Jo, and C.S. Liu, *J. Am. Chem. Soc.*, **94**, 7572 (1972).
12. F.J. Padden, Jr., H.D. Keith, and G. Giannoni, *Biopolymers*, **7**, 793 (1969).
13. P. Stein and W.L. Peticolas, unpublished results.
14. K.G. Brown, E.J. Kiser, and W.L. Peticolas, *Biopolymers*, **11**, 1855 (1972).
15. H.T. Miles, T.P. Lewis, E.D. Becker, and J. Frazier, *J. Biol. Chem.*, **248**, 1115 (1973).
16. E.R. Blout and H. Lenormant, *Nature*, **179**, 960 (1957).

INFRARED STUDIES OF POLYPEPTIDE SOLVATION DURING HELIX-COIL TRANSITIONS

P. COMBELAS, M. AVIGNON, C. GARRIGOU-LAGRANGE, Centre de Recherches P. Pascal, Domaine Universitaire, 33405 — Talence, France, and **J. LASCOMBE,** Laboratoire de Spectroscopie Infrarouge, Associé au CNRS, Université de Bordeaux I, 33405 — Talence, France

SYNOPSIS: Infrared spectra of poly(L-alanine), poly(γ-benzyl-L-glutamate), alternating poly(γ-benzyl-DL-glutamate), and poly(β-benzyl-L-aspartate) dissolved in mixtures of trifluoroacetic acid (TFA) and $CHCl_3$ were investigated for the purpose of analyzing polymer-acid-specific interactions responsible for the helix–random coil transition of polymer chains. In particular, characteristic IR absorptions, including those of amide A, I, II, (backbone), $\nu C=O$ ester (side chains) and acid $\nu C=O$, νOH, as well as solvents that are sensitive to molecular interactions were studied. Previous experimental investigations and assumptions of a classical theoretical model for polypeptide-proton donor interaction are discussed.

We examined the peptide backbone–acid interaction by comparing the spectra of polypeptides and model compounds under the same conditions. The spectra show backbone-acid hydrogen bonding; that is,

$$-NH-CHR-\underset{|}{C}=O \cdots HOOCCF_3 \ (I)$$

for all acid concentrations. Moreover, although the model compound is protonated in moderately and highly acidic solutions according to the exothermic equilibrium

$$>N-\underset{|}{C}=O \cdots HOOCCF_3 \ (I) \rightleftharpoons \left\{>N \stackrel{\cdots}{\cdots} C \stackrel{\cdots}{\cdots} OH\right\}^{\oplus} {}^{\ominus}\left\{OOCCF_3\right\} \ (II)$$

the polypeptide does not show such proton transfer because the amidinium absorption characteristic of protonation is not observed. By quantitative analysis of this association (I), we localized the peptide binding sites of the polymers and found that they belong only to random coil and terminal helical regions. Furthermore, our results have shown that the main binding site is the carbonyl group ($K_{NH} \cong 0$ or $\ll K_{CO}$).

Finally we analyzed the behavior of the polar side chains during the transition. Similarly to the peptide backbone, they bind acid by hydrogen bonding:

$$-(CH_2)_x-\underset{\underset{OBzl}{|}}{C}=O \cdots HOOCCF_3 \ (III)$$

Moreover, this association is more important when the side chains are localized in the coil regions than in the helical regions. We propose, therefore, to accommodate this result by introducing into helix-coil transition theory two more association constants for polar side chains, namely k_1 for the helical regions and $k_2 > k_1$ for the coil regions.

INTRODUCTION

The experimental and theoretical aspects of the role of the proton-donor solvent in the mechanism of the helix–random coil transition of polypeptides with unionizable side chains has been extensively analyzed in recent years. Particular attention has been paid to the transition induced by solvent mixtures containing a halogenated carboxylic acid. The conclusions of these studies are still controversial among investigators who use different techniques. Some explain helix disruption as a consequence of the protonation of the peptide functions (1-11); others propose instead a hydrogen-bond interaction between the polymer backbone and the acid (12-17). The latter hypothesis is generally more consistent with the theoretical model based on competition between the formation of intramolecular (helix) and intermolecular (polymer–proton donor) hydrogen bonds (18-21).

The solvation of the backbone has been studied extensively, but the side-chain interaction with the medium during the conformational transition rarely has been considered (18). It is possible that the basic properties of side chains containing polar functional groups (ester, thioether, etc.) also have an impact on helix disruption.

Infrared (IR) spectroscopy is a suitable technique by which to study molecular interactions. It can be used to investigate these problems when the absorptions of the various species (complexes and ions) are well known from preliminary studies of monomeric model compounds.

EXPERIMENTAL METHOD

Poly(L-alanine) (PLA), $\overline{MW} \cong 5000$, and poly(γ-benzyl-L-glutamate) (PBLG), $\overline{MW} \cong 75000$, were purchased from Miles-Yeda Ltd., Lots AL19 and GL112, respectively. Alternating poly(γ-benzyl-DL-glutamate) (PBDLG), $\overline{MW} \cong 16000$, was synthesized by Heitz and Spach (22). Poly(β-benzyl-L-aspartate) (PBLA) (A: low $\overline{MW}$, B: $\overline{MW}$, ~150,000, Pilot Chemicals, Lot A37) were supplied by E. Marchal.

Model compounds (amides and esters), spectroscopic solvents ($CHCl_3$, methylene chloride, and high-purity TFA) were commercial products that generally were distilled and dehydrated before use.

A Perkin-Elmer 225 spectrometer equipped with air dryer was used. The preservation of products, the preparation of solutions, and the filling of cells were achieved inside a dry box. The thermostatted cells used had CaF_2 windows.

In order to limit errors, spectra were recorded without direct compensation of the solvent's absorption on a reference beam. Two spectra were always recorded: one for a binary mixture of solvents, the other for a tertiary amide-solvents mixture. The difference spectra was then obtained (optical density units). Frequency precision is about ±1 to ±4 cm^{-1} according to half-width of bands.

RESULTS AND DISCUSSION

Analysis of backbone-acid interaction during the helix-coil transition. A preliminary study of a secondary amide model, *N*-methylacetamide (NMA), dissolved in TFA-CHCl$_3$ mixtures of different acid molar fractions (x) was necessary to understand the polymer's behavior. A similar investigation of tertiary amides (23) had shown the steps of amide-acid interaction when x is increased to be

$$XH + B \rightleftharpoons XH \cdots B \rightleftharpoons X^- \cdots HB^+ \rightleftharpoons X^- + HB^+$$

With a secondary amide the same phenomena are likely, but they should be complicated by the presence of the NH group.

The spectra in Figs. 1 and 2 show amide and acid bands' evolution when x increases and amide monomer concentration is kept constant. In NMA-CHCl$_3$ the frequency of the νC=O amide band is at 1672 cm^{-1} [Fig. 1(B)]. When TFA is added $(x \leqslant 0.007)$ the intensity of this band decreases with concomitant increase of a new, large band near 1622 cm^{-1}. This band is characteristic of the νC=O vibration of the 1:1 amide-acid hydrogen-bonded complexes:

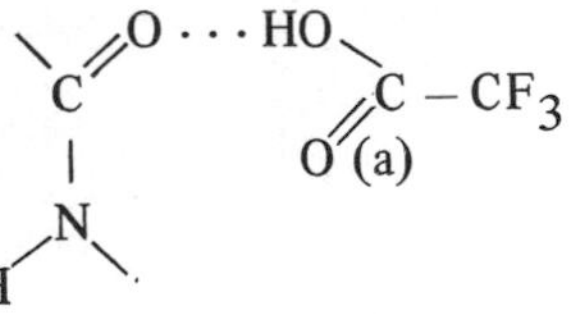

(A)

This interpretation is confirmed by the νOH [Fig. 1(D)] and νC=O [Fig. 1(C)] bands of TFA. The first ($\sim$3480 cm^{-1}) disappears on the spectra of ternary mixtures [Fig. 1(D)] and the νC=O band of acid monomer ($\sim$1805 cm^{-1}), shown in Fig. 1(C), shifts to an absorption at 1775 cm^{-1} corresponding to νC=O (a) vibrators (24, 25).

The intensity and frequency of the νNH band of NMA ($\sim$3468 cm^{-1}), shown in Fig. 1(D), do not vary significantly. This is due to the absence of an interaction of the type

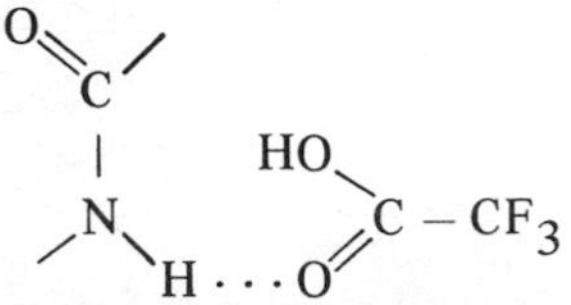

(B)

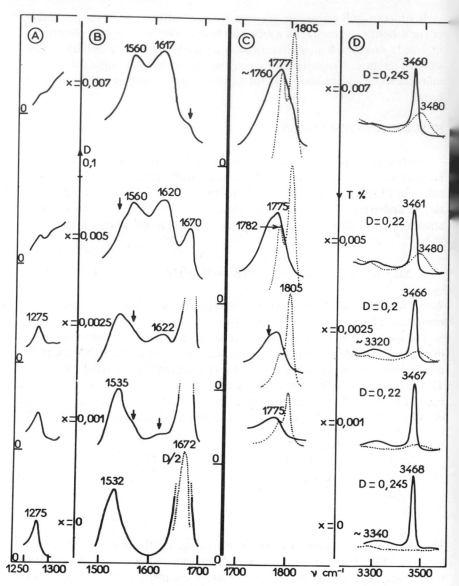

Figure 1. Spectra of mixtures of NMA (0.08 M)-TFA solvent (A, B, D: chloroform; C: methylene chloride) of weak-acid molar fractions (x). (A) Amide III, (B) amide I and II, (C) νC=O TFA, (D) amide A and νOH TFA. The dotted line shows the mixture of solvents without amide. Thermostatted (25°C) cell of ~200 μ thickness.

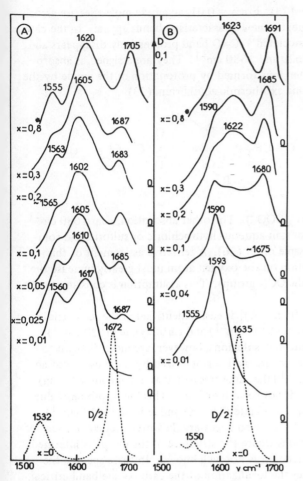

Figure 2. Evolution of the amide-I and -II absorption bands of NMA (A) and the $\nu C=O$ band of DMA (B) dissolved (0.08 M) in TFA-CHCl$_3$ mixtures of strong acid fractions (x). Thickness of the cell $\cong 200\ \mu$, amide concentration $\cong 0.2\ M$.

During the type-A association of NMA and TFA the frequencies of the amide-II and -III bands increase appreciably. The absence of a type-B interaction is explained by the participation of the νCN motion in these modes. Indeed, the interaction on carbonyl amide bond increases the electronic delocalization of OCN group and the double-bond character of the CN bond becomes stronger (26).

For high acid concentrations ($x > 0.01$) considerable alterations in the spectra are observed (Fig. 2). We compared, under the same conditions, the behavior of NMA and N,N-dimethylacetamide (DMA); the last compound had been previously investigated in such mixtures (23). For $x = 0.01$, all amide molecules are associated with TFA. When x is greater, new characteristic bands appear. In the case of DMA [Fig. 2(B)], the "associated" νC=O band progressively disappears and two other bands appear at 1620 and 1680 cm^{-1}. They are assigned as absorptions of acetate and amidinium ions formed by protonation of the amide by the acid according to the following exothermic equilibrium (23).

The band near 1620 cm^{-1} is ν_a(COO$^-$). The other absorption, near 1680 cm^{-1}, is due to the amidinium cations; in aqueous hydrochloric or sulfuric acid solutions its frequency is of the same order (26). A tentative assignment of this vibration is νC=N$^+$ (protonation on the oxygen atom must greatly increase the electronic delocalization of the OCN group). This assumption is consistent with other studies (27, 28).

The protonation of NMA [Fig. 2(A)] is easily identified in the most acidic solution ($x = 0.8$) by amidinium (1705 cm^{-1}) and carboxylate (1620 cm^{-1}) bands. Spectra of intermediate acid solutions, however, are more difficult to interpret. The protonated species are well characterized by the presence of an amidinium band near 1685 cm^{-1} (the same frequency as in aqueous solution), but the ν_a (COO$^-$) absorption is not so well defined. This fact is probably due in part to the proximity of the "associated" νC=O amide band (1617 cm^{-1}), which overlaps the anion band. It is also necessary, however, to take account of the complexity of these solutions where the structure of the ion pairs intermediate between complexes and free ions is not well known. In these species, the anion-cation interaction makes the localization of the carboxylate band critical. It is probable that the trend to form ion pairs is greater for NMA than for DMA. It is worthwhile to note for the interpretation of polypeptide spectra that the protonated peptide species are better characterized by the amidinium band than by the carboxylate band ν_a(COO$^-$).

The helix-coil transition of polypeptides was examined under the aforementioned conditions. Fig. 3 shows the spectra of poly(L-alanine) (PLA) (29). PLA is soluble in CHCl$_3$:a few percent TFA. In such solutions the helix is characterized by the amide-I absorption band near 1655 cm^{-1} [Fig. 3(A)]. Further addition of TFA decreases the intensity of this band in favor of a new, wider band near 1610 cm^{-1}. It is likely that this last band results from a specific polymer-acid interaction. The parallel study of νC=O and νOH acid bands supports this fact. As for NMA, we assign this interaction to hydrogen bonding of the type

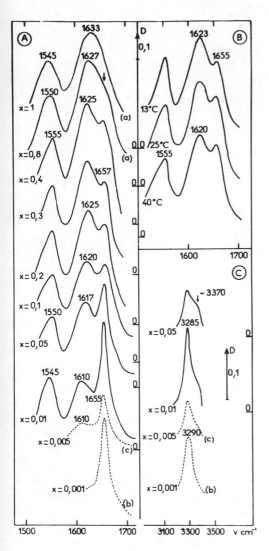

Figure 3. Spectra of PLA dissolved (∼0.08 M-g/l) in mixtures of TFA-CHCl₃.
Amide I and II (A) and amide A (C): variable acid molar fraction (x); amide I
and II (B): variable temperature (PLA = 0.08 M-g/l; x (TFA) = 0.3). Thickness
of the cell ≅200μ. (A) PLA ≅ 0.3 M-g/l, cell ≅ 50 μ. (B) PLA ≅ 0.007 M-g/l,
cell ≅2 mm. (C) PLA ≅ 0.03 M-g/l, cell ≅200 μ.

$$\begin{array}{l} \vdots \\ | \\ \text{NH} \\ | \\ \text{CHR} \\ | \\ \text{C=O} \cdots \text{HOOCCF}_3 \\ | \\ \vdots \end{array} \qquad \text{(I)}$$

This interaction occurs with some of the CONH units and competes with intramolecular association.

At higher acid concentration the PLA-TFA association is much slower than NMA-TFA associations [cf. $x = 0.01$ in Figs. 2(A) and 4(A)]. This feature is in agreement with the difference of basicity between an isolated CONH group and a CONH group in a chain. Only in pure TFA does the helical form disappear [Fig. 3(A)]; this phenomenon is related to the great stability of the right-handed helix of PLA. Indeed, the optical rotatory dispersion (ORD) curve of PLA (30) shows helix disruption in solutions containing more than 70% TFA.

The examination of amide-II [Fig. 3(A)] and amide-A [Fig. 3(C)] vibrations does not lead to any conclusion on a supplementary interaction of type B. The shoulder at 3370 cm^{-1} [Fig. 3(C)] on the 3290 cm^{-1} νNH band of the helix corresponds to the random-coil forms [Fig. 4(B)]. Such a νNH frequency is relatively low compared with the isolated model amide [Fig. 1(D)], but it is unlikely that the NH groups are involved in an association of type B because this interaction was not detected with NMA [Fig. 1(D)]. The frequency of the amide-II vibration of PLA [Fig. 3(A)] does not change much due to interaction with the peptide functions in contrast with that of NMA [Fig. 1(D)]. This observation is related to the high initial frequency of this absorption in helical structure.

Finally, in contrast to NMA, the spectra of PLA in high acid concentration [Fig. 3(A)] do not show alterations due to polymer protonation. First, the evolution of the "associated" νC=O band (1610–1630 cm^{-1}) of PLA in contrast to NMA, indicates the existence of a simple equilibrium (the frequency rise is probably due to a medium effect). Furthermore, the main argument is the absence of a new high-frequency band characteristic of amidinium cation. Previous investigations (23, 26, 31) have shown that such a band appears between 1660 and 1710 cm^{-1}. In the present case, the spectra up to a frequency of 1725 cm^{-1} do not exhibit any change.

The absence of protonation was confirmed by a study of the temperature dependence on the spectra [Fig. 3(B)]. The temperature variation was found to shift the association equilibriums (intra- and intermolecular), but was never found to induce proton transfer.

Our conclusion from these infrared data is in agreement with the hypothesis of a polypeptide-carboxylic acid hydrogen-bond interaction proposed in some experimental and theoretical studies (12, 14–16, 18–21, 32). We disagree with those (1–11) who interpret their results in terms of the protonation of the polypeptide chains (10 to 60% of the residues, according to the authors). Our analysis is based on the absence of the amidinium band in the spectra of polypeptides. Indeed, we believe that protonation can hardly be proposed, because the spectra show a band near 1620 cm^{-1} that could be a carboxylate band (5). We have shown that different species, in particular, hydrogen-bonded complexes (I), can appear near this frequency. In other respects, we believe that the overtone 2 νNH band is not well adapted to elucidate this problem. On one hand, the assignment of overtone regions is not clear for these systems; on the other hand, amide-acid interaction changes the NH group very little and changes mainly the OCN skeleton. Finally, it is surprising to note that the proposal that protonation occurs at low acid concentrations (1–5) and sometimes according to an endothermic process (5) is inconsistent with our previous results concerning the hydrogen-bond–proton transfer equilibrium in amide (or polyamide)–carboxylic acid systems (23, 31).

In contrast to ORD results (30), the evolution of the PLA-TFA association (I) [Fig. 3(A)] indicates a progressive helix-coil transition without apparent cooperativity phenomenon. This is due to the low molecular weight (~5000) and the polydispersity of the sample used. We have studied another sample of a higher degree of polymerization ($\overline{MW}$, ~35000) and compared IR and ORD results (33). There is a good correlation between IR polymer-acid association and the ORD helicity parameter (b_0). So the strong initial complexation (I) that appears in Fig. 3(A) corresponds to the transition of the shortest chains, which have weak helix stability and which offer to TFA a great number of attack sites. Previous observations of the influence of the molecular weight on the cooperativity of the transition (34–36) confirm this finding.

The backbone-acid interaction can be analyzed more precisely by the study of samples of lower polydispersity in order to localize the "binding sites" of the chains (37). The transition of PBLG is shown in Figs. 4 (by variation of acid molar fraction) and 5 (by temperature variation). In contrast to PLA, the strong cooperativity of the transformation is clearly demonstrated: the two amide-I bands [1652 cm^{-1} for the helix and 1610 cm^{-1} for the complexes (I)] are abruptly replaced by a simple, larger band near 1640 cm^{-1} for $0.08 \leqslant x \leqslant 0.1$ [Fig. 4(A)] or for $20 \leqslant t°C \leqslant 25$ [Fig. 5(A)]. This band is the νC–O amide band of the solvated coil whose carbonyl groups are associated with TFA. The frequency of the complexes is higher (1640 cm^{-1}) in this last conformation than in the ordered ones (1610 cm^{-1}) due to the abrupt change in environment and interaction state of these complexes when the transition occurs. The behavior of the amide-A [Figs. 4(B) and 5(B)] and amide-II [Fig. 4(A)] bands also shows the break-up of the intramolecular hydrogen-bonding system.

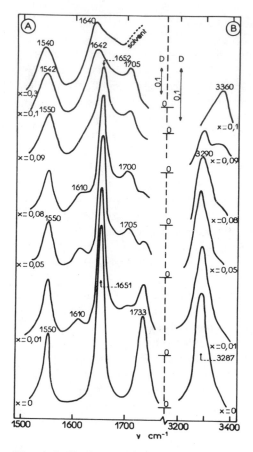

Figure 4. Evolution of $\nu C{=}O$ (A) and νNH (B) bands of PBLG dissolved (0.04 M-g/l) in TFA-CHCl$_3$ mixtures of variable acid molar fractions (x). Thermostatted (25°C) cell of ~200 μ thickness.

Before the transition, the number of complexes (I) (1610 cm^{-1}) that form [Fig. 4(A)] seems relatively low and does not increase appreciably with x—in contrast to PLA [Fig. 3(A)]. We titrated that part of the carbonyl amide group that is associated with TFA when $0 \leqslant x \leqslant 0.09$ (Fig. 6), after having verified that the amide-I band of PBLG obeys the Beer-Lambert law. If C_0 is the initial polymer concentration (referred to the repetitive unit) and C the amount of "free" polymer at equilibrium (obtained by optical density measurements), the value C/C_0 for each acidification gives the association state of the chains. Fig. 6 shows that the first acid additions to polymer solution induce the association of a significant number of amide groups: 25% of peptide units when $x = 0.02$. However, this interaction is largely stabilized for $0.02 \leqslant x \leqslant 0.07$ and increases strongly for $x > 0.07$ when helix disruption occurs. So if we consider the

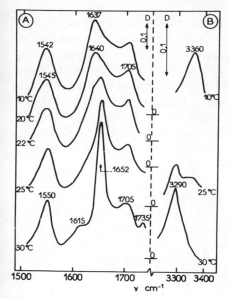

Figure 5. Effect of temperature on the spectra of PBLG dissolved (0.04 M-g/l in TFA ($x = 0.09$)-CHCl$_3$. (A) νC=O, (B) νNH.

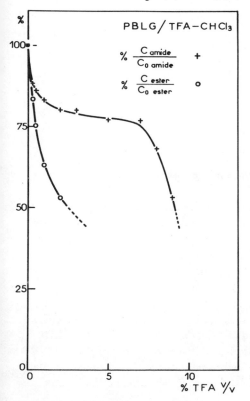

Figure 6. Measurements of amide and ester complexation amounts of PBLG dissolved (0.03 M-g/l) in TFA-CHCl$_3$ at 25°C. Cell ≅ 200 μ.

backbone-acid association before the transition, we must take into account that the acid interacts only with some amide groups. In agreement with theory (18–21), we can consider that these are peptide functions belonging to a limited number of random-coil and terminal-helical regions that have "free" carbonyl groups.

The curve in Fig. 6 can be compared with the transition curve obtained by viscosity measurements (38). Indeed, there is an initial decrease in viscosity when TFA is first added to PBLG solution, well before the transition point. This decrease is attributed to the break-up of polymer aggregates that exist in the $CHCl_3$ solution. During this process, the $-CO \cdots HN-$ bonds formed between different aggregated helical chains are replaced by $-CO \cdots HOOCCF_3$ bonds.

Analysis of polar side-chain–acid interaction during the helix-coil transition. The side chains of PBLG are characterized by the $\nu C=O$ ester absorption band at 1733 cm^{-1} [Fig. 4(A)]. Progressive additions of TFA reveal, in addition to the backbone-acid interaction, another, similar association on these side chains. Indeed, the $\nu C=O$ ester band is replaced by a band near 1705 cm^{-1}. This behavior is typical of the hydrogen-bond interaction

$$\vdots$$
$$|$$
$$NH$$
$$|$$
$$CH-(CH_2)_2-C \overset{\displaystyle O \cdots HOOCCF_3}{\underset{\displaystyle OBzl}{\big\langle}} \qquad (III)$$
$$|$$
$$CO$$
$$|$$
$$\vdots$$

which is similar to the one that takes place in the case of model compounds, such as benzyl acetate or propionate (33).

In contrast to complexes (I), the number of the complexes (III) grows according to a simple law: There is an isobestic point and quantitative measurements on the $\nu C=O$ "free" ester band (Fig. 6) show that C/C_0 decreases simply when x increases. Above $x = 0.02$, optical density determinations are inaccurate because of growing acid absorption, but the qualitative spectra (Fig. 4) show that for $x = 0.08 - 0.09$ (before the transition) most of the side chains are solvated by TFA.

The transition of PBDLG is visible in the spectra shown in Fig. 7. In $CHCl_3$ this alternating meso-copolymer studied by Heitz and Spach (22) takes a left-handed helical conformation whose stability is weaker than that of PBLG (39) (helix disruption occurs at $x = 0.04$). The behavior of the side chains is qualitatively the same as previously described. If, however, we compare the spectra

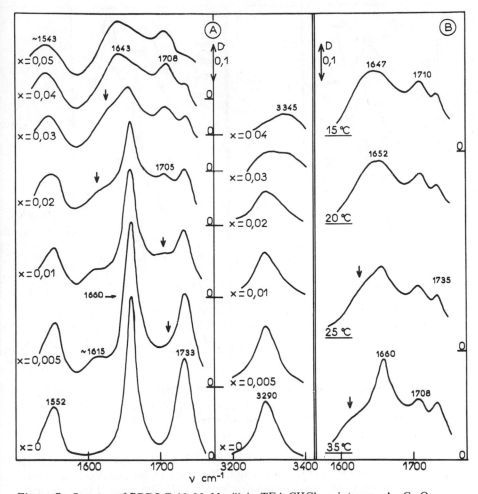

Figure 7. Spectra of PBDLG (0.03 M-g/l) in TFA-CHCl$_3$ mixtures. A νC=O and νNH regions (A): variable acid molar fractions (x) at 25°C and νC=O region (B): variable temperature [x (TFA) = 0.03]. Cell $\cong$ 2 μ.

obtained at x = 0.05 for PBLG in helical form [Fig. 4(A)] and for PBDLG in the coil state [Fig. 7(A)], we see that the number of solvated side chains seems greater for the copolymer than for the homopolymer. This difference could be due to the difference in the residue configurations in . . . L-L-L . . . and . . . D-L-D . . . sequences. Perhaps it is related also to the conformation (helix or coil) of the backbone. In this case, the proton-accepting power of the side chains would be dependent on the secondary structure of the polymer. Support for this point was obtained by the study of two PBLA samples of low and high molecular weights (Fig. 8). The left-handed helix of this polymer is not very stable: 2% TFA is enough to induce the coil. The difference in cooperativity between the transitions of the two samples is clearly apparent in the amide-I bands.

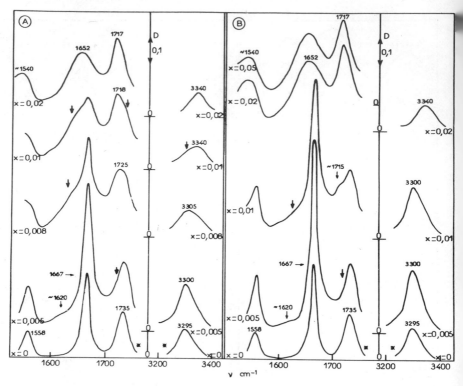

Figure 8. Spectra of PBLA (0.04 M-g/l) in TFA-CHCl$_3$ mixtures of variable acid molar fractions (x) in νC=O and νNH regions. (A) Low $\overline{MW}$, (B) $\overline{MW}$, ~150,000. Thermostatted (25°C) cell of ~200 μ thickness. *PBLA $\cong$ 0.02 M-g/l.

The aspartate side chains also interact with TFA by association of type (III). Furthermore, this interaction seems directly related to the helix-coil transition, as a comparison of the spectra obtained for x = 0.01 illustrates [Figs. 8(A) and (B)]. Indeed, we see that the amount of ester association is a function of the coil rate: For A, in which most helices are broken, the majority of the side chains are associated with TFA, but for B, which is mainly helical, the side chains are essentially free. Thus the proton-accepting power of the side chains seems particularly related to the conformational state of the backbone; it seems to be more significant when the side chains are in coil regions of the polymer rather than in helix segments. Furthermore, the shorter the side chain, the greater the effect, as shown by the glutamate-aspartate comparison. This behavior is perhaps due to the structure of side chains around the helix of polymers. Various experimental and theoretical studies (40–44) have been done to determine side-chain conformations as a function of the possible interactions (backbone-side-chain or side-chain–side-chain). These interactions might decrease the complexing power of the carbonyl ester groups belonging to helical regions. When helix disruption occurs, the conformation of the side chains is destroyed and their association with TFA is increased.

CONCLUSION

An IR study was carried out on the solvation of polypeptides dissolved in TFA-CHCl$_3$ mixtures that induce helix-coil transitions. The analysis of the fundamental amide and acid vibrations, which are sensitive to conformation and molecular interactions, indicates complexing properties not only for the backbone, but also for polar side chains.

We first compared the acido-basic properties of the peptide backbone and a monomer model compound (N-methylacetamide). The results show that the transition between helical chains and coiled chains is induced by the formation of polymer-acid hydrogen-bonded complexes that compete with intramolecular association. Moreover, there is no protonation of the peptide functions by acid, as there appears to be with the model amide. This result is interesting in view of the controversy about this point and the prominent part played by protonation in the mechanism of the transition proposed by many investigators. We concur with the assumption of the theory that only the peptide units belonging to random-coil and terminal-helical regions are capable of interaction with the acid (the regions inside the helix seem unable to form complexes). Although it is generally admitted that the CO and NH sites of these units have identical complexing powers ($K_{CO} = K_{NH}$), our results indicate that the main association takes place on the amide carbonyl groups (K_{CO}); the NH groups do not seem to interact ($K_{NH} \cong 0$ or $\ll K_{CO}$).

The behavior of the polar side chains during the transition shows that they interact with the acid in a process similar to that of the backbone: They form hydrogen-bonded complexes on their basic sites. Furthermore, the association amount is related to the secondary structure of the polymer. Indeed, the association constant seems to be more important in coil regions of the polypeptide than in helix regions. This difference depends on the length of the side chain, which is perhaps due to its conformation and various interaction types that can exist in the ordered structure. The analogous behavior of side chains and backbone with respect to the acid has not been considered previously. Therefore, a precise theoretical development of the helix-coil transition induced by preferential association should take into account this result by introducing two more association constants in the model (k_1 for the helix regions and $k_2 > k_1$ for the coil regions).

ACKNOWLEDGMENTS

We are grateful to E. Marchal (Centre de Recherches sur les Macromolécules, Strasbourg), F. Heitz and G. Spach (Centre de Biophysique Moléculaire, Orléans) and Mr. Marraud (Laboratoire de Chimie Physique Macromoléculaire, Nancy) who supplied most of the samples studied in this work and made ORD measurements of PLA.

REFERENCES

1. S. Hanlon, S.F. Russo, and I.M. Klotz, *J. Am. Chem. Soc.*, **85**, 2024 (1963).
2. S. Hanlon and I.M. Klotz, *Biochemistry*, **4**, 37 (1965).
3. S. Hanlon, *Biochemistry*, **5**, 2049 (1966).
4. S. Hanlon and I.M. Klotz, *Developments in Appl. Spectros.*, **6**, 219 (1968).
5. B.Z. Volchek and A.V. Purkina, *Vysokomol. Soed.*, **A9**, 1257 (1967) and **A11**, 1563 (1969).
6. J.W.O. Tam and I.M. Klotz, *J. Am. Chem. Soc.*, **93**, 1313 (1971).
7. J.H. Bradbury and M.D. Fenn, *Aust. J. Chem.*, **22**, 357 (1969).
8. J.H. Bradbury and B.E. Chapman, *J. Macromol. Sci. Chem.*, **A4**, 1137 (1970).
9. J.C. Haylock and H.M. Rydon, *Colloq. Int. Centre Nat. Rech. Sci.*, **175**, 19 (1968).
10. M.A. Stake and I.M. Klotz, *Biochemistry*, **5**, 1726 (1966).
11. H. Watanabe, K. Yoshioka, and A. Wada, *Biopolymers*, **4**, 43 (1966) and **2**, 91 (1964).
12. W.E. Stewart, L. Mandelkern, and R.E. Glick, *Biochemistry*, **6**, 143, 150 (1967).
13. A. Takahashi, L. Mandelkern, and R.E. Glick, *Biochemistry*, **8**, 1673 (1969).
14. E.M. Bradbury and H.W.E. Rattle, *Polymer*, **9**, 201 (1968).
15. J.A. Ferretti and L. Paolillo, *Biopolymers*, **7**, 155 (1969).
16. F.A. Bovey, *Pure Appl. Chem.*, **16**, 417 (1968).
17. F. Quadrifoglio and D.W. Urry, *J. Phys. Chem.*, **71**, 2364 (1967).
18. J.A. Schellmann, *Compt. Rend. Trav. Lab. Carlsberg, Ser. Chim.*, **29**, 230 (1955) and *J. Phys. Chem.*, **62**, 1485 (1958).
19. M. Bixon and S. Lifson, *Biopolymers*, **4**, 815 (1966).
20. L. Peller, *J. Phys. Chem.*, **63**, 1194, 1199 (1959).
21. J.H. Gibbs and E.A. Dimarzio, *J. Chem. Phys.*, **30**, 271 (1959).
22. F. Heitz and G. Spach, *Macromolecules*, **4**, 429 (1971).
23. P. Combelas, F. Cruege, J. Lascombe, C. Quivoron, and M. Rey-Lafon, *J. Chim. Phys.*, **66**, 668 (1969).
24. M. Haurie, Thèses Doctorat (Bordeaux), 1962 and 1966.
25. M. Kirszenbaum, Thèse 3ème cycle (Paris), 1971.
26. C. de Loze, P. Combelas, P. Bacelon, and C. Garrigou-Lagrange, *J. Chim. Phys.*, **69**, 397 (1972).
27. R. Stewart and L.J. Muenster, *Chem. and Ind.*, 1906 (1961).
28. D. Cook, *Canad. J. Chem.*, **42**, 2721 (1964).
29. P. Combelas, C. Garrigou-Lagrange, and J. Lascombe, *Biopolymers*, **12**, 611 (1973).
30. G.D. Fasman, in *Polyamino Acids, Polypeptides and Proteins*, M.A. Stahmann, ed., Univ. Wisconsin Press, Madison, Wis., 1962, p. 221.
31. P. Combelas, F. Cruege, J. Lascombe, C. Quivoron, M. Rey-Lafon, and B. Sebille, *Spectrochim. Acta*, **26A**, 1323 (1970).
32. C. Lapp and J. Marchal, *J. Chim. Phys.*, **62**, 1032 (1965).
33. P. Combelas, Thèse de Doctorat d'Etat, C.N.R.S. AO 5481 (1973).

34. T.M. Birshtein and O.B. Ptitsyn, *Conformations of Macromolecules,* Wiley(Interscience), New York, 1966, Chapter 9.
35. B.H. Zimm and J.K. Bragg, *J. Chem. Phys.,* **31**, 526 (1959).
36. S. Lifson and A. Roig, *J. Chem. Phys.,* **34**, 1963 (1961).
37. P. Combelas, C. Garrigou-Lagrange, and J. Lascombe, *Biopolymers,* in press.
38. P. Doty, *Collection Czech. Chem. Commun.,* **22**, 5, 11 (1957).
39. F.A. Bovey, J.J. Ryan, G. Spach, and F. Heitz, *Macromolecules,* **4**, 433 (1971).
40. J.F. Yan, G. Vanderkooi, and H.A. Scheraga, *J. Chem. Phys.,* **49**, 2713 (1968).
41. G. Govil, *J. Ind. Chem. Soc.,* **48**, 731 (1971).
42. D.N. Silverman and H.A. Scheraga, *Biochemistry,* **10**, 1340 (1971).
43. D.N. Silverman, G.T. Taylor, and H.A. Scheraga, *Arch. Biochem. Biophys.,* **146**, 587 (1971).
44. E.M. Bradbury, B.G. Carpenter, C. Crane-Robinson and H. Goldman, *Nature* (London), **225**, 64 (1970).

AN EXPERIMENTAL STUDY OF THE INTERNAL ROTATION POTENTIALS ABOUT THE N–C_α AND C_α–C' AXES OF THE PEPTIDE BACKBONE

YASUSHI KOYAMA, Department of Chemistry, Faculty of Science, Kwansei Gakuin University, Nishinomiya, Japan, and TAKEHIKO SHIMANOUCHI, Department of Chemistry, Faculty of Science, University of Tokyo, Hongo, Tokyo, Japan

SYNOPSIS: The molecular conformations of N-ethylacetamide ($CH_3C'ONH–CH_2CH_3$), N-methylpropionamide ($CH_3CH_2–C'ONHCH_3$), N-methylchloro-acetamide ($ClCH_2–C'ONHCH_3$), N-acetylglycine methylamide ($CH_3C'ONH–CH_2–C'ONHCH_3$), and N-acetylalanine methylamide [$CH_3C'ONH–CH(CH_3)–C'ONHCH_3$] were studied by means of electron and x-ray diffraction methods and infrared and Raman spectroscopy. Internal rotation potentials about the N–C_α and C_α–C' axes of the peptide backbone are discussed, based on the results of those studies. The following conclusions are drawn: The N–C_α axis has a potential minimum near the gauche position; the C_α–C' axis has two potential minima, one at *cis* and the other at *trans* position; and the energy difference between the minima is small and dependent on intermolecular interactions.

INTRODUCTION

Studies were designed to clarify the forces determining the conformation of protein molecules on a purely experimental basis. It was necessary, therefore, to select a number of systems in which different types of forces predominate, that is, to select suitable model compounds and environments. In this paper, attention is confined mainly to the following types of intramolecular forces: internal rotation potentials inherent in the rotational axes, which depend on the electronic configuration of the molecule; van der Waals interaction between non-bonded atoms; electrostatic interaction between bond dipoles; and intramolecular hydrogen bonds. The importance of intermolecular interactions also is discussed.

The following model compounds were selected: N-ethylacetamide ($CH_3C'ONH–CH_2CH_3$), which represents molecules having the rotational axis N–C_α, N-methylpropionamide ($CH_3CH_2–C'ONHCH_3$) and N-methylchloro-acetamide ($ClCH_2–C'ONHCH_3$), which represent molecules having the rotational axis C_α–C', and N-acetylglycine methylamide ($CH_3C'ONH–CH_2–C'ONHCH_3$) and N-acetylalanine methylamide [$CH_3C'ONH–CH(CH_3)–C'ONHCH_3$], which contain both of these rotational axes.

Experimental techniques employed for the determination of the molecular conformation of the model compounds are: electron-diffraction analysis, which applies to the gaseous phase of molecules with a low molecular weight and provides rotational angles free from the influence of intermolecular interactions; x-ray-diffraction analysis, which gives the rotational angles in the crystalline state under the influence of intermolecular forces; infrared and Raman spectroscopy, which are applicable to the liquid and crystalline states, as well as in solution, and provide information on the existence of rotational isomers and their conformations; and calculation of normal vibrations of the molecules, which affords approximate rotational angles when the force constants employed are reliable. Thus, definite information concerning the position of the minima in the internal rotation potential is obtained by diffraction methods, whereas data concerning the positions of the minima and the energy difference between them derives from vibrational spectroscopy.

In this paper, the internal rotation angles about the $N-C_\alpha$ and $C_\alpha-C'$ axes are specified by the ϕ and ψ defined by Edsall *et al.* (1) to maintain consistency with our previous papers (2–4). In the case of the characteristic vibrations of the amide group, the abbreviations presented by Miyazawa (5) for *N*-methylacetamide are employed.

MOLECULAR CONFORMATION OF *N*-ETHYLACETAMIDE

Liquid and solid states. The conformation of *N*-ethylacetamide in the liquid and solid states has been studied by means of infrared spectroscopy and calculation of normal vibrations (6). No significant differences were observed between the spectra for the solid and liquid states at various temperatures. The number of observed bands was just that expected for one conformer. Thus, the molecule takes only one conformation in the liquid and solid phases. Spectra for the liquid state at room temperature are shown in Fig. 1. The bands at 936, 892, and 773 cm^{-1} are associated with the methylene-rocking and skeletal-stretching vibrations. The band at 773 cm^{-1}, which is overlapped by the amide-V band, appears clearly in the spectrum of the N-deuterated compound (Fig. 2). The band at 617 cm^{-1} (shoulder) was assigned to amide IV and the band at 602 cm^{-1} to amide VI because the latter disappears upon N-deuteration, whereas the former remains at 616 cm^{-1}. Upon C$'$-methyl deuteration, these bands shift and split into bands at 584 and 537 cm^{-1}, respectively. The bands at 464, 382, and 303 cm^{-1} are related to the three skeletal-deformation vibrations ($C-C'-N$, $C'-N-C$, and $N-C-C$).

The calculations of normal vibrations were carried out according to Wilson's GF matrix method (7) for models with the rotational angle $\phi = 0°$ (*trans*), $60°$, $120°$ (gauche), and $180°$ (*cis*). The potential energy matrix F was always fixed and only the kinetic energy matrix G was varied. The force constants were transferred from *N*-methylacetamide (8) and propane (9) molecules. They were determined for the liquid state and involve the latent effect of intermolecular

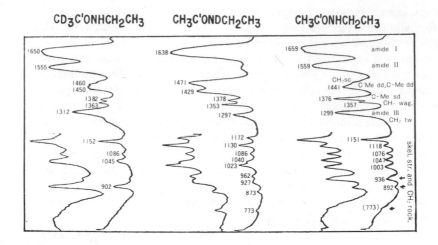

Figure 1. Infrared spectra of N-ethylacetamide and its deuterium homologs in the 1750–650 cm^{-1} region. (Reprinted with permission. Copyright by Kodansha Ltd., Tokyo.)

interactions. The frequencies of the observed conformation-sensitive bands are compared with those calculated for the above models in Fig. 3(a). The absolute values of the calculated frequencies themselves cannot be compared with the observed frequencies, because the force constants used in the calculation were values transferred from other molecules without further refinement. We can, however, compare the relative arrangement of the calculated and observed frequencies that are associated with skeletal vibrations, as determined by the geometry of the molecule through the coupling of vibrations.

The model with the rotational angle $\phi = 120°$ was selected for N-methyl-acetamide because the coupling between the methylene-rocking and skeletal-stretching vibrations depends strongly on the molecular conformation. The model with $\phi = 120°$ best fits the general pattern of the observed band arrangement. The amide-IV, amide-VI, and three skeletal-deformation vibrations are also coupled with each other and are sensitive to conformation. For models with $\phi = 0°$ and $180°$, the frequency associated with the lowest skeletal-deformation vibration appears below 200 cm^{-1}. These models were therefore rejected. For the model with $\phi = 120°$, the separations of the calculated frequencies correspond roughly with those observed. If the molecule takes a *trans* or *cis* conformation, vibration coupling between CH$_2$ wagging and amide III would be expected because the former vibration occurs parallel to the amide plane. As Fig. 1 shows, the strong amide-III band appears at 1299 cm^{-1} and is overlapped by the CH$_2$ twisting band, which appears at 1297 cm^{-1} in the spectrum of the N-deuterated compound. The CH$_2$-wagging band in question appears at 1357 cm^{-1} and remains at 1353 cm^{-1}, unaffected by N-deuteration,

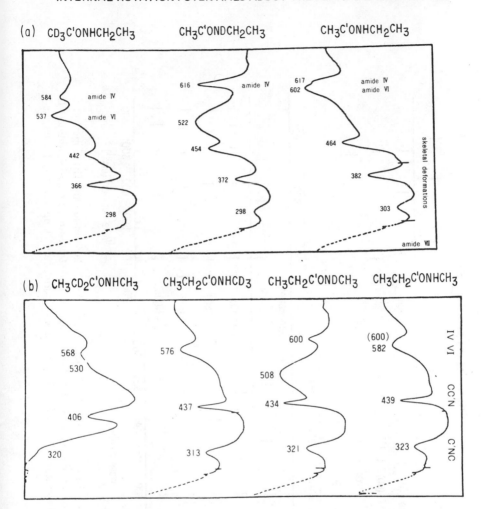

Figure 2. Infrared spectra of (a) *N*-ethylacetamide, (b) *N*-methylpropionamide, and their deuterium homologs in the 700–200 cm^{-1} region. (Reprinted with permission. Copyright by Kodansha Ltd., Tokyo.)

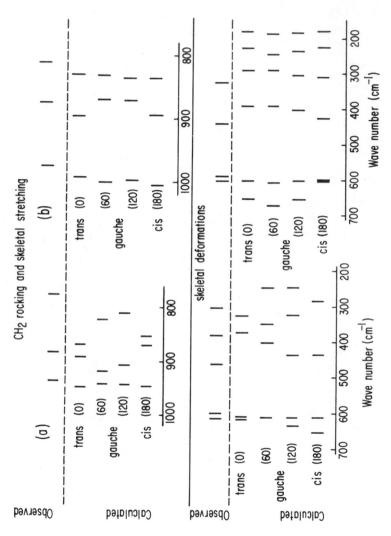

Figure 3. Observed and calculated frequencies for (a) *N*-ethylacetamide and (b) *N*-methyl-propionamide, which are related with CH_2-rocking, skeletal-stretching, amide-IV, amide-VI, and skeletal-deformation vibrations. (Reprinted with permission. Copyright by Kodansha Ltd., Tokyo.)

although the amide III band shifts to as low as ~1000 cm^{-1}. These facts indicate that no such coupling occurs and that the molecule is not planar.

In conclusion, it can be said that this molecule takes a gauche conformation with the rotational angle ϕ near $120°$ in the liquid and solid states. The angle may deviate to smaller angles, however, and the model with $\phi = 60°$ represents a possible second choice.

Gaseous state. The conformation of the N-ethylacetamide molecule in the gaseous state has been studied by electron diffraction. The sample was heated to about $150°$C and then led through a high temperature nozzle (10). The diffraction photographs were taken with 40 KV electrons at camera distances of 112.70 mm (short) and 247.83 mm (long), and the intensity was measured by a digital microphotometer. The molecular intensity data obtained in the ranges $q = 7-65$ Å^{-1} (long) and 25-90 Å^{-1} (short) were joined at $q = 33$ Å^{-1} and a radial distribution curve was then derived. Theoretical radial distribution curves were calculated also for the four models with $\phi = 0°, 60°, 120°,$ and $180°$. The mean amplitudes and shrinkage effect of the molecular vibrations were calculated for each model with the Lx matrix obtained in the calculation of normal vibrations.

The results of these calculations indicate that the radial distribution curves in the 3.0-5.5 Å region are very sensitive to conformation. The observed and calculated curves in this region are compared in Fig. 4(a). The models with the rotational angles $\phi = 0°, 60°$ are rejected because three distinct peaks would be expected to appear in this region. The remaining models with $\phi = 120°, 180°$ have two peaks for C–C$_\alpha$ and C–C atom pairs. The observed radial distribution curve is satisfactorily explained in terms of the model with $\phi = 120°$ (gauche). Further analysis revealed that the rotational angle ϕ is smaller than $120°$. Refinement of the molecular parameters is now in progress, although it appears certain that the molecule takes a gauche conformation in the gaseous state.

MOLECULAR CONFORMATION OF N-METHYLPROPIONAMIDE

Liquid and solid states. The conformation of N-methylpropionamide in the liquid and solid states has been studied by means of infrared (IR) spectroscopy and the calculation of normal vibrations (6). The molecule is believed to take only one conformation in the liquid and solid states based on the reasoning elaborated above for N-ethylacetamide. The IR spectra for the liquid state, including the deuterium homologs CH_3CD_2–$C'ONHCH_3$, CH_3CH_2–$C'ONHCD_3$, and CH_3CH_2–$C'ONDCH_3$, at room temperature are shown in Figs. 2(b) and 5. The bands at 971, 869, and 805 cm^{-1} are associated with methylene-rocking and skeletal-stretching vibrations. The broad, strong band near 582 cm^{-1} is considered an overlap of the amide-IV and amide-VI bands. These bands split into two bands at 568 and 530 cm^{-1} upon methylene deuteration. A sharp, weak band remaining in the spectrum of the N-deuterated homolog was assigned to amide IV. The bands at 439 and 323 cm^{-1} are associated

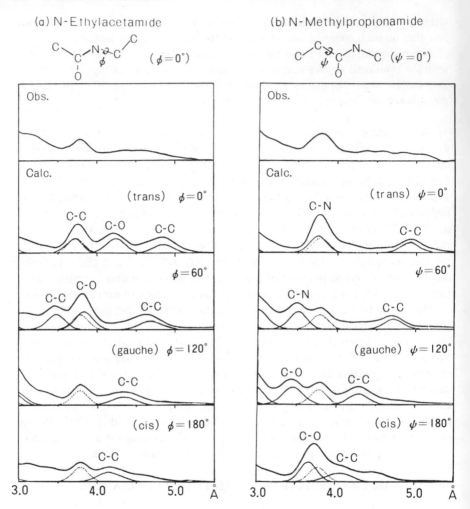

Figure 4. Observed and calculated radial distribution curves for (a) *N*-ethylacet-amide and (b) *N*-methylpropionamide in the 3.0–5.0 Å region. (Reprinted with permission. Copyright by Kodansha Ltd., Tokyo.)

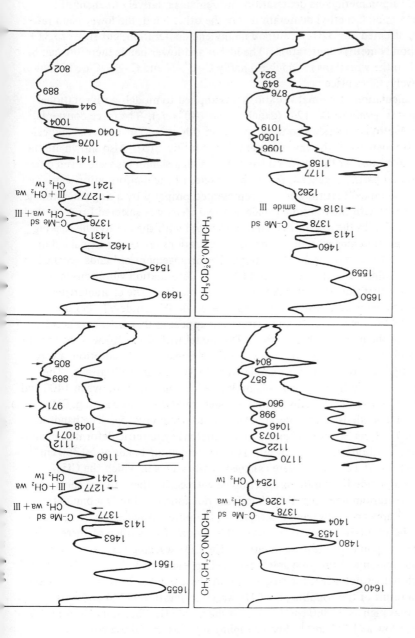

Figure 5. Infrared spectra of *N*-methylpropionamide and its deuterium homologs in the 1750–650 cm^{-1} region. (Reprinted with permission. Copyright by Kodansha Ltd., Tokyo.)

with the skeletal-deformation vibrations. The higher band shifts to as low as 406 cm^{-1} upon methylene deuteration, but remains relatively unchanged at 437 cm^{-1} upon C-methyl deuteration. On the other hand, the lower band remains unaffected at ~320 cm^{-1} upon methylene deuteration, but shifts to 313 cm^{-1} upon N-methyl deuteration. The higher and lower bands therefore can be assigned to the vibrations in which primarily C–C′–N and C′–N–C deformation, respectively, takes place.

The calculation of normal vibrations was applied to models with rotational angles $\psi = 0°$ (*trans*), $60°$, $120°$ (gauche), and $180°$ (*cis*). The force constants used for N-ethylacetamide also were applied in this case. The observed and calculated frequencies of the vibrations are shown in Fig. 3(b). Also shown are the vibrational modes for the observed bands to which assignments were given on an experimental basis. In general, the normal modes of the conformation-sensitive bands are so mixed that they are not represented properly by a single coordinate.

The model with the rotational angle $\psi = 180°$ (*cis*) was selected for N-methylpropionamide. In the case of the bands associated with the methylene-rocking and skeletal-stretching vibrations, the band separations are better explained in models with the $\psi = 0°$ or $\psi = 180°$ (*cis*). The arrangement of bands assigned to amide-IV, amide-VI, and C–C′–N and C′–N–C deformation vibrations indicates clearly that the molecule takes a *cis* conformation. The characteristics of the observed band arrangement (viz., that the amide-IV, amide-VI, and the C–C′–N and C′–N–C deformation vibrations appear in this order) are accommodated by the model with $\psi = 180°$. The model with $\psi = 120°$ was the second choice, and models with $\psi = 0°$ and $\psi = 60°$ were rejected. In addition, a very strong coupling was observed for amide-III and CH$_2$-wagging vibrations. The undeuterated N-methylpropionamide molecule has four bands between 1400 and 1200 cm^{-1}, where the C-methyl symmetric-deformation, CH$_2$-wagging, CH$_2$-twisting, and amide-III vibrations are expected to occur (Fig. 5). The bands at 1377 and 1241 cm^{-1} were assigned to the C-methyl symmetric-deformation and CH$_2$-twisting vibrations, respectively. The two remaining bands at ~1360 cm^{-1} (shoulder) and at 1277 cm^{-1} were assigned to vibrations in which the CH$_2$-wagging and amide-III vibrations are coupled with each other. In the spectrum of the N-deuterium homolog, these bands are not found and only a band at 1326 cm^{-1} appears. This band is assigned to the pure CH$_2$-wagging vibration, because the amide-III band moves out of this region. Decoupling of these vibrations occurs upon methylene deuteration. The CD$_2$-wagging band then moves out of this region and the pure amide-III band appears at 1318 cm^{-1}. N-methyl deuteration does not affect the frequencies of these bands. These facts indicate that, as a result of the coupling, the CH$_2$-wagging vibration (1326 cm^{-1}) is pushed to as high as ~1360 cm^{-1}, whereas the amide-III vibration (1318 cm^{-1}) shifts to as low as 1277 cm^{-1}. Such coupling suggests that these vibrations should take place in the same plane; in other words, the rotational angle should be $0°$ or $180°$.

All of the above results are consistent with each other and indicate that the N-methylpropionamide molecule takes a *cis* conformation with the rotational angle ψ near $180°$. However, a certain deviation to smaller angles is naturally to be expected.

Gaseous state. The conformation of N-methylpropionamide in the gaseous state was studied by electron diffraction. The procedure was the same described above for N-ethylacetamide. The observed and calculated radial distribution curves in the 3.0–5.5 Å region are compared in Fig. 4(b). On the basis of the theoretical radial distribution curves calculated for $\psi = 0°, 60°, 120°$, and $180°$, the models with $\psi = 60°$ and $120°$ were rejected because they do not have three distinct peaks of atom pairs [C–N (C–O), C_α–C, and C–C for $\psi = 60°$ (120°)] in the 3.0–5.0 Å region. The model with $\psi = 0°$ (180°) has a strong peak near 3.7 Å, which arises from the atom pairs C_α–C and C–N (C_α–C, C–O, and C–C). The profile of the curves for these models corresponds to that observed. Thus, a choice between the models with $\psi = 0°$ and $\psi = 180°$ could not be made on the basis of their curve profiles. The possibility remained of a coexistence of *trans* and *cis* rotational isomers. Further analysis showed, however, that the observed curve could be explained satisfactorily in terms of the model with $\psi = 150°$. Refinement of the molecular parameters is still in progress.

MOLECULAR CONFORMATION OF N-METHYLCHLOROACETAMIDE

Crystalline state at room temperature. Full discussions of the x-ray analysis have been published elsewhere (3) and only the results are summarized here. The crystal has a monoclinic space group $P2_1/n$ and lattice constants $a = 7.55$, $b = 27.26, c = 5.11$ Å and $\beta = 106.1°$. The intensities were measured visually and corrected for the Lorentz and polarization factors. The crystal structure was solved by the Patterson function method with 513 independent structure factors. Refinement of the structure was based on the full-matrix least-squares method, in which the individual anisotropic temperature factors were included. The final R factor was 0.09.

The crystal contains two crystallographically independent molecules in the asymmetric unit, which we denote as molecules I and II. The bond lengths and bond angles for the molecules I and II are given in Fig. 6. The molecular parameters involved in the peptide group are in agreement with the normal values, within the limits of experimental error, except that the values for the length of the C'–N bonds (1.39 and 1.37 Å) are rather large. The values of the bond angles of Cl–C_α–C' (115°) are also somewhat large. These anomalies may be ascribed to intramolecular van der Waals repulsions due to the *cis* conformation described below.

The rotational angles about the C–C' axes are $\psi = 179°$ for molecule I and $\psi = 160°$ for molecule II. These values indicate that the molecules I and II are both *cis* and that the difference between their molecular conformations is small; short intermolecular contacts are probably responsible for the difference.

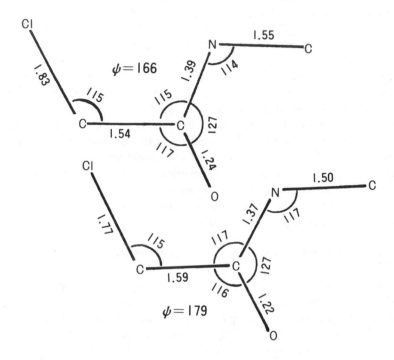

Figure 6. Bond lengths, bond angles and the angle ψ for molecules I and II of *N*-methylchloroacetamide. (Reprinted with permission. Copyright by Kodansha Ltd., Tokyo.)

Gaseous state. The conformation of *N*-methylchloroacetamide in the gaseous state was studied by electron diffraction. The observed and calculated radial distribution curves are shown in Fig. 7. Theoretical radial distribution curves were calculated for the models with the rotational angles $\psi = 0°, 60°, 120°$, and $180°$. The average bond angles and bond lengths were transferred from the crystalline form of molecules I and II. As Fig. 7 shows, only the model with $\psi = 180°$ explains the observed radial distribution curve. Thus, the molecule takes a *cis* conformation. The difference between the observed and calculated curves in the 1.0–3.0 Å region is attributed to differences in bond lengths for the crystal and the gas phase. The bond lengths for the C′–N and C′=O bonds in the gas phase are near 1.39 and 1.21 Å, which are in agreement with those for

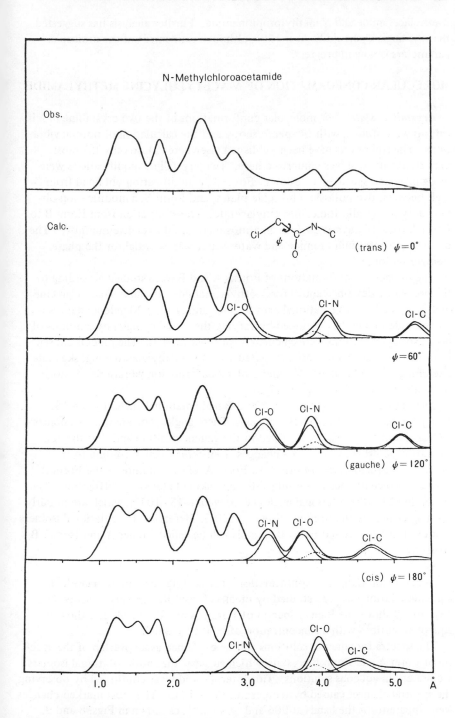

Figure 7. Observed and calculated radial distribution curves for N-methyl-chloroacetamide in the 0.0–5.5 Å region. (Reprinted with permission. Copyright by Kodansha Ltd., Tokyo.)

N-ethylacetamide and *N*-methylpropionamide. Further analysis has suggested that the angle ψ should be near 150°, although refinement of the molecular parameters is now in progress.

MOLECULAR CONFORMATION OF *N*-ACETYLGLYCINE METHYLAMIDE

Crystalline state. The molecular conformations of the two crystalline modifications were studied with IR spectroscopy and the calculation of normal vibrations. The full results have been published elsewhere (2), so only the most important ones will be summarized here. Two crystalline modifications were recognized and specified as follows: Form A, a modification obtained from a melt between two potassium bromide plates; and Form B, a modification obtained by recrystallization from ethylacetate. Transformation from Form B to Form A took 10 days at 20°C in an atmosphere of 50% relative humidity. The presence of potassium bromide and water vapor was essential for the phase transformation.

The molecular conformations of Forms A and B were studied according to the procedure described above for *N*-ethylacetamide. Because *N*-acetylglycine methylamide has two rotational axes (C_α–C' and N–C_α) 20 independent isomeric models could be developed by varying the ψ and ϕ angles by amounts of 60°. Normal vibrations were calculated for these models. The models chosen for our study had frequencies associated with the methylene-rocking, skeletal-stretching, amide-IV, amide-VI, and skeletal-deformation vibrations. Isotope shifts also were used in this case (2).

The models selected for Forms A and B had rotational angles ($\psi,\phi = 180°$, 120°) and ($\psi,\phi = 0°, 120°$), respectively. These angles indicate that the conformation about the N–C_α axis is fixed in the gauche position and that the conformation about the C_α–C' axis changes from *trans* to *cis* according to the phase transformation from Form B to Form A. The structure of the Form B crystal has recently been determined by Iwasaki (11) by x-ray diffraction. The values given for the rotational angles (ψ,ϕ) were (-15°,103°), which are in fairly good agreement with our prediction (0°,120°). Hereafter, the rotational isomers that exist in Form A and Form B crystals will be called Isomer A and Isomer B, respectively.

Aqueous solution. The conformation of *N*-acetylglycine methylamide in aqueous solution has been studied by means of laser Raman spectroscopy (12, 13). Fig. 8 shows the Raman spectra of Forms A and B crystals and those of aqueous solutions with a concentration of ~30% by weight.

The spectra for aqueous solutions can be explained as an overlap of the spectra of Form A and Form B crystals, which suggests that both rotational isomers are present in aqueous solutions. This interpretation was confirmed by observing the spectral changes caused by temperature variations. The most marked changes were apparent for the bands at 906 and 869 cm^{-1}, as shown in Figs. 8 and 9.

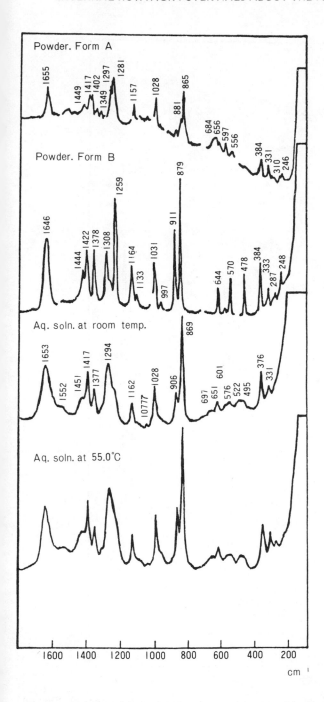

Figure 8. Raman spectra of crystalline Forms A and B and aqueous solutions of
N-acetylglycine methylamide. (Reprinted with permission. Copyright by
Kodansha Ltd., Tokyo.)

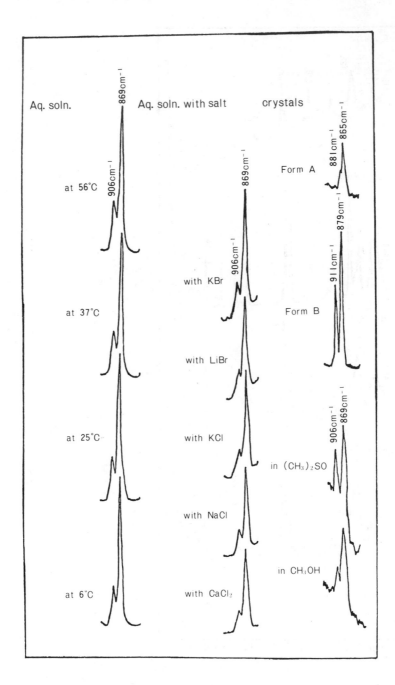

Figure 9. Temperature dependence and solvent effect of the bands at 906 and 869 cm^{-1} for solutions of N-acetylglycine methylamide. (Reprinted with permission. Copyright by Kodansha Ltd., Tokyo.)

An increase in temperature causes the ratio of the 906 cm^{-1} band intensity to that of the 869 cm^{-1} band to increase. This phenomenon can be explained in terms of a change in the proportions of the two isomers. Form A has a strong band at 865 cm^{-1} with a shoulder at 881 cm^{-1}, whereas Form B has two widely separated bands at 911 and 879 cm^{-1}. The band at 906 cm^{-1} for the solution must correspond to the 911 cm^{-1} band of Form B. The solution also has a broad and strong band at 869 cm^{-1}, which may be composed of the remaining bands (879 cm^{-1} of Form B and 865 and 881 cm^{-1} of Form A). If this assignment is correct, the intensity ratio should increase when the amount of Isomer B increases relative to that of Isomer A. Therefore, the above results indicate that the proportion of Isomer B increases at higher temperatures; in other words, Isomer B is less stable than Isomer A. The spectral changes suggest also that the energy difference between the two rotational isomers is small.

The intensity ratio of the 906 cm^{-1} band to the 869 cm^{-1} band increases in less polar solvents such as dimethylsulfoxide or methanol. This increase indicates that the proportion of Isomer B increases in less polar solvents, a reasonable conclusion in view of the fact that the magnitude of the dipole moment of Isomer B should be smaller than that of Isomer A. It suggests also that interactions with the solvent molecules may affect the energy difference between the isomers. Spectral changes were found also when salts such as potassium bromide, potassium chloride, sodium bromide, sodium chloride and calcium chloride were added to the aqueous solution. The relative intensity of the 906 cm^{-1} band diminishes, suggesting that Isomer A is stabilized by the addition of salts.

MOLECULAR CONFORMATION OF *N*-ACETYLALANINE METHYLAMIDE

Crystalline state. Infrared spectra of crystalline *N*-acetyl amino acid methylamides (CH$_3$C$'$ONH–CHR–C$'$ONHCH$_3$) with various side chains, R, have been analyzed and discussed in relation to hydrogen bond formation in crystals (4). Assignments for the observed bands also have been described (4).

The existence of two crystalline modifications for *N*-acetylalanine methylamide was found spectroscopically (Fig. 10) and confirmed by differential thermal analysis. The two forms, so-called Forms A and B, may be specified as follows: Form A, the form that is stable from 128°C to 182°C, the melting point; and Form B, the form obtained by crystallization from ethyl acetate and methanol that is stable at room temperatures. Spectral differences between the two forms are conspicuous in the region where the skeletal-stretching and skeletal-deformation vibrations appear, so this transition must be accompanied by some rotational isomerism. Crystalline *N*-acetyl-DL-alanine methylamide shows a Form-B spectrum and does not have such a transition up to the melting point, 162°C.

As predicted by Ramachandran *et al.* (14, 15) and other investigators (16), the interaction between the β-carbon atom and atoms constituting the peptide groups restricts the rotational angles (ψ,ϕ) to narrow ranges. Calculations of

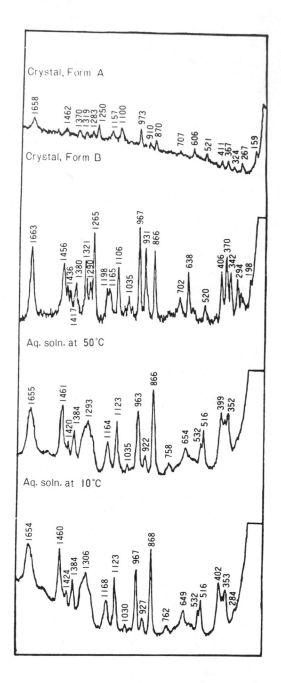

Figure 10. Raman spectra of crystalline Forms A and B and aqueous solutions of *N*-acetylalanine methylamide. (Reprinted with permission. Copyright by Kodansha Ltd., Tokyo.)

normal vibrations were attempted for the configurational models with rotational angles $(\psi,\phi) = (120°,120°)$, $(300°,60°)$, and $(240°,240°)$, which represent the three sterically allowed regions. The frequencies of the amide-IV, amide-VI, and skeletal-deformation vibrations are compared with those of the observed bands in Fig. 11. The frequencies for vibrations that are related primarily to deformation of the $C_\alpha-C_\beta$ bond were not considered because they are missing from the observed spectra; they would be expected to appear near 450 cm^{-1}. The comparison shows that the possibility of taking a conformation in which the angles (ψ,ϕ) are near $(240°,240°)$ can be negated both for Form A and for Form B. This conformation corresponds to that of the left-handed α helix of polypeptide chains and must be sterically unstable. Assignment of the remaining two models to Forms A and B cannot be made definitely. The frequencies suggest that the molecule may take α-helixlike $(120°,120°)$ and β-sheetlike $(300°,60°)$ conformations, respectively, in the crystal Forms A and B.

The x-ray crystal structure of N-acetylalanine methylamide (Form B) has recently been determined by Harada and Iitaka (17). The molecule takes a β-sheetlike conformation in both the L and DL crystals at room temperature, the angles (ψ,ϕ) being near $(340°,100°)$.

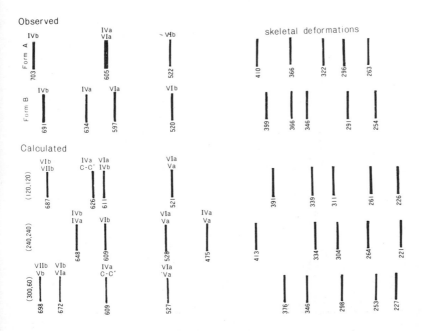

Figure 11. Observed and calculated frequencies of the amides IV and VI and skeletal deformation vibrations of N-acetylalanine methylamide. (Reprinted with permission. Copyright by Kodansha Ltd., Tokyo.)

Aqueous solution. The conformation of *N*-acetylalanine methylamide in aqueous solution has been studied by means of laser Raman spectroscopy (12, 14). Fig. 10 shows the Raman spectra for the two crystal forms and those for aqueous solutions. It was concluded from these spectra that two rotational isomers exist in aqueous solutions. This conclusion is based on the temperature dependence and the solvent effect of the relative peak intensity of the bands at 532 and 516 cm^{-1} in the spectra for aqueous solutions. These bands can be assigned to amide-VIb. They correspond to the bands at 521 and 520 cm^{-1} for Forms A and B. As Fig. 12 shows, the ratio of the intensity of the band at 516 cm^{-1} to that of the band at 532 cm^{-1} increases when the temperature is raised, the relative intensities being 1.6, 1.7, and 1.8 at 25°, 38°, and 55°C, respectively. Thus, the isomer having the amide-VIb band at 516 cm^{-1} is less stable than the isomer having the amide-VIb band at 532 cm^{-1}. The spectral change suggests that the enthalpy difference (ΔH) between these isomers may be

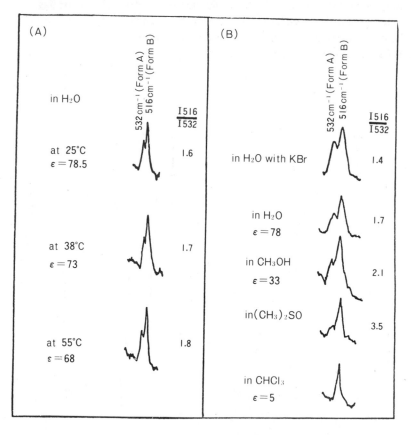

Figure 12. Temperature dependence and solvent effect of the bands at 532 and 516 cm^{-1} for solutions of *N*-acetylalanine methylamide. (Reprinted with permission. Copyright by Kodansha Ltd., Tokyo.)

~0.5 kcal/mole. Fig. 12 shows also that the relative intensity varies when the solvent is changed. The ratios of the peak heights are 2.1 and 3.5 for methanol and dimethylsulfoxide solutions, respectively, and the band at 532 cm^{-1} is virtually indistinguishable in the spectrum for chloroform solution. The relative intensity increases when the value of the dielectric constant for the solvents increases. Therefore the band at 516 cm^{-1} should be assigned to the less polar isomer and the band at 532 cm^{-1} to the more polar isomer.

The dipole moment should be very large for the isomer of Form A (Isomer A) whose amide C=O bonds are almost parallel and in the same direction. The isomer of Form B (Isomer B) also has amide C=O bonds that are almost parallel, but they are in the opposite direction. The dipole moment of Isomer B therefore should be very small. These considerations led to the following general conclusions: The bands at 532 and 516 cm^{-1} can be assigned to the Isomers A and B, respectively; and Isomer A is more stable than Isomer B in aqueous solutions. In addition, the intensity ratio decreases from 1.7 to 1.5 when potassium bromide is added to aqueous solutions, as shown in Fig. 12. Thus, the salt may stabilize Isomer A.

INTERNAL ROTATION POTENTIALS ABOUT THE N–C$_\alpha$ AND C$_\alpha$–C' AXES (18)

Stable conformations and potential minima. Only the gauche conformation ($\phi = 120°$) has been found for the N–C$_\alpha$ axis. The N-ethylacetamide molecule takes a gauche conformation in the gaseous, liquid, and solid states. (The value of the rotational angle ϕ has not yet been established, but it may be a little less than 120°.) The N-acetylglycine methylamide molecule has two rotational isomers, both of which take a gauche conformation with respect to the N–C$_\alpha$ axis. This conformation is stable also in solutions.

The above findings indicate that the internal rotation potentials about the N–C$_\alpha$ axis, in general, have a minimum near the gauche position ($\phi = 120°$). It is probable that steric interactions between nonbonded atoms push the potential minimum toward smaller angles. It is noteworthy that the molecules do not take the *trans* conformation, in which no severe interactions between nonbonded atoms occur. The inherent internal rotation potential about the axis, which originates from the electronic configuration of the peptide group, must have a minimum at $\phi = 120°$ and must play an important role in maintaining the stability of the gauche conformation.

Both *cis* and *trans* conformations have been found for the C$_\alpha$–C' axis. The N-methylpropionamide molecule takes a *cis* conformation in the gaseous, liquid, and solid states. (The rotational angle ψ may be near to 150°.) The N-methylchloroacetamide molecule takes a *cis* conformation. The rotational angle ψ is near 150° in the gaseous state, although it is larger (166° and 179°) in the crystal, probably because of intermolecular van der Waals interactions. In the case of N-acetylglycine methylamide, Isomers A and B take *cis* and *trans* conformations, respectively. The existence of two crystalline modifications is attributable to rotational isomerism about the C_α–C' axis. The two rotational isomers coexist in solution.

The above results indicate that the internal rotation potential about the C_α–C' axis has two minima, one at *cis* and the other at *trans*. Severe steric interactions between nonbonded atoms are expected for the *cis* conformation. These restrictions may push the minimum toward smaller angles as low as ~150°. In other words, the inherent internal rotation potential about the C_α–C' axis must have one minimum at $\psi = 180°$ and must play an important role in stabilizing the *cis* conformation. The exact position of the second minimum is not certain.

The stable conformations of the peptide backbone suggested by the present investigations are indicated by dashed lines in Fig. 13. The points in the figure show the sets of rotational angles (ψ, ϕ) associated with the glycine residues in myoglobin (18), lysozyme (19), α-lactalbumin (20), caboxypeptidase (21), α-chymotrypsin (22), parvalbumin (23), and cytochrome C (24). The values for residues involved in the α-helical and β-sheet structures are excluded from the plots because regular hydrogen bond formation may affect the stable conformations. The observed points clearly cluster around the intersections of the regions formed by the dashed lines. This feature strongly suggests that the forces controlling the conformation of model compounds also should govern those of the peptide backbones of proteins to a considerable extent. It is of interest also that conformations with the angle $\psi = 0°$, which are unstable for the model compounds, appear in the figure. These conformations must be impossible without intramolecular and intermolecular interactions of the long polypeptide chains, and their existence suggests that the inherent potential about the N–C_α axis should be threefold. The points appearing in the $\psi = 0°$ region imply that the inherent potential about the C_α–C' axis should be twofold.

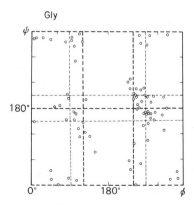

Figure 13. Stable conformations determined by this investigation and the peptide-backbone conformations associated with the glycine residues in the random parts of proteins. (Reprinted with permission. Copyright by Kodansha Ltd., Tokyo.)

Rotational isomerism and energy difference. *Cis-trans* rotational isomerism is common to model compounds having a C_α–C' axis, although the conformation around the N–C_α axis always is fixed to the gauche position. This characteristic suggests that the energy difference between the potential minima (*cis* and *trans*) is very small. The enthalpy difference between two rotational isomers of *N*-acetylalanine methylamide were estimated by observing the temperature dependence of the relative intensity of amide-VIb bands to be 0.5, 0.5, 0.3 and –1.0 kcal/mole for water, methanol, ethanol, and dimethylsulfoxide solutions, respectively. These values contain the latent effect of the variation of the dielectric constants of the solution and the effect of various kinds of intermolecular interactions. However, it is clear at this stage that intermolecular interactions affect the energy difference.

REFERENCES

1. J.T. Edsall, P.J. Flory, J.C. Kendrew, A.M. Liquori, G. Némethy, G.N. Ramachandran, and H.A. Sheraga, *Biopolymers,* **4**, 121 (1966).
2. Y. Koyama and T. Shimanouchi, *Biopolymers,* **6**, 1037 (1968).
3. Y. Koyama, T. Shimanouchi, and Y. Iitaka, *Acta Cryst.,* **B27**, 940 (1971).
4. Y. Koyama, T. Shimanouchi, M. Sato, and T. Tatsuno, *Biopolymers,* **10**, 1059 (1971).
5. T. Miyazawa, T. Shimanouchi, and S. Mizushima, *J. Chem. Phys.,* **29**, 611 (1958).
6. Y. Koyama and T. Shimanouchi, *21st Meeting of the Chemical Society of Japan,* Osaka, Japan, 1968.
7. E.B. Wilson, Jr., J.C. Decius, and P.C. Cross, *Molecular Vibrations,* McGraw-Hill, New York, 1955.
8. K. Ito, Thesis, University of Tokyo (1966).
9. T. Shimanouchi, *Tables of Force Constants,* Molecular Structure Symposium, Osaka, Japan, 1966.
10. A. Yokozeki and K. Kuchitsu, *Bull. Chem. Soc. Japan,* **44**, 72 (1971).
11. F. Iwasaki, *9th International Congress of Crystallography,* Kyoto, Japan, 1972.
12. Y. Koyama, K. Habuchi, and K. Kamei, *12th Meeting of the Biophysical Society of Japan,* Tokyo, 1973.
13. M. Avignon, C.C. Garrigou-Lagrange, and P. Bothorel, *Biopolymers,* **12**, 1651 (1973).
14. G.N. Ramachandran, C.M. Venkatachalam, and S. Krimm, *Biophys. J.,* **6**, 849 (1966).
15. G.N. Ramachandran and V. Sasisekharan, *Adv. Protein Chem.,* **23**, 283 (1968).
16. S.J. Leach, G. Némethy, and H.A. Sheraga, *Biopolymers,* **4**, 369 (1966).
17. Y. Harada and Y. Iitaka, to be published.
18. For detailed discussion see T. Shimanouchi, Y. Koyama, and K. Ito, *Progress in Polymer Science* (Japan), Vol. 7, Kodansha Ltd., Tokyo, 1974.
19. H.C. Watson, *Prog. Stereochem.,* **4**, 299 (1969).
20. D.C. Phillips, *Proc. Nat. Acad. Sci. U.S.,* **57**, 484 (1967).

21.   W.J. Browne, A.C.T. North, and D.C. Phillips, *J. Mol. Biol.*, **42**, 65 (1969).
22.   F.A. Quiocho and W.N. Lipscomb, *Adv. Protein Chem.*, **25**, 1 (1971).
23.   J.J. Birktoft and D.M. Blow, *J. Mol. Biol.*, **68**, 187 (1972).
24.   R.H. Kretsinger and C.E. Nockolds, *J. Biol. Chem.*, **248**, 3313 (1973).
25.   J.N. Brown, T. Takano, and R.E. Dickerson, to be published.

PART V.   INTERMOLECULAR INTERACTIONS

MECHANISM OF THE TRIPLE HELIX ⇌ COIL TRANSITION OF SHORT PEPTIDES WITH COLLAGENLIKE SEQUENCE

H. WEIDNER[*] and J. ENGEL,[**] Department of Biophysical Chemistry, Biozentrum der Universität Basel, Switzerland, and P. FIETZEK, Department of Connective Tissue Research, Max-Planck-Institut für Biochemie, Martinsried, Germany

SYNOPSIS: The kinetics of the triple helix ⇌ coil transition of a collagen-derived peptide with the general formula $(Gly-X-Y)_{12}$, where X and Y stand for residues other than glycine, is governed by more than one relaxation time. This factor rules out the all or none mechanism, although this simple model was successfully applied to the equilibrium transition data. A proposed mechanism is based on the idea that side products with staggered chains can form and that readjustment can take place only by a complete dissociation into single chains. This model describes the equilibrium transition data with the following parameters: enthalpy $\Delta H_P = -2.9$ kcal/mole tripeptide units and entropy $\Delta S_P = 8.7$ e.u./mole tripeptide units of the propagation step and entropy of nucleation $\Delta S_N = -21.5$ e.u./mole single chains. The nucleation parameter $\beta = 2 \times 10^{-5} M^{-2}$ was determined by a fit of the model to published transition data on $(Pro-Pro-Gly)_n$. The mechanism accounts also for the observed relaxation spectrum. A fit of the theoretical to the experimental dependence of the reciprocal mean relaxation time on the degree of conversion yields a negative apparent energy of activation, $E_a = -13$ kcal/mole, which may be explained by fast preequilibria that lead to a nucleus made up of about six tripeptide units. In the framework of this kinetic model, the apparent rate constant of nucleation is $k_N = 1 M^{-2}$ sec^{-1} and the transfer of a tripeptide unit from the coiled to the triple helical state proceeds with a rate constant of propagation $k_P = 4 \times 10^3$ sec^{-1} at $18°$C.

INTRODUCTION[***]

The structure of collagen is simple in one respect: It is linear and (with the exception of the end regions) the same structure extends along the whole rodlike molecule (1). In contrast to transitions of globular proteins in which interactions occur in all directions, cooperative transitions of linear systems are accessible to a quantitative treatment based on relatively simple reaction models.

[*]This paper is based on a thesis submitted by H. Weidner to the University of Munich in partial fulfillment of the requirements for the Ph.D. degree (September 1973). The current address of H. Weidner is: Department of Chemistry, Yale University, New Haven, Conn. 06520.

[**]Correspondence should be addressed to J. Engel, Biozentrum der Universität, CH 4056 Basel, Klingelbergstrasse 70, Switzerland.

[***]Abbreviations used throughout this paper are: AON, all or none; SZ, staggering zipper; SSZ, simplified staggering zipper.

For example, the equilibrium and the kinetic properties of the α-helix $\rightleftharpoons$ coil transition and of the double-helix $\rightleftharpoons$ coil conversion of polynucleotides have been successfully described on the basis of the linear Ising model (2, 3).

Compared with the latter systems, the transition of collagen is more complex for a number of reasons. Three segments of different chains must come together in order to form a nucleus for triple-helix formation. This type of nucleus formation is now generally accepted (4) and an earlier proposed mechanism that postulated a rate-determining nucleation within single chains (5) has been ruled out mainly on the basis that initial rates of triple-helix reformation in small peptides [(6); see also this paper] depend on the third power of chain concentration. Real native collagen molecules can reform only if the nucleus is formed in such a way that the three chains are in proper register. For long chains (each of the three chains of collagen contains about 1000 amino acid residues) many "wrong" nucleations can occur between mutually staggered chains. In such cases, only stretches of triple helix can develop and chain ends that are unable to participate in structure formation within the same molecule will protrude. They can, however, combine with other free chains to form aggregates of much higher molecular weight than collagen. Such side products have been demonstrated by several methods (7, 8, 9, 10). Furthermore, it has been shown that at low chain concentrations the formation of triple-helical nuclei within a single chain folded back on itself becomes a prominent side reaction (4, 10).

In view of the many possible side reactions, it is not surprising that the products are usually kinetically controlled and that real equilibrium is achieved only after experimentally unfeasible times. Even if this difficulty could be overcome, the model would probably contain so many parameters that a quantitative decision about its validity could not be obtained by comparison with experimental results. In particular, it appears to be impossible to extract equilibrium and kinetic data on the elementary steps (nucleation and propagation) from studies with collagen alone.

Thus, the discovery of Piez and Sherman (6, 11) that a small, 36-residue peptide (CB2) obtained from collagen by cyanogen bromide digests of the α_1 chain re-forms the triple helix in a reversible way. The authors were able to reconcile their equilibrium data with an all or none (AON) model, in which all intermediates and side products are neglected. In their kinetic studies, however, they found serious deviations from the predictions of this simplest model. The transition curves of polytripeptides (Pro-Pro-Gly)$_n$ with n = 10, 15, and 20 (12) also fit the AON model reasonably well; no kinetic data have been published for this system. In this paper, the existing data on initial rates of CB2 helix re-formation (6) are supplemented by relaxation studies. A model mechanism is formulated that, in contrast to the simple AON model, can explain both the equilibrium data and the kinetics.

EXPERIMENTAL SECTION

Materials. The peptide CB2 was prepared from calf-skin collagen according to Rauterberg and Kühn (14). It is chemically identical to the rat-skin CB2 (15) used by Piez and Sherman (6). The sequence of the general formula $(Gly-X-Y)_{12}$ is known (14, 15); X and Y stand for residues other than glycine. All measurements were performed in 0.25 M sodium citrate buffer, pH 3.7, which contained 0.02 g/100 ml sodium azide for protection against bacterial growth. The concentration of the stock solution was determined by amino acid analysis.

Methods. Optical rotations α were determined by a Cary-Varian Model 60 spectropolarimeter in thermostatted cells (Perkin Elmer Bodenseewerk, Ueberlingen, Germany) with path lengths $\ell = 1$ mm. Specific optical rotations $[\alpha]$ were calculated according to $[\alpha] = 10^4/c \cdot 1$ where c is the peptide concentration in g/100 ml. Relaxation curves were followed by optical rotation after a small rise or drop of temperature (3°C) within the transition range. A special cell was constructed in which the new temperature was reached within approximately 10 sec. About 1 ml solution contained in a trough of thin platinum foil (3 mm broad, 8 mm deep, optical path length 1 cm) was thermostatted by circulating water. The circulation was quickly switched from one water bath to another by a set of magnetic valves (16). Small corrections were employed to accommodate the temperature dependencies of the pure forms when the degree of conversion θ (fraction of helical segments) was calculated. The calculation and plotting of theoretical functions was performed with a Hewlett Packard Model 10 calculator equipped with a Model 9862 A plotter. The least square fits of theoretical functions to experimental data were performed at the University Computing Center, Basel, on a Univac 1108 computer using the Gaushaus program (17), an algorithm that combines the method of steepest descent with the Gauss method of linear expansion (18).

EXPERIMENTAL RESULTS

Fig. 1 shows the temperature-dependent equilibrium transition curves of CB2 at different concentrations. The limiting rotation for the coiled chains at high temperatures is known with good accuracy. Unfortunately, the corresponding value for the triple-helical state at low temperatures is uncertain because no clear plateau is reached. The transition curves shown are theoretical curves obtained by a Gaushaus fit of the parameters of the simplified staggering zipper (SSZ) model (see next section) to the experimental data. Because of the experimental uncertainty outlined above, the optical rotation of fully helical CB2 also was evaluated by the fitting procedure. With this degree of freedom it is possible to achieve an equally good fit ay another reasonable model, such as the staggering zipper (SZ) model or even the AON model. The fitted transition curves for all three models are almost superimpossible, but significant differences appear in the upper plateau value.

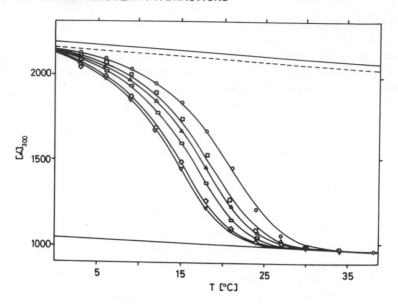

Figure 1. Temperature dependence of optical rotation at 300 nm of the collagen peptide CB2. Peptide concentrations were 1.49×10^{-3} M (∇), 1.71×10^{-3} M ($\Diamond$), 2.53×10^{-3} M ($\square$), 3.46×10^{-3} M ($\triangle$), 4.41×10^{-3} M ($\square$), and 7.85×10^{-3} M ($\bigcirc$). Data obtained by a stepwise increase of temperature are identical with those measured in the reverse way. The lower straight line indicates the temperature dependence of the optical rotation of the coiled state. The transition curves shown are obtained by a fit of the AON or SSZ model to the data (see Table 1 for parameters). The dashed and solid lines above the transition curves indicate the theoretical limiting rotations for fully helical CB2 in the cases of the AON and SSZ models, respectively. The slope of these lines is taken from the temperature dependence of the optical rotation of collagen.

All relaxation curves exhibited more than one phase at various degrees of conversion θ and at different temperatures. This clearly rules out an AON mechanism. Fig. 2 shows a logarithmic plot of $(\theta - \theta_\infty)/(\theta_0 - \theta_\infty)$ versus time measured after a temperature jump of $3°$ in the transition range, somewhat above the midpoint of the transition. The total change in the degree of conversion $\theta_0 - \theta_\infty$ is about 15% in this instance. A clear dependence on the direction of the temperature jump is observed. In accordance with relaxation theory (19), this dependence disappears as the amount of perturbation decreases. It is important to note that a positive curvature is observed that is independent of the direction of the jump. According to Ikai and Tanford (20), this finding rules out a mechanism of the type $3A \rightleftharpoons I \rightleftharpoons B$ where A denotes the free chains, B denotes the triple helix, and I is an intermediate that accumulates in concentrations comparable to those of A and B.

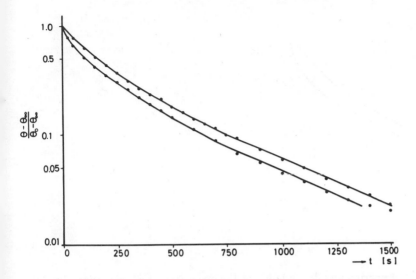

Figure 2. Experimental relaxation curves measured after a temperature jump from 15° to 18°C (upper curve) and from 21° to 18°C (lower curve) at a CB2 concentration of 7.85×10^{-3} M. The degree of conversion measured at time t is θ, and θ_0 and θ_∞ denote the equilibrium θ values before and after perturbation.

It was not possible to derive from the relaxation curves an unambiguous set of relaxation times τ_i and corresponding weight factors β_i Most likely we are dealing with a spectrum of many relaxation times that are not well separated. Therefore the mean reciprocal relaxation time τ^*

$$\frac{1}{\tau^*} = -\frac{1}{\theta_0 - \theta_\infty} \frac{d\theta}{dt}\bigg|_{t=0} = \sum_i \frac{\beta_i}{\tau_i} \tag{1}$$

was evaluated from the initial slope of each curve. Experimentally τ^* can be determined with good precision and it can be calculated from the models with relative ease. Compared with initial rates determined from the overall transition of the pure forms (6), τ^* contains more information because, at least in principle, it is influenced by all τ_i and β_i of the system. Of particular significance is the dependence of τ^* on θ. A plot of $\tau^* c_0^2$ versus θ is shown in Fig. 3. The normalization with the square of the total chain concentration c_0^2 is performed because it is expected from the model calculations that $\tau^* c_0^2$ should be essentially concentration independent. Perturbation in the direction of low to high degrees of conversion yield smaller τ^* values than those in the opposite direction.

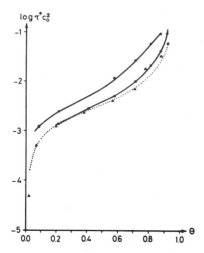

Figure 3. Experimental dependence of the mean reciprocal relaxation time τ^* on the degree of conversion θ. Log $\tau^*c_0^2$ is plotted versus the new equilibrium value θ after perturbation. The concentrations of CB2 were $c_0 = 7.85 \times 10^{-3}\ M$ (shown by darkened and open circles) and $3.46 \times 10^{-3}\ M$ (shown by triangles). The temperature was increased (open symbols) or decreased (darkened symbols) by $3°$ in each relaxation experiment.

Measurements of the initial rates of helix formation from the pure coiled chains (CB2 solutions of different concentrations were cooled from $38°$ to $18°$) obeyed the equation

$$\left.\frac{d\theta}{dt}\right|_{t=0} = kc_A{}^\alpha$$

with the reaction order $\alpha = 2.9 \pm 0.2$, confirming earlier measurements by Piez and Sherman (6).

THEORY

Nucleation and propagation. A conformational transition is cooperative when the formation of the first stretch of structure (nucleation) is thermodynamically more difficult than the growth of structure via propagation steps. For our system it is useful to define as the elementary step of structure formation the transfer of a tripeptide unit Gly-X-Y from the randomly coiled state to the conformation and mode of binding that is characteristic of the triple helix. This step is associated with the formation of a hydrogen bond between Gly and Y of a neighboring chain (1). A helix with r such hydrogen bonds will be designated B_r.

Three chains of n tripeptide units can form a triple helix with a maximum number of $3n - 1$ hydrogen bonds each. The process

$$B_r \underset{k_P'}{\overset{k_P}{\rightleftharpoons}} B_r + 1$$

with $r > x$ (see below) is called a propagation step. In all propagation steps, essentially the same process is repeated. It will be assumed, therefore, that the rate constants k_P and k_P' and the equilibrium constant $s = k_P/k_P'$ do not depend on r. The equilibrium constant s is associated with the enthalpy and entropy changes that occur during propagation:

$$s = e^{-\Delta H_P/RT} e^{\Delta S_P/R} \tag{2}$$

A tripeptide unit that undergoes propagation is near its partners; it is properly oriented and has lost some of its conformational freedom because of the steric restrictions imposed by the other chains. These effects must be introduced during nucleation, thereby giving rise to an entropy change of nucleation ΔS_N, which is negative and renders nucleation difficult. A nucleus B_x is defined as a first piece of triple helix from which further helix formation proceeds via propagation steps. The first x steps may have kinetic and equilibrium constants that are quite different from those of the propagation step. The number of hydrogen bonds in the nucleus is not known a priori. Because these hydrogen bonds and the other interactions are identical in B_x and in any other part of the helix formed via propagation steps, the same ΔH_P and ΔS_P contributions per tripeptide unit are expected. Therefore it is possible to write the equilibrium constant of nucleation as a product:

$$\frac{k_N}{k_N'} = \beta s^x = e^{-x\Delta H_P/RT} e^{(x\Delta S_P + \Delta S_N)/R} \tag{3a}$$

where

$$\beta = e^{\Delta S_N/R} \tag{3b}$$

is called the nucleation parameter.

In accordance with the simplified picture that the free chains are random coils that penetrate each other freely, it is possible to estimate the probability that three tripeptide units of different chains meet within a certain reaction radius and are aligned in the same orientation. Then a factor β_1 of the nucleation parameter can be estimated to account for the meeting and orientation. If the distance between two C_α atoms in a peptide chain (3.8 Å) is arbitrarily chosen to be the reaction radius, $\beta_1 \cong 10^{-4} M^{-2}$ results for chain lengths $n =$ 10–30 (21). The loss of conformational freedom that is imposed on the remaining segments of three chains after a nucleus has been formed has been estimated from Monte Carlo simulations of all possible arrangements of the free-chain ends for the sequence (Gly-Pro-Pro)$_n$ (21). For $n \cong 10$, the number of conformations that are possible after nucleation divided by the number of conformations accessible to the free chains was $\beta_2 \cong 0.1$. The effect of β_2 remains constant only as long as the free ends are still long compared to the range in which significant steric restrictions occur. Thus, propagation becomes easier at the ends of a molecule. A total $\beta = \beta_1 \beta_2 \cong 10^{-5} M^{-2}$ is estimated by this crude treatment, which, among other simplifications, neglects possible enthalpy differences between nucleation and propagation.

According to the above picture of nucleation, most of the difficulty in forming B_x is seen in its formation reaction, whereas little physical difference is expected between the destruction of a nucleus and the back reaction of a propagation step. For the formulation of the kinetic models, therefore, it is assumed that $k'_N = k'_P$, which, according to Eq. (3) is equivalent to

$$k_N = \beta s^x k'_P \quad \text{or} \quad k_N = \beta s^{x-1} k_P \tag{4}$$

The formation of B_x may proceed via a series of fast preequilibria. This process may be demonstrated by the dimer intermediate B_2, which must be postulated because the direct formation of B_3 by a collision of three segments has an extremely low probability

$$3A \underset{k'_1}{\overset{k_1}{\rightleftharpoons}} B_2 + A \underset{k'_2}{\overset{k_2}{\rightleftharpoons}} B_3$$

This mechanism is consistent only with the experimentally observed cubic dependence of the initial rates of helix formation from the purely coiled chains if the first step is a fast preequilibrium of the second and if the concentration of B_2 is always negligible compared to those of A and B_3. Thus, $k'_1 \gg k_2 c_A$; $K_1 c_A \ll 1$ and $K_2 c_A \gg 1$ with $K_1 = k_1/k'_1$ and $K_2 = k_2/k'_2$. Then a single apparent rate constant $k_N = K_1 k_2$ is observed. In the complete series of steps, if similar conditions are assumed for the first $x - 1$ steps ($k'_2 \gg k_3, \ldots, k'_{x-1} \gg k_P$), we can write

$$k_N = k_P \prod_{r=1}^{x-1} K_r$$

and according to our simplified view of the nucleation process

$$\prod_{r=1}^{x-1} K_r = \beta s^{x-1}$$

A similar mechanism was postulated for the double helix ⇌ coil conversion of poly(nucleic acids) (22).

Model mechanisms: AON. The simplest mechanism is derived with the AON assumption that all intermediates between A and B_{3n-1} occur in sufficiently low concentrations that they can be neglected. It can be shown by a population analysis that this situation is approached when cooperativity is high and when the chain length is small:

$$3A \underset{k_R}{\overset{k_F}{\rightleftharpoons}} B_{3n-1} \tag{5}$$

The equilibrium is described by

$$\frac{\theta}{3(1-\theta)^3} = c_0^2 \beta s^{3n-1} \tag{6}$$

where c_0 is the total concentration of peptide chains. The degree of conversion, θ = fraction of helical segments, is in this case $3c_{B,3n-1}/c_0$. There is a single relaxation time (which in this case, of course, is identical to τ^*):

$$\tau = (gk_F c_A^2 + k_R)^{-1} \tag{7}$$

where c_A denotes the equilibrium concentration of single chains. The overall rate constants k_F and k_R can be expressed by the rate constants of the individual reaction steps (see below).
The dependence of τ on θ is given by

$$\tau k_F c_0^2 = \frac{\theta}{3(1-\theta)^2(2\theta+1)} \tag{8}$$

Although the AON model has been ruled out by the observation of more than one relaxation time, it will be used for comparison with the other models. If the AON behavior is a reasonable assumption for a reaction chain, it is possible to calculate its kinetics if a steady state is assumed for all intermediates. For a simple reaction chain in which the nucleation occurs at one end only it is found (22) that

$$k_F = k'_P \beta s^{3n-1} \frac{s-1}{s^{3n-x}-1} \tag{9a}$$

$$k_R = k'_P \frac{s-1}{s^{3n-x}-1} \tag{9b}$$

In the case of small chains in which $s \gg 1$ in the transition range (2) Eq. (9a) simplifies to

$$k_F = k'_P \beta s^x = k_N \tag{10}$$

it is more realistic to allow for nucleations at any tripeptide unit in a chain. Assuming that propagation, which starts from a nucleus somewhere in the middle of a chain, will proceed at the same rate toward both ends, we obtain

$$k_F = \frac{2k'_P}{\displaystyle\sum_{i=x}^{3n-1} [1/(3n-i)\beta s^i]} \tag{11a}$$

$$k_R = \frac{2k'_P}{\displaystyle\sum_{i=0}^{3n-x-1} (s^i/i+1)} \tag{11b}$$

Model mechanisms: the starlike model. There is experimental evidence (see Introduction) that the nucleation of collagen structure can occur at many sites on the chain and in such a way that part of the chain cannot participate in structure formation. Possible products for $n = 4$ are shown in Fig. 4. One possible formation is the complete triple helix with $r = 3n - 1$ hydrogen bonds. Once a wrong species has been formed, it cannot align to the complete triple helix via a sliding of the chains. Such a mechanism is unlikely because of the intimate mutual stabilization of the three chains. The only possibility is seen in a complete dissociation into chains A, which can again undergo nucleation with a perhaps better result. This idea leads to the formulation of a starlike mechanism in which the various reaction chains are coupled only via A (Fig. 5). The equilibrium transition curves predicted by this model were calculated by the method

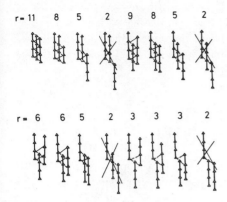

Figure 4. Staggering of chains in a triple helix. All possible arrangements with the maximum number of hydrogen bonds that can form at a given degree of staggering are shown.

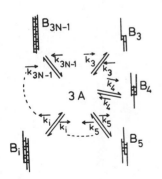

Figure 5. Starlike model for the kinetics of triple-helix formation when staggering of chains is possible.

used by Applequist and Damle (23) for the double-helix ⇌ coil transition, but with other weighting factors, of course. If we consider only the species with the maximum number of hydrogen bonds r that are compatible with a given degree of staggering (Fig. 4), we arrive at a simplified staggering zipper (SSZ) model. The weighting factors by which the equilibrium constants βs^r must be multiplied are

$$g_r = \begin{cases} n-v, & u=0, 2 \\ 0, & u=1 \end{cases} \quad \text{for } r = 3v+u \quad \text{with} \begin{cases} v=1,\ldots,n-1 \\ u=0, 1, 2 \end{cases}$$

The full staggering zipper (SZ), which resembles that of Applequist and Damle (23) is obtained when this limitation is dropped. The weighting factors for the equilibrium constants βs^r are then

$$f_r = (n - v + 1)^{2-u}(n - v)^{1+u}$$

Theoretical transition curves calculated for the AON, SSZ, and SZ models are compared in Fig. 6. At low values of s the SSZ and especially the SZ predict higher θ values than the AON model because of the increased number of ways in which helices can form. Because more possibilities for coiled states exist in the zipper model at high θ values than in the AON model, $\theta = 1$ is approached much more gradually than in the AON case.

For the kinetics calculations of the starlike model a steady state was assumed in each of its reaction chains. The overall rate constant k_r was expressed by Eq. (10) (model 1), Eq. (9a) (model 2), or Eq. (11a) (model 3). Model 3 is certainly the most realistic one, although model 1 is the simplest to calculate. Model 1 was used to calculate the individual rate constants τ_i in addition to the mean relaxation time τ^*. With $x = 3$ we obtain:

$$\frac{1}{\tau^*} = \frac{3}{(3n - 1) c_0 (\theta_0 - \theta_\infty)} \sum_{r=3}^{3n-1} r \frac{dc_\tau}{dt} \bigg|_{t=0}$$

$$= k_N c_0^2 (n - 1)(6n^2 + 3n + 4) \left[s(sQ'c_0^2)' - (sQ'c_0)^2 \frac{9c_A^2}{c_0^2(1+9 c_A^2 Q)} \right]^{-1}$$

with

$$Q = \sum_{r=3}^{3n-1} g_r \beta s^r$$

(Primes indicate the first derivative with respect to s.)

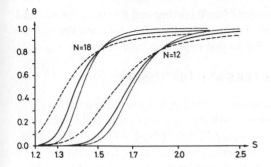

Figure 6. Equilibrium transition curves calculated for the AON ($\cdots$), SSZ (—), and SZ (– –) models with $\beta c_0{}^2 = 10^{-8}$ and $n = 12$ and 18.

The expression for τ^* in models 2 and 3 can be derived in a similar way (21). The dependence of τ^* on θ is shown in Fig. 7 for the three versions of the star-like model. The AON model [Eq. (8)] is shown for comparison. As expected, τ^* decreases with increasing complexity of the model. The dependence of the longest relaxation time τ_{3n-1} on θ also is shown in Fig. 7 for the starlike model 1. This relaxation time is greater than τ^* by a factor of 10–100.

The above relaxation times were calculated for small perturbations for which the linearization of the differential equations is possible. Because the experimentally useful perturbations are of a size at which a dependence on the direction

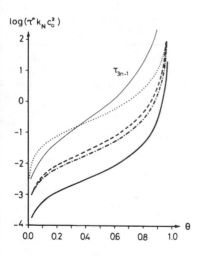

Figure 7. Plot of $\tau^* k_N c_0{}^2$ versus θ for the AON model ($\cdots$) and for the starlike models 1 (– –), 2 (–·–), and 3 (—). The dependence of the longest individual relaxation time τ_{3n-1} on θ is plotted for the starlike model 1 (thin curve). All calculations were performed for $n = 12$ and $\beta c_0{}^2 = 10^{-8}$.

of the temperature jump is still apparent (see Experimental Results), the model-1 relaxation times were calculated also for larger jumps. The predicted dependence on the direction of the jump is similar to that experimentally observed.

EVALUATION OF THE PARAMETERS AND DISCUSSION

The nucleation parameter β can be determined only from chain-length-dependent transition data for polymers with identical tripeptide units. The only useful data of this type have been published (12) for (Pro-Pro-Gly)$_n$. A value of β between 10^{-2} and $10^{-6} M^{-2}$ has been estimated from these data with the AON model (13) (see Table 1). We have performed a least square fit of the SZ and SSZ model to these data. The latter is shown in Fig. 8 and the parameters are summarized in Table 1. The value of β agrees surprisingly well with the very crude estimate of meeting probabilities. The corresponding entropy of nucleation was calculated from Eq. (3b). The fit also yields values for ΔS_P and ΔH_P [Eq. (2)]. The overall entropy change $\Delta S_N + (3n - 1)\Delta S_P$ was calculated for $n = 12$.

If it is assumed that the same β is valid for CB2, reasonable data for the other parameters are obtained by fitting the models to the CB2 transition data (Table 1). The entropy of propagation is more negative for CB2 than for (Pro-Pro-Gly)$_n$. This phenomenon may be explained by the higher conformational freedom due to the lower content of imino acids in CB2. A more negative conformational

Table 1. Equilibrium parameters for CB2 and (Pro-Pro-Gly)$_n$ obtained from least-square fits of the AON, SSZ, and SZ models to experimental data

Model	CB2			(Gly-Pro-Pro)$_n$[a]		
	AON	SSZ	SZ	AON[b]	SSZ	SZ
β[c]	—	4×10^{-4}[g]	2×10^{-5}[h]	10^{-2}–10^{-6}	4×10^{-4}	2×10^{-5}
S_N[d]	—	-15.6[g]	-21.5[h]	-9–-28	-15.6	-21.5
S_P[e]	—	-8.12	-8.72	-5.0	-6.0	-6.7
$S_N + 35 S_P$[d]	-276	-300	-327	-192	-226	-255
H_S[f]	-2.48	-2.66	-2.86	-2.0	-2.20	-2.54

[a] Transition curves taken from Kobayashi et al. (12); see Fig. 8.
[b] Analysis by N. Go and Y. Suezaki (13).
[c] In liter^2 (mole single chains)$^{-2}$.
[d] In e.u./mole single chains.
[e] In e.u./mole tripeptide units.
[f] In kcal/mole tripeptide units.
[g] Value taken from column 5.
[h] Value taken from column 6.

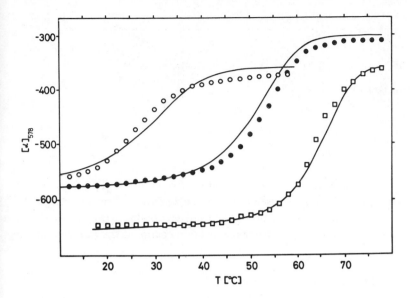

Figure 8. Least-square fit of the SSZ model to experimental transition data (12) of (Pro-Pro-Gly)$_n$ with n = 10 (○), n = 15 (◉), and n = 20 (□). The theoretical transition curves shown were determined by the parameters β, ΔS_p, and ΔH_p (see Table 1) for the lowest least-square sum.

entropy is expected, therefore, for the fixation of tripeptide units in a propagation step. The enthalpy of propagation ΔH_p is found to be only slightly more negative than that of (Pro-Pro-Gly)$_n$. The values are comparable for enthalpies and entropies that have been obtained for collagens of different origins by direct calorimetric measurements (24).

The rate constant of nucleation can be obtained by comparing the experimental mean relaxation times τ^* (Fig. 3) and the $k_N\tau^*$ values calculated from the starlike model mechanism. A comparison of the experimental and theoretical dependencies of τ^* on θ is hampered by the unknown temperature dependence of k_N, which will significantly change the results because θ changes from 0.1 to 0.9 in the broad temperature range of 5° to 25°C. The activation energy E_a in the Arrhenius equation,

$$k_N = k_N°e^{-E_a/RT}$$

was taken as an additional parameter in the fitting procedure. The results are shown in Table 2.

Table 2. Kinetic parameters for CB2 at 18°

Model	k_N $(M^{-2} sec^{-1})$	E_a (kcal/mole)	k_P (sec^{-1})	k'_P (sec^{-1})
Starlike model 1	8	-17	–	–
Starlike model 2	6.3	-14	–	–
Starlike model 3	1	-13	4×10^3	2.6×10^3

The activation energy is negative even for starlike model 3. This finding suggests a fast preequilibrium because only then can a negative apparent energy of activation result from the summation of a negative enthalpy ΔH of the preequilibrium and the always positive real activation energy of the rate-determining kinetic step

$$E_a = \Delta H + E$$

It follows from Eq. (4) that for our system

$$E_a = (x - 1)\Delta H_P + E k_P$$

Assuming that the activation energy $E k_P$, which belongs to the rate constant of propagation, is zero or slightly positive, we can estimate the number of hydrogen bonds in a nucleus to be $x = 6$. It is interesting to note that from a similar experimental result and by a similar argument, $x = 3$ was derived for the number of base pairs that must combine before the first stable nucleus of the double helix is formed (22). With $x = 6$ and Eq. (4) (s is known from Table 1) k_P and k'_P can be determined (see Table 2).

The value obtained for the rate constant of propagation is surprisingly low. The propagation steps in the α-helix $\rightleftharpoons$ coil conversion (3) ($k_P \cong 10^{10} sec^{-1}$) and in the double-helix $\rightleftharpoons$ coil conversion (23) ($k_P \cong 10^7 sec^{-1}$) are much faster. The more complex propagation step in the triple-helix $\rightleftharpoons$ coil conversion may involve a higher entropy of activation (lower frequency factor in the Arrhenius equation) than the corresponding step in the formation of a double helix. Also the activation enthalpy may be higher because more bonds must be adjusted, each of which has a rotational barrier of 1–2 kcal/mole. Another explanation for the small value of k_P is that the nucleation parameter for CB2 may have been overestimated. This and other questions can be resolved only on the basis of chain-length-dependent kinetic studies on synthetic peptides with collagenlike sequence; such work is being performed in our laboratory now.

REFERENCES

1. W. Traub and K.A. Piez, *Adv. in Protein Chem.*, **25**, 243 (1971).
2. J. Engel and G. Schwarz, *Angew. Chem.*, **82**, 468 (1970); *Angew. Chem. Int. Ed.*, **9**, 389 (1970).
3. G. Schwarz and J. Engel, *Angew. Chem.*, **84**, 615 (1972); *Angew. Chem. Int. Ed.*, **11**, 568 (1972).
4. W.F. Harrington and N.V. Rao, *Biochemistry*, **9**, 3714 (1970).
5. P.J. Flory and E.S. Weaver, *J. Am. Chem. Soc.*, **82**, 4518 (1960).
6. K.A. Piez and M.R. Sherman, *Biochemistry*, **9**, 4134 (1970).
7. J. Engel, *Arch. Biochem. Biophys.*, **97**, 150 (1962).
8. G. Beier and J. Engel, *Biochemistry*, **5**, 2744 (1966).
9. P.V. Hauschka and W.F. Harrington, *Biochemistry*, **9**, 3745 (1970).
10. M.P. Drake and A. Veis, *Biochemistry*, **3**, 135 (1964).
11. K.A. Piez and M.R. Sherman, *Biochemistry*, **9**, 4129 (1970).
12. Y. Kobayashi, R. Sakai, K. Kakiuchi, and T. Isemura, *Biopolymers*, **9**, 415 (1970).
13. N. Go and Y. Suezaki, *Biopolymers*, **12**, 1927 (1973).
14. J. Rauterberg and K. Kühn, *Eur. J. Biochem.*, **19**, 398 (1971).
15. P. Bornstein, *Biochemistry*, **6**, 3082 (1967).
16. F.M. Pohl, *Eur. J. Biochem.*, **4**, 373 (1968).
17. The Gaushaus Program: Wisconsin Computing Center, Vol. IV/B, Jan. 1966.
18. D.L. Marquardt, *J. Soc. Indust. Appl. Math.*, **2**, 431 (1963).
19. M. Eigen and L. DeMaeyer in *Techniques of Organic Chemistry*, Vol. 8/2, A. Weissberger, ed., Wiley(Interscience), New York, 1963, p. 890.
20. A. Ikai and Ch. Tanford, *Nature* (London), **230**, 100 (1971).
21. H. Weidner, Ph.D. Thesis, University of Munich, 1973.
22. M. Eigen and D. Poerschke, *J. Mol. Biol.*, **53**, 123 (1970).
23. J. Applequist and V. Damle, *J. Am. Chem. Soc.*, **87**, 1450 (1965).
24. P.L. Privalov and E.I. Tiktopulo, *Biopolymers*, **9**, 127 (1970).

A ^{13}C MAGNETIC RESONANCE STUDY OF THE HELIX AND COIL STATES OF THE COLLAGEN PEPTIDE α1-CB2

D.A. TORCHIA and **J.R. LYERLA, JR.**, National Bureau of Standards, Washington, D.C. 20234, and **A.J. QUATTRONE**, Laboratory of Biochemistry, National Institute of Dental Research, Bethesda, Md.

SYNOPSIS: Carbon-13 chemical shifts, spin-lattice (T_1) and spin-spin (T_2) relaxation times and ^{13}C-^{1}H nuclear Overhauser enhancements (NOE) have been determined for the coil and triple-helical states of the α1-CB2 fragment of rat-skin collagen. Assignment of the C_α resonances of this 36-residue peptide in the random-coil state (30°C) has been achieved with the aid of model polypeptides containing pyrrolidine residues. The T_1 measurements show that the interior backbone carbons of the coil are characterized by effective rotational correlation times (τ_{eff}) of ~0.045 nanosec, whereas the near-terminal backbone τ_{eff} values are two to four times shorter. These results, along with the narrow natural line widths (3–5 Hz) and maximum NOE values (3.0), demonstrate the high degree of backbone mobility, attributable to segmental motion in the unordered state of the peptide. By contrast, the broad lines (50–80 Hz) and low NOE values (1.3) for the α-carbons in the helical state (2°C) suggest much slower motion. These results together with the T_1 values (0.025–0.040 sec) are consistent with a model in which the motion of the ordered state of α1-CB2 is described by rotational diffusion of an axially symmetric rigid ellipsoid having dimensions approximating those expected for a collagenlike triple-helical aggregate of three α1-CB2 chains.

INTRODUCTION

Collagen is the primary protein constituent in bone, skin, cartilage, cornea, and tendon. The primary biological role of this protein is structural and the structural unit common to all collagens is the triple-stranded helix. Each of the three α chains that comprise the helix consists of approximately 1000 amino acids. The stability of the helical structure results from the regular amino acid sequence of the α chains, which consist of Gly-X-Y triplets where X and Y are often pyrrolidine residues. The regularity of the α-chain sequence has prompted many model studies (1) with synthetic peptides of the form poly(Gly-X-Y). Recent advances in collagen chemistry have permitted model studies with fragments of the α chains themselves. One such fragment, α1-CB2, obtained from the α1 chain of rat-skin collagen has the sequence (2)[1]

---

[1]Due to incomplete hydroxylation, rat-skin α1-CB2 contains about 5 Hyp residues (2). Amino acid analysis showed that the sample used in this work contained 5.5 Hyp and 6.5 Pro residues.

Gly-Pro-Ser-Gly-Pro-Arg-Gly-Leu-Hyp-Gly-Pro-Hyp-

Gly-Ala-Hyp-Gly-Pro-Gln-Gly-Phe-Gln-Gly-Pro-Hyp-

Gly-Glu-Hyp-Gly-Glu-Hyp-Gly-Ala-Ser-Gly-Pro-Hse

and undergoes a temperature-induced reversible transition from a random-coil to a triple-stranded structure (trimer) of high (~90%) helix content (3, 4).

It was anticipated that the conformations of this interesting peptide could be further elucidated using ^{13}C nuclear magnetic resonance (NMR) spectroscopy, because resonances attributable to carbons in different residues can be resolved in complex peptides (5). Although signals from the low abundance (1.1%) ^{13}C nuclei are weak, sensitivity enhancement is provided by Fourier transform techniques (6), which also facilitate measurement of spin-lattice relaxation times (T_1) (7). Furthermore, spin-spin relaxation times (T_2) can be estimated from ^{13}C line widths and the nuclear Overhauser enhancement (8) (NOE) can be obtained from a comparison of ^{13}C intensities with and without proton decoupling. These NMR parameters are sensitive functions of the rotational reorientation of individual CH (internuclear) vectors and there is a theoretical framework (9–14) that allows rotational correlation times to be calculated from the measured parameters $(T_1$, line width, and NOE). Hence, the ^{13}C spectra can provide information on the motion at numerous sites in both helical and coil conformations of α1-CB2. In this paper we shall discuss primarily the results obtained for the backbone carbons.

EXPERIMENTAL SECTION

The manner in which α1-CB2 was prepared was similar to that described by Piez and Sherman (3). Amino acid analyses showed that, except for a small amount (~1% by weight) of α1-CB1, the sample was free of other peptide or protein content. A 7.5 × 10^{-3} M solution of α1-CB2 (35 mg in 1.4 ml of 0.15 M AcNa, pH = 4.8) containing 4 λ of 90% enriched ^{13}CH$_3$CN (as an internal reference) was prepared in a 10 mm tube and plugged to prevent vortexing. The temperature of the sample was maintained at 2 ± 1.5°C or 30 ± 1°C in a Dewared probe by a commercial temperature controller. Spectra were obtained with a homebuilt Fourier transform spectrometer, described previously (15, 16), operating at 15.08 MHz and equipped with proton noise decoupling. Partially relaxed spectra were obtained using 180° – t – 90° pulse sequences (7) with a recycle time of about $4T_1$.

RESULTS AND DISCUSSION

Random-coil assignments. The upfield portion of the ^{13}C Fourier transform spectrum of α1-CB2 at 30°C [Fig. 1(a)] resembles that of denatured calf-tendon collagen (17) and has narrow resonances attributable to the rapid segmental motion in the unstructured chains. Most resonances can be readily assigned on the

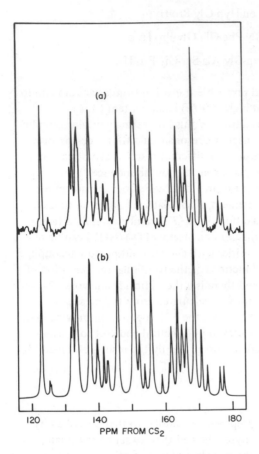

Figure 1. Comparison of experimental and calculated random-coil spectra in the aliphatic region: (a) experimental spectrum at 30°C, 64K scans of the FID were accumulated, protein concentration 25 mg/ml in 0.15 M AcNa buffer at pH = 4.8; (b) computer simulation. Chemical shifts in ppm from external CS_2.

basis of results in the literature (16, 18, 19) because aliphatic and aromatic carbon chemical shifts in random-coil chains are virtually independent of sequence. The one exception to this statement occurs when a residue precedes an *imino* residue such as Pro or Hyp. In this case, spectra of a series of model polypeptides containing pyrrolidine residues show that the C_α and C_β chemical shifts (20) are systematically upfield of their corresponding values when the given residue precedes an *amino* acid residue (21). The known sequence of α1-CB2 was used to determine the residues that precede Pro or Hyp, and these residues were assigned using the results obtained with the model polypeptides. Spectra of N-acetyl homoserine and N-acetyl homoserine lactone were used to assign the carbons in the C-terminal residues. The aliphatic spectrum calculated using these assignments [Fig. 1(b)] is an excellent simulation of the experimental spectrum (Fig. 1a). The α1-CB2 random-coil C_α assignments are listed in Table 1. Individual C' resonances (chemical shifts, 18–24 ppm) in the α1-CB2 spectrum cannot be assigned to carbonyl carbons in specific residues on the basis of available model peptide data.

Random-coil T_1 values and correlation times. Random-coil C_α T_1 values (Table 1), obtained from inversion-recovery (7) Fourier transform spectra (Fig. 2), have estimated uncertainties of 10–25%. The larger error is associated with the resonances of low intensity. Only three inversion-recovery spectra were obtained for the C' resonances because their large T_1 values ($\sim$1.4 sec) necessitated long accumulation times. Hence, the uncertainties in the C' T_1 values are 25–30%.

Comparison of proton-coupled and -decoupled $90° - t - 90°$ spectra showed that the C' NOE was maximal (8, 13, 14) because the area under the C' resonances decreased by a factor of 2.8 ± 0.3 when the C' spectrum was obtained without decoupling. A maximal NOE was deduced also for the aliphatic carbons because the ratio of areas of aliphatic to C' resonances (both measured with

Table 1. Random-coil chemical shifts, assignments, spin-lattice relaxation times, and effective correlation times

Chemical Shift[a]	Relative Intensity	Position of Carbon Residue in Chain	T_1 (sec)	τ_{eff} (nanosec)
151.9	1.0	C_α-Arg-(6)	0.17	0.14
151.9	1.0	C_α-Gly-(1)	0.17	0.14
150.6–150.9	6.0	C_α-Gly-(4,10,16,22,34)	0.065	0.36
150.0–150.3	5.0	C_α-Gly-(7,13,19,25,28,31)	0.065	0.36
145.0–145.7	6.5	C_δ-Pro-(2,5,11,17,23,35)	0.085	0.28
145.0–145.7	1.0	C_α-Ala-(14)	0.085	0.28
145.0–145.7	0.4	C_α-Hselactone-(36)	0.085	0.28
142.3–142.7	1.0	C_α-Ala-(32)	0.012	0.39
142.3–142.7	1.0	C_α-Leu-(8)	0.012	0.39
141.3	2.0	C_α-Glu-(26,29)	0.11	0.43
139.3–139.9	2.0	C_α-Gln-(18,21)	0.12	0.39
139.3–139.9	1.0	C_α-Arg-(6)	0.12	0.39
139.3–139.9	0.6	C_α-Hse-(36)	0.12	0.39
136.9–137.5	5.5	C_δ-Hyp-(9,12,15,24,27,30)	0.07	0.34
136.9–137.5	2.0	C_α-Ser-(3,33)	0.07	0.34
136.9–137.5	1.0	C_α-Phe-(20)	0.07	0.34
132.8–133.8	5.5	C_α-Hyp-(9,12,15,24,27,30)	0.105	0.45
132.8–133.8	2.0	C_α-Pro-(11,23)	0.105	0.45
132.8–133.8	0.6	C_γ-Hse-(36)	0.105	0.45
132.0–132.1	4.5	C_α-Pro-(2,5,17,35)	0.15	0.32

[a]In ppm from external CS_2.

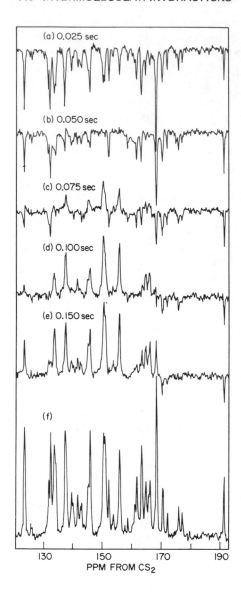

Figure 2. Partially relaxed random-coil spectra of α1-CB2 in the aliphatic region obtained using $180° - t - 90°$ pulse sequences with (a) $t = 0.025$ sec; (b) $t = 0.050$ sec; (c) $t = 0.075$ sec; (d) $t = 0.10$ sec; and (e) $t = 0.15$ sec. In each case 65K scans of the FID were accumulated with a 0.75 sec delay between the 90° and 180° pulses. For the $90° - t - 90°$ spectrum (f) 96K FID were accumulated and the vertical amplitude was multiplied by two-thirds to facilitate comparison with the partially relaxed spectra. Protein concentration, temperature, solvent, and chemical-shift scale were the same as in Fig. 1.

proton decoupling and $t = 4.5$ sec) was found to be 2.7 ± 0.3, in agreement with the theoretical value (2.61) calculated from the amino acid composition. Consistent with the finding that all coil resonances have equal (maximal) NOE values, the intensity of each resonance in the simulation is proportional to the number of carbons assigned to the resonance.

The fact that the NOE is maximal has two important consequences (8). First, the carbon relaxation times are determined solely by the ^{13}C-^{1}H dipole-dipole interaction, and second, the motional narrowing condition is valid; that is, $(\omega_H + \omega_C)^2 \tau_r^2 \ll 1$, where τ_r is a rotational correlation time and ω_C and ω_H are the respective Larmor precession frequencies of the ^{13}C and ^{1}H nuclei. Under these conditions, an effective rotational correlation time, τ_{eff}, for the vector connecting directly bonded C and H atoms is given by (11, 14)

$$\tau_{eff} = \frac{r_{CH}^{6}}{KT_1 N_H} \tag{1}$$

where r_{CH} is the internuclear distance (1.09 Å), T_1 is the spin-lattice relaxation time, N_H is the number of attached protons, and K is a constant equal to 3.56×10^{10} Å^6 sec^{-2}.

Some comment regarding the physical meaning of τ_{eff} is in order. The simplest model for α-CH rotational motion assumes that backbone segmental motion results in isotropic rotational diffusion of the α-CH internuclear vector. In this case, $\tau_{eff} = 1/(6\mathcal{R})$, where $\mathcal{R}$ is the rotational diffusion constant characterizing the α-CH reorientation (9). Two or more correlation times are required to characterize anisotropic reorientation (11, 12), and the τ_{eff} value calculated from Eq. (1) is a weighted average of these correlation times; that is, $\tau_{eff} = \Sigma c_i \tau_i$, where the c_i are orientation-dependent coefficients. The τ_i are functions of the various molecular rotation rates and quantitative values of these rates can be obtained from the τ_{eff} value only if a detailed model of the (complex) molecular motion is assumed (11, 12).

Although precise interpretation of τ_{eff} is difficult when molecular motion is complex, the τ_{eff} values in Table 1 do provide a reasonable description of the mobility of various α-carbons in the α1-CB2 random coil. For instance, of the 12 pyrrolidine C_α carbons in α1-CB2, 4.5 are Pro C_α that resonate at 132.1 ppm, whereas the remaining 7.5 pyrrolidine C_α resonate at 132.8–133.8 ppm. It is significant that three of the Pro C_α that resonate at 132 ppm and have $\tau_{eff} = 0.32$ nanosec are in residues whose position is within five residues of the chain termini, whereas the higher-field pyrrolidine α-carbons have $\tau_{eff} = 0.45$ nanosec and are in residues that are located at least seven residues from the chain ends. The various C_α that resonate in the 139–143 ppm range are also in residues away from the chain ends and have τ_{eff} values ($\sim$0.4 nanosec) closer to those of the higher-field pyrrolidine C_α. The N-terminal Gly residue has the smallest $C_\alpha \tau_{eff}$ value (0.14 nanosec), and the remaining Gly residues, distributed uniformly along the chain, have an average τ_{eff} value of 0.36 nanosec. Taken together, these results show that the position of a residue in the chain, rather than the residue type, determines τ_{eff}. This conclusion is consistent with the idea that cooperative segmental motion reorients the CH vectors in the backbone. Clearly, the motion of residues at or near chain termini requires cooperative movement of fewer atoms, resulting in smaller τ_{eff} values. These conclusions regarding the C_α motion in the α1-CB2 random coil are consistent with those obtained for backbone motion in random-coil sequential polypeptides composed of Pro and Gly residues (16).

NMR parameters in the helical state. Assignment of resonances in the aliphatic portion of the 2°C spectrum [Fig. 3(e)] is complicated by the presence of random-coil conformations. Separate spectra are expected for random-coil and helical conformations because the activation energy for the helix-coil transition is 18 kcal/mole (4). At 2°C and 7.5 mM, approximately 3% of α1-CB2 is random coil (4), and as noted, the solution contained (~1% by weight) the nonhelical fragment α1-CB1. Although the fraction of random coil is small, its spectrum [calculated in Fig. 3(a)] is a measurable component of the 2°C spectrum because of the smaller line width and larger NOE values of the coil resonances.

Coil resonances are narrow because of the rapid reorientation of the flexible polypeptide backbone. It is possible for local backbone motion to occur in the terminal triplets of the α1-CB2 trimer because the individual chains are staggered (1), the chain alignment may be imperfect, and terminal H bonds may be broken. Indeed, it was found that the experimental spectrum [Fig. 3(e)] could be satisfactorily simulated [Fig. 3(d)] only when the line widths of residues in the terminal trimers [calculated spectrum, Fig. 3(b)] were assumed to be two to five times smaller than the line widths (Table 2) used to calculate the spectrum [Fig. 3(c)] of these residues in the helical portion of the trimer. The sharper lines required for the near-terminal residues confirms the conclusion, inferred from optical rotation data (4), that the terminal regions of the trimer are not helical.

A comparison of areas of the C' resonances in the helix, with and without proton decoupling, yielded a C' NOE of 1.4 ± 0.3. The same average NOE was deduced for the aliphatic carbons, because the measured ratio of areas of aliphatic and C' resonances was equal to the theoretical value, known from the amino acid composition. The helix NOE values were also obtained by comparing areas of helix C' and aliphatic resonances with respective areas measured in random-coil spectra. This method yielded NOE values of 1.4 ± 0.3 for both C' and aliphatic carbons in the helix. Because the NOE generally shows (13, 14) a monotonic increase with increased molecular mobility, NOE values of 1.5 and 1.35 were assumed in the simulations for the flexible (near-terminal) and rigid carbons, respectively.

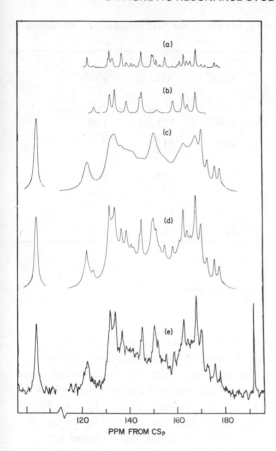

Figure 3. Comparison of α1-CB2 aromatic and aliphatic low-temperature (2°C) spectra: (a) calculated spectra of α1-CB2 and α1-CB1 random coil; (b) calculated spectra of near-terminal helix residues; (c) calculated spectrum of residues in the rigid portion of the helix; (d) sum of calculated spectra (a), (b), (c); (e) experimental spectrum at 2°C, 400K scans of the FID were accumulated, concentration 25 mg/ml in 0.15 M AcNa buffer at pH = 4.8. Chemical shifts in ppm from external CS_2.

The C_α NOE values in the helix are significantly smaller than in the coil, as are the C_α helix T_1 values (Table 2), obtained from inversion-recovery spectra (Fig. 4). The reduction in these C_α NMR parameters and the corresponding increase in C_α line widths (Table 2) that occur when the temperature is reduced from 30°C to 2°C suggest the presence of a rigid macromolecule at 2°C. It will now be shown that the low-temperature data are consistent with those expected for a rigid ellipsoid of revolution having dimensions approximating those of the α1-CB2 triple helix.

Table 2. Chemical shifts, assignments, T_1, line width, and NOE values of α1-CB2 carbons in the triple helix

Chemical Shift[a]	Carbon Residue	T_1[b] (sec)	Line Width[c] (Hz)	NOE[d]
150.1–150.9	C_α-Gly	0.025	90	1.35
145–146	C_δ-Pro, C_α-Ala	0.04	125, 54	1.35
139–143	C_α-Gln, Arg, Glu, Leu, Ala	0.04	54	1.35
137–137.5	C_δ-Hyp	0.03	150	1.35
137–137.5	C_α-Ser, Phe	0.04	54	1.35
132.8–133.8	C_α-Hyp, Pro	0.04	54	1.35
131.5–132.1	C_α-Pro, C_β-Ser	0.04	54, 60	1.35

[a] In ppm from external CS_2.
[b] Uncertainty about ± 30%.
[c] Obtained from computer simulation; uncertainty about ± 40%.
[d] Uncertainty ± 30%.

Model for the motion of the helix. It is supposed that in solution at 2°C α1-CB2 behaves like a rigid (prolate) ellipsoid of revolution undergoing rotational diffusion. Model ellipsoid dimensions were calculated for anhydrous and hydrated helices. An ideal anhydrous α1-CB2 helix has a volume $V_H = 1.14 \times 10^4$ Å^3, total length $\ell = 113$ Å, and an average diameter (3) $d = 11.4$ Å. The model ellipsoid major and minor axes, $2a$ and $2b$, were specified by the conditions $b/a = d/\ell$ and $V_H = 4\pi ab^2/3$. In the second model, bound water was assumed to increase the dimensions of the anhydrous helix by 3 Å, that is, $\ell = 116$ Å, $d = 14.4$ Å. This value of d corresponds closely to the separation of the collagen helices in hydrated fibers (1). The dimensions of the corresponding ellipsoid (Table 3) were calculated in the manner described for the anhydrous model.

Using the dimensions thus obtained, the two rotational diffusion constants that describe the reorientation of the ellipsoid can be calculated (11, 22). These diffusion constants are designated $\mathcal{R}_1$ and $\mathcal{R}_2$ and they are, respectively, the diffusion constants for reorientation about the long axis of the ellipsoid and an axis perpendicular to the long axis. Using the dimensions of the anhydrous and hydrated ellipsoid models of the helix, values of $\mathcal{R}_2$ and $\sigma = \mathcal{R}_1/\mathcal{R}_2$ were calculated (Table 3) and as expected, $\sigma \gg 1$, because both ellipsoid models have rodlike shapes.

Woessner (11) has developed formulas for the spectral densities, $J(\omega)$, needed to calculate nuclear spin relaxation rates in rigid ellipsoids. Combining Woessner's relations for $J(\omega)$, with Solomon's dipolar relaxation theory (10) yields equations that show that T_1, T_2, and NOE depend on only three variables,

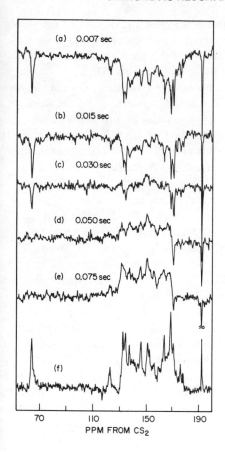

Figure 4. Low-temperature (2°C) partially relaxed spectra of α1-CB2 in the aromatic and aliphatic regions, obtained using 180° – t – 90° pulse sequences with (a) t = 0.007 sec; (b) t = 0.015 sec; (c) t = 0.030 sec; (d) t = 0.050 sec; (e) t = 0.075 sec (a 0.4 sec delay was used between the 90° and 180° pulses); (f) a 90° – t – 90° spectrum with t = 0.4 sec. In each case, 192K FID were accumulated and protein concentration, temperature, solvent, and chemical-shift scale were the same as in Fig. 3.

Table 3. Dimensions and rotational diffusion constants of α1-CB2 ellipsoid models

	Anhydrous Model	Hydrated Model
Major axis (2a)	130 A	132 A
Minor axis (2b)	13 A	16.4 A
σ (= R_1/R_2)	20	14
R_2 (theory)	2.5×10^6 sec^{-1}	2.2×10^6 sec^{-1}
R_2 (experiment)	1.0×10^6 sec^{-1}	1.3×10^6 sec^{-1}

R_2, σ, and $\cos^2\theta$, where θ is the angle between the given CH vector and the long axis of the helix. Plots of the NMR parameters as functions of R_2 and θ, with σ held constant, illustrate the results of the theory. Such plots are shown in Fig. 5 and were calculated assuming $\sigma = 14$ for $\theta = 0°, 30°, 60°$, and $90°$.

It is clear that the curves in Fig. 5 provide a reliable estimate of R_2 only if the values of $\theta_{\alpha,H}$, the angles between the α-CH internuclear vectors and the long axis of the helix are known. Because atomic coordinates are not available for the $\alpha 1$-CB2 helix, the structure of the poly(Gly-Pro-Pro) triple helix is assumed to approximate that of $\alpha 1$-CB2. From the poly(Gly-Pro-Pro) atomic coordinates (1) the following angles were calculated for the Gly-X-Y residues in the $\alpha 1$-CB2 triplets; $\theta_{\alpha 2,H}$ (Gly) $= 60°$, $\theta_{\alpha 1,H}$ (Gly) $= 120°$, $\theta_{\alpha,H}(X) = 105°$, and $\theta_{\alpha,H}(Y) = 90°$. Although the Gly C_α is bonded to two hydrogens, the fact that $\cos^2(120°) = \cos^2(60°)$ means that the $\theta = 60°$ curves in Fig. 5 can be used to determine τ_A from the Gly relaxation data, provided that the measured values of the Gly T_1 and line width are multiplied respectively by 2 and 0.5. Since $2T_1$ for the Gly C_α is 0.045 sec, two values of τ_A (6 or 170 nanosec) are possible. However, the small value of the Gly C_α NOE (1.35 + 0.3) rules out the smaller value of τ_A. In concert with this result, the value of the Gly C_α line width/2 (45 Hz) corresponds to $\tau_A = 120$ nanosec. Combining T_1 and line width data for the three types of residues in the triplet yields an average τ_A of 130 nanosec for the hydrated ellipsoid and $\tau_A = 170$ nanosec for the ellipsoid model, assuming no bound water. A comparison of the corresponding values of R_2 for both models obtained from these τ_A values $[R_2 = (6\tau_A)^{-1}]$ (Table 3) and the theoretical values of R_2 reveals that the theoretical values of R_2 do not differ greatly because R_2 is not a sensitive function of b. Considering the uncertainties in both theory and experiment, there is reasonable agreement between the theoretical and experimental values of R_2 for both models, indicating that at $2°C$, $\alpha 1$-CB2 diffuses in solution as a rigid rod of dimensions 11.5-14.5 Å by 110-120 Å.

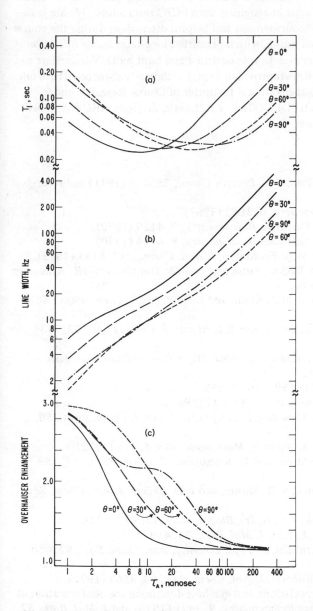

Figure 5. Calculated ^{13}C NMR parameters: (a) T_1; (b) line width = $1/(\pi T_2)$; (c) NOE, plotted as a function of $\tau_A = (6\Re_2)^{-1}$ and θ, the angle between the CH bond and the ellipsoid symmetry axis, for $\theta = 0°$ (—), $\theta = 30°$ (····), $\theta = 60°$ (—–), and $\theta = 90°$ (–·–·). The calculations assume that one hydrogen is bonded to the carbon ($r_{CH} = 1.09$ Å) and that $\sigma = 14$, where $\sigma = \Re_1/\Re_2$.

ACKNOWLEDGMENTS

We are indebted to D.F. DeTar and E.R. Blout for kindly providing the model polypeptides that were crucial in assigning the $\alpha 1$-CB2 resonances. We are grateful also to K.A. Piez for encouragement and helpful discussions during the course of this work. The Dewared probe, which permitted long-term stable operation of the spectrometer system at $2°C$, was designed and built by D. VanderHart and H.M. McIntyre, and D.E. Brown provided expert technical assistance. The work was supported, in part, by the National Institutes of Dental Research under interagency agreement #Y01-DE-30013. J.R. Lyerla, Jr., is an NRC-NAS Postdoctoral Research Associate.

REFERENCES

1. W. Traub and K.A. Piez, *Adv. Protein Chem.*, **25**, 243 (1971) and references therein.
2. P. Bornstein, *Biochemistry*, **6**, 3082 (1967).
3. K.A. Piez and M.R. Sherman, *Biochemistry*, **9**, 4129 (1970).
4. K.A. Piez and M.R. Sherman, *Biochemistry*, **9**, 4134 (1970).
5. J.R. Lyerla, Jr., and M.H. Freedman, *J. Biol. Chem.*, **247**, 8183 (1972).
6. T.C. Farrar and E.D. Becker, *Pulse and Fourier Transform NMR*, Academic Press, New York, 1970.
7. R.L. Vold, J.S. Waugh, M.P. Klein, and D.E. Phelps, *J. Chem. Phys.*, **48**, 3831 (1968).
8. K.F. Kuhlmann, D.M. Grant, and R.K. Harris, *J. Chem. Phys.*, **52**, 3439 (1970).
9. A. Abragam, *The Principles of Nuclear Magnetism*, Oxford University Press, London, 1961.
10. I. Solomon, *Phys. Rev.*, **99**, 559 (1955).
11. D.E. Woessner, *J. Chem. Phys.*, **37**, 647 (1962).
12. D.E. Woessner, B.S. Snowden, Jr., and G.H. Meyer, *J. Chem. Phys.*, **50**, 719 (1969).
13. J. Schaefer and D.F.S. Natusch, *Macromolecules*, **5**, 416 (1972).
14. A. Allerhand, D. Doddrell, and R. Komoroski, *J. Chem. Phys.*, **55**, 189 (1971).
15. T.C. Farrar, S.J. Druck, R.R. Shoup, and E.D. Becker, *J. Am. Chem. Soc.*, **94**, 699 (1971).
16. D.A. Torchia and J.R. Lyerla, Jr., *Biopolymers*, **13**, 97 (1974).
17. D.A. Torchia and K.A. Piez, *J. Mol. Biol.*, **76**, 419 (1973).
18. W.J. Horsley, H. Sternlicht, and J.S. Cohen, *J. Am. Chem. Soc.*, **92**, 680 (1970).
19. M. Christl and J.D. Roberts, *J. Am. Chem. Soc.*, **94**, 4565 (1972).
20. For IUPAC-IUB abbreviations and symbols describing the conformation of polypeptide chains see *Biochemistry*, **9**, 3471 (1970) and *J. Mol. Biol.*, **52**, 1 (1970).
21. D.A. Torchia, J.R. Lyerla, Jr., and A.J. Quattrone, in preparation.
22. F. Perrin, *J. Phys. Radium*, **5**, 497 (1936).

ORDERED WATER IN COLLAGEN

E. SUZUKI and **R.D.B. FRASER**, CSIRO Division of Protein Chemistry, 343 Royal Parade, Parkville, Victoria 3052, Australia

SYNOPSIS: Measurements of infrared dichroism in partly hydrated sections of kangaroo-tail tendon are described. The spectra were placed on a molar absorptivity basis and the water content was estimated by reference to the spectra of model compounds. The spectrum in the amide A region is overlaid by a broad absorption band caused by sorbed water, which exhibits perpendicular dichroism, indicating that ordered water is present. The value of the dichroic ratio, 3450 cm^{-1}, which corresponds to the maximum of the band associated with the antisymmetrical stretching vibration of the water molecule, is such that an estimate of $75 \pm 15°$ can be made for the angle between the fiber axis and the line joining the two hydrogen atoms in the water molecule. The dichroisms measured at 3250 and 6800 cm^{-1} also are perpendicular, and because the absorption in these regions is likely to contain significant contributions from vibrations in which the associated transition moment is parallel to the symmetry axis of the water molecule, it seems likely that this axis is also preferentially oriented perpendicular to the fiber axis.

INTRODUCTION

Collagen and collagenlike sequential polypeptides have a conformation in which it is not possible for all the amide groups to participate in intramolecular or intermolecular hydrogen bonds (1). Evidence for the presence of ordered water molecules in collagen has been obtained from a study of the dichroism of combination bands in the infrared spectrum (2) and from measurements of nuclear magnetic resonance (NMR) spectra (3-5), and it seems likely that the major part of this ordered water is immobilized through the formation of hydrogen bonds with the amide groups of the polypeptide main chain. There has been a considerable amount of speculation about the possible location of ordered water in collagen and collagenlike sequential polypeptides (6-9), and if sufficiently precise x-ray-diffraction and spectroscopic data were available it would be possible to carry out objective tests of these and alternative models. In the present communication we report a quantitative study of the infrared dichroism exhibited by ordered water in collagen; an account of the x-ray-diffraction studies will be reserved for a later publication.

EXPERIMENTAL PROCEDURE

Tendons from the tail of a mature female specimen of Bennett's Wallaby (*Macropus rufogriseus fruticus*) were washed in distilled water, stretched approximately 6% in length and air-dried while held in the extended state.

For measurements in the fundamental region, longitudinal sections 1.5–2 μm thick were cut from the tendon with a microtome fitted with a glass knife and mounted under tension over a rectangular aperture of dimensions $180 \times 5000\,\mu$m cut in platinum foil. The assembly was mounted in an evacuable temperature-controlled cell fitted with barium fluoride windows. Sections were dried by evacuating the cell to a pressure of 2.10^{-6} torr, or partly hydrated by connecting the cell to a water reservoir maintained at $25°$C. In both cases the cell temperature was maintained at $35°$C. Spectra were recorded with a Beckman IR9 spectrophotometer equipped with a selenium transmission polarizer and refracting beam condensors in both the sample and the reference beams.

For measurements in the overtone region, individual tendons were flattened between glass plates while being held under tension. Pieces of the flattened tendon were equilibrated to the desired water content and then mounted in a silica cell with a path length of 1 mm. The unoccupied volume was filled with hexachlorobutadiene to reduce scattering. Spectra were measured with a Beckman DK2 spectrophotometer and the radiation incident on the specimen was polarized by means of a suitably oriented sheet of Polaroid HR polarizing film. A similar sheet was placed in the reference beam.

Spectra of the model compounds, N-ethylpropionamide and N-propionylpyrrolidine, were obtained in the fundamental region using the IR9 spectrophotometer. Both compounds, which are liquids at $25°$C, were introduced into fixed-thickness cells with barium fluoride windows.

DETERMINATION OF MOLAR ABSORPTIVITY

In order that spectra from specimens with different water contents can be compared quantitatively and correlated with spectra obtained from model compounds, it is necessary to convert measured absorbances to molar absorptivities. The molar absorptivity ϵ is conveniently expressed in units of mole^{-1}-liter-cm^{-1}, in which case

$$\epsilon = \frac{AM}{1000\,\rho t} \tag{1}$$

where A is absorbance, M is molecular (or residue) weight, ρ is the density in g-cm^{-3}, and t is the section or cell thickness in cm.

In order that Eq. (1) may be applied to partially hydrated collagen it is necessary to know the values of ρ and t that were obtained during measurement of the spectrum. It would be difficult to make such measurements directly, so the alternative procedure of using the integrated area of the CH stretching bands about 2900 cm^{-1} as an internal standard was chosen. An initial calibration was carried out by comparing the integrated area of the amide-I band in dried tendon with the integrated molar absorptivities of the amide-I bands in N-ethylpropion-amide (a) and N-propionylpyrrolidine (b),

$$C_2H_5-NHCO-C_2H_5$$

$$\begin{array}{c} CH_2-CH_2 \\ | \quad\quad\quad\quad NCO-C_2H_5 \\ CH_2-CH_2 \end{array}$$

(a)   (b)

which serve as models for amino acid residues and imino acid residues, respectively. The integrated molar absorptivities of the amide-I bands in (a) and (b) were combined in the ratio appropriate to tendon collagen (0.784:0.216) as a basis for determining the value of the product ρt appropriate to the dried-tendon spectrum. Because tendon is optically anisotropic, it was necessary to calculate the effective isotropic absorbance A from the values A_π and A_σ, measured with the electric vector vibrating, respectively, parallel and perpendicular to the optic axis, using the formula

$$A = \frac{1}{3}(A_\pi + 2A_\sigma) \tag{2}$$

Once the product ρt was known for one specimen, the values of ρt for other specimens could be determined by comparing the integrated areas of the CH stretching band. The values so obtained will be in error by about 5–10% because of neglect of the contribution by side-chain groups to the measured area of the amide-I band in collagen.

DETERMINATION OF WATER CONTENT

It is extremely difficult to measure infrared spectra under conditions of controlled humidity because the specimen temperature will be higher than that of the cell because of the heating effect of the radiation. The relative humidity in the immediate vicinity of the specimen, therefore, will be some unknown amount lower than that in the main body of the cell. The quantity of interest in structural studies is generally the number of water molecules sorbed per residue, and even if the relative humidity in the vicinity of the specimen is known accurately, the water content must be derived from sorption isotherms measured on bulk specimens.

Because the spectra obtained from tendon had been placed on a molar absorptivity basis, as outlined in the previous section, it was possible to estimate the water content in the partially hydrated specimen from the intensity of the broad band centered around 3450 cm^{-1} (Fig. 1) that may be assigned to sorbed water (10). When water is sorbed onto polymers the spectrum differs from that of

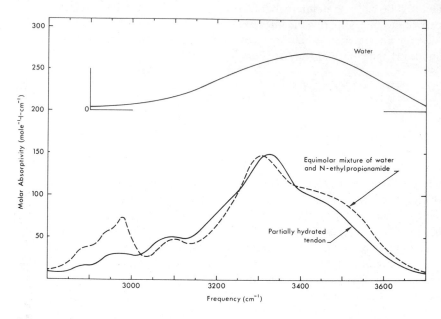

Figure 1. Infrared spectra of liquid water and an equimolar mixture of
N-ethylpropionamide and water. In the latter spectrum the absorptivity refers
to 1 mole of N-ethylpropionamide. The lower full curve is the spectrum of
iostropic partly hydrated tendon calculated from the polarized spectra given
in Fig. 2 using Eq. (2).

liquid water (11) (Fig. 1). A better estimate of the water content of partly
hydrated tendon was obtained by comparing the spectrum with that obtained
from N-ethylpropionamide and water in known molar proportions. The intensity of the water band per amide group in an equimolar mixture of water and
N-ethylpropionamide was found to be close to that of the intensity per residue
in the partly hydrated collagen specimen. It was estimated that the water content of the specimen was 1.0 ± 0.2 water molecules per residue, equivalent to
about 20 g of water per 100 g of dry tendon. According to the sorption iso-
therm given by Rougvie and Bear (12), this ratio corresponds to a relative
humidity of about 50%. In the experimental arrangement used the water
reservoir was maintained at a temperature of approximately 25°C, and 50%
relative humidity would correspond to a specimen temperature of 37.2°C. As
noted earlier, the specimen cell was maintained at 35°C during the measurement.

INTERPRETATION OF DICHROISM

Fig. 2 shows the spectrum of partially hydrated tendon containing approximately one water molecule per residue. It was obtained using polarized radiation. The amide-A band, centered at 3330 cm^{-1} in the perpendicular spectrum

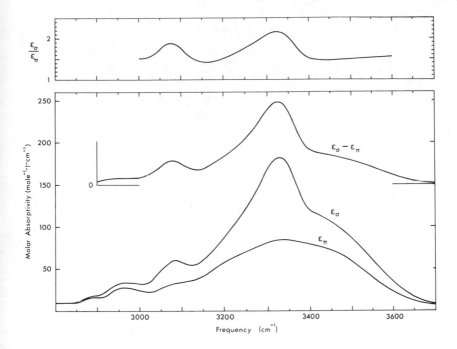

Figure 2. Infrared spectrum of partly hydrated tendon section, which contains about one water molecule per residue, obtained with the electric vector vibrating parallel to the fiber axis (ϵ_π) and perpendicular to the fiber axis (ϵ_σ).

(ϵ_σ), is superimposed on a broad absorption band with a maximum about 3450 cm^{-1}. On the basis of earlier studies, this band may be assigned to water (10, 11) together with a small contribution from side-chain groups (13). The amide-A band is very weak in the parallel spectrum (ϵ_π) and the dichroic ratio, after correction for overlap by water absorption, would appear to be in the range 1:10–1:20.

The water band appears to be more intense in the perpendicular spectrum. This observation is confirmed by the difference spectrum $\epsilon_\sigma - \epsilon_\pi$ shown in Fig. 2. The perpendicular dichroism in the range 3400–3600 cm^{-1} may be attributed to the presence of ordered water in the structure.

In the vapor state, the water molecule has three normal modes of vibration (14), which are shown in Fig. 3 together with the associated transition moments. In the liquid state, the absorption bands are broadened because of the formation of hydrogen bonds and the absorption maximum associated with ν_3 is reduced in frequency (15), the value being about 3450 cm^{-1}. No distinct band associated with ν_1 is apparent, but it has been suggested that the shoulder at 3285 cm^{-1} (Fig. 1) is associated with this mode (15).

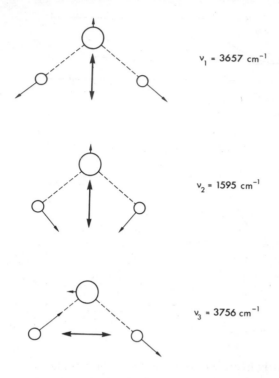

$v_1 = 3657$ cm^{-1}

$v_2 = 1595$ cm^{-1}

$v_3 = 3756$ cm^{-1}

Figure 3. Normal modes of vibration of an isolated water molecule together with the corresponding frequencies observed in the vapor state and the directions of the associated transition moments (bold double-headed arrows).

If the perpendicular dichroism in the range 3400–3600 cm^{-1} (Fig. 2) is attributed to absorption associated with v_3, it is possible to draw conclusions about the preferred orientation of the ordered water. The ratio $\epsilon_\sigma/\epsilon_\pi$ (Fig. 2) is essentially constant in the range 3400–3600 cm^{-1} and the dichroic ratio R, defined as $\epsilon_\pi/\epsilon_\sigma$, has a value close to 1/1.5. The transition-moment direction associated with v_3 is parallel to the line joining the two hydrogen atoms (Fig. 3), and if all the water molecules were equivalent and the ordering perfect, the inclination θ_a of the HH direction to the fiber axis could be calculated from the expression (16)

$$\theta_a = \cot^{-1}(\tfrac{1}{2}R)^{1/2} = 60°$$

(3)

Because it cannot be assumed that the ordering is complete, it is preferable to restate the result in the form of the inequality relationship for perpendicular dichroism (16)

$$\cot^{-1}(\tfrac{1}{2}R)^{\frac{1}{2}} \leqslant \theta_a \leqslant 90° \tag{4}$$

which gives an estimate for θ_a of $75 \pm 15°$ (Fig. 4).

The transition-moment direction associated with ν_1 is parallel to the twofold symmetry axis of the water molecule (Fig. 3). If it is assumed that the low-frequency portion of the OH stretching band (Fig. 1) arises from ν_1, the perpendicular dichroism in partly hydrated collagen (Fig. 2) indicates that the ordered water molecules are oriented in such a manner that θ_s (Fig. 4) lies in the range $54.7°$–$90°$ (16). To check this conclusion, we measured the dichroism in the prominent water band about 6800 cm^{-1} and found it to be perpendicular across the entire breadth of the band. Three binary combinations might be expected to contribute to the absorption in this region. These are, in ascending order of frequency: $2\nu_1, (\nu_1 + \nu_3)$, and $2\nu_3$. The transition moment associated with $(\nu_1 + \nu_3)$ has the same direction as that for ν_3 (Fig. 3), but the other two have transition-moment directions parallel to the twofold axis. The uniformly perpendicular character of the 6800 cm^{-1} band lends support to the conclusion that both transition-moment directions are preferentially oriented perpendicular to the fiber axis, and from Eq. (3) it may be concluded that $54.7° < \theta_a \leqslant 90°$ and $54.7° < \theta_s \leqslant 90°$. It is not feasible to use the measured values of the dichroic ratio in the 6800 cm^{-1} band to set further limits on θ_a and θ_s because the contributions of the three binary combinations to the 6800 cm^{-1} band are not clearly resolved.

The spectrum of tendon dried at $35°$C *in vacuo* is shown in Fig. 5. The major portion of the sorbed water has been removed, but the difference curve $\epsilon_\sigma - \epsilon_\pi$

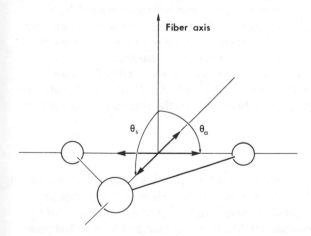

Figure 4. Preferred orientation of the ordered water molecules in partly hydrated tendon, deduced from the observed perpendicular dichroisms of the absorption band associated with $\nu_3(\theta_a)$ and $2\nu_3(\theta_s)$.

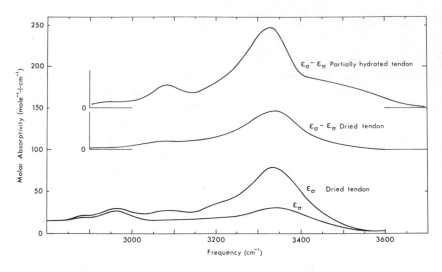

Figure 5. Infrared spectrum of dried tendon section obtained with the electric vector vibrating parallel (ϵ_π) and perpendicular (ϵ_σ) to the fiber axis. The difference spectrum ($\epsilon_\sigma - \epsilon_\pi$) is compared with that of a partly hydrated specimen containing about one water molecule per residue.

shows some evidence of residual perpendicular dichroism in the vicinity of the maximum of the water band. It seems likely, therefore, that the ordered water is not entirely removed by drying at 35°C. This result is consistent with findings of Rougvie and Bear (12), but it is contrary to the conclusions reached by Susi *et al.* (17, 18). Drying at 105°C *in vacuo* produced a slight reduction in intensity at 3450 cm^{-1}. The effect was small, however, and did not exceed the experimental error by a sufficient margin to warrant quantitative analysis.

The dichroism of the amide-A band is reduced and the band width is increased in the spectrum of tendon dried at 35°C (Fig. 5). Both of these observations are consistent with a progressive disordering of the structure as the ordered water is removed.

DISCUSSION

In an earlier study of ordered water in tendon (2) it was found that the combination band at 5150 cm^{-1} caused by sorbed water exhibited perpendicular dichroism. If the 5150 cm^{-1} band is assumed to be due to a combination of ν_2 and ν_3 (Fig. 3), the transition-moment direction is the same as for ν_3. The present result, in which a perpendicular dichroism associated with ν_3 was found for the 3450 cm^{-1} band is therefore consistent with the earlier study of the combination band at 5150 cm^{-1}.

The infrared data does not permit the precise location of the ordered water to be determined, but it places severe restrictions on possible models. Some suggested models involve the formation of a single hydrogen bond between a water molecule and a main-chain amide group, but, in general, this situation would not account for appreciable preferential orientation of the HH direction. A more likely explanation is that the ordered water molecules act as donors in the formation of hydrogen bonds to two CO groups. Specific models in which water molecules form pairs of hydrogen bonds between CO groups in adjacent chains in the collagen molecule have been suggested by Burge *et al.* (6) and by Ramachandran and Chandrasekharan (8). In both cases, the water molecule is presumed to form a bridge between O_3 of one chain and O_1 of another chain in a conformation topologically similar to that of the collagen-II model described by Rich and Crick (7) (Fig. 6). The observed perpendicular dichroisms of the bands caused by sorbed water are consistent with such an orientation.

Yonath and Traub (9) proposed a detailed model for the structure of the sequential polypeptide poly(propyl-Gly-Pro), which adopts a collagenlike conformation, and suggested the presence of two ordered water molecules per asymmetric unit of three residues. Although coordinates were not given for the hydrogen atoms, reasonable assumptions about their location lead to the prediction that the 3450 cm^{-1} absorption band will exhibit pronounced parallel

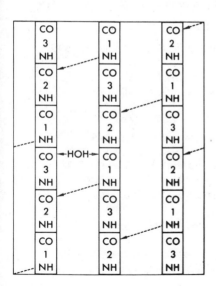

Figure 6. Topological arrangement of hydrogen bonds in the collagen-II model described by Rich and Crick (7). Residue 1 is always glycine. One hydrogen bond per asymmetric unit of 3 residues is formed between the NH of residue 1 and the CO group of residue 2 in an adjacent chain. Possible sites for an ordered water molecule are indicated.

dichroism. The observed dichroism in tendon is not consistent with the presence of two ordered water molecules per asymmetric unit in the positions suggested for the sequential polypeptide.

ACKNOWLEDGMENTS

We are grateful to the CSIRO Division of Wildlife Research for the supply of a wallaby tail and to F.H.C. Stewart for supplying the model compounds. We are also indebted to E.F. Woods for measuring the densities of the model compounds and to E.G. Bendit and T.P. MacRae for helpful discussions.

REFERENCES

1. R.D.B. Fraser and T.P. MacRae, *Conformation in Fibrous Proteins,* Academic Press, New York, 1973.
2. R.D.B. Fraser and T.P. MacRae, *Nature* (London), **183**, 179 (1959).
3. H.J.C. Berendsen, *J. Chem. Phys.,* **36**, 3297 (1962).
4. H.J.C. Berendsen and C. Migchelson, *Ann. N.Y. Acad. Sci.,* **125**, 365 (1965).
5. G.E. Chapman and K.A. McLauchlan, *Proc. Roy. Soc.* (London), **B173**, 223 (1969).
6. R.E. Burge, P.M. Cowan, and S. McGavin, in *Recent Advances in Gelatin and Glue Research,* G. Stainsby, ed., Pergamon Press, New York, 1958, p. 25.
7. A. Rich and F.H.C. Crick, *J. Mol. Biol.,* **3**, 483 (1961).
8. G.N. Ramachandran and R. Chandrasekharan, *Biopolymers,* **6**, 1649 (1968).
9. A. Yonath and W. Traub, *J. Mol. Biol.,* **43**, 461 (1969).
10. E.M. Bradbury, R.E. Burge, J.T. Randall, and G.R. Wilkinson, *Disc. Faraday Soc.,* **25**, 173 (1958).
11. C. de Lozé and M.L. Josien, *Biopolymers,* **8**, 449 (1969).
12. M.A. Rougvie and R.S. Bear, *J. Am. Leather Chem. Assoc.,* **48**, 735 (1953).
13. E.G. Bendit, in *Symposium on Fibrous Proteins Australia 1967,* W.G. Crewther, ed., Butterworths (Australia), Sydney, 1968, p. 386.
14. D. Eisenberg and W. Kauzmann, *The Structure and Properties of Water,* Oxford University Press, London, 1969.
15. J.G. Bayly, V.B. Kartha, and W.H. Stevens, *Infrared Physics,* **3**, 211 (1963).
16. R.D.B. Fraser, in *Analytical Methods of Protein Chemistry,* Vol. 2, P. Alexander and R.J. Block, eds., Pergamon Press, New York, 1960, p. 287.
17. H. Susi, J.S. Ard, and R.J. Carroll, *Biopolymers,* **10**, 1597 (1971).
18. H. Susi, J.S. Ard, and R.J. Carroll, *J. Am. Leather Chem. Assoc.,* **66**, 508 (1971).

THE STEREOCHEMISTRY OF THE ACTINOMYCIN D–DNA COMPLEX IN SOLUTION

DINSHAW J. PATEL, Bell Laboratories, Murray Hill, N.J. 07974

SYNOPSIS: The structures of 1:2 complexes of actinomycin D with d-pG, d-pGpC, and d-ApTpGpCpApT in aqueous solution have been investigated by nuclear magnetic resonance (NMR) spectroscopy. The stereochemical relationship between the stacked nucleic acid bases and the phenoxazone ring of the antibiotic have been evaluated by comparison of the experimental complexation chemical shifts and those predicted from ring-current calculations. Phosphorus-31 NMR spectroscopy can be a powerful probe for locating the antibiotic intercalation site in a double-stranded nucleotide sequence because the internucleotide phosphates at the intercalation site in d-pGpC and d-ApTpGpCpApT exhibit downfield-shifted ^{31}P resonances on complexation with the antibiotic. The guanine 2-NH$_2$ protons of d-pG and d-pGpC shift downfield upon complex formation with actinomycin D in aqueous solution, in agreement with their participation in intermolecular hydrogen bond formation in aqueous solution. The NMR resonance of the guanine-N$_1$H and thymine-N$_3$H Watson–Crick hydrogen-bonded protons are used as probes for the double-stranded structure of d-ApTpGpCpApT in aqueous solution to demonstrate that the phenoxazone ring of the antibiotic intercalates between the (GC)$_{central}$ base pairs. The NMR data in solution support the model of the Act-D–DNA complex put forward by Sobell and Jain.

INTRODUCTION

Several detailed stereochemical models incorporating the principles of intermolecular hydrogen bonding (1) and intercalation (2, 3) have been proposed for the complex of actinomycin D (Act-D), a peptide antibiotic, with double-stranded DNA (4–7). The structure of actinomycin D is shown in Fig. 1. The Sobell–Jain model (8) is based on an x-ray–crystallographic analysis of the complex of Act-D with two deoxyguanosines (9). From this crystallographic data, Sobell and Jain proposed models for the 1:2 Act-D:d-pGpC and 1:2 Act-D:d-ApTpGpCpApT complexes (8). In their model, the phenoxazone ring intercalates between GC and CG Watson–Crick base pairs with the pentapeptide lactone rings extending over two base pairs on either side of the intercalation site. The symmetry axis relating the Act-D pentapeptide lactone rings coincides with the symmetry axis relating the sugar-phosphate backbone and base sequence of the double helix (10).

The Sobell–Jain proposal for complex formation between actinomycin D and DNA (10) led to a systematic spectral investigation of complex formation between Act-D and d-pG (11–13), d-pGpC (14–17), and d-ApTpGpCpApT (18) in

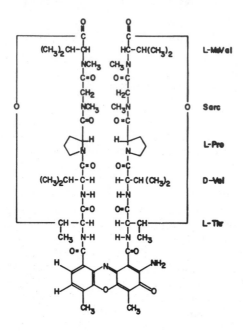

Figure 1. Chemical structure of actinomycin D.

aqueous solution. Experiments in our laboratory (13, 17, 18) were initiated to determine whether high-resolution NMR could be used to monitor the intercalation and hydrogen-bond interactions proposed to account for the specificity and stability of polypeptide–nucleic acid complexes (10).

1:2 ACT-D:d-pG AND 1:2 ACT-D:d-pGpC COMPLEXES IN SOLUTION

Complex formation. The proton and phosphorus NMR studies were undertaken by adding Act-D gradually to d-pGpC (or d-pG) in aqueous solution. Beyond a 1:2 ratio of Act-D:d-pGpC (or Act-D:d-pG), the antibiotic precipitated. Because d-pGpC (or d-pG) is water soluble and the antibiotic sparingly so, the data suggest that the stoichiometry of the water-soluble complex is 1 equivalent of antibiotic to 2 equivalents of dinucleotide (or mononucleotide).

The proton NMR resonances of d-pG shifted as average resonances upon addition of Act-D in aqueous solution. The shift suggests rapid exchange of the antibiotic between free and complexed forms. The lifetime of this exchange process is estimated to be $\ll 2$ msec.

The ^{31}P spectrum of a mixture of 1 equivalent of Act-D and 4 equivalents of d-pGpC in H_2O (pH 8) has been studied as a function of temperature. The terminal phosphorus atom exhibits a narrow resonance at 9°C, but the internucleotide phosphorus atom (which shifts 1.6 ppm downfield on complex formation) is considerably broadened. When the temperature is raised, this resonance undergoes exchange narrowing. The data suggest that the exchange process

$$2\, d\text{-}pGpC + 1:2\ \text{Act-D}:d\text{-pGpC} \underset{k_{-1}}{\overset{\kappa_1}{\rightleftharpoons}} 2\ d\text{-pGpC} + 1:2\ \text{Act-D}:d\text{-}pGpC \qquad (1)$$

exhibits intermediate rates on the NMR time scale at 9°C. The exchange process was simulated for two sites of equal population separated by 63 Hz (i.e., 1.6 ppm at 40 MHz) and undergoing rapid exchange with lifetime broadening. The average lifetime, τ, was determined to be 2.6 msec and 0.23 msec at 9° and 27°C, respectively. The rate constants, k_1 and k_{-1} for process (1) are equal to $\frac{1}{2}\tau$, permitting an estimation of the free energy of activation at 9°C to be 13.6 kcal/mole. kcal/mole.

Stacking interactions (1:2 Act-D:d-pG). Upfield proton (11–13) and carbon (13) chemical shifts of the purine ring of d-pG upon complexation with Act-D in solution suggest that the purine and phenoxazone rings stack in the 1:2 Act-D:d-pG complex. The stacking geometry in the complex has been evaluated based on the ring currents of the phenoxazone ring on the purine carbon chemical shifts of d-pG and the ring currents of the purine ring (19) on the phenoxazone proton chemical shifts. Compared to a unique orientation of stacked phenoxazone and purine rings in the structure of the complex in the crystal (9), a range of stacking geometries are suggested in aqueous solution (13).

Stacking interactions (1:2 Act-D:d-pGpC). The 300 MHz proton NMR spectra of 1:2 Act-D:d-pGpC in D_2O (pH 5.8) at 26°C have been recorded and many of the resonances of the peptide and nucleotide identified by spin decoupling techniques. Selective phenoxazone CH_3 and H chemical shifts for the complex in solution are compared with those reported by Angerman et al. (20) for Act-D in aqueous solution under conditions in which dimerization is negligible. Neglecting shifts of <0.1 ppm, the CH_3 and H protons of the phenoxazone ring of the antibiotic are shifted upfield on complexation (16, 17).

Giessner-Prettre and Pullman (19) have evaluated the ring-current shifts in a plane parallel to and at a distance of 3.4 Å from the guanosine purine and cytosine pyrimidine ring. The shielding ring-current contours are presented in Fig. 2. The CH_3-4 and CH_3-6 groups in 1:2 Act-D:d-pGpC are predicted to be upfield-shifted 0.2 ppm by the ring currents of the stacked base rings in the Sobell–Jain model of the complex (8). The experimental upfield shifts are 0.46 and 0.37 ppm for the CH_3 resonances. The predicted upfield shifts of the phenoxazone H-7 and H-8 from the ring currents of the GC base pairs stacked on either side (Fig. 2) are 0.5 and 0.2 ppm, respectively. Experimentally, upfield shifts of 0.68 and 0.23 ppm are observed for these benzenoid protons on complex formation. Thus, the experimental proton complexation shifts are consistent with the Sobell–Jain model (8) of the stacking geometries in the 1:2 Act-D:d-pGpC complex.

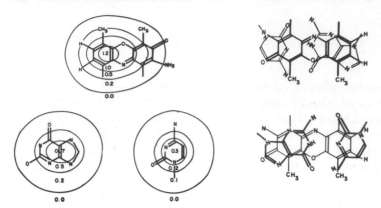

Figure 2. (Left) Ring current contours (in ppm) for the guanine, cytosine, and phenoxazone rings in a plane parallel to and at a distance of 3.4 Å from the ring systems. (Right) The stacking orientations of the Watson–Crick GC base pairs above and below the phenoxazone ring of the antibiotic.

Intercalation. For 1:2 Act-D:d-pGpC, the terminal and internucleotide phosphorus atoms are distant from and in the plane of the phenoxazone ring, respectively (8). On complex formation, the terminal phosphorus of d-pGpC undergoes an 0.1 ppm upfield shift and the internucleotide phosphorus atom undergoes a 1.7 ppm downfield shift (Fig. 3).

Schematic representations of the structures of 1:2 Act-D:d-pCpG and 1:2 Act-D:d-pGpC derived from the Sobell–Jain proposal (8) and supported by spectral measurements (14–16) are presented in Fig. 4. For the complex with d-pCpG, only the purine rings stack with the phenoxazone ring, but in the complex with d-pGpC, the phenoxazone ring is intercalated between GC and CG Watson–Crick base pairs. In contrast to the 1.6 ppm downfield shift for the internucleotide phosphate of d-pGpC on complex formation with Act-D (Fig. 3), no shift is observed for the internucleotide phosphate of d-pCpG upon complex formation with Act-D. It appears, therefore, that intercalation results in a large ^{31}P downfield chemical shift of the phosphate resonances at the intercalating site (17).

Hydrogen bonding (1:2 Act-D:d-pG). From the intercalation geometry of the complex in the crystal (9) and the phenoxazone ring-current contours, an average upfield shift of ~0.2 ppm is predicted for the $G-NH_2$ protons resulting from the ring currents of the phenoxazone ring. Experimentally, the $G-NH_2$ protons shift downfield by 0.1 ppm from 6.29 to 6.39 ppm upon complexation with Act-D in aqueous solution at 26°C (Fig. 5). Correcting for the ring-current contribution, the ~0.3 ppm downfield shift is assigned to hydrogen-bond formation between the $G-NH_2$ proton(s) and accepter group(s) on the Act-D (13).

The $G-NH_2$ protons and the exchangeable phenoxazone amino protons ($A-NH_2$) are selectively broadened out in the 1:2 Act-D:d-pG complex when the temperature is lowered from 37°C to 2.5°C. Thus, it is proposed that the rates of rotation of the amino groups of $A-NH_2$ and $G-NH_2$ about the C—N

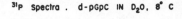

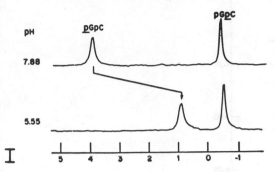

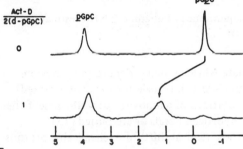

Figure 3. The 40 MHz ^{31}P NMR spectra relative to 16% phosphoric acid. (I) 0.05 M d-pGpC in D_2O (pH 7.88 and 5.55) at 8°C; (II) 0.05 M d-pGpC in D_2O (pH 7.88) at 9°C and 0.025 M 1:2 Act-D:d-pGpC in D_2O (pH 7.88) at 8°C.

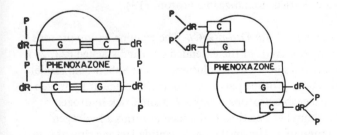

Figure 4. (Left) Intercalation of the phenoxazone ring between GpC sequences in 1:2 Act-D:d-pGpC complex. (Right) The stacking of the guanosine rings with the phenoxazone ring in the 1:2 Act-D:d-pCpG complex.

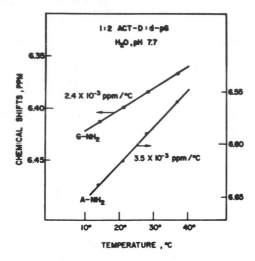

Figure 5. Plot of the temperature dependence of guanine 2-NH$_2$ resonance in d-pG and 1:2 Act-D:d-pG in H$_2$O (pH 7.7).

bonds are decreased to an intermediate NMR exchange condition in the complex, compared to fast rotation on the NMR time scale for the uncomplexed species in solution. Such restricted rotations of amino groups could arise if their exchangeable protons participate in hydrogen bonds or are buried (13).

The temperature dependence of the chemical shifts of exchangeable protons (21) has been utilized extensively in polypeptide conformational studies to determine whether the proton is hydrogen-bonded or buried, on the one hand ($\leqslant 2 \times 10^{-3}$ ppm/°C), or exposed to solvent, on the other hand ($\geqslant 6 \times 10^{-3}$ ppm/°C). The temperature coefficient of the G-NH$_2$ protons at 6.4 ppm in the 1:2 Act-D:d-pG complex has a value of 2.4×10^{-3} ppm/°C, which is characteristic of hydrogen-bonded or buried exchangeable protons (13).

Hydrogen-bond formation (1:2 Act-D:d-pGpC). The guanine amino protons (G-NH$_2$) are exposed to water solvent in the extended (22) and stacked (23, 24) conformation(s) of dinucleotides in aqueous solution. In the Sobell–Jain model of the complex (8), both amino protons participate in intermolecular hydrogen bonding. One of the exchangeable protons forms a Watson–Crick hydrogen bond with the cytosine carbonyl at position 1, the second forms a hydrogen bond with the carbonyl group of L-Thr on the pentapeptide lactone ring of Act-D.

From the intercalation geometry of the Sobell–Jain model of 1:2 Act-D:d-pGpC (8) and the phenoxazone ring-current contours, an average upfield shift of ~0.2 ppm is predicted for the G-NH$_2$ protons, resulting from the ring currents of the phenoxazone ring. Experimentally, the G-NH$_2$ proton of d-pGpC undergoes a downfield shift (0.6 ppm at 30°C) on complexation with Act-D in aqueous solution (Fig. 6). This shift, which persists over a temperature range of

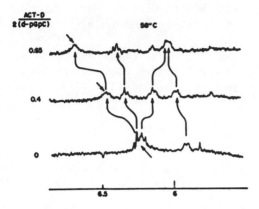

ADDITION OF ACT-D TO d-pGpC IN H₂O, pH 5.8

Figure 6. The 300 MHz ^{1}H NMR spectra of the gradual addition of Act-D to 0.025 M d-pGpC at 58°C. The straight arrows designate the chemical shifts of the G-NH$_2$ resonance.

10° to 50°C, is attributed to the participation of the G-NH$_2$ protons in intermolecular hydrogen bonding with accepter groups (17).

1:2 ACT-D:d-ApTpGpCpApT COMPLEX IN SOLUTION

Complex formation between Act-D and the hexanucleotide d-ApTpGpCpApT has been investigated by ^{1}H and ^{31}P NMR spectroscopy (18). The proton spectra monitor the Watson–Crick base pairs (G-N$_1$H and T-N$_3$H resonances), and the phosphorus spectra monitor the sugar-phosphate backbone.

In the all-or-none approximation, the transition from a fully Watson–Crick hydrogen-bonded double helix to separated single strands is referred to as the *melting transition*. For short oligonucleotide sequences, such as hexanucleotides, there is, in addition, a finite probability that the terminal base pairs are nonhydrogen-bonded in the Watson–Crick double helix. This so-called *fraying* at the ends of the double helix occurs much faster than the melting of the fully hydrogen-bonded double helix.

G-N$_1$H and T-N$_3$H. Earlier investigations (25, 26) established that one exchangeable proton per base pair [in this case, the guanine N-1 proton (G-N$_1$H) and thymine N-3 proton (T-N$_3$H)] is observed between 11 to 15 ppm downfield from standard DSS in the high-resolution proton NMR spectra of transfer RNA in aqueous solution.

Hydrogen-bond formation shifts the G-N$_1$H and T-N$_3$H resonances downfield and slows their exchange rate with water. Furthermore, these ring NH resonances experience upfield ring-current shifts from the ring currents of nearest-neighbor base pairs (21, 28) and are shielded by them from the solvent. The chemical shift and line width of the ring NH resonances depends on their location (i.e., terminal versus internal and nearest-neighbor sequence) and the stability of the double helix (i.e., fraying of the helix at its ends and on the concentration of double helix and single strands at a particular temperature) (29, 30).

d-ApTpGpCpApT. The Watson–Crick double-helical form of the deoxyoligonucleotide d-ApTpGpCpApT exhibits a twofold symmetry axis such that only three base pairs need be considered. They are designated $(AT)_{terminal}$, $(AT)_{internal}$, and $(GC)_{central}$.

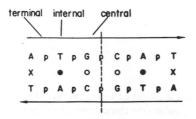

Fig. 7 shows the high-resolution proton NMR spectrum of d-ApTpGpCpApT in H_2O between 11.5 and 14.5 ppm as a function of temperature. Earlier studies (25, 26) predict the observation of three resonances, each having an area of two protons per double-stranded helix, corresponding to the ring NH resonances of the $(AT)_{terminal}$, $(AT)_{internal}$, and $(GC)_{central}$ base pairs. At $3°C$, two distinct resonances at 13.8 and 12.7 ppm and a very broad resonance at ~13.15 ppm are observed. At $14°C$, the resonance at 13.8 ppm broadens, and at $25°C$, the resonance at 12.7 ppm broadens. Three distinct resonances of equal area are observed in the high-resolution proton NMR spectrum of d-ApTpGpCpApT in H_2O:MeOH (3:2) at $-11°C$ between 12 and 14 ppm from DSS (Fig. 7).

The assignments of the resonances between 12 and 14 ppm in the spectrum of d-ApTpGpCpApT in H_2O at $3°C$ are: $T-N_3H$, $(AT)_{terminal}$, ~13.15 ppm; $T-N_3H$, $(AT)_{internal}$, 13.8 ppm; $G-N_1H$, $(GC)_{central}$, 12.7 ppm. Raising the temperature of the hexanucleotide solution has two effects on the stability of the fully hydrogen-bonded double-helical structure. The melting process, which involves the conversion of the double-stranded structure to separate single strands can be monitored by the $G-N_1H$ resonances of the $(GC)_{central}$ base pairs. Rapid fraying at the ends of the double-stranded structure can be monitored by the $T-N_3H$ resonances of the $(AT)_{terminal}$ base pairs between $-11°$ and $0°C$ and the $(AT)_{internal}$ base pairs between $0°$ and $14°C$. In all cases, upfield shifts and line broadening reflect increasing populations of the nonhydrogen-bonded forms (18, 30). The fraying process is fast on the NMR time scale because the ring NH resonances shift as average signals, giving a lifetime $\tau < 0.2$ msecs (30).

d-ApTpGpCpApT IN H₂O, pH 7

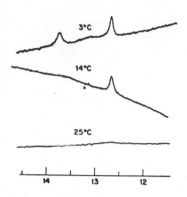

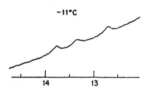

d-ApTpGpCpApT   H₂O: MeOH = 3:2

Figure 7. The temperature dependence of the high-resolution 300 MHz proton NMR spectrum of d-ApTpGpCpApT between 12 and 14 ppm in H$_2$O (pH 7) at 3°, 14°, and 25°C and H$_2$O:MeOH = 3:2 at –11°C.

1:2 Act-D:d-ApTpGpCpApT. The proton NMR spectral changes resulting from the gradual addition of Act-D to d-ApTpGpCpApT in H$_2$O (pH 7) at 0°C are shown in Fig. 8. With the addition of 0.5 equivalent of the antibiotic (i.e., one Act-D molecule per double-stranded hexanucleotide), chemically shifted resonances are observed having the twofold symmetry of the oligonucleotide double helix removed in the spectrum of the complex. The Act-D lacks an exact twofold symmetry axis due to asymmetric substitution of the phenoxazone ring and, therefore, a tightly bound complex of Act-D and double-helical d-ApTpGpCpApT also lacks this element of symmetry. Addition of Act-D greater than 0.5 equivalent results in no additional spectral changes. The addition of 0.3 equivalent of Act-D (i.e., 0.6 Act-D molecule per double-stranded hexanucleotide) to d-ApTpGpCpApT gives a spectrum that is a superposition of the spectra of the hexanucleotide and the complex (Fig. 8). This effect suggests that the exchange of the antibiotic between double-helical hexanucleotides is slow on the NMR time scale ($\tau > 2$ msecs) at 0°C. Complex stability, therefore, increases on the order 1:2 Act-D:d-pG $<$ 1:2 Act-D:d-pGpC $<$ 1:2 Act-D:d-ApTpGpCpApT.

d–ApTpGpCpApT IN H$_2$O, pH 7, 0°C
ADDITION OF Act–D

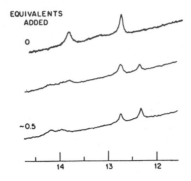

d–ApTpGpCpApT + Act–D, H$_2$O:MeOH = 3:2

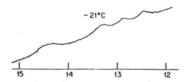

Figure 8. (Top) The high-resolution 300 MHz proton NMR spectra of d-ApTpGpCpApT in H$_2$O (pH 7) at 0°C as a function of Act-D concentration. (Bottom) 1:2 Act-D:d-ApTpGpCpApT in H$_2$O:MeOH = 3:2 at –21°C.

As shown in Fig. 8, the ring N protons of the $(GC)_{central}$ and $(AT)_{internal}$ base pairs of d-ApTpGpCpApT can be traced through gradual addition of Act-D in H$_2$O at 0°C, and H$_2$O:MeOH = 3:2 is used to follow the $(AT)_{terminal}$ base pairs below 0°C. Because these resonances have been identified with specific base pairs in d-ApTpGpCpApT, they can be readily identified also in the complex (see Table 1). Upon complex formation, the largest nonequivalence, 0.4 ppm, is observed for the $G-N_1H$ protons of the $(GC)_{central}$ base pairs; 0.2 ppm nonequivalence is observed for the $T-N_3H$ protons of the $(AT)_{internal}$ base pairs, and the smallest nonequivalence is observed for the $T-N_3H$ protons of the $(AT)_{terminal}$ base pairs (18). The data strongly suggest that the asymmetrically substituted phenoxazone ring of Act-D binds at the $(GC)_{central}$ Watson–Crick base pairs in double-helical d-ApTpGpCpApT (Fig. 9).

Sobell and Jain (8) proposed intercalation of the antibiotic between GpC sequences in d-ApTpGpCpApT. In this intercalation model, the $G-N_1H$ resonance of one GC base pair lies over the benzenoid ring and the $G-N_1H$ resonance of the other GC base pair lies over the quinonoid ring of the phenoxazone. The experimental nonequivalence of the $G-N_1H$ resonances of the GC base pairs in the complex could result from the different magnitudes of the upfield

Table 1. Chemical-shift nonequivalence of symmetry-related d-ApTpGpCpApT ring NH resonances on complexation with Act-D

	d-ApTpGpCpApT H$_2$O:MeOH = 3:2 at –11°C	Chemical-Shift Nonequivalence in 1:2 Act-D:d-ApTpGpCpApT[a]
T-N$_3$H, (AT)$_{terminal}$	13.35 ppm	~0.0 ppm
T-N$_3$H, (AT)$_{internal}$	13.8 ppm	0.2 ppm
T-N$_1$H, (GC)$_{central}$	12.7 ppm	0.4 ppm

[a]Temperature-independent between –20° and 0°C in H$_2$O:MeOH = 3:2 and 0° to 25°C in H$_2$O (pH 7).

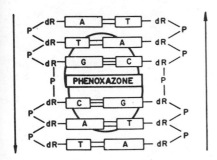

Figure 9. A schematic of the 1:2 Act-D:d-ApTpGpCpApT complex with intercalation of the antibiotic between the (GC)$_{central}$ base pairs.

ring-current contribution from the phenoxazone ring. The chemical-shift changes that occur upon complexation also reflect the perturbed ring-current contributions of neighboring base pairs when the helix unwinds to incorporate the intercalating residue.

The 1:2 Act-D:d-ApTpGpCpApT complex melts at a higher temperature than double-stranded d-ApTpGpCpApT because the G-N$_1$H resonances of the (GC)$_{central}$ base pairs observed in the spectrum of the complex at 25°C are broadened at this temperature in the spectrum of the hexanucleotide.

DNA helix type. It has been proposed that the specificity of Act-D complexation to DNA double helices and its lack of complexation to RNA and DNA-RNA double helices (31–33) indicates a preference for the B rather than the A form of double-helical polyribonucleotides. Because NMR studies indicate complex formation between Act-D and d-ApTpGpCpApT in aqueous solution, double-helical d-ApTpGpCpApT in H$_2$O is probably in the DNA-B form at the concentrations (25 mg single strand/1 ml) utilized in this study.

Intercalation monitored by ^{31}P NMR spectra. The 40 MHz ^{31}P NMR spectrum of d-ApTpGpCpApT in aqueous solution exhibits several poorly resolved resonances between 0 and –1 ppm. In the >1:2 Act-D:d-ApTpGpCpApT complex in H_2O (pH 7) at 10°C, three new ^{31}P resonances are observed at 2.2, 1.2, and 0.55 ppm with an area of one phosphorus each to approximately seven phosphorus resonances between 0 and –1 ppm (Fig. 10) (18). The temperature-independent lines at 1.2 and 2.2 ppm are assigned to the internucleotide phosphate atoms at the intercalating site. Based on the Sobell–Jain model (8) and proton NMR data, the assignment of the downfield-shifted resonances at 1.2 and 2.2 ppm is made to the internucleotide GpC phosphate groups (18). The

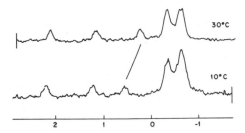

^{31}P SPECTRA
>1:2 Act-D: d-ApTpGpCpApT
0.1M NaCl, 0.01M EDTA, D$_2$O, pH7

30°C

10°C

Figure 10. The 40 MHz ^{31}P NMR spectra of >1:2 Act-D:d-ApTpGpCpApT in 0.1 M NaCl-D$_2$O in 0.01 M EDTA (pH 7) at 30° and 10°C.

slow exchange of Act-D between free and complexed d-ApGpCpApApT indicates a tightly bound complex. One of the GpC phosphates faces the nonpolar benzenoid ring of the phenoxazone and the GpC phosphate on the other chain faces the polar quinonoid ring of the phenoxazone. A hydrogen bond through a water molecule has been proposed for the 2-amino group of the phenoxazone ring and its neighboring phosphate group (8). The different environments of the GpC phosphates on the two chains of the double-stranded hexanucleotide in the complex due to asymmetric substitution of the intercalating phenoxazone ring could account for the chemical shifts at 2.2 and 1.2 ppm. The ^{31}P resonance at 0.55 ppm (10°C) and 0.2 ppm (30°C) in the spectrum of >1:2 Act-D:d-ApTpGpCpApT cannot be readily assigned because there should be only one intercalation binding site per hexanucleotide. The presence of a third downfield resonance could arise from weak surface binding.

ACKNOWLEDGMENT

I am pleased to acknowledge the expert technical assistance of L.L. Canuel.

REFERENCES

1. E. Reich, I.H. Goldberg, and M. Rabinowitz, *Nature,* **196**, 743 (1962).
2. L.S. Lerman, *Proc. Nat. Acad. Sci. U.S.,* **49**, 94 (1963).
3. M. Waring, *J. Mol. Biol.,* **54**, 247 (1970).
4. L.D. Hamilton, W. Fuller, and E. Reich, *Nature,* **198**, 538 (1963).
5. G.V. Gursky, *Biofizika,* **11**, 737 (1966).
6. W. Müller and D.M. Crothers, *J. Mol. Biol.,* **35**, 251 (1968).
7. H.M. Sobell, S.C. Jain, T.D. Sakore, and C.E. Nordman, *Nat. New Biol.,* **231**, 200 (1971).
8. H.M. Sobell and S.C. Jain, *J. Mol. Biol.,* **68**, 21 (1972).
9. S.C. Jain and H.M. Sobell, *J. Mol. Biol.,* **68**, 1 (1972).
10. H.M. Sobell, *Prog. Nucl. Acid Res.,* **13**, 153 (1973).
11. B.H. Arison and K. Hoogsteen, *Biochemistry,* **9**, 3976 (1970).
12. T.R. Krugh and J.W. Neely, *Biochemistry,* **12**, 4418 (1973).
13. D.J. Patel, *Biochemistry,* **13**, 1476 (1974).
14. T.R. Krugh, *Proc. Nat. Acad. Sci. U.S.,* **69**, 1911 (1972).
15. R. Schara and W. Müller, *Eur. J. Biochemistry,* **29**, 210 (1972).
16. T.R. Krugh and J.W. Neely, *Biochemistry,* **12**, 4418 (1973).
17. D.J. Patel, *Biochemistry,* **13**, 2388 (1974).
18. D.J. Patel, *Biochemistry,* **13**, 2396 (1974).
19. C. Giessner-Prettre and B. Pullman, *J. Theor. Biol.,* **27**, 87 (1970).
20. N.S. Angerman, T.A. Victor, C.L. Bell, and S.S. Danyluk, *Biochemistry,* **11**, 2402 (1972).
21. K.D. Kopple, M. Ohnishi and A. Go, *J. Am. Chem. Soc.,* **91**, 4264 (1969).
22. C.D. Barry, J.A. Glasel, A.C.T. North, R.J.P. Williams, and A.V. Xavier, *Biochem. Biophys. Acta,* **262**, 101 (1972).
23. P.O.P. T'so, H.S. Kondo, M.P. Schweizer, and D.P. Hollis, *Biochemistry,* **8**, 997 (1969).
24. C. Altona, J.H. van Boom, J.D. Jager, H.J. Koeners, and G. van Binst, *Nature,* **247**, 558 (1974).
25. D.R. Kearns, D.J. Patel, and R.G. Shulman, *Nature,* **229**, 338 (1971).
26. D.R. Kearns, D.J. Patel, R.G. Shulman, and T. Yamane, *J. Mol. Biol.,* **61**, 265 (1971).
27. R.G. Shulman, C.W. Hilbers, Y.P. Wong, K.L. Wong, D.R. Lightfoot, B.R. Reed, and D.R. Kearns, *Proc. Nat. Acad. Sci. U.S.,* **70**, 2042 (1973).
28. D.J. Patel and A.E. Tonelli, *Proc. Nat. Acad. Sci. U.S.,* **71**, (1974).
29. D.J. Patel and A.E. Tonelli, *Biopolymers,* in press.
30. D.J. Patel and C.W. Hilbers, to be submitted.
31. E. Reich, in *The Role of Chromosomes in Development,* M. Locke, ed., Academic Press, New York, 1964.
32. R. Haselkorn, *Science,* **143**, 682 (1964).
33. M. Gellert, C.E. Smith, D. Neville, and G. Felsenfeld, *J. Mol. Biol.,* **11**, 445 (1965).

DESIGN AND PROPERTIES OF α-HELIX COILED-COIL MODEL POLYPEPTIDES

ALAN G. WALTON and **KATHLEEN P. SCHODT**, Department of Macromolecular Science, Case Western Reserve University, Cleveland, Ohio 44106

SYNOPSIS: Design criteria for the preparation of α-helix coiled-coil models for fibrous proteins are explored. Important criteria include sequence, intramolecular stability, conformational-directing properties of the constituent peptides, specific hydrophobic and hydrophilic interactions, and molecular weight.

Some of these parameters are investigated with four polyheptapeptides: poly[Glu-(OBzl)-Ala-Ala-Glu-(OBzl)-Ala-Ala-Ala], designated [A]; poly(Glu-Ala-Ala-Glu-Ala-Ala-Ala), designated [B]; poly[ε-Z-Lys-Ala-Ala-Glu-(OBzl)-Ala-Ala-Ala], or [C], and poly(Lys-Ala-Ala-Glu-Ala-Ala-Ala), or [D]. Data for the conformations of these polypeptides in a variety of solvents and film cast from solvents are reported. All four materials show predominantly α-helical conformation in the solid state, although for some solvents, significant β-sheet or random components were observed. On the other hand, [A] and [C] show only moderate amounts of α-helical conformation in nonaqueous solvents, and [B] and [D] are essentially random in aqueous solution under all conditions. To date, no definitive x-ray-diffraction data have been obtained that would indicate the presence of a supercoiled structure. The conformational implications of these polypeptides are explored in terms of improved design of polyheptapeptide α-helix chain models.

INTRODUCTION

Our understanding of the conformational properties of proteins, particularly fibrous proteins, has developed to a large extent from model polypeptides, of which the poly(amino acids) were the first to be put to such use. Studies of poly(α-amino acids) and their derivatives in the solid state revealed the predominance of two major discrete conformations, the α helix and β sheet, along with several less frequent conformations, such as the poly(L-proline)-I and -II helices, the polyglycine-II helix and the ω helix. In addition, the solution conformations of ionized poly(α-amino acids) have been referred to as random, random-coil, extended-coil, and distorted polyglycine helices. Although these conformations have provided a useful framework for conformational analysis, fibrous proteins rarely have conformations that are as symmetrical as those of the poly(amino acids), basically because of the nature of the peptide repeat. Perhaps the closest native analogs to homopolymers of amino acids are some of the silk fibroins that possess antiparallel or cross-β structures that are analogous in many ways to similar structures in poly(L-valine), and so on.

In nature, coiled or supercoiled forms generally strengthen the fibrous proteins whose main function is mechanical. Thus, supercoiled α helices (tropomyosin, α keratin, fibrinogen), supercoiled β structure (β keratin), and supercoiled poly(L-proline)-II or polyglycine (left-handed) helices (collagen) are well documented (1).

The collagen structure, now elucidated in considerable detail, is known to be derived from the polytripeptide repeating sequence in which glycine and proline play dominant roles. The production of synthetic glycyl polytripeptides with collagenlike triple-coiled-coil helices represented a considerable advance in our ability to synthesize model proteins and understand the features controlling conformation. Indeed, perhaps the most detailed model for the structure of collagen has been developed from x-ray-diffraction studies of the polytripeptide poly(Gly-Pro-Pro) (2).

So far no serious attempts have been made to model α-helix coiled coils, a problem of much greater complexity than the elucidation of the structure of collagen. Poly(α-amino acids) that are α-helical in the solid state generally pack hexagonally with no supercoiling. Uniform supercoiling, therefore, must originate in a repeat pattern of peptides. Various numbers of strands may be imagined for the supercoiled chain structure, but for the present purpose, it is convenient to study only the two-stranded α-helix coiled coil. It would seem likely at first sight that shorter repeating sequences would lead to a more tightly wound chain, and to some extent they do. Thus, for example, imaginary supercoiled double α helices with repeating sequences of three, four, seven, or eleven peptides would possess a superhelix repeat distance of approximately 30 Å, 55 Å, 190 Å, and 300 Å, respectively. The first two probably are not sterically feasible, but they would have left-handed and right-handed superhelices, respectively. The latter two do appear to be sterically feasible and also would have left-handed and right-handed superhelices [based on a right-handed sense of the minor (α) helix].

Experimentally, the observations of Astbury (3–5) concerning x-ray diffraction from the k-m-e-f groups of proteins were interpreted by Crick (6, 7) and Pauling and Corey (8) in terms of coiled α helices. Often, low-angle x-ray diffraction shows α-coiled superhelix repeat distances of ~190 Å, suggesting that perhaps the heptad repeat is basic to such structures.

The factors determining the stability of such coiled coils may be expected to be derived from suitable interactions within the coiled-coil structure and with the environment, as well as from the steric packing restrictions previously noted. Most of the α-helix coiled-coil proteins have a large content of polar amino acids, often 30–50%, usually with the acidic peptides glutamate and aspartate predominant. Tropomyosin, for example, contains 36 residue-percent glutamate and aspartate and 18 residue-percent lysine and arginine (9).

The interesting question concerning electrostatic interaction between these charged groups in solution relates to the stability of glutamate and lysine sequential polypeptides in solution. Until recently the presence of large quantities of glutamate groups in a polypeptide could be related only to the properties of poly(L-glutamic acid), which had been described as random. However, several glutamate- and lysine-containing sequential polypeptides that have now been prepared and characterized show varying degrees of compatibility of charged groups with α helices. Because much of this material has been presented only recently or is currently in press a summary is presented in Table 1.

It is noteworthy that the glutamate polypeptides poly(Glu-Ala) and poly(Glu-Ala-Ala) are not α-helical in aqueous (neutral pH) solution (10, 11), probably due to a combination of solvation and charge-repulsion effects (14), but readily form α helices in the solid state. The lysine polymers poly(Lys-Ala-Ala) and poly(Lys-Ala-Ala-Ala), on the other hand, are much more prone to form α helices in neutral aqueous solution (12, 13) and readily form α helices in the solid state. Thus, it may be expected that one of the difficulties in synthesizing α-helix coiled coils will be the subtle conformation-destroying interactions between charged peptide side groups (at least in solution).

With the preceding information in mind, it is evident that polymers of the type $(A_6B)_n$, where both A and B are α-helix-formers would be the simplest α-helix coiled-coil models to design. One of these residues should be hydrophobic and one hydrophilic. If A were to be an ionizable peptide such as glutamate, a disrupted structure would undoubtedly be produced, whereas no such effect should occur if B were the ionizable peptide. Cohen has suggested (15) that stabilization of a coiled-coil chain structure derives, in part, from interaction between nonpolar residues on the interior of the chains. This proposal led to a predicted repeating-dyad/triad-type of sequence, that is, $[X-N-X-X-N-X-X]_n$, where N is a nonpolar residue. In view of this suggestion, we have concentrated on polymers of the type $(BA_2BA_3)_n$, where A is L-alanine and B is L-glutamate or its benzyl ester. In addition, in case of conformational disrupting effects between glutamate groups in solution, we have replaced one of the B peptides

Table 1. Conformation of glutamate and lysine sequential polypeptides

Polypeptide	MW	Conformation in Solution (pH 7)	Solid State	Reference
poly(Ala-Glu)		E.C.[a]	α or β	10
poly(Ala-Ala-Glu)	12,000	E.C. + random	α	11
poly(Ala-Ala-Lys)	13,000	E.C. + 20% α		12
poly(Ala-Ala-Ala-Lys)	3,000	0% α		
	10,000	20% α		
	21,000	50% α	α	13

[a] E.C. is correlated with the poly(L-glutamic acid) structure by CD spectroscopy and is denoted E.C. (extended coil) to differentiate its conformation from the random (gelatin) CD pattern.

in the heptad, $(CA_2BA_3)_n$, with L-lysine or its benzyloxycarbonyl (Z) derivative. The four polymers described in this paper are poly[Glu-(OBzl)-Ala-Ala-Glu-(OBzl)-Ala-Ala-Ala], [A]; poly(Glu-Ala-Ala-Glu-Ala-Ala-Ala), [B]; poly[ε-Z-Lys-Ala-Ala-Glu-(OBzl)-Ala-Ala-Ala], [C]; and poly(Lys-Ala-Ala-Glu-Ala-Ala-Ala), [D].

SYNTHESIS AND PRELIMINARY CHARACTERIZATION

One of the major difficulties of synthesizing sequential polypeptides with optically active residues is the possibility of racemization at various stages of the reaction. Methodology devised by Jones *et al.* (16, 17) has facilitated synthesis of sequential polymers containing only optically active residues, essentially without racemization. The four polyheptapeptides studied in our laboratory were synthesized by L. Treiber. The method was stepwise synthesis from the C terminus using active esters (18). Molecular weights of water-soluble [B] and [D] were obtained on Sephadex G50 calibrated with known polypeptide markers. Because [B] was obtained from [A] and [D] from [C], it is reasonable to assume that the molecular weights obtained for [B] and [D] represent minimum values for [A] and [C].

The average molecular weight of [B] was 8600 daltons, with a maximum observable MW of 14,000. The molecular weight of [D] was considerably lower, with a mean value of ~3500 and a maximum observable value of 6000. Improved methods of polymerization perhaps might have doubled the molecular weight attainable, but clearly the production of high-molecular-weight polyheptapeptides is, at least in the present case, difficult. Because of the disappointingly low molecular weights, it is appropriate to explore the molecular weight required for supercoiling to become evident. The molecular weight of the chains of tropomyosin is ~30,000, which is unusually small for a fibrous protein and only a factor of four or so higher than our synthetic analogs.

By analogy, poly(Pro-Pro-Gly) forms supercoiled (collagenlike) triple helices when the molecular weight is as low as 3000 (19). The collagen superhelix repeat distance (~86 Å) is much smaller, however, than that of the α-coiled coil (~190 Å). Nevertheless, about 130 peptides would be required to form one superhelix repeat in a single strand of an α-helical chain, in contrast to about 30 in a strand of triple polyglycine-II helices. Thus, molecular weights of the order of 12,000 or so seem to be required in order for supercoiling to become evident. At present, there is no clear cut evidence, however, for a lower limit of required molecular weight.

RESULTS

The materials were subjected to circular dichroism (CD) and infrared (IR) (including Fourier Transform) spectroscopy and, in some cases, x-ray diffraction (powder patterns) and electron microscopy. A search through conformations produced in and from various solvents has been directed toward finding appropriate crystallization media.

In view of the rather extensive information gathered on these four polypeptides in different solvents (20) under different conditions and also in the solid state, only a brief summary of the salient points will be made here.

SOLUTION MEASUREMENTS

Solution studies were carried out using CD spectroscopy only. Unfractionated poly[Glu-(OBzl)-Ala-Ala-Glu-(OBzl)-Ala-Ala-Ala] [A] was insoluble in water, trifluoroethanol (TFE), and only slightly soluble in formic acid (FA). It was readily soluble in hexafluoroisopropanol (HFIP), where it showed 65% α-helix content at 5°C [based on ellipticities of poly(Glu-Ala)] (21). The polypeptide "melts" linearly with increasing temperature, presumably indicating the lower α-helix stability of the lowest-molecular-weight component.

Poly(Glu-Ala-Ala-Glu-Ala-Ala-Ala) [B] is water soluble, but it showed a random conformation at all pHs and temperatures. Up to 20% α-helix formation could be induced by addition of methanol.

Poly[ϵ-Z-Lys-Ala-Ala-Glu-(OBzl)-Ala-Ala-Ala] [C] is soluble in TFE, where its α-helix content is ~40%, the remainder being random. The helix content decreases with increasing temperature in much the same manner as [A].

Poly(Lys-Ala-Ala-Glu-Ala-Ala-Ala) [D] behaves in much the same fashion as [B] in aqueous solution. The polymer appeared random at all pHs and temperatures and developed a small amount of α-helix content upon addition of methanol or TFE. The α helix may be melted out at elevated temperatures.

SOLID STATE

The main methods of evaluation in the solid state were by IR and CD spectroscopy of films. In all cases, there was a clear correlation between the two methods in terms of conformations observed, but precise quantitation of the various conformers was limited by the methodology.

Poly[Glu-(OBzl)-Ala-Ala-Glu-(OBzl)-Ala-Ala-Ala] [A]. This polymer was examined "as polymerized" (by IR only) and film cast from dichloroacetic acid (DCA), trifluoracetic acid (TFA), and hexafluoroisopropanol (HFIP). Of these preparations, the first two showed essentially equal proportions of α helix and antiparallel β sheet. Virtually no random component was identifiable (sharp IR bands and CD pattern). The latter two preparations showed high proportions of α helix. In particular, HFIP films showed virtually no random or β-sheet component.

Attempts to orient films for IR dichroism measurements by stroking induced a strong β component in all cases, presumably by mechanical deformation. X-ray powder-diffraction patterns of the "as polymerized" sample revealed lines at 8.05 Å, 4.62 Å, 3.69 Å, 3.33 Å, 2.23 Å, and 2.09 Å. The line at 4.62 Å probably is related to the distance between chains that are hydrogen-bonded together in a β sheet. There was no evidence of the 5.4 Å pseudorepeat of the α helix or the 5.1 Å repeat of the α-coiled structure.

Similarly, electron micrographs of low-molecular-weight material (1400–2000) film cast from DCA, showed fibers that were apparently cross-β structure.

Poly(Glu-Ala-Ala-Glu-Ala-Ala-Ala) [B]. Films of this polypeptide were prepared from FA, TFA, DCA, and HFIP. Films were not readily prepared from aqueous solution. Despite the high random conformational content of [B] found in solution, the solid films were predominantly α helix with various amounts of β sheet present. The α-helix content increased in the order DCA < FA < TFA < HFIP. Films formed from the last two solvents showed essentially no conformation other than α helix. Once again, attempts to orient films for IR dichroism studies induced an α-helix–β-sheet transition. In addition attempts to precipitate material from these solvents also produced material that was revealed by x-ray diffraction to have a major β-sheet content.

Poly(ϵ-Z-Lys-Ala-Ala-Glu(OBzl)-Ala-Ala-Ala) [C]. The solid state conformation of [C] was examined in the "as polymerized" form and film cast from FA, TFA, and HFIP. The results were virtually identical with those obtained for [A] under similar circumstances, namely, equal amounts of β sheet and α helix for the first two preparations and virtually 100% α helix for the last two. This result is perhaps a little surprising because the molecular weight of [C] is much less than that of [A], and consequently, a greater tendency to produce β sheet might have been predicted. Again the α-helix–β-sheet transition occurred when orientation was attempted.

Poly(Lys-Ala-Ala-Glu-Ala-Ala-Ala) [D]. Films that had been prepared from water at pH 5.0 and from TFA, DCA, and FA were examined. The predominant conformation in all cases was α helix, although a certain amount of random material was also evident (slightly broadened IR bands and slightly modified CD curves). The random component was apparently rather small for the first two preparations and very small for the last two. Once again α–β transitions were induced by attempts to orient samples. No interpretable x-ray-diffraction patterns have yet been obtained from this polypeptide.

DISCUSSION

Although we have failed to detect α-supercoiling in x-ray-diffraction patterns, there are several encouraging features in the work done so far. Perhaps most encouraging is the fact that in the solid state, the polypeptides may be formed into α-helical films with essentially no other detectable conformation. Unfortunately, manipulation of material that is α-helical in film form readily converts the structure to β sheet. Low-molecular-weight poly(L-alanine) is crystallizable in the β-sheet form and the lower-molecular-weight fractions of the polyheptapeptides discussed here are probably β conformation. It seems that a higher molecular weight might be helpful in obtaining the appropriate α form. Because we do not yet have a diffraction pattern of the α-helical form, it is not possible to say whether supercoiling occurred. Current efforts are underway to obtain more crystalline specimens by precipitation with divalent cations, which, as with tropomyosin, may produce a definitive solid state structure. X-ray-fiber diagrams are not currently feasible because of mechanical deformation.

In solution, the polypeptides behave in a manner that appears to be a logical extrapolation from simpler sequential analogs described in the introduction. As may be seen from Table 1, there is a strong tendency to form random- or extended-coil forms in solution for both alanine-glutamate and alanine-lysine polymers, but in the solid state, all of the polymers are readily produced in α-helical form.

It would seem, therefore, that destabilizing (strong) interactions with the solvent disrupt the α-helical form. Although nothing is known of sequential leucine-glutamate or leucine-lysine polypeptides, the presence of strongly hydrophobic leucine groups, along with higher molecular weights, may produce a higher α-helical content in aqueous solution.

The role of α-helix-directing residues is important in the above context. Normally, L-leucine and L-alanine are both considered to be α-helix-directing residues. The available evidence (22, 23) indicates that poly(L-alanine) forms a more stable α helix in solution than poly(L-leucine) and thus supports the concept that alanine is the more likely candidate for the chain model. Nevertheless, the increased hydrophobic nature of the leucine residue also makes it a strong candidate.

It may be that hydrophobic interchain interactions play a much larger part in dictating the α-helix coiled-coil regime than was supposed in the original design of this work. Since this work was begun, certain α-helix coiled-coil sequences have been elucidated, and it is now possible to assess in more detail the requirements for the primary structure. Fig. 1 compares the repeating heptad sequence of a portion of the α-helix chain structure in α keratin (merino wool) (24) with the repeat sequence in two of our model polymers. The most noticeable feature is the much higher polar-residue content of the native material. More detail can be derived from the data of Hodges *et al.* (25, 26) for the sequence of the C-terminal portion (150 residues) of tropomyosin. It is apparent from an analysis of their data that three of the seven positions are essentially invariant, that is,

$$1\ 2\ 3\ 4\ 5\ \ 6\ \ 7$$
$$X\text{-}N\text{-}X\text{-}X\text{-}N\text{-}\ominus\text{-}X$$

where N is a nonpolar residue and $\ominus$ is glutamate or aspartate. Positions 1, 3, and 7 are strongly polar with acidic groups slightly predominant in positions 3 and 7 and basic groups slightly predominant in position 1. There is rarely a sequence of seven successive peptides that does not contain one or more basic groups and thus perhaps the most representative heptad sequence would be

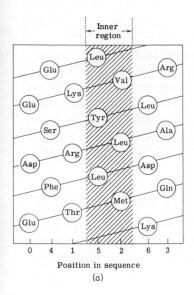

Position in sequence

(a)

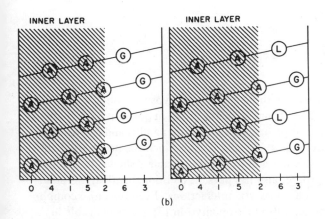

(b)

Figure 1. (a) Schematic projection diagram of α-helix coiled coil from α-keratin sequence (21) showing the hydrophobic nature of residues 2 and 5 in a repeating heptad sequence. (b) Schematic projection diagrams of poly(Ala-Ala-Glu-Ala-Ala-Ala-Glu) (left) and poly(Ala-Ala-Lys-Ala-Ala-Ala-Glu) (right) represented as repeating heptads in a supercoil (3.5 residues per turn) showing 2/5 hydrophobic residues in a hydrophobic core and a hydrophilic exterior.

```
 1  2  3  45 6   7
-⊕ -N-⊖ -N-N-⊖ -⊖ -
```

It is interesting that the basic groups do seem to mediate the disruptive charge repulsion of acidic groups in the native material. Thus new and more sophisticated models might be directed at increasing the charge density from ~30%, as in the present case, to 57% for the above model. There are, however, certain interesting details of chain interaction that result from such models. As shown in the following section, the most suitable model polypeptide for studying α-helix coiled coils in fact differs slightly from the sequence shown above and would have 43% charged residues.

ROLE OF INTERCHAIN INTERACTIONS

Whereas it is clear that certain sequences are necessary to avoid interchain conformation disruption, attention must be paid also to the nature of interchain forces. If the two charged polyheptapeptide models studied are represented as

```
 1 2  3  45 6  7
-N-N-⊖ -N-N-⊖ -N-
```

and

```
 1 2  3  45 6  7
-N-N-⊖ N-N- ⊕ -N-
```

it can be seen that the first has the three main tropomyosin features with non-polar residues in positions 2 and 5 and an acidic group in position 6. If hydrophobic forces hold the chains together, however, both parallel and antiparallel helices are feasible for both models, [1] and [2], in which the necessary 2 and 5 residue positions (or equivalent 1 and 4) interact. Table 2 shows that model [2], which contains basic groups, also has possible additional antiparallel stabilizing forces originating from electrostatic interactions. Even the more complicated heptad sequence representative of tropomyosin [3] seems plausible in both the parallel and antiparallel arrays with hydrophobic interactions in positions 2 and 5.

It seems evident from the preceding considerations that the formation of a parallel or antiparallel array of α-coiled-coil helices must be strongly dependent upon specific interchain hydrophobic forces. The suggestion of Sodek *et al.* (26) that a parallel chain phasing may occur, bringing into juxtaposition certain pairs of hydrophobic residues, is particularly interesting in this context. (We note that alanine-leucine pairs are complementary hydrophobes.)

Based on all of the above considerations, we have concluded that a rather complex set of criteria are required to effect the unique combination of interactions required to allow the formation of α helices, to dictate parallel chains, to cause supercoiling, to effect the appropriate balance of basic and acidic

Table 2. Interchain interactions for polyheptapeptides[a]

Residue[e]	Model 1[b] Parallel[f]	Antiparallel[f]	Model 2[c] Parallel[f]	Antiparallel[f]	Model 3[d] Parallel[f]	Antiparallel[f]
1†	↑N ↑N	↓N	↑N ↑N	↓N	↑+ ↑+	−
2*	N N	N	N N	N	N N	N
3	− −	−	+ +	−	− −	↓N
4†	N N	N	N N	N	N N	−
5*	N N	N	N N	N	N N	N
6	N N	N	N N	N	− −	+
7	− −	−	− −	+	− −	−
8†	N N	N	N N	N	+ +	−
9*	N N	N	N N	N	N N	N
10	− −	−	+ +	−	− −	N
11†	N N	N	N N	N	N N	−
12*	N N	N	N N	N	N N	N
13	N N	N	N N	N	− −	+
14	− −	−	− −	+	− −	−

[a] Most favorable arrangement of parallel and antiparallel arrays showing the required 1/4 or 2/5 hydrophobic peptide interactions.
[b] Poly(Ala-Ala-Glu-Ala-Ala-Ala-Glu).
[c] Poly(Ala-Ala-Lys-Ala-Ala-Ala-Glu).
[d] The hypothetical tropomyosin model.
[e] The dagger (†) and asterisk (*) denote required nonpolar combinations (one or the other is necessary).
[f] N, nonpolar residues; −, acidic residues; +, basic residues.

residues, and to provide interlocking hydrophobic residues. It does not appear possible to satisfy all these criteria with a simple model polyheptapeptide, but the most promising sequence appears to be poly(Ala-Ala-Glu-Ala-Leu-Lys-Glu). [Equally plausible is the complementary sequence poly(Leu-Leu-Glu-Leu-Ala-Lys-Glu).] Although all of the required criteria are fulfilled by this polymer, there appears to be an equal possibility of parallel (phase-staggered chains) and antiparallel helices, as shown in Table 3. More quantitive conformational analysis will reveal the probability of parallel or antiparallel chain sense in such a model. Nevertheless we believe that poly(Ala-Ala-Glu-Ala-Leu-Lys-Glu) is the simplest heptad sequence in which virtually all of the native design elements are combined with the rationale derived from current studies of charged sequential polypeptides. It is noteworthy that our poly(Ala-Ala-Glu-Ala-Ala-Ala-Glu) model differs in only two of the residues from that of the prospective model. Those two residues, however, provide the two important elements of lock and key hydrophobic interactions and intrachain stabilization, which apparently are missing from the current models.

In any event, the successful synthesis of α-helix coiled-coil model polypeptides not only should provide a detailed understanding of the forces dictating

Table 3. Interchain interactions for proposed model polyheptapeptides[a]

Residue	Parallel[b]		Antiparallel[b]	
1	A	A	A	L+
2	A	Le	A	Le
3	G –	L +	G –	A
4	A	G –	A	G –
5	Le	A	Le	A
6	L +	A	L +	A
7	G –	G –	G –	G –
8	A	A	A	L+
9	A	Le	A	Le
10	G –	L +	G –	A
11	A	G –	A	G –
12	Le	A	Le	A
13	L +	A	L +	A
14	G –	G –	G –	G –

[a] Poly(Ala-Ala-Glu-Ala-Leu-Lys-Glu) [α-helix coiled coil (2 strands)]
showing plausible parallel (chain-staggered) and antiparallel arrays.
[b] Boxes represent specific desirable hydrophobic interactions; the
parallel chain couplet is staggered by four residues.

such a conformation, but also may provide substrates for the binding of troponin or actin. Thus, such models may generate important new insight into the complex function of muscle proteins.

CONCLUSIONS

The reported studies indicate that design criteria for α-helix coiled-coil model polypeptides are:

1. The model must be a sequential heptapeptide.
2. The sequence must contain α-helix-directing residues.
3. An alternation of hydrophilic and hydrophobic residues must be strategically placed.

4. It must contain a sequence of charged residues (basic and acidic) that minimize intrachain disruption.
5. It must contain hydrophobic residues that interact in a complementary fashion with the adjacent chain.

In addition, molecular weights of ~12,000 daltons or greater are preferred.

The four polyheptapeptides reported in this paper meet most, but not all, of the above criteria. In particular, they lack strategic hydrophobic interacting residues. Nevertheless, we have isolated films of polyheptapeptide in the α-helical conformation that might be potentially recrystallizable in the conformation of an α-helix coiled coil.

It is suggested that the simplest sequence that essentially meets all of the above criteria is poly(Ala-Ala-Glu-Ala-Leu-Lys-Glu).

ACKNOWLEDGMENTS

We are pleased to acknowledge the financial support of this work by the Molecular Biology Division of the U.S. National Science Foundation and appreciate the extensive synthesis support provided by L. Treiber and G. Sigler. In addition, we would like to thank W.B. Rippon for his interest and helpful discussion.

REFERENCES

1. A.G. Walton and J. Blackwell, *Biopolymers,* Academic Press, New York, 1973, Chapter 10.
2. A. Yonath and W. Traub, *J. Mol. Biol.,* **43**, 461 (1969).
3. W.T. Astbury and H.J. Woods, *Nature* (London), **126**, 913 (1930).
4. W.T. Astbury and H.J. Woods, *Phil. Trans. Roy. Soc.* (London), A**232**, 333 (1933).
5. W.T. Astbury, *Proc. Roy. Soc.* (London), B**134**, 303 (1947).
6. F.H.C. Crick, *Nature* (London), **170**, 882 (1952).
7. F.H.C. Crick, *Acta Cryst.,* **6**, 685, 689 (1953).
8. L. Pauling and R.B. Corey, *Nature* (London), **171**, 59 (1953).
9. S. Seifter and P.M. Gallop in *The Proteins,* Vol. 4, H. Neurath, ed., Academic Press, New York, 1966.
10. W.B. Rippon, H.H. Chen, and A.G. Walton, *J. Mol. Biol.,* **75**, 369 (1973).
11. R.R. Gruetzmacher, L.R. Treiber, and A.G. Walton, *Biopolymers,* in press.
12. A. Yaron, N. Tal, and A. Berger, *Biopolymers,* **11**, 2461 (1972).
13. D.B. Wender, L.R. Treiber, H.B. Bensusan, and A.G. Walton, *Biopolymers,* in press.
14. W.A. Hiltner, A.J. Hopfinger, and A.G. Walton, *Biopolymers,* **12**, 157 (1973).
15. C. Cohen in *Principles of Biomolecular Organization,* G.E.W. Wolstenholme and M. O'Conner, eds., Churchill, London, 1966, p. 101.
16. J.H. Jones and G.T. Young, *J. Chem. Soc.* (C) (London), 436 (1968).
17. R.D. Cowell and J.H. Jones, *J. Chem. Soc.* (London), 1814 (1972).

18.   L.R. Treiber and A.G. Walton, in preparation.
19.   Y. Kobayashi, R. Sakai, K. Kakiuchi, and T. Isemura, *Biopolymers,* **9,** 415 (1970).
20.   K.P. Schodt, M.S. Thesis, Case Western Reserve University, Cleveland, Ohio (1974).
21.   W.B. Rippon and W.A. Hiltner, *Macromolecules,* **6,** 282 (1973).
22.   G.D. Fasman in *Polyamino Acids, Peptides and Proteins,* M.A. Stahmann, ed., University of Wisconsin Press, Madison, Wis., 1962, p. 221.
23.   A.G. Walton and J. Blackwell, *Biopolymers,* Academic Press, New York, 1973, p. 386.
24.   I.J. O'Connell, *Aust. J. Biol. Sci.,* **22,** 471 (1969).
25.   R.S. Hodges and L.B. Smillie, *Can. J. Biochem.,* **50,** 330 (1972).
26.   J. Sodek, R.S. Hodges, L.B. Smillie, and L. Jurasek, *Proc. Nat. Acad. Sci. U.S.,* **69,** 3800 (1972).

STRUCTURAL INFLUENCE ON THE DYNAMIC MECHANICAL PROPERTIES OF POLYPEPTIDES AND COLLAGEN

A. HILTNER, S. NOMURA, and E. BAER, Department of Macromolecular Science, Case Western Reserve University, Cleveland, Ohio 44106

SYNOPSIS: Wet dura-mater collagen and various ionic homo- and copolyamino acids have similar dispersion processes below physiological temperature that are dependent on specific interactions between the macromolecule and water. Specifically, three strong relaxation maxima were observed: T_{H_2O}, β_1, and β_2 around 270°, 200°, and 150°K, respectively. The T_{H_2O} process, which occurs only when the water content exceeds 50%, is due to devitrification. Both the β_1 and β_2 processes decrease in temperature with increasing water content. Since β_2 occurs first, this water must be tightly bound to the macromolecular substrate, and it is suggested that this dispersion process is similar to the "L1" defect mechanism in ice. It follows that the water associated with the β_1 process, which occurs at a higher water content, is more mobile and could act as a plasticizing medium to enhance side-group mobility. This paper shows that dynamic mechanical spectroscopy is a powerful tool that can be used to compare the behavior of water in collagen and in model poly(α-amino acids) in our search for prosthetic materials that mechanically resemble the fibrous proteins.

EXPERIMENTAL METHOD

Dynamic mechanical measurements were made with a free-oscillating, inverted torsional pendulum at about 1 Hz over the 80–300°K temperature range (1). Because the poly(α-amino acids) either formed intractable films or were available in small quantities, the torsional braid technique was used (2). In this method the polymer is cast from solution onto a viscoelastically inert glass braid support. In this way, measurements can be made with as little as 5 mg of the material.

Poly(L-glutamic acid) (MW 50,000–80,000) was obtained from Pierce Chemical Company and was cast from dimethylformamide solution. Poly(L-glutamic acid):Na (MW 50,000–80,000) also was obtained from Pierce Chemical Company and was cast from water. Glutamic acid:Na-leucine copolymer was obtained by dissolving the unionized copolymer (3) in an equivalent amount of base and was cast directly from this solution. Poly(L-lysine):HBr was provided by J.M. Anderson and was cast from aqueous solution. Poly(β-benzyl-L-aspartate) (MW 250,000) was obtained from Pilot Chemicals and was cast from chloroform. Poly(S-Z-methyl-L-cysteine) was provided by G.D. Fasman and was cast from trifluoroacetic acid solution. All specimens were dried in vacuo for several days at 50°C.

The torsional braid specimens were hydrated by exposure to an atmosphere of known humidity. The desired relative humidities were obtained with saturated salt solutions at 25°C. Salts used were $CaCl_2$, 32%; $NaNO_2$, 66% and $NaClO_3$, 75%.

Frozen human dura mater (HDM: age, 29 yr) was thawed in a buffered saline solution and cut into 6.0 × 0.5 cm strips. Residual blood was removed from the strips in a saline solution and the specimens were dehydrated by freeze-drying. Rehydration was achieved by exposing specimens to an atmosphere of constant humidity controlled over saturated salt solutions. Water content of the rehydrated dura is calculated on the basis of a completely dry vacuum-dried specimen, although some residual water is certainly present even after several days in vacuo at 25°C.

POLY(α-AMINO ACIDS)

Side chain effect. Poly(α-amino acids) for which relaxation behavior has previously been described exhibit primarily side-group processes below physiological temperature. Polymers with blocked functional groups (Group 1, Table 1), especially the esters of glutamic acid, were the first to be studied in detail. The most intense relaxation is evident in the side-group peak near room temperature (β_{sc}). McKinnon and Tobolsky (4) observed a discontinuity in the slope of the specific volume-temperature relationship for poly(benzyl-L-glutamate) at 290°K. This behavior is typical of a glasslike transition. Kaneko *et al.* (5) found that the frequency dependence of β_{sc} for homo- and copolymers of benzyl-L-glutamate and methyl-L-glutamate does not follow an Arrhenius relationship. Rather the data fit the WLF equation, which is generally applicable only to the glass transition. They likened these polymers to a two-phase system with a rigid α-helical core in an amorphous side-chain region that can undergo a glasslike transition. The poly(α-amino acids) with aliphatic side groups show a side-chain peak (γ) at a much lower temperature of about 120°K.

Table 1.   Side-group relaxations in poly(α-amino acids)

Group 1	γ	β_{sc}	α	Ref.
Poly(γ-methyl-L-glutamate)		262°K[a]	415(110 Hz)	4
Poly(γ-benzyl-L-glutamate)		300	396(110 Hz)	4
Poly(β-benzyl-L-aspartate)[b]		300		
Poly(ε-Z-L-lysine)		303		4
Poly(S-Z-methyl-L-cysteine)[b]		285	325	
Group 2				
Poly(L-alanine)	120			3
Poly(L-valine)	120			3
Poly(L-leucine)	120			4

[a]1 Hz.

[b]This work.

Conformation has not been considered as a variable that might affect relaxation behavior. The poly(α-amino acids) in Table 1 assume rigid hydrogen-bonded conformations (α-helix or β-sheet) and the backbone is thought to contribute little, if at all, to the γ and β_{sc} processes. It is worth summarizing the experimental findings that support this conclusion.

In polyglycine, where only backbone mechanisms are operative, Krug and Gillham (6) observed no relaxations below room temperature. Available data suggest that structural changes that affect backbone flexibility do not alter the activation energy and temperature position of the γ and β_{sc} processes. A comparison of poly(α-olefins) with identical side groups with poly(amino acids) [poly(4-methyl-pentene-1) and poly(L-leucine), poly(3-methyl-butene-1) and poly(L-valine)], shows the side-chain γ peaks at the same temperature rather than at a lower temperature for the polyolefin. Hiltner et al. (7) found that copolymers of benzyl-L-glutamate and L-leucine show the two side-chain peaks (γ at 120° and β_{sc} at 300°K) characteristic of homopolymers, even though leucine is reported to have a helix-stabilizing effect on poly(benzyl-L-glutamate) (8). In this study it was found also that the intensity of the benzyl glutamate peak is proportional to the composition, but that of the leucine peak is not. One interpretation is that cooperative effects, which affect intensity but not activation energy, are important in the side-group motions of leucine.

Effect of water on nonionic hydrophilic polymers. It is known that the relaxation spectra of many polymers with polar groups exhibit features that can be associated with the presence of water. Polymers with the amide linkage are of particular relevance in a discussion of proteins and poly(α-amino acids). Sorbed water in nylon is of two types, bound and free. The bound water, about 2% by weight, has by far the greatest effect on the relaxation spectrum (9, 10). It is associated with a large decrease in the temperature of the glass transition, 362° to 317°K, and the appearance of a β process, which shifts from 230° to 190°K as water content increases from 0 to 2%. Water in excess of 2% is free water that has little additional effect on the relaxation behavior.

Changes in the relaxation spectra of poly(α-amino acids) caused by sorbed water have recently been reported. Hiltner et al. (3) found that water decreases the intensity, but does not alter the temperature of the γ peak in poly(L-leucine). It gives rise to a new peak at about 180°K (β). A similar water peak is also apparent in the relaxation spectra of wet poly(L-alanine) and polyglycine reported by Krug and Gillham (6).

Additional structural considerations are introduced when poly(α-amino acids) with polar or ionic side groups are considered because the barriers to reorientation are affected by polar interactions between side chains. These polymers are also hydrophilic. Water can reduce polar side-chain–side-chain interactions, thereby acting as a plasticizer. Water is also known to produce conformational changes in salts of poly(L-lysine) (11) and poly(L-glutamic acid) (12, 13), a feature that may further complicate the relaxation behavior.

The side-chain peak for vacuum dried poly(L-glutamic acid) was found at
$340°K$ (1 Hz) in this study. In contrast to its effect on aliphatic-side-chain peaks,
water appears to have a plasticizing effect on the glutamic acid side chain; the
$340°K$ peak has been reported 45° lower, at $295°K$ (110 Hz), for a specimen at
ambient humidity (14).

Polar interactions between side chains are also reduced by introducing a
comonomer with a nonpolar side chain. In two copolymers of glutamic acid
and leucine, the leucine peak has been reported at $120°K$, the same temperature
as the homopolymer, but the glutamic acid peak was at $278°K$, about 60° below
that of the homopolymer (3). Water suppresses the intensity, but causes no
additional change in the temperature of the side-group processes in the copolymers. Water also gives rise to a peak at about $175°K$ (β). For a 1:1 copolymer,
this β peak increases in intensity up to about 1% water, but at higher water content the intensity of the β remained constant and a broad devitrification peak
was observed at about $260°K$.

Salts of poly(L-glutamic acid) and poly(L-lysine). The sodium salt of poly-
(glutamic acid) is viscoelastically inactive when dry. In the presence of water,
two processes are distinguished: one at about $185°K$ (β_2) that does not change
in intensity but shifts about 15°, from 190° to $175°K$, over the range of water
contents studied, and another that shifts from above room temperature to about
$250°K$ (β_1) with increasing water content (Fig. 1). The absence of a devitrification peak in all but the wettest specimen is evidence of the sorption of a significant amount of water, equal to several molecules per peptide, that does not have
the properties of bulk water.

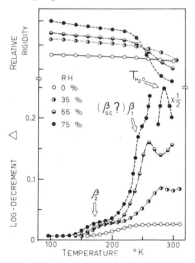

Figure 1. Mechanical relaxation spectra of the sodium salt of poly(glutamic
acid) as a function of water content showing two β processes.

Measurements were obtained also on a less hydrophilic copolymer of sodium glutamic acid and leucine (1:1). The leucine peak is again at 120°K, but it is quite weak and, as with the homopolymer, there is no glutamic acid peak in the dry specimen. Water has essentially the same effect as in the homopolymer: The β_2 peak shifts from 200° to 185°K and β_1 is apparent but is less intense than for the homopolymer.

Poly(L-lysine):hydrobromide behaves similarly (Fig. 2). A wet specimen shows β_2 and β_1 peaks at 175° and 240°K, both 10° lower than the analogous peaks in sodium poly(L-glutamic acid).

The β_1 peak may be the water-plasticized side-chain peak (β_{sc}) for the ionized poly(α-amino acids). The drop in rigidity and the peak intensity are comparable with those observed for other side-group processes. The temperature of the water-plasticized process in sodium poly(L-glutamic acid) (250°K) is also close to that of the acid form when polar interactions are minimized by water or a comonomer.

DURA-MATER COLLAGEN

The structure and role of water in collagen has been discussed by many authors who used wide- and small-angle x-ray scattering (15–19), infrared absorption (20–22), NMR (23–26), and various other techniques (27, 28) for their investigations. By analyzing x-ray diffraction patterns of collagen specimens containing different levels of hydration, Rougvie and Bear (15) distinguished several kinds of water bound in different degrees to the macromolecular system. It was

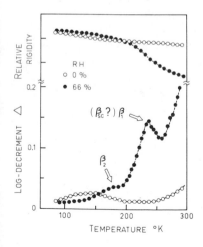

Figure 2. Mechanical relaxation spectra of polylysine:HBr.

found that about 13% of the water was primarily or tightly bound within both the band and interband regions.

The dynamic mechanical behavior of polymers has been studied extensively in attempts to interrelate the supermolecular structure in the solid state to their mechanical properties. Recently, several investigators have reported the dynamic mechanical behavior of native and reconstituted collagen (29–33), and at least three kinds of relaxations that are sensitive to the water content (31) have been shown to exist. To date, the effect of hydration on the relaxation behavior of collagen has not been examined in detail, so we present in this paper a detailed study of the behavior of native human dura mater (HDM), which was chosen because of its very high collagen content.

The effects of water. Native wet HDM collagen has three mechanical relaxation maxima below physiological temperature. They occur around 270°, 200°, and 150°K and are labeled in Figs. 3–6 as the T_{H_2O}, β_1 and β_2 relaxation processes. These results are similar to those of Papir and Baer (31) for human-diaphragm tendon and Chien and Chang (32) who used rat-tail tendon. Rehydration of freeze-dried specimens of dura in either saturated water vapor or saline solution caused relaxation spectra essentially identical to that of the native tissue.

Figs. 3 and 4 show changes in the transitions at various water contents. The dehydrated sample has two small relaxations at 260° and 140°K, namely, the β_2 and γ processes shown in Fig. 3. The γ mechanism has previously been related to various side-chain motions observed in poly(α-amino acids), as described above. Both the β_1 and β_2 processes are shifted to lower temperatures with increasing water content and reach constant temperatures of 200° and 150°K at water contents of about 50% and 15%, respectively. The T_{H_2O} mechanism appears only above 50% water at about 270°K, which, of course, is about the freezing point of water (Fig. 5).

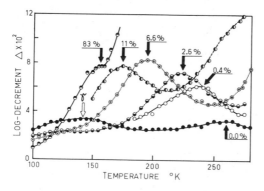

Figure 3. Effect of water on a β_2 process in collagen showing a decrease in the relaxation temperature with increasing water content.

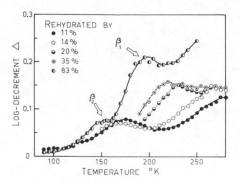

Figure 4. Effect of water on β_1 process in collagen again showing a decrease in this relaxation temperature with increasing water content.

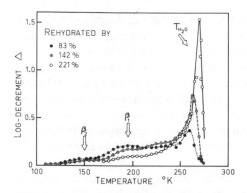

Figure 5. Relaxation spectra of collagen showing devitrification process T_{H_2O}, observed only at high water content.

Changes in relaxation behavior with hydration are summarized in Fig. 6. Rougvie and Bear (15) have shown that the initial 13% water is strongly bound to the collagen. Our results on the behavior of β_2 relaxation also showed a concentration dependency in this range, so it follows that the β_2 process is affected by the concentration and structure of bound water. The water associated with β_1 also must be interacting with the protein because evidence for devitrification appears only at higher water contents.

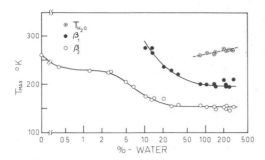

Figure 6. Summary of the effect of water on the three relaxation processes in collagen.

DISCUSSION

There are strong similarities in the behavior of the poly(α-amino acids) with ionic side groups and dura-mater collagen. Both adsorb nearly 50% water that does not have the properties of bulk water as evidenced by the lack of devitrification. On the basis of repeated observations of the β_1 and β_2 relaxation maxima, which occur at similar temperatures for all cases studied, at least two types of water must be interacting with the macromolecule. In every case, these relaxations decrease in temperature with increasing water content. The fact that the β_2 process is always observed first and reaches a limiting temperature before the β_1 process implies that the β_2 mechanism is associated with the most tightly bound water.

Only the most general suggestions can be made in regard to the mechanisms of the β_1 and β_2 processes. The β_1 process is observed in all wet poly(α-amino acids) studied to date; it begins with water sorption, it is difficult to remove completely, and above a certain water content, the intensity is constant. A similar peak has been reported in other water-absorbing polymers, including polyamides, polymethylmethacrylate, polyhydroxyethylmethacrylate, and poly(vinyl alcohol). Because the β_2 process is observed so frequently in these wet polymers, the mechanism may be traceable to the rearrangement of water molecules by the breaking and reforming of hydrogen bonds with the macromolecular substrate. This is analogous to the "L1" defect mechanism in ice, which has been used to explain the β mechanism in wet nylon 6 (34).

The higher-temperature β_1 process is observed at higher water content than the β_2 process and thus involves less tightly bound, more mobile water. The mechanism is not clear, but it may include motions of both the macromolecular side chains and the water molecules. For example, the β_1 process may be water-plasticized motion of the ionic or polar side chains analogous to the glasslike transition in poly(benzyl-L-glutamate).

The sensitivity of the dynamic mechanical technique to interaction of water with proteins and model poly(α-amino acids) has been clearly demonstrated. A probe with this capability is applicable to problems involving the relationship between structural and mechanical changes, such as those observed during aging, and in the development of synthetic materials with mechanical properties resembling those of the fibrous proteins for prosthetic application.

ACKNOWLEDGMENTS

We wish to thank J.M. Anderson and G.D. Fasman for kindly providing some of the poly(α-amino acids), and R.R. Kohn for specimens of human dura mater. We also wish to thank R. Woods and E. Borkowski for competent technical assistance.

This work was generously supported by the National Institutes of Health (Grant No. HD-00669-13).

REFERENCES

1. C.D. Armeniades, I. Kuriyama, J.M. Roe, and E. Baer, *J. Macromol. Sci.-Phys.*, **B1**, 777 (1967).
2. J.K. Gillham, *Crit. Revs. Macromol. Sci.*, **1**, 83 (1972).
3. A. Hiltner, J.M. Anderson, and E. Baer, *J. Macromol. Sci.-Phys.*, **B8**, 449 (1973).
4. A.J. McKinnon and A.V. Tobolsky, *J. Phys. Chem.*, **70**, 1453 (1966).
5. M. Kaneko, K. Hikichi, A. Tsutsumi, Y. Yamashita, M. Kanke, and N. Matsushima, *J. Macromol. Sci.-Phys.*, in press.
6. R.R. Krug and J.K. Gillham, *J. Appl. Polym. Sci.*, **17**, 2089 (1973).
7. A. Hiltner, E. Borkowski, and J.M. Anderson, *Macromolecules*, **5**, 446 (1972).
8. G.D. Fasman, C. Lindblow, and E. Bodenheimer, *Biochem.*, **3**, 155 (1964).
9. R. Puffr and J. Sebenda, *J. Polym. Sci.*, **16C**, 79 (1967).
10. Y.S. Papir, S. Kapur, C.E. Rogers, and E. Baer, *J. Polym. Sci.*, A-2, **10**, 1305 (1972).
11. U. Shmueli and W. Traub, *J. Mol. Biol.*, **12**, 205 (1965).
12. G.D. Fasman, H. Hoving, and S.N. Timasheff, *Biochem.*, **9**, 3316 (1970).
13. H. Lenormant, A. Baudras, and E.R. Blout, *J. Am. Chem. Soc.*, **80**, 6191, (1958).
14. A. Nguyen, B.T. Vu, and G.L. Wilkes, *J. Macromol. Sci.-Phys.*, in press.
15. M.A. Rougvie and R.S. Bear, *J. Am. Leather Chemists Assoc.*, **48**, 735 (1953).
16. R.E. Burge, *J. Mol. Biol.*, **7**, 213 (1963).
17. W. Traub, A. Yonath, and D.M. Segal, *Nature* (London), **221**, 914 (1964).
18. V.G. Tumanyan, *Biopolymers*, **9**, 995 (1970).
19. A. Miller and J.S. Wray, *Nature* (London), **230**, 437 (1971).
20. E.M. Bradbury, R.E. Burge, J.T. Randall, and G.R. Wilkinson, *Discussion Faraday Soc.*, **25**, 173 (1958).
21. R.D.B. Fraser and T.P. MacRae, *Nature* (London), **183**, 179 (1959).

22. H. Strassmair, J. Engel, and G. Zundel, *Biopolymers,* **8**, 237 (1969).
23. H.J.C. Berendsen, *J. Chem. Phys.*, **36**, 3297 (1962).
24. H.J.C. Berendsen and C. Migehelsen, *Proc. Fed. Am. Soc. Exp. Biol.*, **25**, 998 (1966).
25. R.E. Dehl and C.A.J. Hoeve, *J. Chem. Phys.,* **50**, 3245 (1969).
26. G.E. Chapman, I.D. Campbell, and K.A. McLauchlen, *Nature* (London), **225**, 639 (1970).
27. R.E. Burge and R.D. Hynes, *J. Mol. Biol.,* **1**, 155 (1959).
28. J. Schnell and H. Zahn, *Makromol. Chem.*, **84**, 192 (1965).
29. J.K. Kolske and J.A. Faucher, *J. Phys. Chem.*, **69**, 4040 (1965).
30. P. Mason and J. Unsworth, *Kolloid-Z. Z. Polym.,* **249**, 1101 (1971).
31. E. Baer, R. Kohn, and Y.S. Papir, *J. Macromol. Sci.-Phys.,* **B6**, 761 (1972).
32. J.C.W. Chien and E.P. Chang, *Biopolymers,* **11**, 2015 (1972).
33. H. Stefanou, A.E. Woodward, and D. Morrow, *Biophysical J.,* **13**, 772 (1973).
34. S. Kapur, C.E. Rogers, and E. Baer, *J. Polym. Sci.-Polym. Phys. Ed.,* **10**, 2297 (1972).

SYNTHETIC ANTIGENS AND RECENT PROGRESS IN IMMUNOLOGY

MICHAEL SELA, Department of Chemical Immunology, The Weizmann Institute of Science, Rehovot, Israel

SYNOPSIS: Designing an antigen molecule to incorporate structural features desirable for the elucidation of defined immunological problems has been proven worthwhile by studies using synthetic poly(amino acids). The availability of synthetic antigens has led not only to the elucidation of much of the molecular basis of antigenicity (e.g., the role of steric conformation), but also to a better understanding of such disparate immunological phenomena as the inverse relationship between the net charge on an antigen and the antibodies it provokes, the molecular basis of purely cellular immune response, immunological tolerance, antigenic competition, cell cooperation in efficient induction of immune response, and the genetic control of immunological responsiveness.

INTRODUCTION

Undoubtedly, one of the most important uses of poly(α-amino acids) as protein models has been in the field of immunology. Many of them have been shown to be antigenic; in fact, the use of synthetic antigens as highly efficient tools for the elucidation of the molecular basis of antigenicity was described at the symposium on poly(amino acids) in Madison 13 years ago (1). Since that time, synthetic antigens have been used successfully in many laboratories, and through proper antigen design we have learned a lot about immunological phenomena. It is, therefore, not an exaggeration to state that the availability of synthetic antigens has been of crucial importance for the development of modern immunology.

Several review articles describing this work in general terms (2-5) and in terms of specific phenomena (6-10), as well as published lectures (11-20), have appeared in recent years. I shall not try to discuss every aspect here but I shall try to illustrate the importance of poly(amino acids) in immunology with several examples on both the molecular and the cellular level, mainly from our laboratory in Rehovot.

MOLECULAR BASIS OF ANTIGENICITY

My personal interest in immunology is partially a consequence of the synthesis of poly(L-tyrosine) (21) and of polypeptidic azo dyes (22). The latter, which were used extensively by Landsteiner in his classical studies of antigenic specificity (23), could serve as a synthetic model for azoproteins. The synthesis of polytyrosine and the availability of N-carboxy-L-tyrosine anhydride permitted experimental investigation of the speculation that the lack of antigenicity of gelatin is due to its lack of tyrosine. We prepared polytyrosyl gelatin (24) and demonstrated that the attachment of tyrosine peptides to gelatin converted it to a potent immunogen (25, 26). Depending on the extent of enrichment with tyrosine, the tyrosylated gelatin preparations led mainly to the formation of antibodies specific for gelatin (at lesser enrichment) or to antibodies directed exclusively to tyrosine peptides (at greater enrichment) (27). Today it is well known that gelatin is not completely devoid of tyrosine, nor is it completely incapable of provoking an immune response.

As a result of these studies, it became desirable to distinguish clearly between the notion of *immunogenicity* (i.e., capacity to provoke an immune response) and the notion of *antigenic specificity*, which defines the capacity of a molecule to react specifically with an antibody (3, 4). Thus, immunogenicity is independent of specificity, whereas the capacity of a molecule to react with an antibody does not presume at all that the molecule would be capable of stimulating the production of a specific antibody.

A detailed investigation of the immunochemical properties of many polypeptidyl gelatins (26, 27) led to the preparation of a synthetic antigen, a multichain synthetic polypeptide in which poly(DL-alanine) side chains were attached to the ϵ-amino groups of a poly(L-lysine) backbone and the polyalanyl chains, in turn, were elongated with peptides containing L-tyrosine and L-glutamic acid (28, 29). We called this polymer (T, G)-A--Lys. The branched polymer induced the formation of antibodies specific for the peptides of tyrosine and glutamic acid. Around the same time, the antigenicity of some synthetic polypeptides was reported from other laboratories (2, 30).

The synthetic approach offers the advantage that, once the immunogenicity of one synthetic material has been unequivocally demonstrated, tens of analogs may be prepared and tested. Knowing the chemistry of these compounds, we believed it possible to arrive at conclusions concerning the roles of various structural features in their antigenic functions through a systematic study of copolymers showing only limited variations in their chemical structures. The problems considered over the years include the roles of shape, size, composition, and electrical charge of the macromolecule, of the locus in the molecule of the area important for immunogenicity, as well as of the optical configuration of its component amino acids and of the conformation of the immunogenic macromolecule.

Homopolymers of amino acids are very poor immunogens. Immunogenicity increases with increased variations in composition, so macromolecular substances are more reliably immunogenic, although low-molecular-weight peptide conjugates may be immunogenic provided they have the right composition (31). The presence of electrical charges on a macromolecule is not a requirement for immunogenicity (4), but when the antigen is charged, an inverse relationship exists between the net electrical charge on the immunogen and that on the antibodies it provokes (32). In order for antibodies to be produced, the immunogenically important area of the antigen must be readily accessible and cannot be hidden in the interior of the molecule (3, 29).

Peptides or sugars of opposite optical configuration are exquisitely recognized by specific antibodies, which are usually stereospecific (i.e., totally unreactive towards the other optical isomer). Attachment of D-tyrosine peptides enhances the immunogenicity of gelatin as efficiently as does the attachment of peptides of L-tyrosine (33). Polymers containing exclusively D-amino acids are poor immunogens and easily induce immunological tolerance unless administered in very small doses (4). Thus it seems that the inefficient formation of antibodies to polymers of D-amino acids is due to their slow and incomplete catabolism (6).

ROLE OF CONFORMATION

The conformation of antigenic determinants seems especially important when we consider the tremendously specific interaction of the determinant group on the antigen with the combining site on the antibody. This interaction of two macromolecules represents a uniquely specific pattern of recognition on a molecular level.

In protein and polypeptide antigens it is possible to distinguish between "sequential" determinants and conformation-dependent determinants (4, 12). I call a sequential determinant one that implies a given amino acid sequence in the polymeric chain. Antibodies to such a determinant are expected to react with a peptide of identical or similar sequence. On the other hand, a conformational determinant results from the conformation of the immunogenic macromolecule and leads to antibodies that do not necessarily react with peptides derived from that area of the molecule. It seems that antibodies to native globular proteins are directed mostly against conformational rather than sequential determinants, and the same is true for at least some fibrilar proteins, such as collagen. By the way, similar argumentation makes it possible to distinguish between sequential and conformational determinants in other macromolecules, such as nucleic acids.

Studies of the immunological properties of native proteins have shown conclusively that the antibodies obtained are directed mostly, and in several well-documented cases exclusively, against conformation-dependent determinants. Thus, for example, antisera produced against native bovine pancreatic ribonuclease do not react with the denatured, open-chain ribonuclease, in which all the disulfide bridges have been severed, nor do antibodies against the open chain react with the native enzyme (34). Rabbit immunoglobulin G is another example. Antibodies prepared in goats are not able to react with a derivative in which all the disulfide bridges in the immunoglobulin G have been opened (12, 35).

In recent years we have tried to gain a better understanding of the role of conformation in antigenicity by building synthetic models. In one example, the tripeptide Tyr-Ala-Glu was attached to a branched polymer of alanine. It behaved as a sequential determinant. In a second case Tyr-Ala-Glu was polymerized to a high-molecular-weight periodic polymer that exists in an α-helical form under physiological conditions (36). I have discussed this particular antigenic system extensively elsewhere (4, 11, 12, 14, 17) and would like to note here only that antibodies against the helical poly(Tyr-Ala-Glu), may transform a smaller, not yet helical, polymer of identical amino acid sequence into a helical shape, followed by circular dichroism (37). Thus, we could measure an induced fit occurring upon the interaction of two specific sites present on two biologically active macromolecules.

The synthetic polypeptide poly(Pro-Gly-Pro) is another interesting example. This ordered sequence polypeptide, previously shown to have physical properties similar to those of collagen, was found to be immunogenic in guinea pigs and rabbits (38). We have now investigated in detail the immunological properties of the ordered poly(Pro-Gly-Pro), which possesses the collagenlike triple-stranded structure, and compared them with the properties of poly(Pro^{66}Gly^{34}), which possesses a ratio of proline to glycine residues similar to that found in the ordered polymer, but is prepared by a random polymerization technique (from N-carboxy-L-proline anhydride and N-carboxyglycine anhydride) and does not resemble collagen in its conformation (39, 40).

Specific antibodies to poly(Pro-Gly-Pro) were obtained in guinea pigs, rabbits and goats by immunization with the free polymer or with covalent conjugates of the polymer with a carrier protein. Antibodies to the ordered polymer were shown to be specific to the unique collagenlike conformation of the immunogen, as they cross-react well with other collagenlike polyhexapeptides and to a much smaller extent with the random copolymer poly(Pro^{66}Gly^{34}), which is not collagenlike. Antibodies to poly(Pro-Gly-Pro) cross-react with various collagens, and anticollagen antibodies cross-react with the ordered synthetic collagen model. Sara Fuchs will discuss the immunologic properties of the synthetic collagen model at this symposium (41).

In view of the findings pertinent to the role of conformation in the antigenic specificity of proteins and synthetic polypeptides described above, the conclusion is unavoidable that no significant splitting by proteolytic enzymes may occur between the moment an immunogen is administered and the moment it is recognized at the biosynthetic site (4). Such proteolysis would result in the destruction of the conformation of most protein determinants. Additional evidence comes from a study (42) of the "loop" region of hen egg-white lysozyme. The loop contains a disulfide bridge, and we have shown that antibodies prepared either against the natural loop—isolated from lysozyme (43)—or against the loop that we synthesized in the laboratory (44) [in both cases the loop is first covalently linked either to multichain poly(DL-alanine) or to multichain poly(L-proline)], react with the native loop, but not with the open chain derived from it after reduction and carboxymethylation. We have thus succeeded in preparing a completely synthetic immunogen that in experimental animals stimulates production of specific antibodies that react with a native protein (the lysozyme) through a unique region that is conformation-dependent. From here the road is open, conceptually, to the development of synthetic vaccines in which unique regions of viral coated proteins (20, 45) might be similarly used.

SYNTHETIC ANTIGENS CROSS-REACTIVE WITH NATURAL SUBSTANCES

We have shown already that we can make synthetic nucleoside-containing antigens that provoke antibodies that react with denatured DNA and RNA (46). We have produced synthetic antigens leading to antibodies that react with cytolipin H (47), as well as with brain lipids devoid of sugar moieties (48). The sphingomyelin-specific antibodies reacted with the natural sphingomyelin present in sheep red-blood-cell membrane. On the other hand, the red-blood-cell membranes of a guinea pig have a low level of sphingomyelin and they did not undergo hemolysis with our antisera. Synthetic copolymers of α-amino acids have also been described as capable of provoking antibodies that cross-react with a bacterial cell wall (49) and with the basic protein of the myelin sheath (50). It is thus possible to conclude that we can prepare materials that will lead to antibodies capable of reacting with all kinds of natural substances.

INVERSE RELATIONSHIP BETWEEN NET CHARGE ON ANTIGEN AND ANTIBODY

The inverse relationship between the net electrical charge of immunogenic macromolecules and the net charge of the antibodies they provoke has been demonstrated for antibodies produced in several animal species and is valid for antibodies of IgG, IgM, and IgE classes (17). Much of the initial work in this field has been performed using synthetic polypeptides or their haptenic conjugates as immunogens. Antihapten antibodies obtained upon immunization with conjugates of negatively or positively charged carriers differ in their net charge, but not in their specificity or affinity, as measured by using small molecules related chemically to the original haptens (51).

Cell fractionation on glass beads (52) has been used to establish that the net charge correlation has a cellular basis. If a population of potentially immunocompetent cells could select immunogens on the basis of net antigenic charge, as well as on the basis of determinant specificity, then more positively charged immunocompetent precursor cells would be expected to react preferentially with more acidic immunogens. Thus, more acidic immunocompetent cells should be more readily eluted from columns of glass beads because glass is acidic, whereas more basic cells should adhere to such surfaces. Consequently, cells collected after glass bead filtration should be preferentially stimulated by the more positively charged immunogen. Using dinitrophenylated copolymers of L-tyrosine, L-glutamic acid, and L-lysine that were acidic or basic, depending on their relative content of glutamic acid and lysine, we have indeed proved experimentally that glass-bead columns reduced the number of immunocompetent spleen cells preferentially reactive with more acidic immunogens.

We have also used glass beads coated with poly(L-lysine) to reverse their net electrical charge (53). In contrast to uncoated glass beads, which bind the more basic cells, polylysine-coated glass beads were expected to bind selectively the more acidic cells (i.e., those reactive with basic antigens). Indeed, fractionation of spleen cells on polylysine-coated columns, followed by their inoculation with antigens into irradiated mice, resulted in a marked reduction in the percentage of responders with the basic dinitrophenylated copolymer, whereas no change was observed in response to the acidic dinitrophenylated copolymer.

In accord with presently accepted views on the role of cell cooperation between T (thymus-derived) and B (Bursa-analog- or bone-marrow-derived) cells, we have found that for successful transfer experiments into irradiated mice, both T cells and B cells are needed to provoke antibody formation with the synthetic dinitrophenyl conjugates of the L-amino acid copolymers mentioned above. It was thus possible to enquire which of the two cell types is mainly responsible for the net charge effect (53). Similarly to the results obtained with spleen cells, it was observed that cell suspensions containing a mixture of thymocytes fractionated on glass beads and unfractionated bone-marrow cells showed a reduced capacity to respond to the dinitrophenyl group on the acidic copolymer, but not on the basic immunogen. In contrast, no reduction was detected in the ability of cell mixtures containing fractionated bone-marrow cells and unfractionated thymus cells to respond to the dinitrophenyl hapten on either copolymer.

Similar experiments were carried out with T cells and B cells fractionated on polylysine-coated glass beads (53). Again, cell mixtures of fractionated thymocytes and unfractionated marrow cells reduced the percentage of responders to the basic copolymer, but not to the acidic one; and no reduction occurred in the response frequency to the dinitrophenyl hapten on either carrier when mixtures of fractionated marrow cells and unfractionated thymocytes were used. In conclusion, relevant cells may be fractionated either on negatively or on positively charged columns. The results obtained verify and extend the cellular basis for the inverse net-charge relationship and raise the possibility that there may be populations of thymocytes that recognize immunogens on the basis of their overall electrical charge.

We are currently studying the immune response to analogs of the above immunogens that are built exclusively of D-amino acids (54). Dinitrophenyl conjugates of the basic and acidic copolymers of D-tyrosine, D-glutamic acid, and D-lysine were used for immunization, and the anti-dinitrophenyl antibodies they provoked were thymus-independent. Cellular fractionation studies indicate that fractionation of the various types of cells (spleen, thymus, bone marrow) does not affect the immune response, suggesting that the inverse relationship between the net electrical charge on antigen and on the antibodies it provokes does not hold for thymus-independent antigens.

THYMUS-INDEPENDENT ANTIGENS

I would like to illustrate proper antigen design for another cellular immunological problem. For most immunogens, cooperation between thymus-derived and bone-marrow-derived lymphocytes seems necessary and no antibodies are formed—at least in the mouse—in the absence of the thymus (55, 56). Nevertheless, several thymus-independent immunogens have been described. They include polymerized flagellin, pneumococcal polysaccharide, *E. coli* lipopolysaccharide, and polyvinylpyrrolidone. It has been suggested that the common characteristic of all these thymus-independent antigens is that they possess repeating antigenic determinants. This condition may be necessary but not sufficient for their thymus independence, for several synthetic polypeptide antigens comprised of repeating antigenic determinants have been shown to need both T and B cells for an efficient immune response in mice. The thymus-dependent synthetic antigens include the multichain copolymers poly(Phe, Glu)-poly(Pro)--poly(Lys), poly(Phe, Glu)-poly(DLAla)--poly(Lys), and poly(Tyr, Glu)-poly(DLAla)--poly(Lys).

It seemed possible that in addition to repeating antigenic determinants, slow metabolism might be a reason for efficient responses in mice to some antigens in the absence of thymus cells (11). To check this hypothesis, we used immunogens composed of D-amino acids, which are known to be metabolized slowly and incompletely (6). The four multichain copolymers that were prepared had the general structure poly(Tyr,Glu)-poly(Pro)--poly(Lys) [abbreviated (T,G)-Pro--Lys], but the optical activity of the component amino acids of each was different (57). Each of these polymers is composed of two moieties: the *outside* determinants, peptides of tyrosine and glutamic acid; and the *inside* area, a multichain polyproline (i.e., polyproline side chains attached to a polylysine backbone). Thus, the following four enantiomers of (T,G)-Pro--Lys were synthesized: (a) all L; (b) all D; (c) L outside and D inside; and (d) D outside and L inside. It is of interest that the polymer that was D outside and L inside (more than 90% of its amino acid residues were of the L configuration) behaved similarly to the all-D polymer in that its immunogenicity in rabbits is expressed only at very small doses (57) and it is slowly metabolized (58). This similar behavior is due to the lack of endopeptidases capable of splitting peptide bonds between two L-proline residues. Every poly(L-proline) chain in the multichain copolymer (DTyr,DGlu)-poly(Pro)--poly(Lys) is linked to an ε-amino group of lysine on one end and to a peptide composed of D-amino acids on the other end, so no significant digestion of this polymer can occur within the animal body, a situation that is similar to that of copolymers composed exclusively of D-amino acids.

The four enantiomorphs of (T, G)-Pro--Lys were tested for their efficiency in provoking an immune response in the irradiated nonthymectomized and thymectomized mice into which various combinations of cells were transferred (59). The all-L polymer was found to be thymus-dependent. The other three polymers were thymus-independent. The relation of thymus independence to the slow metabolism of the antigen would be an adequate explanation for the all-D polymer and for the polymer that is D outside and L inside. The polymer that is L outside and D inside is expected to be easily digested at least in its outside L moiety. The reason for its thymus independence is that this particular polymer provokes in SJL/J mice mainly an anti-polyprolyl response, but no anti-(Tyr, Glu) response, and thus the relevant determinants are composed of D-proline. To verify this explanation, we investigated another polymer built of L-amino acids on the outside and D-amino acids inside. This time we used the multichain copolymer poly(Phe,Glu)-poly(DPro)--poly(DLys) in the DBA/1 strain of mice, which are good responders to both moieties of this immunogen. It is of interest that, whereas mice of the DBA/1 strain are good responders to the poly-(Phe,Glu) moiety and poor responders to the poly(Pro)--poly(Lys) moiety (60) when immunized with the all-L polymer, they respond well to both moieties when immunized with the polymer that is L outside and D inside (61). In this case, the anti-poly(DPro) response is thymus-independent, whereas the anti-(Phe,Glu) response was found to be thymus-dependent.

It may be concluded that there is a direct correlation between metabolizability and thymus dependence in the response toward these polymers in mice. Moreover, this correlation seems to be valid at the level of unique immunopotent determinants because synthetic immunogens can be prepared that are composed of thymus-dependent and thymus-independent regions. The results support the hypothesis that poorly metabolized immunogens, as well as poorly metabolized antigenic regions of immunogenic macromolecules, are thymus-independent. Also indicated is the relation of the thymus independence of the response to immunogens with similar size, shape, and amino acid composition to the optical configuration of some or all of their amino acid components. Moreover, because thymus independence of the antibody responses for the five above-mentioned, related multichain polypeptides was correlated with the metabolizability of the immunogens or their various moieties, it is likely that for the category of thymus-independent immunogens possessing repeating antigenic determinants, slow metabolism may be a necessary requirement for a multipoint binding of antigenic determinants to the lymphocyte, which, in turn, is needed to induce an immune response.

It is of interest to mention that recently, using experimental procedures similar to those described above, we found the ordered collagenlike periodic polymer poly(Pro-Gly-Pro) to be thymus-independent, whereas the random copolymer poly(Pro^{66}Gly^{34}) is thymus-dependent (62). Similarly, rat-tail collagen is thymus-independent, but its gelatin derivative is a thymus-dependent immunogen.

GENETIC CONTROL OF IMMUNE RESPONSE

One area of immunology in which antigen design and, consequently, the use of synthetic antigens have been of paramount importance is the genetic control of immune response (9, 10, 11). The immune state of an individual is not itself an inherited characteristic. Nevertheless, the ability of an animal to produce an immune response to a specific immunogen is subject to genetically determined factors. Investigations reported during the last decade indicate that the immune response potentials of several rodent species to natural and synthetic immunogens are genetically regulated. Significant progress in elucidating the mechanisms responsible for generating genetic variations in the immune systems has been made using immunogens possessing a restricted number of antigenic determinants. Thus, immunological studies using synthetic polypeptides and haptens coupled to synthetic polypeptides have opened the way for molecular and cellular studies of genetic control of immunological responsiveness.

A single dominant autosomal gene determines whether or not guinea pigs can form antibodies to hapten-poly(L-lysine) conjugates (9). In this case, the ability to respond depends not on the nature of the hapten, but on the nature of the polymeric backbone. Evidence for determinant-specific genetic control of antibody response in inbred strains of mice has been obtained with synthetic multichain polypeptide antigens in which short peptides containing glutamic acid and tyrosine, histidine, or phenylalanine were attached to the amino acid termini of multichain poly(DL-alanine) (63, 64). For example, C57 black mice are good producers of antibodies against the tyrosine-containing polymer and they respond poorly to the histidine-containing polymer; the situation is completely reversed in CBA mice. In short, genetic factors can cause discrimination among tyrosine, histidine, and phenylalanine in the determinant. The genetic differences are dominant, unigenic, quantitative, and determinant-specific.

The ability of mice to respond to the above antigens is a genetic trait that can be transferred with "responder" spleen cells and that is closely associated with the major histocompatibility (H-2) locus in the ninth mouse linkage group (9). All strains of the same H-2 type exhibit the same pattern of immune response toward the above-mentioned three multichain antigens independently of the remainder of a given strain's genetic background. When multichain polymers in which polyproline chains replaced the poly(DL-alanine) side chains were built and their immunogenicity in inbred strains of mice was tested, the response was different (and not linked to the H-2 locus), even though the same short sequences of tyrosine or phenylalanine and glutamic acid were attached to the polypeptide side chains in both series (60, 65).

Different inbred strains of mice may produce similar amounts of antibodies against the same protein, but the reason may be the complexity of the multi-determinant antigen, so that the specificity of the antibodies formed may differ. For example, two different mouse strains (DBA/1 and SJL) immunized with the antigen poly(Phe, Glu)-poly(Pro)--poly(Lys) responded equally well, but the antisera they produced had markedly different specificity. Antibodies formed in the DBA/1 strain cross-reacted well with poly(Phe, Glu)-poly(DLAla)--poly-(Lys) and only weakly with poly(Tyr, Glu)-poly(Pro)--poly(Lys). The opposite was true of the antibodies produced in the SJL strain. Thus, either the specificity of the antibodies produced or the recognition of antigenic determinants is under direct genetic control.

Direct evidence for genetic control at the level of unique antigenic determinants of native proteins came from a recent study of the loop peptide of lysozyme. Although most inbred mice responded well to hen egg-white lysozyme, some strains did not respond to the loop region of this protein. This was true independently of whether the synthetic conjugate of the loop with multichain poly(DL-alanine) was used for immunization or whether lysozyme itself was the immunogen (66).

In order to better understand the cellular aspects of the genetic control of immune response, we used both the "limiting dilution" technique and carefully controlled allogeneic transfers. In the limiting dilution technique, graded and limiting numbers of cells were transferred with the antigen into lethally irradiated recipients. The first results indicated that the genetic control of immunity to the synthetic polypeptide antigens investigated is directly correlated to the relative number of precursor cells reactive with the immunogen in high- and low-responder strains (67, 68). Studies of the relative importance of T and B cells showed that the genetic defect is similar for antigens based on multichain poly-proline and those based on multichain polyalanine at the level of spleen cell and bone-marrow cell, but not at the level of thymocyte. Thus, a major difference in the nature of the genetic control of immune response seems to be due to the particular chemical nature of the whole immunogenic macromolecule. On the other hand, the mouse strain also seems very important in the role assigned to various cell types in the genetic control. The same (T, G)-A--Lys antigen for which a clear genetic defect has been observed in the thymus cells and in the bone-marrow cells of the poor responder SJL/J mouse strain does not reveal any genetic defect in thymocytes when tested in the poor responder C3H/HeJ strain (69).

Interestingly, the SJL/J strain is a poor responder toward any determinant attached to multichain poly(DL-alanine) [the determinants tested to date are (Phe, G), (T, G), and lysozyme loop], whereas the strain C3H/HeJ is a poor responder toward (T, G)-A--Lys, but not toward (Phe, G)-A--Lys. Thus, it seems that the T cell is of crucial importance as far as the genetic defect is concerned in those cases in which the defect is at the *carrier* level, whereas the defect at the *determinant* level is reflected mainly in the B cells (11).

Very recently, we have reported evidence that a carrier-dependent strain defect in immune response is, indeed, reflected only in thymocytes (70). The SWR strain of mice does not produce antibodies to determinants such as peptides of L-phenylalanine and L-glutamic acid or to the loop peptide of lysozyme when attached to multichain poly(L-proline), although they respond well to the same antigenic determinants when conjugated to multichain poly(DL-alanine).

Transfer experiments have been carried out in which irradiated SWR recipients were injected with an excess of DBA/1 thymocytes [which do not exhibit a defect in response to poly(L-proline)] mixed with graded numbers of SWR marrow cells, prior to immunization with poly(Tyr, Glu)-poly(Pro)--poly(Lys). The results indicate that the poor response potential of SWR mice to polyproline is not reflected in their bone-marrow cells. Allogeneic transfers in which irradiated mice were injected with mixtures of thymocytes and marrow cells from high and low responders and then immunized with poly(Tyr, Glu)-poly(Pro)--poly(Lys) or poly(Phe, Glu)-poly(Pro)--poly(Lys) have revealed a clear defect in the thymus-derived population of SWR mice when the response potential to determinants attached to multichain polyproline was tested (70).

These results are compatible with two alternative interpretations. One is simply that the genetic defect is indeed in the thymus cells when it is at the carrier level and in the bone-marrow cells when it is at the determinant level. The alternative interpretation is that, at least for histocompatibility-linked genetic responses, the defect is always in the thymocytes, but it expresses itself in thymus and marrow cells depending whether the genetic defect is at the carrier or determinant level, respectively. In order to accept the second interpretation, it is necessary to explain why low response potential reflects itself sometimes in thymocytes and at other times in the bone-marrow cells.

As mentioned previously, the immune response to the multichain antigen poly(Tyr, Glu)-poly(DLAla)--poly(Lys), denoted (T, G)-A--Lys, is determinant-specific (63) and closely associated with the H-2 histocompatibility locus of mice (9). The gene controlling the antigen has been called the Ir-1 gene. It has been generally postulated that the genetic control is expressed at the level of thymocytes when the ability to respond is H-2 linked (71), and there is indirect evidence to support this contention (72). On the other hand, our cellular studies implicated a genetic defect for the Ir-1 gene at the level of B cell (69). This finding has been recently confirmed by a study in which T cells were replaced by a thymus-derived antigen-specific cell-free factor (73).

Our results showed that educated T cells of both high- and low-responder origin produce active cooperative factors to (T, G)-A--Lys, and no differences between the strains, in respect to production of T-cell factors, could be found. Such factors, whether of high- or low-responder origin, cooperated efficiently with B cells of high-responder origin only, and hardly at all with B cells of low-responder origin. As far as could be discerned by the methods used, no T-cell defect existed in low-responder mice, and the expression of the controlling (determinant-specific) Ir-1 gene was solely at the level of the B cells (73).

The (T, G) moiety in (T, G)-A--Lys is a short random copolymer of L-tyrosine and L-glutamic acid. Better-defined immunogens that have recently been prepared in our laboratory under strict genetic control (74) will be discussed at this symposium by Edna Mozes (75). The availability of immunogens with determinants of defined amino acid sequence that are H-2 linked should permit within the foreseeable future a much better understanding of the nature of this most interesting and fascinating—yet so obscure—phenomenon of the genetic control of the immune response.

CONTRIBUTION TO OTHER AREAS OF IMMUNOLOGY

Synthetic antigens may be prepared that stimulate the production of specific antibodies against almost any moiety, such as "classical" haptens, sugars, nucleosides, pyridoxal, folic acid and methotrexate, a phytoestrogen, ferrocene, the glycolipid cytolipin H, brain lipids, angiotensin, bradykinin, glucagon, triiodothyronine, and prostaglandin [for literature, see (4 and 76)].

Amino acid copolymers have also been used in connection with experimental allergic encephalomyelitis, an autoimmune disease of possible relation to multiple schlerosis. A synthetic copolymer, Cop 1, which is composed of L-alanine, L-glutamic acid, L-lysine, and L-tyrosine in a residue molar ratio of 6.0:1.9:4.7: 1.0, with an average molecular weight of 23,000, is cross-reactive immunologically with the basic protein of the myelin sheath (50). Cop 1 can efficiently suppress the onset of the experimental disease in guinea pigs (77) and rabbits (78), it can protect the surviving animals against future exposure to the natural basic encephalitogen (79), and it can suppress the disease in monkeys after clinical symptoms have already appeared (80).

Synthetic antigens have also been helpful in investigations of antigen metabolism (6), delayed hypersensitivity (7), antigenic competition (8), and immunological tolerance (81–84).

CONCLUSIONS

I have tried to give several typical illustrations of our use of synthetic antigens to probe the molecular and cellular aspects of immunological phenomena. Undoubtedly, the availability of precisely designed immunogens and haptens permits a better understanding of these phenomena. The relative simplicity of the synthetic molecules facilitates the interpretation of the results obtained with them and sometimes permits the detection of effects, such as genetic variations in immune response, that are not easily observable with complex natural antigens.

The success encountered in the use of poly(α-amino acids) as synthetic antigens led to the synthesis of molecules capable of providing antibodies that can react with unique, conformation-dependent antigenic determinants on native proteins. This synthesis, in turn, may lead to the development of new synthetic vaccines. Any developments in this direction will have to take into account both chemical and genetic parameters, especially in view of the apparent close genetic link, in several species, between good immune response to certain antigens and the major histocompatibility locus.

ACKNOWLEDGMENTS

Studies from the author's laboratory discussed in this paper were supported by grants from the National Institutes of Health, U.S. Public Health Service, Bethesda, Maryland, and the Minerva Foundation, Germany. The paper was prepared while the author was a Fogarty Scholar-in-Residence at the Fogarty International Center, National Institutes of Health.

REFERENCES

1. M.A. Stahmann, ed., *Polyamino Acids, Polypeptides and Proteins,* University of Wisconsin Press, Madison, Wisconsin, 1962.
2. P.H. Maurer, *Progress in Allergy,* 8, 1 (1964).
3. M. Sela, *Advan. Immunol.,* 5, 29 (1966).
4. M. Sela, *Science,* 166, 1365 (1969).
5. T.J. Gill III, in *Immunogenicity,* F. Borek, ed., North-Holland, Amsterdam, 1972.
6. T.J. Gill III, *Current Topics in Microbiology and Immunology,* 54, 19 (1971).
7. F. Borek, *Current Topics in Microbiology and Immunology,* 43, 126 (1968).
8. M.J. Taussig, *Current Topics in Microbiology and Immunology,* 60, 125 (1973).
9. H.O. McDevitt and B. Benacerraf, *Advan. Immunol.,* 11, 31 (1969).
10. E. Mozes and G.M. Shearer, *Current Topics in Microbiology and Immunology,* 59, 167 (1972).
11. M. Sela, *Harvey Lectures,* 67, 213 (1973).
12. M. Sela, B. Schechter, I. Schechter, and F. Borek, *Cold Spring Harbor Symposia on Quantitative Biology,* 32, 537 (1967).
13. M. Sela, in *Gamma Globulins,* J. Killander, ed., Almqvist and Wiksell, Stockholm, 1967, p. 455.
14. M. Sela, *Naturwissenschaften,* 56, 206 (1969).
15. M. Sela, *Ann. N.Y. Acad. Sci.,* 169, 23 (1970).

16. M. Sela, I. Schechter, B. Schechter, and A. Conway-Jacobs, in *Homologies in Enzymes and Metabolic Pathways and Metabolic Alterations in Cancer*, Vol. 1, Miami Winter Symposia, W.J. Whelan and J. Schultz, eds., North-Holland, Amsterdam, 1970, p. 382.
17. M. Sela, *Ann. N.Y. Acad. Sci.*, **190**, 181 (1971).
18. M. Sela, in *Immunoglobulins: Cell Bound Receptors and Humoral Antibodies*, Vol. 26, FEBS Proc., R.E. Ballieux, M. Gruber, and H.G. Seijen, eds., North-Holland/American Elsevier, Amsterdam and New York, 1972, p. 87.
19. M. Sela, *Behring Institute Mitteilungen*, **53**, 1 (1973).
20. M. Sela, *Bull. Inst. Pasteur*, **72**, 1 (1974).
21. E. Katchalski and M. Sela, *J. Am. Chem. Soc.*, **75**, 5284 (1953).
22. M. Sela and E. Katchalski, *J. Am. Chem. Soc.*, **77**, 3662 (1955).
23. K. Landsteiner, *The Specificity of Serological Reactions*, Harvard University Press, Cambridge, Mass., 1945.
24. M. Sela, *Bull. Research Counc. Israel*, **4**, 109 (1954).
25. M. Sela, E. Katchalski, and A.L. Olitzki, *Science*, **123**, 1129 (1956).
26. M. Sela and R. Arnon, *Biochem. J.*, **75**, 91 (1960).
27. R. Arnon and M. Sela, *Biochem. J.*, **75**, 103 (1960).
28. M. Sela and R. Arnon, *Biochim. Biophys. Acta*, **40**, 382 (1960).
29. M. Sela, S. Fuchs, and R. Arnon, *Biochem. J.*, **85**, 223 (1962).
30. T.J. Gill III and P. Doty, *J. Mol. Biol.*, **2**, 65 (1960).
31. A.L. de Weck, in *The Antigens*, Vol. 2, M. Sela, ed., Academic Press, New York, in press.
32. M. Sela and E. Mozes, *Proc. Nat. Acad. Sci. U.S.*, **55**, 445 (1966).
33. M. Sela and S. Fuchs, *Proc. Symp. on Molecular and Cellular Basis of Antibody Formation*, Czechoslovak Academy of Sciences, Prague, 1965, p. 43.
34. R.K. Brown, *Ann. N.Y. Acad. Sci.*, **103**, 754 (1963).
35. M.H. Freedman and M. Sela, *J. Biol. Chem.*, **241**, 2383 (1966).
36. B. Schechter, I. Schechter, J. Ramachandran, A. Conway-Jacobs, M. Sela, E. Benjamini, and M. Shinizu, *European J. Biochem.*, **20**, 309 (1971).
37. B. Schechter, A. Conway-Jacobs, and M. Sela, *European J. Biochem.*, **20**, 321 (1971).
38. F. Borek, J. Kurtz, and M. Sela, *Biochim. Biophys. Acta*, **188**, 314 (1969).
39. A. Maoz, S. Fuchs, and M. Sela, *Biochemistry*, **9**, 4238 (1973).
40. A. Maoz, S. Fuchs, and M. Sela, *Biochemistry*, **9**, 4246 (1973).
41. S. Fuchs, in *Peptides, Polypeptides, and Proteins*, Proceedings of the Rehovot Symposium 1974, Wiley(Interscience), New York, 1974.
42. R. Arnon, in *Peptides, Polypeptides, and Proteins*, Proceedings of the Rehovot Symposium 1974, Wiley (Interscience), New York, 1974.
43. R. Arnon and M. Sela, *Proc. Nat. Acad. Sci. U.S.*, **62**, 163 (1969).
44. R. Arnon, E. Maron, M. Sela, and C.B. Anfinsen, *Proc. Nat. Acad. Sci. U.S.*, **68**, 1450 (1971).
45. R. Arnon, in *Immunity in Viral and Rickettsial Diseases*, A. Kohn, and M.A. Klingberg, eds., Plenum, New York, 1972, p. 209.
46. H. Ungar-Waron, E. Hurwitz, J.-C. Jaton, and M. Sela, *Biochim. Biophys. Acta*, **138**, 513 (1967).
47. R. Arnon, M. Sela, E.S. Rachaman, and D. Shapiro, *European J. Biochem.*, **2**, 79 (1967).

48. D. Teitelbaum, R. Arnon, M. Sela, Y. Rabinsohn, and D. Shapiro, *Immunochemistry*, **10**, 735 (1973).
49. A.R. Zeiger and P.H. Maurer, *Biochemistry*, **12**, 3387 (1973).
50. C. Webb, D. Teitelbaum, R. Arnon, and M. Sela, *European J. Immunol.*, **3**, 279 (1973).
51. A. Licht, B. Schechter, and M. Sela, *European J. Immunol.*, **1**, 351 (1971).
52. M. Sela, E. Mozes, G.M. Shearer, and Y. Karniely, *Proc. Nat. Acad. Sci. U.S.*, **67**, 1288 (1970).
53. Y. Karniely, E. Mozes, G.M. Shearer, and M. Sela, *J. Exp. Med.*, **137**, 183 (1973).
54. F. Falkenberg, Y. Karniely, E. Mozes, and M. Sela, unpublished data.
55. H.N. Claman, E.A. Chaperon, and R.F. Triplett, *J. Immunol.*, **97**, 828 (1966).
56. G.F. Mitchell and J.F.A.P. Miller, *J. Exp. Med.*, **128**, 821 (1968).
57. J.-C. Jaton and M. Sela, *J. Biol Chem.*, **243**, 5616 (1968).
58. J. Medlin, J.H. Humphrey, and M. Sela, *Folia Biologica*, **16**, 156 (1970).
59. M. Sela, E. Mozes, and G.M. Shearer, *Proc. Nat. Acad. Sci. U.S.*, **69**, 2696, (1972).
60. E. Mozes, H.O. McDevitt, J.-C. Jaton, and M. Sela, *J. Exp. Med.*, **130**, 1263 (1969).
61. E. Mozes, H.O. McDevitt, and M. Sela, *European J. Immunol.*, **3**, 1 (1973).
62. S. Fuchs, E. Mozes, A. Maoz, and M. Sela, *J. Exp. Med.*, **139**, 148 (1974).
63. H.O. McDevitt and M. Sela, *J. Exp. Med.*, **122**, 517 (1965).
64. H.O. McDevitt and M. Sela, *J. Exp. Med.*, **126**, 969 (1967).
65. E. Mozes, H.O. McDevitt, J.-C. Jaton, and M. Sela, *J. Exp. Med.*, **130**, 493 (1969).
66. E. Maron, H.I. Scher, E. Mozes, R. Arnon, and M. Sela, *J. Immunol.*, **111**, 101 (1973).
67. E. Mozes, G.M. Shearer, and M. Sela, *J. Exp. Med.*, **132**, 613 (1970).
68. G.M. Shearer, E. Mozes, and M. Sela, *J. Exp. Med.*, **135**, 1009 (1972).
69. L. Lichtenberg, E. Mozes, G.M. Shearer, and M. Sela, *European J. Immunol.*, in press.
70. E. Mozes and M. Sela, *Proc. Nat. Acad. Sci. U.S.*, in press.
71. B. Benacerraf and H.O. McDevitt, *Science*, **175**, 273 (1972).
72. G.J. Hammerling and H.O. McDevitt, *Behring Institute Mitteilungen*, **53**, 28 (1973).
73. M.J. Taussig, E. Mozes, and R. Isac, Submitted for publication.
74. E. Mozes, M. Schwartz, and M. Sela, *J. Exp. Med.*, in press.
75. E. Mozes, in *Peptides, Polypeptides, and Proteins*, Proceedings of the Rehovot Symposium 1974, Wiley (Interscience), New York, 1974.
76. M. Sela, *FEBS Letters*, in press.
77. D. Teitelbaum, A. Meshorer, T. Hirshfeld, R. Arnon, and M. Sela, *European J. Immunol.*, **1**, 242 (1971).
78. D. Teitelbaum, C. Webb, A. Meshorer, R. Arnon, and M. Sela, *European J. Immunol.*, **3**, 273 (1973).
79. D. Teitelbaum, C. Webb, A. Meshorer, R. Arnon, and M. Sela, *Nature*, **240**, 564 (1972).
80. D. Teitelbaum, C. Webb, M. Bree, A. Meshorer, R. Arnon, and M. Sela, submitted for publication.

81. C.A. Janeway, Jr., and J.H. Humphrey, *Israel J. Med. Sci.*, **5**, 185 (1969).
82. S. Bauminger and M. Sela, *Israel J. Med. Sci.*, **5**, 177 (1969).
83. M. Goldman and S. Leskowitz, *J. Immunol.*, **104**, 874 (1970).
84. R.K. Gershon, P.H. Maurer, and C.F. Merryman, *Proc. Nat. Acad. Sci. U.S.*, **70**, 250 (1973).

STUDIES ON THE CHEMICAL AND GENETIC BASES OF IMMUNOGENICITY AND ANTIGENIC REACTIVITY

THOMAS J. GILL III, HEINZ W. KUNZ, and SANDRA K. RUSCETTI,
Department of Pathology, University of Pittsburgh School of Medicine,
Pittsburgh, Pa. 15261

SYNOPSIS: Synthetic polypeptide antigens have been used to study the chemical factors influencing the ability of the antigen to elicit an immune response, the metabolic fate of the antigen, the structure of antigenic sites and the genetic control mechanisms influencing the ability of the animal to respond. The properties of a molecule minimally necessary to induce antibody formation are those quite general to macromolecules. Quantitative control of antibody formation depends upon both the general properties and the detailed composition of the antigen. The topography of a macromolecule is crucial to its ability to provide sites reactive with an antibody, and all levels of organization in polypeptides and proteins can provide structural specificity for antigenic sites. The composition of the antigenic sites does not reflect the overall composition of the molecule, but preferentially contains certain amino acid residues.

The antibody response of genetically inbred rats is controlled by a complex polygenic system. There are at least two autosomal genes and a sex influence, which also may be genetically determined. The genetic control of the quantity, binding constant, and specificity of the antibody formed is linked to the major histocompatibility locus. All of the low responders make a uniformly low antibody response, whereas there are quantitative differences among responders. Factors other than the major genetic ones and the sex influence control the quantity of antibody formed, because animals of the same genotype can make significantly different amounts of antibody, depending upon the crosses by which they acquire the major histocompatibility alleles.

These studies have led to formulation of a hypothesis that seeks to explain antigenic stimulation or tolerance on the basis of the same sequence of events that is governed by the mass-action law and whose outcome depends upon multiple, interrelated equilibria.

The major thrust of our research program has been the investigation of the mode of action of antigen and the description of the immune response as a quantitative, antigen-driven phenomenon. Studies in our laboratory have utilized synthetic polypeptide antigens, and this paper is a review of these studies. This subject has been reviewed elsewhere (1–4) and in other papers presented at this Symposium. Our approach has been to study the chemical factors influencing the ability of synthetic polypeptide antigens to elicit an immune response (immunogenicity), the metabolic fate of the antigen, the structure of the antigenic sites, and the genetic control mechanisms influencing the ability of an animal to respond.

THE CHEMISTRY OF ANTIGENS AND ITS INFLUENCE ON IMMUNOGENICITY

The ability of an animal to mount an immune response depends upon the interplay of the chemistry of the antigen and the physiological state of the host. The way in which the antigen is presented and the use of adjuvants greatly affect its action. We propose that there is a quantitative balance between the stimulation of an immune response and the induction of tolerance following the introduction of the antigen. This balance varies for each antigen, and the chemistry of the antigen is the crucial factor in determining its immunological activity. The effective amount of antigen depends upon the dose and method of administration and upon the degradation of the antigen *in vivo*. The major role of antigen metabolism is postulated to lie in the regulation of the amount of antigen left intact and capable of stimulating an immune response.

Aromatic amino acids were not needed for a polypeptide to induce an antibody response (5, 6), but they consistently enhanced the amount of antibody formed (5, 7-10). Tyrosine and phenylalanine were equally effective in enhancing immunogenicity, and there was no clear correlation between the amount of tyrosine or phenylalanine in the polypeptide and the amount of antibody formed. This ability to enhance antibody formation was consistent, but not exclusive, because copolymers of glutamic acid, lysine, and alanine could elicit amounts of antibody comparable to those elicited by the tyrosine-containing polypeptides.

The effects of charge on immunogenicity were examined with a series of glutamic acid:lysine and glutamic acid:lysine tyrosine polymers in which the amounts of glutamic acid and lysine were systematically varied (9). The polypeptides containing tyrosine elicited more antibody than their counterparts containing only glutamic acid and lysine. The best immunogens fell in the range +75% to -75% net charge density, and within this range there was no effect of charge on the amount of antibody formed. Highly charged polymers elicited only small amounts of antibody whether or not they contained tyrosine. Polypeptides in which the net charge was zero elicited amounts of antibody comparable to those induced by polypeptides with a net charge between -75% and +75%, and completely uncharged polypeptides (11, 12) also evoked good antibody responses. Therefore, charge is not a requirement for immunogenicity, and over a wide range, it does not influence the amount of antibody elicited; however, excessively high charge depresses the antibody response.

The shape of a polypeptide does not affect its ability to elicit an antibody response, although it can alter the amount of antibody formed, but it greatly changes the specificity of the antibodies. Linear polypeptides, which have no organized conformational structure, can be potent immunogens (9). Ordered sequence polypeptides that have the α-helical conformation [e.g., poly(Tyr-Ala-Glu)] or the collagenlike triple-helical conformation [e.g., poly(Pro-Gly-Pro)] were immunogenic (13, 14). Intramolecularly cross-linked synthetic polypeptides (15), which have an ordered spatial structure, were potent immunogens.

Finally, multichain polymers (16), which are compact, globular molecules, were highly immunogenic. Thus, all the levels of structure and surface topography can be present in immunogenic polypeptides. Presumably, the same conclusion holds true for proteins, and the loss in immunogenic potency with denaturation may be explained in part on grounds other than conformational change, for example, rapid degradation *in vivo*.

Small molecules can be immunogenic, and molecular weight plays an important role in their relative immunogenicity, but above a certain threshold, it does not appear to be a determining factor. This limit is a function both of size and of amino acid composition: for glutamic acid:lysine polymers, the threshold is around 30,000 or 40,000 and for glutamic acid:lysine tyrosine polymers, it is between 10,000 and 20,000. Both types of polypeptides elicited the same antibody response over a threefold variation in molecular weight once the molecular weight threshold was exceeded (9). Because $poly(Glu^{50}Ala^{40}Tyr^{10})$ of molecular weight 4000 was a good immunogen (7), the presence of both tyrosine and alanine appears to lower the molecular-weight threshold even further.

$Poly(DGlu^{55}DLys^{39}DTyr^6)$ elicited about one-fourth as much antibody as the L enantiomorph (17). In an effort to understand the immunogenic differences between the D and L polymers, studies of their metabolism were undertaken in the rabbit (18–21) and in the mouse (22–24). In the rabbit, the metabolic fate of intravenously injected, ^{131}I-labeled $poly(DGlu^{55}DLys^{39}DTyr^6)$ differed markedly from that of its enantiomorph, $poly(Glu^{56}Lys^{33}Tyr^6)$. Large amounts of the D polymer were retained in the liver and in the kidneys. Chromatographic analyses of the dialysable radioactivity in the urine and in the homogenates of liver and kidney showed the presence of labeled peptides of molecular weight ~1000 that derived from the degradation of the D or L polymer. There was no evidence for specific binding between the polypeptides and serum proteins, although there was some nonspecific binding *in vivo* and *in vitro*; hence, the polypeptides alone acted as immunogens. These metabolic studies suggested that the apparent lack of immunogenicity of D-amino acid polymers was due to their prolonged retention in the organs and gradual release over a long period of time: This caused immunological paralysis. In order to verify this hypothesis, small doses of a variety of D polypeptides were injected over a long period of time. All of the polymers consistently elicited an antibody response; continued administration of antigen depressed and then abolished the antibody response (9, 22, 25, 26). Finally, low doses of poly(glutamic acid) or polylysine of both the L and D configurations given over a relatively long period of time elicited an antibody response (9).

The hypothesis developed from the study of D-amino acid polymers (viz., that low doses of antigen over long periods of time can elicit an immune response to poorly immunogenic macromolecules) was tested further with four vinyl polymers, each of which possessed properties of particular interest (27). Polyvinylpyrrolidone (MW = 180,000) is the most peptidelike vinyl polymer because it contains a substituted amide group and a heterocyclic ring, which is a model for aromatic amino acids. Poly(methacrylic acid-2-dimethylaminoethyl methacrylate)

(MW = 360,000) is a polyampholite that resembles proteins in respect to charge distribution; it is the vinyl analog of a synthetic polypeptide containing glutamic acid and lysine. Lastly, poly(methacrylic acid) (MW = 15,000) and polyvinylamine (MW = 43,000) are completely charged vinyl homopolymers (a combination of characteristics that should produce the most unfavorable type of immunogen) and they are the vinyl analogs of poly(glutamic acid) and polylysine, respectively. Polyvinylpyrrolidone and poly(methacrylic acid-2-dimethylaminoethyl methacrylate) consistently elicited a moderate amount of antibody (100–200 μg Ab/ml), and poly(methacrylic acid) and polyvinylamine elicited a low and variable antibody response (5–20 μg Ab/ml). The amounts of antibody elicited by these vinyl polymers were in the same relative proportions as the amounts elicited by their polypeptide analogs. The antibody responses decreased and were eventually abolished as more antigen was administered. The antibodies elicited by the vinyl polymers had the same deficiencies in their biological capabilities as the antibodies elicited by D-amino acid polymers (see below). These findings imply that: the antigen does not have to be fragmented in order to induce antibody formation, because there is no evidence that vinyl polymers can be enzymatically or in any other way physiologically attacked; immunogenicity is probably a general property of most, if not all, macromolecules, because the vinyl polymers that elicited an antibody response are biologically quite alien substances; and the surface topography of the molecule plays a role in its interaction with the immunocompetent cell.

Thus, the studies on synthetic polypeptide antigens have delineated some of the molecular characteristics that influence the induction and magnitude of the antibody response. A summary of these properties is given in Table 1.

ANTIGENIC DETERMINANTS ON SYNTHETIC POLYPEPTIDES

The composition of the antigenic sites on synthetic polypeptides does not reflect the overall composition of the molecule, but preferentially contains certain amino acid residues (9). As a result of studies with synthetic polypeptides and with polypeptidyl proteins, a hierarchy of antigenic potency can be established for amino acids and amino acid combinations. Tyrosine is very potent, lysine moderately so, and glutamic acid and alanine are only modestly effective (28). The antigen reactive sites on macromolecules encompass a discrete part of the molecule, and the sites for reaction in delayed hypersensitivity appear to be larger than those required for combination with antibody. On the other hand, the ability to elicit an immune response, and probably tolerance as well, is a function of the properties of the whole macromolecule or a large segment thereof. There appears to be an inverse relationship between the size of the antigenic site and the potency of the amino acid residues of which it is constituted (29).

All levels of organization in proteins and polypeptides can provide the structural specificity for antigenic sites. Linear synthetic polypeptides have antigenic determinants that depend mainly upon the primary structure of the polypeptide

Table 1. Summary of the chemical properties affecting the immunogenicity of synthetic antigens[a]

Property	Effect on Immunogenicity[b]	Comment
Composition		
three or more different types of amino acids	+	$poly(Glu^{56}Lys^{38}Tyr^6) =$ $poly(Glu^{42}Lys^{28}Ala^{30}) > poly$ $(Glu^{60}Lys^{40}) > poly(Glu) = poly(Lys)$
presence of tyrosine or phenylalanine	+	not necessary, but enhances amount of antibody in most cases
presence of alanine + tyrosine	+	$poly(Glu^{63}Ala^{28}Tyr^9) >$ $poly(Glu^{56}Lys^{38}Tyr^6) =$ $poly(Glu^{42}Lys^{28}Ala^{30})$
Charge		
high	−	completely charged polymers are weakly immunogenic
none	0	uncharged polymers are immunogenic
Shape	0	polypeptides of all conformations are immunogenic
Size	±	molecular size over 5000–20,000, depending upon composition, is not critical
D-amino acids	−	less immunogenic than L enantiomorph
Metabolism	variable	role in regulating the amount of antigen available
Nonbiological components	−	vinyl polymers are weakly immunogenic

[a]Reprinted with permission from Gill, in *Immunogenicity*, F. Borek, ed., North-Holland, Amsterdam, 1972, p. 5.

[b]The symbols are: +, increased immunogenicity; ±, slightly increased; 0, no effect; and −, decreased.

(2, 9). More highly organized polypeptides provide antigenic sites based upon the secondary structure of the molecule: poly(Tyr-Ala-Glu) is α-helical (13), poly(Pro-Gly-Pro) has the triple helix of collagen (14), and polyproline can provide antigenic specificity in either the *cis* or the *trans* conformation (30). Intramolecularly cross-linked synthetic polypeptides, which are models for the tertiary structure of proteins, have antigenic sites that depend upon the spatial organization of the polypeptide (15).

The partial topographical similarities among various chemically and biologically unrelated macromolecules probably accounts for their immunological cross-reactivity (Table 2).

Table 2. Antibody-antigen cross-reactions illustrating the importance of molecular topography in the structure of antigenic sites[a]

Cross-reacting Antibody-Antigen System	Evidence for Cross-Reactivity	Extent of Cross-Reactivity (%)
D and L enantiomorphic synthetic polypeptides	precipitin reaction; passive cutaneous anaphylaxis	2–26 $\leqslant 10$
Enantiomorphic DNP-containing tetrapeptides	binding studies	–
$(Tyr)_n$-gelatin	precipitin reaction	5–15
Enantiomorphic branched chain derivatives of polyproline	precipitin reaction	5–10
Synthetic polypeptide-protein cross-reactions	precipitin reaction	7–28
Poly(Pro-Gly-Pro) antibody with native collagen	precipitin reaction	large
Polyvinylpyrrolidone antibody with poly($Glu^{52}Lys^{33}Tyr^{15}$)	precipitin reaction	20

[a]Reprinted with permission from Gill, in *Specific Receptors of Antibodies, Antigens, and Cells,* Third International Convocation on Immunology, Karger, New York, 1973, p. 136.

The biological properties of antibodies to synthetic polypeptides vary with the chemical composition of the polypeptide and with the species in which the antibodies are made (Table 3). Immunization of rabbits with polypeptides containing only glutamic acid and lysine elicited an antibody that gave relatively weak Arthus, precipitin, and delayed hypersensitivity reactions. In contrast, polymers that contain glutamic acid, lysine and alanine, tyrosine or phenylalanine all gave strong Arthus, precipitin, and delayed hypersensitivity reactions. The antibodies elicited by enantiomorphic polypeptides containing glutamic acid, lysine, and tyrosine differed in their biological properties. The antibody elicited by the L isomer precipitated, gave a strong Arthus reaction, fixed complement, and gave a passive cutaneous anaphylaxis reaction. The antibody to the D isomer precipitated and gave an Arthus reaction, but it fixed complement poorly and did not give a passive cutaneous anaphylaxis reaction. If the D polypeptide was aggregated with methylated bovine serum albumin, however, these reactions were all as strong as those elicited by the antibody to the L-amino acid polymer.

Table 3. Biological properties of rabbit antibodies to synthetic polypeptides[a] (31)

Antigen	Molecular Weight	Precipitin	Arthus	PCA	CF	DH
Poly(Glu, Lys)	50,000–100,000	+	+	+	+	+
Poly(Glu, Lys, Ala)	50,000–100,000	+++	+++	+++	+++	+++
Poly(Glu, Lys, Tyr, or Phe)	50,000–100,000	+++	+++	+++	+++	+++
Poly($Glu^{58}Lys^{36}Tyr^{6}$)	110,000	+++	+++	+++	+++	
Poly($DGlu^{55}DLys^{39}DTyr^{6}$)	93,000	+++	+++	0	+	
Aggregated poly($DGlu^{55}DLys^{39}DTyr^{6}$)[b]		+++	+++	+++	+++	

[a] Reactions scored on a basis of 0 to +++. The abbreviations used are: PCA, passive cutaneous anaphylaxis; CF, complement fixation; and DH, delayed hypersensitivity.

[b] Aggregated with methylated bovine serum albumin.

GENETIC AND CELLULAR FACTORS IN THE IMMUNE RESPONSE

The antibody response to poly(Glu^{52}Lys^{32}Tyr^{15}) was studied in two inbred strains of rats, the highly responding ACI strain and the poorly responding F344 strain (31). The amounts of antibody formed by the parental strains, the F_1 and F_2 generations, and the backcrosses were assayed by a quantitative immunoadsorbent micromethod. Radioimmunoelectrophoretic studies showed that the antibody response was distributed among the IgM, IgG$_1$, and IgG$_2$ immunoglobulin classes in both males and females. The antibody formed by the F344 strain was almost exclusively in the IgG class, whereas the ACI strain formed antibody in all immunoglobulin classes. This finding suggests that an inability to make IgM antibody to poly(Glu^{52}Lys^{33}Tyr^{15}) may be a factor in the poor response of the F344 strain to immunization with that antigen. There was no correlation between the quantity of antibody made and the immunoglobulin class.

The results of the experiments are consistent with a model in which genetic control of the antibody response is exercised by at least two autosomal genes— one for recognition of antigen (R-gene) and one for quantitative control of the amount of antibody formed (Q-gene). In addition, there is a sex influence seen in most of the highly responding strains that is associated with a higher and more heterogeneous antibody response in females. This factor may also have a genetic basis.

Further breeding studies were carried out to test the polygenic model for the control of the antibody response (32). The backcrosses of reciprocally mated F_1 hybrids into both the highly responding ACI strain and the poorly responding F344 strain yielded offspring with low, moderate, and high responses in a ratio compatible with that predicted by the polygenic model. The backcrosses having a low antibody response bred true with inbreeding and with second backcrossing, as predicted, so they apparently have only those genetic factors that lead to a low antibody response. Limited inbreeding studies with the highly responding backcrosses indicated that they also bred true. Inbreeding of moderately responding backcrosses with moderately or highly responding backcrosses gave offspring that showed the whole spectrum of antibody responses, as would be expected for control by multiple genetic factors. Pedigree analyses of the antibody responses in a variety of hybrids of the ACI and F344 strains were performed in order to obtain unexpectedly high responses (mean of the parental responses plus or minus three standard deviations) or minimal responses (< 50 μg antibody/ml) in the offspring of parents that made moderate amounts of antibody. Such unexpected antibody responses were found in all of the crosses examined: F_1 hybrids, F_2 hybrids, F_1 backcrosses, inbred F_1 backcrosses, second F_1 backcrosses, and F_2 backcrosses. These findings indicate that multiple genetic factors are involved in the control of the antibody response. Furthermore, they strongly suggest that genetic recombination has occurred in the animals with the very high or very low antibody responses.

The highly responding ACI strain mounted a strong delayed hypersensitivity response against both free and aggregated poly(Glu^{52}Lys^{33}Tyr^{15}), whereas the poorly responding F344 strain was sensitized only by the aggregated polypeptide. Immunization with 2,4-dinitrophenyl-poly(Glu^{52}Lys^{33}Tyr^{15}) aggregated with methylated bovine serum albumin or with polylysine sensitized the animals to the antigen and the aggregating agent, but not to the hapten (33). These findings are consistent with the hypothesis that a major factor in the action of the aggregate is the prolonged retention and slow release of the antigen (33). There were more differences in the distribution of antibodies among chromatographic subclasses in the F344 strain than in the ACI strain, and these differences did not always correlate with the net charge of the antigen. Most of the differences in the F344 strain were in the antibodies eluting with 0.01 M buffer (IgG$_2$), whereas the differences in the ACI strain were almost entirely in the antibody eluting with 0.20 M buffer (mainly IgG$_1$). Aggregating the antigen caused some differences in the chromatographic subclasses of the antibody compared to those elicited by the antigen alone, and the differences were also strain-dependent.

We postulated that one of the major genes controlling the antibody response influences the size of the cell population initially capable of reacting with the antigen (34, 35). This hypothesis was tested by measuring the fraction of spleen cells that produced antibody after immunization with poly(Glu^{52}Lys^{33}Tyr^{15}) in female rats of the highly responding ACI strain, the poorly responding F344 strain, and a hybrid of the two. The unimmunized animals of each strain generally had less than one antibody-forming cell per million spleen cells. The ACI, F344, and hybrid strains all produced the same small number of plaque-forming cells following primary immunization. Following a second injection of antigen, however, there were marked differences: the ACI strain produced a large number of antibody-forming cells, the F344 strain did not show any increase, and the hybrid strain showed only a modest increase. Because the ACI strain showed a marked secondary response, despite the fact that its response following the primary course of immunization was no different from that of the poorly responding F344 strain or of the hybrid, the primary course of immunization must have sensitized a large number of cells and induced antibody production only in a very few. These sensitized cells were responsible for the increase in antibody-producing cells following the second injection of antigen. In contrast, there was no such effect in the poorly responding F344 strain, presumably because no large population of sensitized cells was available after primary immunization. We suggest, then, that the difference in the ability of the ACI and F344 strains of rats to respond to immunization with poly(Glu^{52}Lys^{33}Tyr^{15}) is due to the presence of more cells capable of being sensitized to poly(Glu^{52}Lys^{33}Tyr^{15}) in the highly responding ACI strain.

Genetic control of the antibody response to poly(Glu^{52}Lys^{33}Tyr^{15}) was studied in 24 strains of inbred rats representing all the major histocompatibility (AgB) groups (36). All of the low responders (AgB-1, 3, and 6) made less than 65 μg antibody per ml. In contrast, there were quantitatively significant differences among responders: the AgB-2 group was a moderate responder (200–500 μg antibody per ml) and the AgB-4 and AgB-5 groups were high responders

(700–1400 μg per ml). Representative data are shown in Fig. 1. Direct linkage studies, and three subsidiary lines of evidence, demonstrated that the control was linked to the major histocompatibility locus. The binding constant of the antibody (Fig. 2) and its specificity also are under genetic control and linked to the major histocompatibility locus. There was no evidence for genetic control of the antibody response to the D isomer of the polymer, and there was no cross-reactivity between the L and D isomers in the induction of the immune response. Aggregation of the L isomer with methylated bovine serum albumin (MeBSA) abolished the pattern of genetic control in most cases by increasing antibody formation in the low responders and by decreasing it in the high responders. Aggregation of the D isomer increased the antibody response of all strains, except the moderately responding WF strain. The antibody response to the MeBSA in the aggregates was uniformly low, although there was a relatively high, genetically controlled response to immunization with MeBSA alone. Factors other than the major genetic one and the sex influence control the quantity of antibody formed: Animals of the same genotype can make significantly different amounts of antibody, depending upon the crosses by which they acquired the major histocompatibility alleles.

CONCLUSIONS

When an animal is injected with an antigen, there is a balance between immunological stimulation and paralysis that depends upon the chemistry of the antigen and the genetic background of the host. The chemical properties of the antigen set the level and range of dosage that can be used to stimulate an antibody response. The antigen acts intact, and the role of antigen catabolism is to function in concert with the original dose to regulate the amount of antigen available to stimulate antibody formation. A poor immunogen is probably a molecule that induces tolerance easily, but this proposition is a difficult one to test because the mechanism of tolerance is not understood.

The properties of a molecule minimally necessary to induce antibody formation are those quite general to macromolecules. Even biologically alien macromolecules, such as vinyl polymers, can elicit antibody formation; hence, all macromolecules can probably elicit antibody formation if given in the proper dosage and by the proper schedule. The quantitative control of antibody formation depends upon both the general properties and the detailed composition of the immunogen. The presence of aromatic amino acids enhances immunogenicity, and there is a particular amount necessary for optimal enhancement. Too many aromatic amino acids residues can shift the specificity of the antigen and finally decrease its ability to elicit antibody formation. Other properties, such as the D configuration or high charge, depress immunogenicity but do not abolish it. Immunogenicity is a property that may involve portions of a macromolecule different from those required for the interaction with antibody, and the antigenic determinant appears to involve a smaller area of the molecule than the portion required for interaction with the immunocompetent cell.

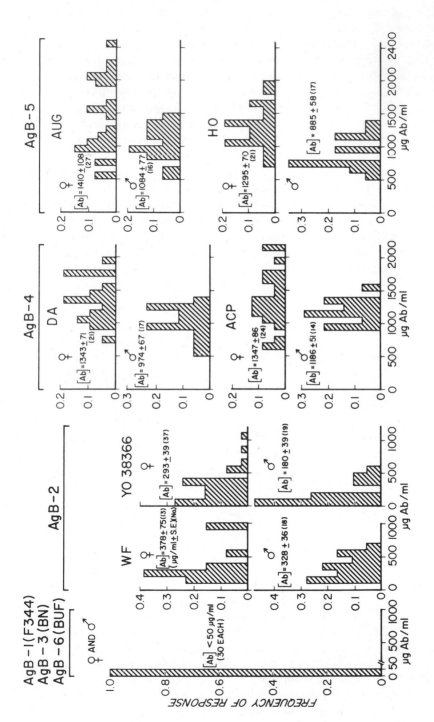

Figure 1. The antibody responses of representative strains from each rat histocompatibility (AgB) group. The animals were immunized by a standard protocol, and their responses are plotted as frequency histograms. The AgB-1, 3, and 6 are low responders. AgB-2 is a moderate responder; and AgB-4 and 5 are high responders.

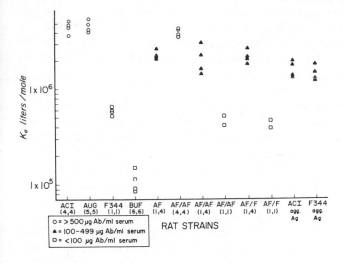

Figure 2. The genetic control of the binding constants of antibodies made in various high responders, low responders, and their hybrids. The abbreviations are A for ACI and F for F344. The hybrids are designated by an abbreviation having the maternal strain first and the paternal strain second. For example, AF is the F_1 hybrid of an ACI female and a F344 male, and AF/AF is the F_2 hybrid. The quantities of antibody made by each group are keyed as indicated in the figure, and the genotypes of each group are shown in parentheses. The binding constants of the high (ACI and AUG) and the low (F344 and BUF) responders are significantly different ($P \leqslant 0.001$). The F_1 hybrid produces antibody with a binding constant intermediate between those of the parental strains and significantly different from them. Linkage of the binding constant to the genotype is seen in the F_2 hybrids and in the backcrosses. Aggregation of the antigen decreases both the amount of antibody formed and its binding constant in the high-responder ACI strain and increases them in the low-responder F344 strain.

The topography of a macromolecule is crucial in its ability to provide sites reactive with antibodies. In theory, the topography of a given site could be made up of several different types of components and still display the same immunochemical behavior. This concept could explain cross-reactivity among ostensibly unrelated antigens and could provide the rationale for the fact that a limited number of antibody specificities exists to deal with an extensive repertoire of antigens. This hypothesis finds some support in the demonstration of cross-reactivity between vinyl polymers and synthetic polypeptides, between synthetic polypeptides and proteins, and between synthetic polypeptides of different optical configurations. The intermolecular interactions involve relatively large areas of the molecules, because the volume change that occurs during the antibody-antigen reaction is quite large (37). Finally, the immunogenicity of vinyl polymers is in the same relative order as that of their synthetic polypeptide analogs; therefore, the van der Waals contour of the molecules, whether they are vinyl polymers or synthetic polypeptides, probably contributes to the ability of the macromolecules to interact with the immunocompetent cell and to elicit antibody formation.

Our hypothesis to explain the action of antigen at the cellular level and the effects of the factors modulating the immune response is shown in Fig. 3 (1). The amount of antigen that is available to stimulate the immunocompetent cell depends upon the dose, the time over which it is given, and the rate at which the antigen is degraded *in vivo*. Aggregating the antigen would protect it from enzymatic degradation and provide for slow release over a long period of time; therefore, it would exert a dosage effect among other effects. Then, the ability of this effective concentration of antigen to stimulate an antibody response or to induce tolerance would depend upon the genetically determined number of immunocompetent cells capable of reacting with a given antigen and the binding affinity of the cellular receptors. The effectors that are produced following stimulation exert a negative feedback control. The population of immunocompetent cells capable of reacting with a given antigen and of producing antibody could be increased by periodic stimulation with the appropriate amount of antigen; in this way tolerance could be avoided. In like manner, immunological memory may be due to the continuous stimulation of the immunocompetent cell population by small amounts of antigen retained in the host. Instead of residing in the lymphoid tissue the retained antigen may be released slowly from a variety of tissues. Thus, this scheme proposes to explain stimulation or tolerance on the basis of the same sequence of events that is governed by the mass-action law and whose outcome depends upon multiple, interrelated equilibria.

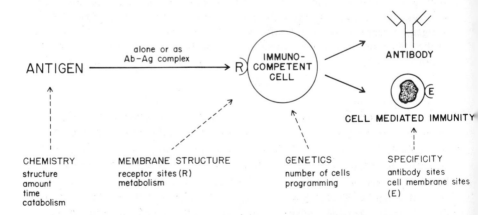

Figure 3. Hypothetical scheme for the action of antigen and for the effects of the factors modulating the immune response. The hypothesis seeks to explain the various aspects of the immune response as a quantitative, antigen-driven phenomenon.

Table 4. Cross-reactivity between antipoly(Glu^{52}Lys^{33}Tyr^{15}) antibodies from various sources and related synthetic polypeptide antigens as determined by the precipitin reaction

Antibody	Poly(Glu^{52}Lys^{33}Tyr^{15}) (%)	Poly(Glu^{56}Lys^{38}Tyr^{6}) (%)	Poly(Glu^{57}Lys^{34}Phe^{9}) (%)
ACI antipoly(Glu^{52}Lys^{33}Tyr^{15})	100	73	21
F344 antipoly(Glu^{52}Lys^{33}Tyr^{15})	100	7	5
BUF antipoly(Glu^{52}Lys^{33}Tyr^{15})	100	6	2

ACKNOWLEDGMENTS

The work in the authors' laboratory has been supported by grants from the National Institutes of Health (AI 11061 and GM 00135), U.S. Army Medical Research and Development Command (DADA 17-73-C-3020), National Science Foundation (GB 30826), Sarah Mellon Scaife Foundation Fellowship Program in Pathology, and the Beaver County Cancer Society.

REFERENCES

1. T.J. Gill, III, in *Immunogenicity*, F. Borek, ed., North-Holland, Amsterdam, 1972, p.5.
2. M. Sela, *Science*, **166**, 1365 (1969).
3. P.H. Maurer, *Progr. Allergy*, **8**, 1 (1964).
4. B. Benacerraf and H.O. McDevitt, *Science*, **175**, 273 (1972).
5. T.J. Gill III, and P. Doty, *J. Biol. Chem.*, **236**, 2677 (1961).
6. P.H. Maurer, *J. Immunol.*, **88**, 330 (1962).
7. M. Sela, S. Fuchs, and R. Arnon, *Biochem. J.*, **85**, 223 (1962).
8. S. Fuchs and M. Sela, *Biochem. J.*, **87**, 70 (1963).
9. T.J. Gill III, H.W. Kunz, and D.S. Papermaster, *J. Biol. Chem.*, **242**, 3308 (1967).
10. P.H. Maurer, *J. Immunol.*, **90**, 493 (1963).
11. M. Sela and S. Fuchs, *Biochim. Biophys. Acta*, **74**, 796 (1963).
12. P.H. Maurer, B.F. Gerulat, and P. Pinchuck, *J. Immunol.*, **97**, 306 (1966).
13. M. Sela, B. Schechter, I. Schechter, and F. Borek, *Cold Spring Harbor Symposia on Quantitative Biology*, **32**, 537 (1967).
14. F. Borek, J. Kurtz, and M. Sela, *Biochim. Biophys. Acta*, **188**, 314 (1969).
15. T.J. Gill III, D.S. Papermaster, H.W. Kunz, and P.S. Marfey, *J. Biol. Chem.*, **243**, 287 (1968).
16. M. Sela, S. Fuchs, and M. Feldman, *Science*, **139**, 342 (1963).
17. T.J. Gill III, H.W. Kunz, H.J. Gould, and P. Doty, *J. Biol. Chem.*, **239**, 1107 (1964).
18. T.J. Gill III, D.S. Papermaster, and J.F. Mowbray, *Nature* (London), **203**, 644 (1964).
19. T.J. Gill III, D.S. Papermaster, and J.F. Mowbray, *J. Immunol.*, **95**, 794 (1965).
20. D.S. Papermaster, T.J. Gill III, and W.F. Anderson, *J. Immunol.*, **95**, 804 (1965).
21. C.B. Carpenter, T.J. Gill III, and L.T. Mann, Jr., *Immunology*, **98**, 236 (1967).
22. C.A. Janeway, Jr. and M. Sela, *Immunology*, **13**, 29 (1967).
23. C.A. Janeway, Jr. and J.H. Humphrey, *Immunology*, **14**, 225 (1968).
24. C.A. Janeway, Jr., *Immunology*, **17**, 715 (1969).
25. Y. Stupp and M. Sela, *Biochim. Biophys. Acta*, **140**, 349 (1967).
26. J.-C. Jaton and M. Sela, *J. Biol. Chem.*, **243**, 5616 (1968).
27. T.J. Gill III and H.W. Kunz, *Proc. Nat. Acad. Sci. U.S.*, **61**, 490 (1968).
28. T.J. Gill III, H.W. Kunz, E. Friedman, and P. Doty, *J. Biol. Chem.*, **238**, 108 (1963).

29.   T.J. Gill III, D.S. Papermaster, H.W. Kunz, and P.S. Marfey, in *Nucleic Acids in Immunology*, O.J. Plescia and W. Braun, eds., Springer-Verlag., New York, 1968, p. 330.
30.   T.J. Gill III, in *Specific Receptors of Antibodies, Antigens and Cells,* Third International Convocation on Immunology, D. Pressman, N. Rose, and T. Tomasi, eds., Karger, New York, 1973, p. 136.
31.   T.J. Gill III, H.W. Kunz, D.J. Stechschulte, and K.F. Austen, *J. Immunol.,* **105**, 14 (1970).
32.   T.J. Gill III and H.W. Kunz, *J. Immunol.,* **106**, 980 (1971).
33.   B.P. Sloan and T.J. Gill III, *J. Immunol.,* **108**, 26 (1972).
34.   T.J. Gill III, J. Enderle, R.N. Germain, and C.T. Ladoulis, *J. Immunol.,* **106**, 1117 (1971).
35.   J. Shonnard, B.K. Davis, C.T. Ladoulis, and T.J. Gill III, submitted for publication.
36.   T.J. Gill III, H.W. Kunz, and B. Borland, *J. Immunogenetics,* submitted for publication.
37.   Y. Ohta, T.J. Gill III, and C.S. Leung, *Biochemistry,* **9**, 2708 (1970).

GENETIC CONTROL OF IMMUNE RESPONSES IN MICE AGAINST RANDOM POLYMERS OF GLUTAMIC ACID, ALANINE, AND TYROSINE

PAUL H. MAURER and **CARMEN F. MERRYMAN**, Department of Biochemistry, Thomas Jefferson University, 1020 Locust Street, Philadelphia, Pa. 19107

SYNOPSIS: Studies on the genetic control of immune responses in inbred mice against random terpolymers of glutamic acid, alanine, and tyrosine are presented. The immune responses against $(\text{Glu}^{60}\text{Ala}^{30}\text{Tyr}^{10})_n$ in mice are controlled by an immune response (Ir) gene located in the mouse's major histocompatibility locus, which is linked to the specific H-2 haplotype. Mice of H-2 haplotype a, b, d, f, k, and r are responders and those of haplotype j, p, q, and s are nonresponders. The SJL mouse (H-2^s), which does not react against GAT^{10}, reacts against polymers of Glu and Ala with limited amounts of tyrosine (GAT^4). The recognition of the GAT^{10} polymer by T cells of nonresponder mice was shown by measurement of the DNA synthetic response of these lymphoid cells. Nonresponsiveness against these polymers, which are "T-cell-dependent antigens," appears to be due *not* to nonrecognition by the T cells, but rather to the interaction of GAT^{10} with "suppressor" T cells, which interferes with the response of the competent B cells. Inhibition studies with anti-GAT^{10} sera indicate that specificities are directed against GA, GT, and "GAT" and that GAT^{10} is a heterogeneous polymer. These findings also explain the studies on immunological tolerance in which GAT^{10} (but not GA or GT) was found to be tolerogenic for subsequent responses of C57BL/6 mice against GAT^{10}.

INTRODUCTION

Since the First International Symposium on Poly(α-Amino Acids) in 1961, there has been a phenomenal growth in the use of synthetic polymers of amino acids to study many of the parameters and questions associated with the immune response (1). Because polymers of amino acids can be tailored to contain a more restricted range of determinants than exists in proteins, one of the major contributions during the past 10 years has been the study of the genetic control of the immune response (2–4).

At the 1961 symposium, we presented data obtained in rabbits and guinea pigs with the copolymers of glutamic acid, lysine, and alanine. It appeared then that there indeed was genetic control of the immune responses of guinea pigs, and we concluded (5),

Although the introduction of tyrosine into copolymers of glutamic acid and lysine appears to be important in determining the *specificity* of the antibody produced, it does not appear to enhance the immunogenicity as measured by the number of positive reactors. The variation in responses among rabbits and guinea pigs may be due to the genetic make-up of the animal which may govern its ability to *recognize* a configuration as foreign.

Since then, our immunochemical studies have dealt with factors that control immunogenicity (6, 7), the nature of antigenic determinants in synthetic polymers (8–15), the mechanism of antigen-antibody reactions as measured by tritium-exchange techniques (16–18) and circular dichroic (CD) methods (19). In this paper, we will present some of our findings obtained in the past few years in the area of genetic control of the immune response and immunological tolerance. Although our studies with different kinds of amino acid polymers have been conducted with rabbits (20), guinea pigs (21–23), mice (24–30), rats (31), and chickens, the subject of this report is our studies with mice and polymers of glutamic acid, alanine, and tyrosine.

MATERIALS AND METHODS

Polymers. Random polymers of amino acids were prepared by polymerizing N-carboxy-α-amino acid anhydrides (NCA) and purified before use (32). Unless indicated otherwise, all polymers consist of α-L-amino acids. Nomenclature is based on the recommendations of the IUPAC and the IUB; for example, poly($Glu^{60}Ala^{30}Tyr^{10}$) represents a linear random terpolymer of α-L-glutamic acid, α-L-alanine, and α-L-tyrosine, in which the superscript indicates the mol percent of each amino acid in the polymer.

The aggregates of GAT^{10} with methylated bovine serum albumin (MeBSA) were prepared as referred to previously (33). The basic procedure was as follows: To a solution of GAT^{10} (1 mg/ml H_2O) containing some ^{125}I-labeled GAT^{10} tracer, MeBSA (5 mg/ml) was added dropwise, and the solution was stirred until maximum flocculation and aggregate formation had occurred. The precipitate was centrifuged in the cold and washed three or four times with 0.15 M NaCl until the washings were free of radioactivity. A suspension of the precipitate containing 200 μg/ml of GAT^{10} was made in saline.

The neutral polymer poly(γ-N-hydroxypropyl glutamide^{60} alanine^{30} tyrosine^{10}) (GAT "amide"), in which the carboxyl groups were modified with propanolamine, was prepared as outlined in a previous publication (15).

The nitration of the GAT^{10} and GAT^4 polymers was accomplished using tetranitromethane (34). About 50% of the tyrosine residues were nitrated with GAT^{10}, whereas about 40% of the residues were modified with GAT^4 (15).

Iodination of polymers. Synthetic polypeptides were labeled with ^{125}I according to the chloramine-T procedure (35), as described previously. Unreacted ^{125}I was removed by extensive dialysis against distilled water until none was detected in the dialyzing medium.

Specific activity of polymers. The ^{125}I content and specific activity of all iodinated polymers were determined on an aliquot of a solution. At least 90% of the polymer-associated ^{125}I was precipitable with homologous rabbit antisera prepared against the noniodinated polymer.

Animals. Most of the inbred and congenic strains of mice were obtained from Jackson Labs, Bar Harbor, Maine. The random-bred mice referred to in the text were obtained from Charles River, Massachusetts.

Immunization. Groups of 5 to 15 mice were immunized with varying amounts of the polymer. For the first injection the polymer was incorporated in complete Freund's adjuvant and injected in the footpad. The "booster" injection was given intraperitoneally with the polymer in solution. In general, the concentration of antigen employed for immunization was 10 μg. In some instances, 25 μg or 100 μg of the appropriate polymer was used.

Immunization with the aggregates of MeBSA followed the same protocol. The "booster" injection was given intraperitoneally as a suspension of the aggregate.

Serum antibody determination. The basic protocol for analysis involved a modification of the antigen-binding assay with ^{125}I labeled polymers (27). All values given in the text and tables refer to the percentage of 0.003 μg N of ^{125}I-labeled polymer bound by 25 μl of 1:2 dilution of mouse antiserum. Generally, the iodinated homologous antigens were used for analyses. For analysis of the responses against the GAT-NO_2 polymers, however, the nonderivatized preparations were used. Analyses of anti-GA and anti-GAL[10] responses employed polymers that were first tyrosylated and then iodinated. The basic procedure for tyrosylating the polymer was as follows (27): The copolymer was precipitated from aqueous solution (pH 1), filtered, washed twice with distilled water and dessicated. To 14.2 mg of the copolymer dissolved in 2 ml distilled dimethylformamide (DMF) were added successively 0.7 mg of tyrosine methyl ester hydrochloride and 0.5 μl triethylamine dissolved in 100 μl DMF and 0.83 mg of dicyclohexylcarbodiimide (DCC) dissolved in 100 μl DMF. The solution was stirred at room temperature for two hours and evaporated to dryness. The residue was suspended in 2 ml phosphate-buffered saline (PBS), which was then adjusted to pH 8.2, centrifuged to remove dicyclohexylurea and dialyzed. Spectroscopy indicated that about 2% tyrosine residues were introduced. Recovery was almost quantitative.

RESULTS AND DISCUSSION

Response patterns of mice against poly(GAT[10]). Table 1 presents the responses against the polymer GAT[10] in inbred and congenic strains of mice. The data show that mice responsive to the polymer are of the H-2 haplotypes a, b, d, f, k, and r; mice of the H-2 haplotypes j, p, q and s were nonresponders. The data also indicate that the responsiveness against GAT[10] is controlled by an immune response (Ir) gene that is linked to the major histocompatibility alleles of the responder strains (H-2 haplotype). The nonimmunogenicity of the "GAT amide" polymer indicates the importance of free glutamyl determinants. In addition, the substitution of D-tyrosine for L-tyrosine in the preparation of the polymer (GAT LLD) does not alter the immune response pattern, although it does alter the level of antibody detected.[1] This finding is additional evidence that the important determinants for polymer *recognition* are associated with poly(Glu, Ala) determinants.

---

[1] It will be shown later that antibodies against GAT[10] are heterogeneous and contain anti-GA, anti-GT and anti-GAT.

Table 1. Antigen-binding capacity of sera of mice immunized with GAT[10] and related polymers and relationship to H-2 haplotype

Strain	H-2 Haplotype	% Antigen-Bound GAT[10]	GAT(LLD)	Strain	H-2 Haplotype	% Antigen-Bound GAT[10]	GAT(LLD)
A	a	67 ± 6	84 ± 4	C3H/He	k	77 ± 1	32 ± 13
B10.A	a	80 ± 2		AKR	k	87 ± 1	
C57BL/6	b	86 ± 3	67 ± 4 (29 ± 9)[a]	B10.BR	k	75 ± 4	
C57L	b	83 ± 5		P	p	2 ± 3	
BALB/c	d	81 ± 8	46 ± 17	C3H.NB	p	1 ± 2	2 ± 4
DBA/2	d	82 ± 2		DBA/1	q	1 ± 1	1 ± 2 (5 ± 4)[a]
A.CA	f	86 ± 3		C3H.Q	q	1 ± 1	
B10.M	f	77 ± 6		B10.R111	r	76 ± 8	
C3H/JK	j	0 ± 0		SJL	s	2 ± 2	3 ± 0 (3 ± 2)[a]
				A.SW	s	0 ± 0	
Random Mice							
Swiss-Webster		(30–88)[b] (16:18)[c]					
CD-1		(39–93)[b] (24:53)[c]					
		0 ± 0 (10)					

[a] Values obtained with (GAT[10])-NO_2.
[b] Range of antigen binding values.
[c] No. responders:total no. of mice immunized.

The data obtained with the random-bred mice show that the response patterns of these mice against the polymer differ; that is, there are responder and nonresponder mice among them. In our laboratory it has been possible to breed these random-bred mice and obtain high-responder and nonresponder lines against the polymer GAT[10].

Response patterns of mice against GAT[4], GAL[10], and GA. The immune response patterns of inbred mice against the polymers GAT[4], GAL[10], and GA were similar. Mice of the H-2[p] and H-2[q] haplotypes, which were nonresponders to GAT[10], also did not respond to these polymers. However, the SJL and A.SW mice, which were nonresponders to GAT[10], responded to the three polymers presented here (percent antigen bound: 41 ± 20 to 68 ± 12).

The observation that the polymer GA is immunogenic in inbred mice is exceptionally important because it was the first time that a random copolymer of two amino acids was shown to be immunogenic in mice (36). A reinvestigation of this problem indicates that GA is immunogenic in mice of H-2 haplotypes a, b, d, f, k, and s, but not in mice of H-2 haplotypes j, p, and q. We have postulated, therefore, that the *recognition* of polymers such as GA, GAT[4], GAT[10], and GAL[10], is under the control of an Ir GA gene that is not present in mice of H-2[j,p,q] haplotypes. The gene controlling the response has been mapped in the I-A locus of the major H-2 complex.

With the exception of C57BL/6 mice, inbred strains of mice do not have the Ir gene for GT (36), although following immunization with GAT[10] they can make antibody with GT specificity; that is, they do have B cells capable of producing antibody against the GT determinants. Therefore GAT[10] can be considered to consist of carrier (GA) and haptenic (GA, GT, or "GAT") determinants. It would follow that for inbred mice (responders to GA), in contrast to random-bred mice (responders to GT), tyrosyl determinants are not important for the recognition mechanism. This feature explains the response patterns against GAT (LLD) and various nitrated derivatives of GAT. In essence, modification of the tyrosine group, or changing the L tyrosine to a D tyrosine, did not change the immune response pattern of the mice.

Conversion of nonresponder mice to responders. The "lesions" in mice that are nonresponsive to some of the random polymers of amino acids appear to be associated with a defect in thymic lymphocyte (T cell) function rather than with a bone-marrow-derived (B cell) defect. Although the exact nature of the lesions is not understood, a method of converting nonresponder animals to responders is to immunize them with an aggregate of the nonimmunogenic polymer complexed with another carrier such as methylated bovine serum albumin. In this situation, the polymer and its determinants function as haptenic groups associated with the new immunogenic carrier. Immunization with the MeBSA complex converts mice of H-2[q] and H-2[s] haplotypes to responders (Table 2). However, the findings for the various strains of mice of H-2[p] haplotype are not too

Table 2. Immune responses of mice against methylated bovine serum albumin aggregates of GAT^{10} ($2°$ response)

Strain	H-2 Haplotype	Immunogen[a]	
		10 μg	25 μg
C57BL/6	b	84 ± 3	85 ± 5
		66 ± 21	
		69 ± 8	
DBA/2	d	71 ± 7	
P	p	0 ± 0 (51)[b]	7 ± 9 (51)[b]
C3H.NB	p	72 ± 4	12 ± 7
BDP	p	0 ± 0	
B10.P	p	32 ± 17	
DBA/1	q	21 ± 19	79 ± 5
		35 ± 4	
		47 ± 16	
SJL	s		87 ± 5

[a] Percent antigen-bound ± S.E.
[b] 1:5 mice responded with value in parentheses. Other mice were nonresponders.

consistent. Although the P and BDP mice do not respond to the GAT^{10}-MeBSA complex, the C3H.NB mice and B10.P mice are responders. The P mice did respond, however, after immunization with a mouse IgG-GAT^{10} complex. These data indicate that the mice of $H-2^{p,q,s}$ haplotypes definitely have B cells that recognize GA and GT determinants, and therefore the absence of response against GAT^{10} indicates a lesion at the T-cell level (33).

Immune response patterns of SJL mice. Although the Ir GA gene, which accounts for responses against GA polymers is present in mice of $H-2^s$ haplotype, SJL mice, these mice are unique nonresponders against GAT^{10}. Moreover, the polymer GAT^{10} can neither prime SJL mice for subsequent immunization with GA nor "boost" SJL mice that have been immunized with GA. This finding is in contrast to data for strains of mice that respond to both GA and GAT^{10}.

An initial explanation that we entertained for nonresponsiveness of SJL mice invoked the concept of steric hindrance, that is, that introduction of more than 4 mol % tyrosine into GA might have created steric structures (29). Our recent findings, however, can be interpreted differently. Because of the polymerization kinetics of the 3 N-carboxyanhydrides of Glu, Ala, and Tyr, in the preparation of GAT^{10}, a mixture of polymers is obtained that contains variable proportions of tyrosine and polymers in which most of the tyrosine residues appear to be concentrated at the end of the molecule. In fact, it has been possible to fractionate GAT^{10} and obtain 14 fractions (37), only some of which are immunogenic.

The explanation we would like to offer for the responses of SJL mice relates to the presence in the heterogeneous GAT^{10} preparation of both immunogenic and "tolerogenic" fractions. For strains responsive to GAT^{10} the entire mixture is immunogenic, and for nonresponder mice of $H-2^q$ and $H-2^p$ haplotype (5), the entire mixture is nonimmunogenic. With respect to the SJL mice, however, there is a correct proportion of tolerogen (nonimmunogen) and immunogen in GAT^{10} that makes the entire preparation nonimmunogenic. It would appear, therefore, that there are unique structures in GAT^{10} that "tolerize" the response of SJL mice.

Specificity of antibody directed against GAT polymers. The nature of the determinants in GAT^{10} against which antibody is produced can be determined by cross-reactions and inhibition experiments with related polymers. The data in Table 3 present "titrations" of the binding of GAT^{10}, GA, and GT at varying dilutions of antiserum. It is evident that considerable specificity is directed against GT determinants (haptenic groups) even though the initiation of the immune response is via the GA determinants (Ir-GA gene).

The ability of the GT determinants to elicit higher levels of antibody than do the GA determinants also is shown in Table 4. Whereas following immunization with GAT^4 most of the antibody is directed against GA, when GAT^{10} is the

Table 3. Effect of dilution of antisera[a] on binding of GAT^{10}, GA, and GT polymers

Serum Dilution	GAT^{10}			GA			GT		
	A	B	C	A	B	C	A	B	C
1:2		93	74		64	57		84	77
1:10	90	82	48	47	48	31	90	83	59
1:50	70	60	16	21	12	4	79	90	23

[a]Pool of sera tested: A, B, C.

Table 4. Homologous reactions and cross-reactions with sera from mice immunized with GAT^4 or GAT^{10}[a]

Mouse Strain	GAT^4 (10 γ)			GAT^{10} (10 γ)		
	GAT^4	GA	GT	GAT^{10}	GA	GT
C57BL/6	60	65	0,0,2,3,13	86	70	82
BALB/c	62	77	23,41,55,60	81	42	86
C57BL/Ks	77	69	0,0,0,10	75	48	76
CBA	51	59	13,14,24,51	87	81	85

[a]Values refer to average antigen binding except for GT reactions with anti-GAT^4 sera. Individual values are given in the latter case.

immunogen, more of the specificity is directed against GT determinants (27). This phenomenon was noted also with random-bred mice and is reminiscent of the phenomenon of antigenic competition at the determinant level (i.e., GA and GT competing within the same molecule).

That additional specificities can exist was determined from inhibition experiments,[1] as shown in Table 5. Good inhibition is obtained with each homologous system; that is, GA inhibits binding of ^{125}I-GA, GT inhibits the binding of ^{125}I-GT, and GAT^{10} inhibits the binding of ^{125}I-GAT^{10}. It is apparent that specificties over and above GA and GT do exist. That GA plus GT could not inhibit the binding of GAT^{10} and that GAT^{10} in some situations could not inhibit totally the binding of GT again indicate that the GAT^{10} is heterogeneous and that there may be a small concentration of highly immunopotent determinants that contain significant amounts of tyrosyl residues.

Although GAT^{10} can inhibit binding of GA, the reverse situation does not obtain. These observations, which indicate that the GAT^{10} polymer has more specificities than the sum of GA plus GT, also help to explain some of the "tolerance" data described below.

Because of the heterogeneity in the GAT^{10} polymer and the "heterogeneity" of immune responses among random-bred mice, several patterns of responses against the polymer have been obtained: GAT^{10}+, GT+, GA+; GAT^{10}+, GT+, GA-; and GAT^{10}+, GA-, GT-.

---

[1] It is important to note that GA and GT are indeed different specificities because no cross-inhibition was noted with these polymers.

Table 5. Inhibition of anti-GAT^{10} by GA, GT, and GAT^{10}

Serum Dilution Studied	Inhibitor Concentration ($\times$ 0.0003 μg N)	Inhibitor Added	% Binding with ^{125}I-Labeled		
			GAT^{10}	GA	GT
1:2	100x	None	93	64	84
		GAT^{10}	1	0	26
		GA	92	5	85
		GT	89	71	20
		GA + GT	70	0	9
1:50	100x	None	60	12	90
		GAT^{10}	20	0	76
		GA	55	0	91
		GT	27	9	33

Tolerogenicity studies. Immunological tolerance, or the inability to respond to an immunogen, can be produced by exposure of a host to high concentrations of the immunogen before immunization. Our studies on tolerance have been carried out with a number of polymers. In a typical experiment, pretreatment of C57BL/6 mice by intravenous injection of 1 mg GAT^{10} reduced totally the subsequent responses against GAT^{10} ($2°$ response: 0 ± 0 binding versus 59 ± 13 binding). Neither GA nor GT (1 mg) were effective tolerogens.[1] In contrast, with the random-bred CD-1 mice, in addition to the expected tolerogenic effect of the GAT^{10} polymer, both GA and GT were tolerogenic. After a secondary immunization, 2:6 mice injected with GA responded, 2:5 animals injected with GT responded, whereas in the control group, 4:5 mice responded. The effectiveness of all three polymers in suppressing the response against GAT^{10} in the random-bred mice, in contrast to the effectiveness of only GAT^{10} in the inbred strain studied, is a phenomenon that is currently under study.

Together with Y. Borel, we investigated whether the tolerogenicity of GAT^{10} would be enhanced when coupled covalently to mouse IgG, as has been noted with other haptenic systems. Coupling of GAT^{10} to a BALB/c γl myeloma IgG was achieved as follows: Approximately 29 mg IgG was mixed with 62 mg of GAT^{10} to create a 6:1 molar ratio. The pH was then adjusted to 5.0 and 250 μl of isobutyl chloroformate was added. The solution was stirred at room temperature for two hours. Another 250 μl of isobutyl chloroformate was added. One hour later, the reaction was stopped by the addition of 42 ml of 66% saturated ammonium sulfate. The mixture was then centrifuged at $4°C$. The pellet was washed several times, and the aggregate was dissolved, dialysed extensively, and lyophilized. The molar ratio of polymer to IgG was determined to be 4:1.

Experiments were conducted in C57BL/6 and in random-bred CD-1 mice. Following pretreatment with IgG-GAT^{10}, there was an enhanced response both by the C57BL/6 and CD-1 mice (Table 6). The enhanced response of the C57BL/6 mice is analogous to our findings with these mice pretreated with GAT^{10} before immunization with GAT^{10}-MeBSA complex.

[1] GA can be tolerogenic for the subsequent responses against GA.

Table 6. Immune responses of mice to 10 μg GAT^{10} following pretreatment with IgG-GAT^{10} complex

Pretreatment	Mouse Strain	% Antigen-Bound
None	C57BL/6	70 ± 7
	CD-1	81 ± 1 (2:10)
500 γ GAT^{10}	C57BL/6	32 ± 9
	CD-1	63 (1:7)
500 γ GAT^{10}/IgG	C57BL/6	71 ± 3
	CD-1	56 ± 15 (8:8)

Some possible explanations for the results in this section are as follows: That neither GA nor GT is tolerogenic for GAT^{10} in inbred mice might be due to the presence in the heterogeneous GAT^{10} preparation of unique specificities that are neither GA nor GT. That the GAT^{10} covalently bound to IgG has failed to induce tolerance in both the inbred and random bred mice might indicate that coupling of GAT^{10} to mouse IgG has altered this protein so that it behaves as an immunogenic carrier for the hapten GAT^{10}. In fact, recent studies in which the IgG-GAT^{10} complex has been shown to be immunogenic (anti GAT^{10} response) in C57BL/6, DBA/1 and P/J mice confirms the above hypothesis that mouse immunoglobulins can function as carriers. A number of studies presently underway are utilizing the autologous model, namely, BALB/c mice injected with BALB/c-IgG-GAT^{10} complex, to confirm or disprove these possibilities.

Recognition of GAT^{10} by lymphoid cells of nonresponder mice. We have examined the interaction of T cells from nonresponder mice with antigen, using a technique for measuring T cell DNA synthesis (38). Ordinarily, thymocytes incorporate little ^{125}I dU into DNA after injection into lethally irradiated syngeneic mice, unless they are stimulated with antigens that elicit a thymus-dependent immune response. The thymus-independent antigens, such as polyvinylpyrolidone and pneumococcal polysaccharide, have not stimulated a significant DNA synthetic response (39). In collaborative experiments with Richard Gershon, we have shown that thymocytes of nonresponder mice synthesize significant amounts of DNA when they meet the antigen in the spleen. Therefore we have suggested that the genetic defect is not in the *recognition* of the immunogen by T cells, but at some other level (33).

The basic procedure that we employed involved intravenous injection of lethally irradiated nonresponder DBA/1 mice with 3–5 × 10^7 syngeneic thymocytes. On the same day the mice were injected intraperitoneally with the test or control polymers. At intervals thereafter, they were injected with ^{125}I dU and the uptake of isotope by the cells was measured. The increase in isotope uptake produced by the GAT^{10} immunization was statistically significant on all assay days. When the same experiment was performed with responder mice of the DBA/2 strain, the DNA synthetic response of their thymocytes was indistinguishable from saline controls. It appeared, therefore, that the thymocytes from nonresponder mice respond to antigen, but those from responder mice do not. As the thymocytes can synthesize significant amounts of DNA during tolerance induction, we investigated the possibility that tolerance was being induced in the thymus cells in the spleens of nonresponder mice. Syngeneic thymocytes were injected into lethally irradiated recipients. Some were then injected with 100 γ of GAT^{10} and others with saline. Eight days later, the spleens were harvested and the cells injected into another group of syngeneic, lethally irradiated mice. The DNA synthetic response of the cells that had passed through the spleens to antigen was again studied. We noted that the immunized thymocytes from nonresponder mice *failed* to respond to a second challenge with GAT^{10}. The nonimmunized thymocytes that had passed through the spleens did not produce a response either. However, the responder thymocytes that had passed through the spleens and had been immunized with GAT^{10} exhibited a striking response to a second immunization with GAT^{10}. We have postulated, therefore, that the results are compatible with the notion that both nonresponder and responder thymus cells recognized GAT^{10}, but that the recognition event leads to paralysis in the nonresponder strain and to immunity in the responders.

Our most recent experiments indicate that the nonresponsiveness of the DBA/1 strain of mouse is related to the interaction of GAT^{10} with suppressor T cells that interfere with or cut off the response to the polymer of the competent B cells (40). In addition, responder mice do not become nonresponders to GAT^{10}-MeBSA after the removal of T cells.

ACKNOWLEDGMENTS

These studies were supported by research Grant AI 07825 from the National Institute of Allergy and Infectious Diseases, Grant GM 15574 from the National Institute of General Medical Sciences, and Research Grant IM-5C from the American Cancer Society.

We wish to thank Allen Zeiger for preparing the tyrosine-containing polymers and Robert Smyth for the iodination of the polymers. Allen Zeiger and M. Frankel prepared the complex of GAT^{10} with BALB/c γl myeloma IgG.

The competent technical assistance of Jeanette Jones and Maria Filinska is gratefully acknowledged.

REFERENCES

1. M. Sela, *Science,* **166,** 1365 (1969).
2. H.O. McDevitt and B. Benacerraf, *Adv. Immunol.,* **11,** 31 (1969).
3. B. Benacerraf and H.O. McDevitt, *Science,* **174,** 272 (1972).
4. E. Mozes and G.M. Shearer, *Current Topics in Microbiology and Immunology,* **59,** 167 (1972).
5. P.H. Maurer, in *Polyamino Acids, Polypeptides and Proteins,* M. Stahmann, ed., Wisconsin University Press, Madison, Wis., 1962, p. 359.
6. P.H. Maurer, *Progress in Allergy,* **8,** 1 (1964).
7. P.H. Maurer, *Med. Clinics of North Amer.,* **49,** 1505 (1965).
8. P.H. Maurer, B.F. Gerulat, and P. Pinchuck, *J. Biol. Chem.,* **239,** 922 (1964).
9. P.H. Maurer, P. Pinchuck, and B.F. Gerulat, *Immunochemistry,* **3,** 403 (1966).
10. L.G. Clark and P.H. Maurer, *Int. Arch. Allergy,* **35,** 58 (1969).
11. P.H. Maurer, L.G. Clark, and P.A. Liberti, *J. Immunol.,* **105,** 567 (1970).
12. H.J. Callahan, P.H. Maurer, and P.A. Liberti, *Biochemistry,* **10,** 3467 (1971).

13. H.C. McDonald, G. Odstrchel, and P.H. Maurer, *Immunochemistry*, **10**, 119 (1973).
14. A.R. Zeiger and P.H. Maurer, *Biochemistry*, **12**, 338 (1973).
15. G. Odstrchel and P.H. Maurer, *Immunochemistry*, **11**, 15 (1974).
16. P.A. Liberti, P.H. Maurer, and L.G. Clark, *Biochemistry*, **10**, 1632 (1971).
17. P.A. Liberti, W.A. Stylos, and P.H. Maurer, *Biochemistry*, **11**, 3312 (1972).
18. P.A. Liberti, W.A. Stylos, P.H. Maurer, and H.J. Callahan, *Biochemistry*, 3321 (1972).
19. H.J. Callahan, P.A. Liberti, and P.H. Maurer, *Immunology*, **25**, 517 (1973).
20. P. Pinchuck and P.H. Maurer, in *Symposium on Regulation of the Antibody Response*, C.C. Thomas, ed., 1966, p. 97.
21. S. Ben-Efraim and P.H. Maurer, *J. Immunol.*, **97**, 577 (1966).
22. H.G. Bluestein, I. Green, P.H. Maurer, and B. Benacerraf, *J. Exp. Med.*, **135**, 98 (1972).
23. P.H. Maurer, G. Odstrchel, and C.F. Merryman, *J. Immunol.*, **111**, 1018 (1973).
24. P. Pinchuck and P.H. Maurer, *J. Exp. Med.*, **122**, 665 (1965).
25. P. Pinchuck and P.H. Maurer, *J. Exp. Med.*, **122**, 673 (1965).
26. P. Pinchuck and P.H. Maurer, *J. Immunol.*, **100**, 384 (1968).
27. C.F. Merryman and P.H. Maurer, *J. Immunol.*, **108**, 135 (1972).
28. C.F. Merryman, P.H. Maurer, and D.W. Bailey, *J. Immunol.*, **108**, 937 (1972).
29. P.H. Maurer and C.F. Merryman, *Ann. Immunol.*, **125C**, 189 (1974).
30. P.H. Maurer and C.F. Merryman, *Immunogenetics*, in press.
31. P.H. Maurer, D. Wilson, and C.F. Merryman, *Fed. Proc.*, **31**, 777 (1971).
32. E. Katchalski and M. Sela, *Adv. Protein Chem.*, **13**, 243 (1958).
33. R.K. Gershon, P.H. Maurer, and C.F. Merryman, *Proc. Nat. Acad. Sci. U.S.*, **70**, 250 (1973).
34. M. Sokalovsky, J.F. Riordan, and B.L. Vallee, *Biochemistry*, **5**, 3583 (1966).
35. W.M. Hunter, in *Handbook of Experimental Immunology*, F.A. Davis Co., 1969, p. 608.
36. C.F. Merryman and P.H. Maurer, *Fed. Proc.*, **32**, Abs. 4372 (1973).
37. C. Vickerman and P. Liberti, unpublished results.
38. R.K. Gershon and R.S. Hencin, *J. Immunol.*, **107**, 1723 (1971).
39. J. Kruger and R.K. Gershon, *J. Immunol.*, **108**, 581 (1972).
40. Unpublished results.

CONFORMATION-DEPENDENT ANTIGENIC DETERMINANTS IN PROTEINS AND SYNTHETIC POLYPEPTIDES

RUTH ARNON, Department of Chemical Immunology, The Weizmann Institute of Science, Rehovot, Israel

SYNOPSIS: In lysozyme, antibodies specific exclusively to a unique conformation-dependent determinant were prepared by the use of a conjugate in which a fragment of the molecule, or *loop*, was attached to a macromolecular carrier. These antibodies recognize the native lysozyme and the disulfide-containing loop fragment, but not its unfolded peptide derivative. A chemically synthesized looplike derivative was proved immunologically identical to the natural fragment, and when forming part of a completely synthetic conjugate, it elicited conformation-specific antibodies reactive with native lysozyme. Synthetic analogs of this loop peptide were used to elucidate the role played by various amino acid residues in the antigenic specificity of this conformational determinant.

The immunological relationship between lysozymes of various species and between lysozyme and α-lactalbumin are shown to depend on the similarities and differences in the amino acid sequence of the compared molecules, namely, on the extent and nature of amino acid replacements in their sequence. The exact character and position of amino acid replacements are of crucial importance, particularly when the molecular species involved are distinguished by a limited number of interchanges. In these cases, replacements that have profound effects on conformation bring about more drastic changes in antigenic reactivity. The conformational determinants are less important for cell-mediated immunity, but they are shown to be essential for humoral immune response and specificity.

INTRODUCTION

The decisive role of conformation in determining the antigenic specificity of protein and polypeptide antigens is widely recognized. A large body of experimental evidence indicates that antigenic properties change drastically upon denaturation of native proteins (by heat or chemical modification) or upon unfolding of their polypeptide chains (1–3). The denatured or unfolded proteins are usually still immunogenic, but their antigenic specificity is totally different from that of the corresponding native proteins. This fact has been convincingly demonstrated for proteins such as ribonuclease (4, 5), papain (6), trypsin (7, 8), lysozyme (9, 10), and albumin (11).

In many instances, more subtle conformational alterations in a protein also are accompanied by a change in antigenic reactivity. The best example of this phenomenon is the change in the antigenicity associated with the removal of the heme group from sperm-whale myoglobin. In the conversion of metmyoglobin to apomyoglobin the loss of heme is associated with only a small structural change and the antigenic properties of the two molecular species are not much different. The precipitate formed between metmyoglobin and antiapomyoglobin, however, is colorless and does not contain the ferriheme group (12); thus the antibodies specific to the heme-free molecule must have induced a conformational change in the metmyoglobin and released the heme group from it during the antigen-antibody interaction.

Not all conformational alterations have a measurable effect upon antigenicity and vice versa. There are, for example, several aberrant hemoglobins for which x-ray crystallography shows distortion in tertiary structure, and yet they cannot be distinguished antigenically from hemoglobin A (13). On the other hand, α and β chains of human hemoglobin do not show any antigenic resemblance in spite of the remarkable similarity in their conformations (14). These are probably exceptions, however, to the general rule that there is an intimate relationship between conformation and antigenicity.

For a better assessment of the role of conformation in determining the antigenic properties of a protein molecule, the structural features that affect antigenic specificity should be analyzed. Antibodies elicited in response to immunization with protein antigens are reactive with various antigenic determinants and may be directed against one or more of the structural aspects of the protein, including the primary structure (the amino acid sequence of the polypeptide chain), the secondary structure (which is dictated by the backbone of the polypeptide, such as α helix and β-pleated sheet), the tertiary structure (conferred by interactions between various groupings in the chain and associated with its folding), and the quaternary structure (which results from specific association of several polypeptide chains to form a multi-subunit protein). The antigenic determinants therefore are divided theoretically into two broad categories (15) according to whether their specificity is due only to stretches of amino acid sequences in the protein (sequential) or to the other structural features mentioned above (conformational). Conformational determinants include those composed of amino acid residues that, even if remote in the unfolded polypeptide chain, are juxtaposed in the native structure.

Examination of the three-dimensional structures of a number of globular proteins reveals that they contain short sequences of adjacent amino acids whose side chains are partially or fully exposed on the surface of the protein. Consequently, these could exist as sequential determinants. In practice, however, it appears that the interference of such short peptide fragments of a protein with the interaction of the native protein and its antibodies is often due to the fact that the peptides are induced by the antibodies to refold into the structure that they hold in the native protein. There are only a few clearcut cases for which sequential determinants can be demonstrated [e.g., the terminal segments of collagen (17) or silk fibroin (18)], but, as mentioned previously, antibodies to native proteins are directed mostly, and in several cases exclusively, against conformation-dependent determinants (19).

In this paper, we will describe several well-documented systems of both native proteins and synthetic polypeptide antigens in which conformational determinants were allocated and characterized. Our studies of a unique conformation-dependent determinant of lysozyme that can be either isolated from the native enzyme or prepared by chemical synthesis will be discussed in more detail, along with the implication of these studies in regard, on the one hand, to the applicability of the immunological approach in the elucidation of structural aspects and, on the other hand, to pursuit of immunological phenomena on a molecular level.

CONFORMATIONAL DETERMINANTS

Our attempts to identify antigenic determinants of proteins usually involved the fractionation of fragments obtained by limited proteolytic digestion or chemical cleavage and the screening of the resultant fractions for immunologically active components. These components, which by definition embody antigenic determinants, were subsequently analyzed and defined in terms of their location in the native structure. In many cases, it was demonstrated that the antigenic determinants were mainly conformational; several of these will be discussed in this section.

Myoglobin. This protein (MW 18,400) has no disulfide bridges. Its structural integrity is due mainly to its high content of ordered structure. Immunologically active fragments, obtained by chymotryptic (20) or tryptic (21) digestion, occupy "corners" in the three-dimensional configuration of the molecule. Similarly, all the antigenic fragments obtained by cyanogen bromide cleavage (21) or by cleavage at the proline (22) or arginine (23) residues, also are included in the same regions of the molecule. Examination of a space-filling molecular model reveals that these corners of the polypeptide chain coincide with the more exposed areas, which are held in a fixed conformation due to the folding of the molecule. The isolated immunologically active fragments were shown to have a random-coil conformation (16). The antibodies, however, are specific to the secondary structure that the peptides assume in the whole native myoglobin.

Most of the antigenic activity of myoglobin may be attributed to its primary and secondary structures. The tertiary structure, brought about by the ferriheme group, also is implicated in the antigenicity, because of the capacity of antibodies to the heme-containing protein to distinguish between the apoprotein and the metmyoglobin (24) and because antibodies to the apoprotein are able to release the heme group from the metmyoglobin (12).

Staphylococcal nuclease. This globular protein also is devoid of disulfide bridges. Its folded structure is held together mainly by hydrophobic interactions, yet examination of the antisera formed to the native protein unequivocally demonstrates that its antigenic determinants possess conformational specificity (25). Fragmentation of the protein by cyanogen bromide or trypsin allowed the antigenicity to be correlated with three regions of the molecule, namely, residues 18–47, 99–126, and 127–149 (25, 26). These regions of the three-dimensional molecular structure again include the corner regions of the folded polypeptide chain that is exposed on the surface of the native protein.

In addition, the existence of conformation-dependent determinants was inferred from the enhanced binding of the fragments to the antibodies when incorporated into a noncovalent, enzymatically active complex, as compared to the binding of the fragments as such. Moreover, antibodies formed by immunization with the fragment 99–149 were compared with antibodies isolated from antisera to the intact protein by using the fragment attached to Sepharose as an immunoadsorbent. These two types of antibodies, both directed apparently to the same 50-amino acid C-terminal region of the molecule, are markedly different. Whereas the anti-whole protein recognized only the native form of the fragment, the antibodies elicited by the fragment recognized only the random unfolded protein. It was concluded that although nuclease has a low helix content and lacks disulfide bonds, its structural conformation influences its antigenic determinants.

Bradykinin. Bradykinin is a nonapeptide with the sequence Arg-Pro-Pro-Gly-Phe-Ser-Pro-Phe-Arg. It is mentioned here only because, in spite of its small size, its specific antibodies are apparently directed towards its whole globular conformation.

Antibodies to bradykinin were obtained by immunizing rabbits with a conjugate of the peptide attached to polylysine (27). The relationship between peptide structure and the antigenic reactivity was evaluated from experiments in which various synthetic analogs of the peptide were used to inhibit binding between the native peptide and antibodies. These experiments (28, 29, 30) revealed that the overall charge of the molecule or the local charges at the termini are not a decisive factor in antigenic specificity, nor are the side chains of residues such as Phe, Arg, and Ser. In contrast, the size of the molecule is crucial, and the presence of Gly in position 4 and all three proline residues (at positions 2, 3, and 7) is essential for efficient inhibitory activity. Their replacement by alanine or by D-proline brought about a profound reduction in activity. Thus, it was concluded that the peptide may exist in a cyclic form as a globular structure, and hence length is important for stability. This assumption, which is corroborated by a molecular model, strongly suggests that the antigenic determinant comprises the entire peptide and that the antibodies recognize the peptide in the preferred globular conformation (28).

Collagen and synthetic collagenlike poly(Pro-Gly-Pro). A better understanding of the role of conformation in antigenicity has been achieved in recent years by building appropriate synthetic models and analyzing their immunochemical properties. One example of this approach is a collagenlike synthetic copolymer. Collagen is a protein that, except for the 10–15 N-terminal amino acid residues, comprises highly repetitive amino acid sequences in which the triplet Gly-Pro (or Hyp)-X is the repeating unit arranged in a unique triple-helical structure (31). Immunization with native collagen elicits mainly antibodies directed to the N- and C-terminal nonhelical regions that exhibit interspecies differences. Only in a few cases (e.g., the immunization of rats with calf collagen) is it possible to obtain antibodies that interact primarily with the triple-helix conformation (32).

The presence of conformational determinants in collagen was indirectly proven by the use of a synthetic analog. Poly(Pro-Gly-Pro), which was shown to have a collagenlike triple-helical structure (33), was found to be immunogenic in guinea pigs and rabbits (34). Immunization with this copolymer elicited antibodies that cross-reacted with collagens of several species (35, 36). On the other hand, a random copolymer of similar composition, poly(Pro^{66}Gly^{34}), did not cross-react with collagens. Thus, the polymer of ordered sequence reacts immunologically with the various collagens by virtue of the triple-helix conformation that is common to both substances and is the major antigenic feature.

Conformation-dependent determinants in a synthetic antigen. The role of conformation in antigenic specificity was most convincingly demonstrated by the use of two synthetic antigens containing the same tyrosyl-alanyl-glutamyl sequence. One was a high-molecular-weight ordered copolymer composed of the repeated sequence of the tripeptide Tyr-Ala-Glu, which had been previously shown to exist as an α helix (37). In the second antigen, the same tripeptide was attached to a branched polymer of alanine, where it exists as a random coil. These two polymers elicited the formation of antibodies with distinct specificities, and no cross-reaction occurs between them (38). Furthermore, the system of the branched polymer was efficiently inhibited by the tripeptide, whereas the system of the helical peptide was not. Inhibition of the latter system was achieved only with oligopeptides of the general formula (Tyr-Ala-Glu)$_n$, with $n = 3$–9. The inhibitory capacity increased with the value of n. The oligopeptide (Tyr-Ala-Glu)$_{13}$ was able to cross-precipitate the antibodies to the helical polymer. Circular dichroism studies (39) showed that the above oligopeptides possess very little helical structure, but upon interaction with the Fab fragment of the antibody to the helical polymer, their α-helix content increases. Thus, it was concluded that the antigenic determinants of the α-helical copolymer are *conformation-dependent*. Moreover, their specific antibodies are capable of inducing a transconformation in the oligopeptides into a structure more similar to that of the high-molecular-weight helical polymer.

ANTIBODIES TO A UNIQUE CONFORMATIONAL DETERMINANT OF LYSOZYME

Properties and specificity of the lysozyme loop. The immunological specificity of lysozyme is almost entirely dependent on its three-dimensional native conformation, as has been implied, by its total lack of cross-reactivity with its unfolded, reduced, and carboxymethylated derivative, on the one hand, and on the other hand, by the isolation of at least two fragments of the native molecule that retain immunological activity and represent independent antigenic determinants. One such fragment consists of two peptides derived from the NH_2-terminus (residues 1–27) and the COOH-terminus (residues 122–129) linked together by a disulfide bond. Forty-seven percent of the antilysozyme antibodies were directed against this fragment (40). Thirty percent of these antibodies were directed against another immunologically active component of lysozyme, namely, the region between residues 57 and 107, which contains two disulfide bridges (41, 42). From this fragment we isolated a smaller immunologically active peptide consisting of the amino acid sequence 60–83 and containing an intrachain disulfide bond. The location of this so-called *loop* (43) region in the three-dimensional structure of lysozyme is shown in Fig. 1 (44).

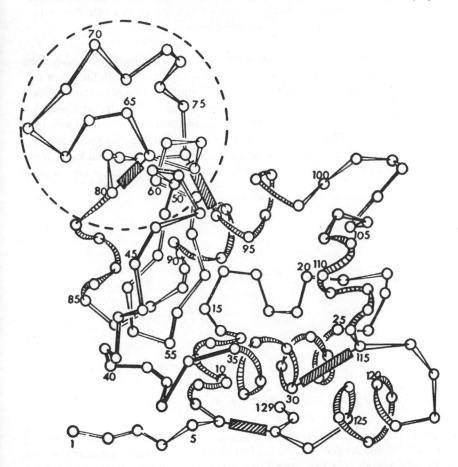

Figure 1. Schematic drawing of the main chain conformation of hen egg-white lysozyme. The area encompassing the loop peptide is encircled. [Adapted from Blake *et al., Nature* (London), **206**, 75 (1965). Copyright by Macmillan Journals Ltd.]

Antibodies specific exclusively to this region were prepared either by selective isolation from antilysozyme serum on a loop immunoadsorbent, or by immunization with a conjugate containing the loop attached to a synthetic carrier. These antiloop antibodies were much less heterogenous than the total antilysozyme antibodies and they reacted efficiently with native lysozyme as well as with the isolated loop peptide derived from it, but not at all with the open-chain loop peptide in which the disulfide bond was disrupted by reduction and alkylation (45). The detailed specificity of these antibodies was investigated by several sensitive techniques, including a fluorometric method, using a loop derivative in which a fluorescent chromophore (dansyl group) was attached (46). The results, shown in Fig. 2, indicate that the loop region is a conformation-dependent antigenic determinant whose specificity is dictated by a spatial structure similar to that which it assumes in the native lysozyme molecule.

   Synthetic approach to the elucidation of antigenic specificity of the lysozyme loop. In order to arrive at a more detailed characterization of this unique antigenic region, a looplike peptide of lysozyme, comprising 19 amino acid residues corresponding to the sequence 64–82 has been synthesized by the solid-phase technique (47). The synthetic material, which differs from the natural loop only in the replacement of Cys_{76} by Ala (in order to avoid ambiguous disulfide bond formation), is identical in its immunological reactivity to the natural loop. When attached to a high-molecular-weight carrier, it elicits antibodies that are similar

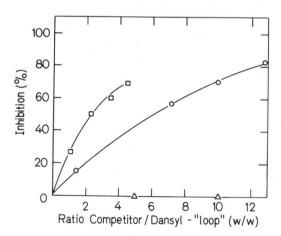

Figure 2. Inhibition of enhanced fluorescence of the mixture of dansyl-loop $(4 \times 10^{-6} M)$ and antiloop antibodies $(0.6 \times 10^{-6} M)$ by varying concentrations of the loop peptide (□), hen egg-white lysozyme (○), and the open-chain peptide obtained by reduction and carboxymethylation of the loop peptide (△). [Reprinted with permission from J. Pecht, E. Maron, R. Arnon, and M. Sela, *Eur. J. Biochem,* **19**, 368 (1971). Copyright by the Federation of European Biochemical Societies.]

in every respect, including specificity, to those elicited by the conjugate of the natural loop. Thus, these antibodies reacted very efficiently with either intact lysozyme or the closed loop fragment, but did not react at all with the unfolded loop peptide.

More subtle changes in the gross conformation of the molecule also affected its antigenic reactivity. We have recently observed (48) that reduction of the synthetic loop followed by conversion of the cysteinyl residues to dehydroalanine to form a linear structure devoid of negative charges at its termini results in a marked decrease in the binding efficiency (Table 1), but not as drastic as that effected by reduction and carboxymethylation. Reduction followed by reoxidation led to a complete reversal of antigenic activity (45). If, however, the reoxidation was performed in the presence of mercuric chloride, the product of the closure of the loop, which contained one mercury atom per molecule, was slightly less active than the intact loop. These data indicate that even a relatively small increase in the size of the loop might change the immunological properties.

One of the advantages offered by a synthetic approach in biochemical and biological research is that once the biological properties of one synthetic material have been unequivocally demonstrated, many analogs may be prepared and tested. Because the chemistry of these analogs is known and controlled, they can increase our understanding of the role of different molecular parameters in conferring specific biological activity. With this in mind we synthesized several analog derivatives of the loop in which one or two amino acids were replaced by alanine (Fig. 3) to test the extent of involvement of residues such as proline, arginine, leucine, and isoleucine in the immunological reactivity of this antigenic region (49). The results are shown in Fig. 4. The derivatives in which either

Table 1. Binding properties of synthetic loop derivatives to antiloop antibodies

Derivative	Observed Binding Activity (%)
Intact synthetic loop (reduced and reoxidized)	100
RCM loop (reduced and carboxymethylated)	0
Linear loop peptide (reduced and cysteines converted to dehydroalanine)	56
Hg loop (reduced and reoxidized in the presence of $HgCl_2$; contains 1 Hg atom per peptide molecule)	88

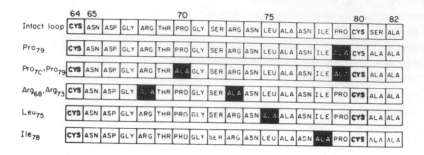

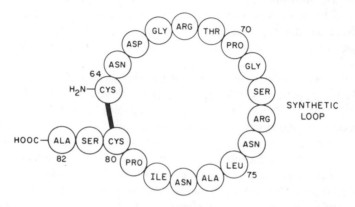

Figure 3. Schematic description of the synthetic loop derivatives. The intact synthetic loop is depicted in the bottom part of the figure. The top part of the figure contains the linear amino acid sequence of the various analogs prepared (all analogs were subsequently oxidized to yield the loop form). The black positions represent those amino acids that were replaced by alanine. [Adapted from Teicher *et al., Immunochemistry,* **10,** 265 (1973). Copyright by Pergamon Press Ltd.]

leucine (residue 75) or isoleucine (residue 78) was replaced by alanine were almost indistinguishable from the intact synthetic loop. On the other hand, replacing the proline residues (70 and 79), or even one proline (residue 70), drastically decreased antigenic activity.

The results just mentioned were analyzed in light of the contribution of each of the respective residues to the β structure in the region of the lysozyme loop, or rather the change in probability of retaining the β structure upon replacement of the particular residue with alanine. The results indicate agreement between the expected probability of retention of the β structure and the observed value for the antigenic reactivity. Accordingly, several additional analogs of the loop

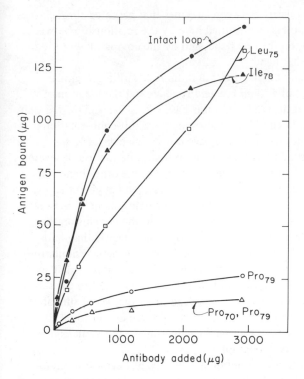

Figure 4. Antigen binding by antiloop antibodies of the various loop analogs. The designation of the derivatives is according to the amino acid(s) replaced by alanine, as shown in Fig. 3. [Reprinted with permission from Teicher *et al.*, *Immunochemistry*, **10**, 265 (1973). Copyright by Pergamon Press Ltd.]

have been synthesized for which the probability of maintaining the conformation of the intact loop was predicted in advance. In these calculations both the involvement of the β structure and the hydrogen bonding known to be effective in lysozyme, were taken into account (50). The results indicate agreement between the predicted and the observed values and corroborate the previous evidence that conformation has a crucial role in determining the antigenic reactivity of the lysozyme loop.

Immunological approach to studies of the evolution and possible relationships of proteins. In view of the role played by conformation in determining antigenicity, it is not surprising that antigenic cross-reactivity has been proposed as a sensitive probe for conformational differences and similarities in related proteins. The immunological approach has often been used in the study of biochemical evolution, especially of enzymes, where specific antibodies can be employed in the search for biological pathways that disappeared in the course of evolution, or to detect the extent of similarity between enzymes that persisted through the ages. This type of investigation is exemplified here by the relationship between lysozyme and α-lactalbumin, on the one hand, and by the comparative study of various lysozymes, on the other hand.

Hen egg-white lysozyme and bovine α-lactalbumin have strikingly similar amino acid sequences; 49 out of 123 amino acid residues and the location of their disulfide bridges are identical (51). On this basis, it has been suggested that they might be functionally related (52) and also might exhibit conformational similarities (53). Our immunological studies have not revealed any cross-reactivity between native lysozyme and α-lactalbumin, despite the use of several extremely sensitive immunochemical techniques (54). On the other hand, as depicted in Fig. 5, a definite cross-reaction has been observed between the reduced and carboxymethylated derivatives of the two proteins, namely, between their unfolded peptide chains (55). These findings lead to the conclusion that antibodies to the native conformation are apparently more specific than antibodies raised against the unfolded peptide chain and fail to recognize similarities in amino acid sequence per se and that lysozyme and α-lactalbumin, which are obviously related, have different "hydrophilic peripheries" exposed to the surrounding medium, namely, those areas that may lead to antibodies of similar specificities. They might still be similar in their overall three-dimensional conformation, which is dictated by the "internal" residues of the molecule and is the feature detected by x-ray crystallography.

As is the case with many other isofunctional enzymes, lysozymes of phylogenetically close species that have similar amino acid sequences are immunologically similar (56, 57). With species that are phylogenetically distant from one another (e.g., human and hen) there is hardly any cross-reaction (58), but various gallinaceous lysozymes are all cross-reactive with each other (56). When the immunological reagents used in these studies included both antilysozyme and antiloop antibodies, it was possible to obtain selective information about the differences in defined regions of the lysozymes of various species (59). In a recent study (60), we observed an extremely interesting phenomenon: In the interaction between hen lysozyme and antiloop antibodies, turkey lysozyme was as efficient as hen lysozyme in interfering with the specific interaction, but quail lysozyme was much less reactive (Fig. 6). The loop regions of these lysozymes differ from hen lysozyme in that lysine replaces arginine, but in different positions (position 68 in the quail and position 73 in the turkey). This finding indicates a stronger influence of the arginine in position 68 than in position 73 on the antigenic properties of this conformation-dependent region and is in accord with x-ray–crystallographic data indicating that Arg_{68} participates to a greater extent in hydrogen bonding with other residues in lysozyme than Arg_{73} does. It may be concluded that comparison of the immunological effects of phylogenetically related proteins can assist in the elucidation of the effects of amino acid substitutions in the local conformation of particular antigenic determinants in a molecule.

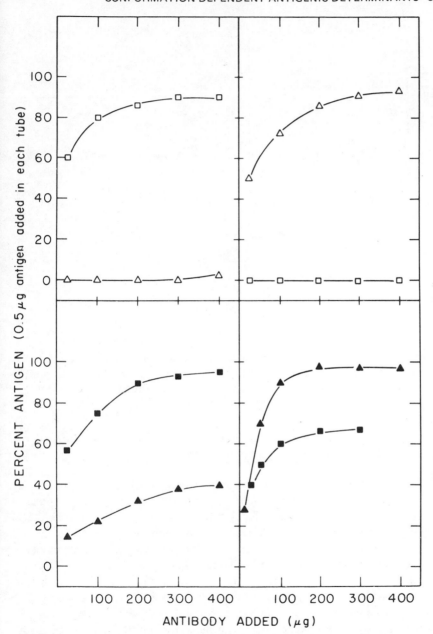

Figure 5. Antigen binding capacity of native and open-chain hen egg-white lysozyme and bovine α-lactalbumin by isolated antibodies against native lysozyme (upper left), native lactalbumin (upper right), reduced carboxymethylated lysozyme (lower left), and reduced carboxymethylated lactalbumin (lower right). The antigens are: □ native lysozyme; △ native lactalbumin; ■ reduced carboxymethylated lysozyme; and ▲ reduced carboxymethylated lactalbumin.

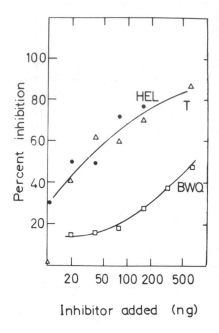

Figure 6. Inhibition of the inactivation of lysozyme-bacteriophage conjugate by antiloop antibodies with the following lysozymes: ● hen egg-white; □ bobwhite quail; and △ turkey. [Reprinted with permission from Fainaru *et al., J. Mol. Biol.*, **84**, 635 (1974). Copyright by Academic Press.]

CELLULAR RECOGNITION AND HUMORAL RESPONSE TO CONFORMATIONAL AND SEQUENTIAL DETERMINANTS

All the data and findings reported to date argue strongly in support of the view that protein antigenic determinants that interact with humoral antibodies are usually conformation specific. This relationship does not appear to hold, however, for the determinants that mediate cellular immunity. Several years ago it was suggested that there are possible differences in both specificity and manifestation of cell-mediated response and antibody production (61). This assumption is borne out by studies of several systems, which are described below.

In contrast to the observation that unfolded (reduced and carboxymethylated) lysozyme failed to cross-react with antiserum to the native enzyme and that the native lysozyme did not react with the antiserum to the unfolded chain, extensive cross-reactivity was detected between native and unfolded lysozymes at the cellular level (62). This reaction was manifested both in vivo, by marked delayed hypersensitivity reaction, and in vitro, by the inhibition of capillary macrophage migration (MIF technique).

A similar phenomenon, namely, a difference between cellular and humoral response, was observed also in cross-reaction between related native proteins. Thus, collagen of *Ascaris lumbricoides*, which possesses a radically different structure than vertebrate collagens (63), appears to have completely distinct antigenic structure and does not show any humoral cross-reactivity with the vertebrate collagens (64). However, a cell-mediated cross-reaction has been observed between Ascaris and human collagens (65).

Hen egg-white lysozyme and bovine α-lactalbumin, which, as discussed earlier, were shown to be completely non cross-reactive at the humoral level, have been reported to exhibit a definite cross-reaction at the cellular level. Thus, both in vivo (delayed hypersensitivity) and in vitro (lymphocyte transformation) a marked cross-reactivity was observed between the two proteins (66), a phenomenon that serves as another indication of the relationship between them.

Various studies with oligopeptides led to similar observations. Thus, oligopeptide fragments of proteins with random conformation induced delayed-type hypersensitivity in guinea pigs, but failed to stimulate the production of circulating antibodies against the whole protein. For example, flagellin fragments of molecular weight greater than 4000 had negligible reactivity with antiflagellin sera, but very effectively induced a state of cell-mediated immunity (67).

It appears that all the observations reported here have a similar molecular basis. It may be concluded, then, that whereas the specificity of humoral antibodies is strict and they recognize and react with mainly conformational antigenic determinants, the determinants mediating cellular immunity are less conformation dependent. In other words, the conformational integrity of a protein molecule, although it is not crucial for interaction with cell-bound antibodies, is essential for eliciting humoral antibodies and reacting with them. Apart from emphasizing the different order of specificity recognized by humoral and cell-bound antibodies, the above results suggest that cell-mediated immunity represents a more informative parameter than humoral immunity in the study of amino acid sequence or phylogenetic relationships, especially between distantly related proteins (3, 65). On the other hand, humoral antibodies are more useful in the study of structural relationships of proteins.

ACKNOWLEDGMENT

This work was supported in part by agreement No. 06-010 with the National Institutes of Health of the U.S. Public Health Service.

REFERENCES

1. R. Arnon, *Current Topics in Microbiology and Immunology,* **54**, 47 (1971).
2. E. Benjamini, D. Michaeli, and J.D. Young, *Current Topics in Microbiology and Immunology,* **58**, 85 (1972).
3. M.J. Crumpton, in *The Antigens,* Vol. 2, M. Sela, ed., Academic Press, New York, in press.

4. R.K. Brown, R. Delaney, L. Levine, and H. Van Vunakis, *J. Biol. Chem.,* **234**, 2043 (1969).
5. H. Neumann, I.Z. Steinberg, J.B. Brown, R.F. Goldberger, and M. Sela, *Eur. J. Biochem.,* **3**, 171 (1967).
6. E. Shapira and R. Arnon, *J. Biol. Chem.,* **244**, 1026 (1969).
7. J.O. Erickson and H. Neurath, *J. Exp. Med.,* **78**, 1 (1943).
8. R. Arnon and H. Neurath, *Immunochemistry,* **7**, 241 (1970).
9. J. Gerwing and K. Thompson, *Biochemistry,* **7**, 3888 (1968).
10. J.D. Young and C.Y. Leung, *Biochemistry,* **9**, 2755 (1970).
11. C. Jacobsen, L. Fundling, N.P.H. Moller, and J. Steengaard, *Eur. J. Biochem.,* **30**, 392 (1972).
12. M.J. Crumpton and J.M. Wilkinson, *Biochem. J.,* **100**, 223 (1966).
13. M. Reichlin, *J. Mol. Biol.,* **64**, 485 (1972).
14. M. Reichlin, E. Bucci, C. Fronticelli, J. Wyman, E. Antonini, C. Luppolo, and A. Rossi-Fanelli, *J. Mol. Biol.,* **17**, 18 (1966).
15. M. Sela, B. Schechter, I. Schechter, and F. Borek, *Cold Spring Harbor Symp. Quant. Biol.,* **32**, 537 (1967).
16. M.J. Crumpton and P.S. Small, *J. Mol. Biol.,* **26**, 143 (1967).
17. U. Becker, R. Timpl, and K. Kuhn, *Eur. J. Biochem.,* **28**, 221 (1972).
18. J.J. Cebra, *J. Immunol.,* **86**, 190 (1961).
19. M. Sela, in *The Harvey Lectures,* Series 67, Academic Press, New York, 1973, p. 213.
20. M.J. Crumpton and J.M. Wilkinson, *Biochem. J.,* **94**, 545 (1965).
21. M.Z. Atassi and B.J. Saplin, *Biochemistry,* **7**, 688 (1968).
22. M.Z. Atassi and R.P. Singhal, *Biochemistry,* **9**, 3854 (1970).
23. R.P. Singhal and M.Z. Atassi, *Biochemistry,* **10**, 1756 (1971).
24. M. Reichlin, M. Hay, and L. Levine, *Biochemistry,* **2**, 971 (1963).
25. D.H. Sachs, A.N. Schechter, A. Eastlake, and C.B. Anfinsen, *Proc. Nat. Acad. Sci. U.S.,* **69**, 3790 (1972).
26. G.S. Omenn, D.A. Ontjes, and C.B. Anfinsen, *Biochemistry,* **9**, 313 (1970).
27. J. Spragg, K.F. Austen, and E. Haber, *J. Immunol.,* **96**, 865 (1966).
28. E. Haber, F.F. Richards, J. Spragg, K.F. Austen, M. Valloton, and L.B. Page in *Cold Spring Harbor Symp. Quant. Biol.,* **32**, 299 (1967).
29. J. Spragg, E. Schroeder, J.M. Stewart, K.F. Austen, and E. Haber, *Biochemistry,* **6**, 3933 (1967).
30. J. Spragg, R.C. Talamo, K. Suzuki, D.M. Appelbaum, and K.F. Austen, *Biochemistry,* **7**, 4086 (1968).
31. K.A. Piez, H.A. Bladen, J.M. Land, E.J. Miller, P. Bornstein, W.T. Butler, and A.H. Kang, in *Brookhaven Symp. Biol.,* Vol. 1, No. 21, 345 (1968).
32. W. Beil, R. Timpl, and H. Furthmayr, *Immunology,* **24**, 13 (1973).
33. J. Engel, J. Kurtz, E. Katchalski, and E. Berger, *J. Mol. Biol.,* **17**, 255 (1966).
34. F. Borek, J. Kurtz, and M. Sela, *Biochim. Biophys. Acta,* **188**, 314 (1969).
35. A. Maoz, S. Fuchs, and M. Sela, *Biochemistry,* **12**, 4238 (1973).
36. A. Maoz, S. Fuchs, and M. Sela, *Biochemistry,* **12**, 4246 (1973).
37. J. Ramachandran, A. Berger, and E. Katchalski, *Biopolymers,* **10**, 1829 (1971).

38. B. Schechter, I. Schechter, J. Ramachandran, A. Conway-Jacobs, M. Sela, E. Benjamini, and M. Shimizu, *Eur. J. Biochem.*, **20**, 309 (1971).
39. B. Schechter, I. Schechter, J. Ramachandran, A. Conway-Jacobs, and M. Sela, *Eur. J. Biochem.*, **20**, 301 (1971).
40. H. Fujio, M. Imanishi, K. Nishioka, and T. Amano, *Biken J.*, **11**, 207 (1968).
41. S. Shinka, M. Imanishi, N. Miyagawa, T. Amano, M. Inouye, and A. Tsugita, *Biken J.*, **10**, 89 (1967).
42. R. Arnon, *Eur. J. Biochem.*, **5**, 583 (1968).
43. R. Arnon and M. Sela, *Proc. Nat. Acad. Sci. U.S.*, **62**, 163 (1969)
44. C.C.F. Blake, D.F. Koenig, G.A. Mair, A.C.T. North, D.C. Phillips, and V.R. Sarma, *Nature* (London), **206**, 757 (1965).
45. E. Maron, C. Shiozawa, R. Arnon, and M. Sela, *Biochemistry*, **10**, 763 (1971).
46. I. Pecht, E. Maron, R. Arnon, and M. Sela, *Eur. J. Biochem.*, **19**, 368 (1971).
47. R. Arnon, E. Maron, M. Sela, and C.B. Anfinsen, *Proc. Nat. Acad. Sci. U.S.*, **68**, 1450 (1971).
48. E. Teicher and R. Arnon, unpublished results.
49. E. Teicher, E. Maron, and R. Arnon, *Immunochemistry*, **10**, 265 (1973).
50. R. Arnon, E. Teicher, and H. Scheraga, *Abstr. 9th International Cong. Biochem. Stockholm*, 88 (1973).
51. K. Brew, F.J. Castellino, T.C. Vanaman, and R.L. Hill, *J. Biol. Chem.*, **245**, 4570 (1970).
52. R.L. Hill, K. Brew, T.C. Vanaman, I.P. Trayer, and P. Mattock, *Brookhaven Symp. Biol.*, Vol. 1, No. 21, 139 (1968).
53. J.W. Browne, A.C.T. North, D.C. Phillips, K. Brew, T.C. Vanaman, and R.L. Hill, *J. Mol. Biol.*, **42**, 65 (1969).
54. R. Arnon and E. Maron, *J. Mol. Biol.*, **51**, 703 (1970).
55. R. Arnon and E. Maron, *J. Mol. Biol.*, **61**, 225 (1971).
56. N. Arnheim and A.C. Wilson, *J. Biol. Chem.*, **242**, 3951 (1967).
57. E.M. Prager and A.C. Wilson, *J. Biol. Chem.*, **246**, 7010 (1971).
58. N. Arnheim, J. Sobel, and R. Canfield, *J. Mol. Biol.*, **61**, 237 (1971).
59. E. Maron, R. Arnon, M. Sela, J.-P. Perrin, and P. Jolles, *Biochim. Biophys. Acta*, **214**, 222 (1970).
60. M. Fainaru, A.C. Wilson, and R. Arnon, *J. Mol. Biol.*, **84**, 635 (1974).
61. G. Senyk, E.B. Williams, D.E. Nitecki, and J.W. Goodman, *J. Exp. Med.*, **133**, 1294 (1971).
62. K. Thompson, M. Harris, E. Benjamini, G. Mitchell, and M. Noble, *Nat. New Biol.*, **238**, 20 (1972).
63. O.W. McBride and W.F. Harrington, *Biochemistry*, **6**, 1484 (1967).
64. S. Fuchs and W.F. Harrington, *Biochim. Biophys. Acta*, **221**, 119 (1970).
65. D. Michaeli, G. Senyk, A. Maoz, and S. Fuchs, *J. Immunol.*, **109**, 103 (1972).
66. E. Maron, C. Webb, D. Teitelbaum, and R. Arnon, *Eur. J. Immunol.*, **2**, 294 (1972).
67. L.B. Ichiki and C.R. Parish, *Immunochemistry*, **9**, 153 (1972).

A SYNTHETIC POLYPEPTIDE AS AN IMMUNOLOGICAL MODEL FOR COLLAGEN

SARA FUCHS, Department of Chemical Immunology, The Weizmann Institute of Science, Rehovot, Israel

SYNOPSIS: The unique three-dimensional conformation of collagen and the availability of synthetic polypeptides representing that specific conformation, enabled us to use these synthetic systems in biological studies of collagen. The synthetic ordered polytripeptide poly(Pro-Gly-Pro) was shown to be an appropriate immunological model for collagen. The specificity of the immune response elicited by this polymer is governed by its unique triple-helical conformation, which is like that of collagen. The extent of cross-reactivity of antibodies to poly(Pro-Gly-Pro) with related synthetic polymers depends both on the size and conformation of the polymers tested.

There is an immunological cross-reaction between poly(Pro-Gly-Pro) and natural collagens of several species. Antibodies to poly(Pro-Gly-Pro), like antibodies to rat-tail collagen, have a cytotoxic activity, which is complement dependent on rat fibroblasts. Under similar conditions, osteoblasts are not affected. Antigen-binding lymphocytes from spleens of mice immunized with poly(Pro-Gly-Pro) or with native collagen can be specifically fractionated by means of different collagen fibers and gels. The unique three-dimensional structure of collagen and the ordered collagenlike polytripeptide consists of repeating antigenic determinants and plays an important role in determining the need for cell-to-cell interaction to elicit an antibody response. Thus, poly(Pro-Gly-Pro) and collagen are thymus-independent immunogens, whereas an efficient immune response to the random noncollagenlike polymer poly($Pro^{66}Gly^{34}$) and to gelatin requires thymus cells.

INTRODUCTION

Research on synthetic antigens by our group stemmed from earlier studies on the immunology of collagen. These investigations into the possible structural basis of the poor immunogenicity of collagen and gelatin brought about the research on polypeptidyl gelatins that began with polytyrosyl gelatins (1-3) and was extended to other polypeptidyl derivatives of gelatin (4, 5). From the studies on polypeptidyl gelatins and on their antigenic specificity came simpler models of *completely* synthetic antigens (6), studies of which contributed much to the understanding of the chemical basis of antigenicity.

The choice of synthetic models for collagen was determined by collagen's unique amino acid composition, which includes a high content of glycine, proline, and hydroxyproline. The immunological properties of polyproline, polyhydroxyproline and several linear random copolymers of glycine, proline, and/or hydroxyproline were studied (7-9). None of the random polymers tested cross-reacted with natural collagens, and it became apparent that they were not appropriate models for collagen because, although they were similar to collagen in amino acid composition, they did not have the right conformation. Thus, it became apparent that in order to use synthetic models for a *certain* protein, more sophisticated structural requirements had to be fulfilled (10-13). For this reason, in order to elucidate the nature of the collagen-fold conformation, collagen models were synthesized in the last decade by polymerization of tri- and hexapeptides (14-17). The resulting ordered polypeptides have in common a sequential occurrence of glycine at every third position along the polypeptide chain and a high content of proline. The availability of these ordered polymers and the relationships between their sequence and physical-chemical properties in solution and in the solid state, permitted the elucidation of the relationship between their chemical nature and their biological properties.

The three-dimensional structure of ordered repeating-sequence polymer poly(Pro-Gly-Pro) was shown to resemble the structure of collagen both in solution (14) and its x-ray-diffraction pattern (18, 19). Borek *et al.* (20) have shown that poly(Pro-Gly-Pro) is immunogenic in guinea pigs and rabbits and demonstrated a weak cross-reaction with several collagens. We extended the immunological characterization of the ordered collagenlike polymer poly(Pro-Gly-Pro) and utilized it as a model for studying several immunological phenomena involving collagen.

DISCUSSION OF RESULTS

Immunogenicity and antigenic specificity of poly(Pro-Gly-Pro). The ordered polypeptide poly(Pro-Gly-Pro) was a poor immunogen in rabbits, goats, and guinea pigs when used for immunization as such (20, 21). In order to increase the antibody production against the collagenlike ordered polymer, we conjugated the polymer to a carrier protein and used such conjugates for immunization. Poly(Pro-Gly-Pro) was covalently conjugated to ovalbumin and to RNase by means of water-soluble carbodiimide (21).

Rabbits were immunized with poly(Pro-Gly-Pro)-ovalbumin. The antisera obtained gave precipitin reactions with both the homologous immunogen and the conjugate of the polymer with an unrelated protein, RNase. The unconjugated ordered polymer inhibited this precipitation. On the other hand, the monomeric tripeptide Pro-Gly-Pro did not inhibit this precipitation, and also, the conjugate prepared between the tripeptide and ovalbumin, Pro-Gly-Pro–ovalbumin, was incapable of precipitating antibodies against the polymer. In the reverse system, however, the ordered polymer poly(Pro-Gly-Pro) inhibited the precipitin reaction of anti-Pro-Gly-Pro–ovalbumin with Pro-Gly-Pro–RNase (21). Poly(Pro-Gly-Pro)-specific antibodies were detected also in the anti-poly(Pro-Gly-Pro)-ovalbumin sera by immunological techniques other than precipitin, such as hemagglutination of poly(Pro-Gly-Pro)-coated sheep erythrocytes, as well as by the modified bacteriophage technique (22, 23).

The specificity of antipoly(Pro-Gly-Pro)-ovalbumin antiserum toward ordered and random polymers of proline and glycine of different molecular weights was tested by the capacity of these polymers to inhibit the immunospecific inactivation of poly(Pro-Gly-Pro)-modified bacteriophage by the antiserum. As can be seen in Fig. 1, polymers of higher molecular weights are more efficient inhibitors in all cases (21). Moreover, within a wide range of molecular weights of the polymers tested, the ordered polymers [poly(Pro-Gly-Pro)] were always much better inhibitors than the random ones [poly($Pro^{66}Gly^{34}$)]. A plot of the inhibitor concentration required for 50% inhibition as a function of the molecular weight of the inhibitor (Fig. 2) revealed that the ordered polymers are better

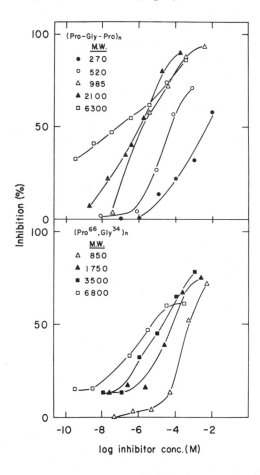

Figure 1. Inhibition of the inactivation of poly(Pro-Gly-Pro)-RNase bacteriophage T4 by rabbit antipoly(Pro-Gly-Pro)-ovalbumin serum with ordered (top) and random (bottom) polymers. [Reprinted with permission from Maoz *et al.*, *Biochemistry*, **12**, 4238 (1973). Copyright by the American Chemical Society.]

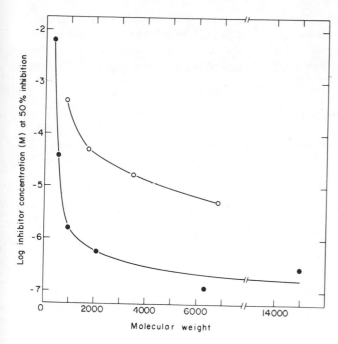

Figure 2. Molecular-weight dependence of the inhibitory capacity of poly(Pro-Gly-Pro) (○) and poly(Pro^{66}Gly^{34}) (●). The concentrations of inhibitor required for 50% inhibition of the inactivation of poly(Pro-Gly-Pro)-RNase-T4 by antipoly(Pro-Gly-Pro) serum were derived from the data presented in Fig. 1. [Reprinted with permission from Maoz *et al.*, *Biochemistry*, **12**, 4238 (1973). Copyright by the American Chemical Society.]

than the random polymers of similar molecular weights by about two orders of magnitude. Thus, the affinity of the antibodies to conformation-dependent determinants can be assumed to be far higher than the affinity to sequential stretches.

The curve describing the dependence of the inhibition on the molecular weight of the ordered polymer (Fig. 2) seems to be composed of two regions. With small polymers there is a strong dependence on the molecular weight. The tripeptide building unit of the polymer shows extremely low binding to these antibodies. With polypeptides of higher molecular weight, the inhibitory capacity is not significantly augmented by the molecular weight. It seems that in the low range of molecular weights, the relatively weak inhibitor does not possess the characteristic three-dimensional structure of collagen.

Additional evidence of the conformation dependence of antibody specificity to the ordered polymer came from studies with collagenase. The ordered and random polymers were digested with collagenase and their inhibition of the inactivation of the modified phage was measured as a function of time of enzymatic digestion (21). The inhibitory capacity of the ordered polymer decreased markedly as a result of collagenase digestion, whereas the inhibitory capacity of the random polymer did not change significantly upon a similar digestion. It seems that the collagenase effect on the ordered polymer is attributable to the loss of its three-dimensional, collagenlike structure.

Immunological cross-reactions between collagens and poly(Pro-Gly-Pro). After it had been shown that the specificity of the antibodies to the ordered polymer is toward the three-dimensional conformation of the polymer, which is like that of collagen, it was of interest to test for cross-reactivity between the immune response to poly(Pro-Gly-Pro) and natural collagens of different species. It was previously reported (20) that there is a weak cross-reactivity by passive cutaneous anaphylaxis reactions between antipoly(Pro-Gly-Pro) and ichthyocol (fish collagen), rat-skin collagen and guinea-pig-skin collagen. A synthetic antigen that will elicit an immune response that is cross-reactive with natural collagens is of special interest and may be helpful in studying structure and function relationships in collagens.

Guinea pigs were immunized with the ordered polypeptide poly(Pro-Gly-Pro) and the random copolymer poly($Pro^{66}Gly^{34}$) and were skin-tested ten days later with both polymers and with natural collagens (24). Guinea pigs that had been immunized with the ordered polymer exhibited delayed type skin reactions with several natural collagens, such as *Ascaris* cuticle collagen, guinea-pig-skin collagen and rat-tail-tendon collagen, whereas animals immunized with the random copolymer exhibited no such cross-reactions. Of special interest was the cross-reaction with collagen from the same species, namely, guinea-pig collagen. Immunization with the synthetic polymer broke the tolerance to self collagen and the animal became sensitive to its own collagen. Such experimentally induced autoimmunity can be helpful in understanding the involvement of collagen in autoimmune diseases.

No humoral cross-reaction between the natural collagens and the synthetic polypeptides was detected by immediate type skin reactions ten days after the immunization. It seems that the cellular type of cross-reaction manifested by the delayed reaction precedes the appearance of cross-reacting antibodies. The immediate reactions were weaker than the delayed ones, even with the homologous antigen. In other experiments, when the animals were skin tested at later times after immunization, immediate cross-reactions with collagens were recorded in addition to the delayed cross-reactions (24) (see Table 1).

The ability of collagens to produce skin reactions in animals immunized with synthetic polymers was tested as a function of the size of the immunogen (24). Guinea pigs immunized with poly(Pro-Gly-Pro) of different molecular weights were skin tested with poly(Pro-Gly-Pro) (MW, 6300) and with three natural collagens. The results show clearly that cross-reactivity with collagens depends on the molecular weight of the immunizing polypeptide (Table 1). No cross-reactions with collagens could be detected in animals immunized with

Table 1. The correlation between the size of the immunogen and the cross-reactions of natural collagens with synthetic collagenlike polytripeptides

| No. of Days after Immunization | Test Antigen | Immunogen[a] | | | | | |
| | | Poly(Pro-Gly-Pro) MW, 915 | | Poly(Pro-Gly-Pro) MW, 1910 | | Poly(Pro-Gly-Pro) MW, 6300 | |
		Immediate Skin Reaction[b]	Delayed Skin Reaction[b]	Immediate Skin Reaction[b]	Delayed Skin Reaction[b]	Immediate Skin Reaction[b]	Delayed Skin Reaction[b]
9	Poly(Pro-Gly-Pro)[c]	0/5[d]	5/5 (9.5)[e]	2/5 (10)[e]	5/8 (13.5)[e]	4/5 (11)[e]	5/5 (14.5)[e]
	RCM Ascaris collagen[f]	0/5	0/5	0/5	3/5 (6)	0/5	5/5 (9.5)
	Guinea-pig-skin collagen	0/5	0/5	0/5	1/5 (10)	0/5	3/5 (6)
	Rat-tail-tendon collagen	0/5	0/5	0/5	4/5 (7)	0/5	4/5 (8)
18	Poly(Pro-Gly-Pro)	2/5 (11)	3/5 (9)	4/5 (10)	5/5 (16.5)	5/5 (20)	5/5 (20)
	RCM Ascaris collagen	0/5	0/5	2/5 (9.5)	5/5 (11.5)	5/5 (15)	5/5 (16)
	Guinea-pig-skin collagen	0/5	0/5	0/5	2/5 (13)	1/5 (10)	4/5 (9)
	Rat-tail-tendon collagen	0/5	0/5	0/5	3/5 (9)	2/5 (10)	3/5 (7)
32	Poly(Pro-Gly-Pro)	1/5 (11)	3/5 (15)	4/5 (19)	4/5 (17)	5/5 (22.5)	5/5 (14)
	RCM Ascaris collagen	0/5	2/5 (15)	4/5 (11)	4/5 (9)	5/5 (15.5)	4/5 (14.5)
	Guinea-pig-skin collagen	0/5	0/5	2/5 (8)	3/5 (9)	5/5 (13.6)	3/5 (11)
	Rat-tail-tendon collagen	0/5	0/5	2/5 (6)	3/5 (7)	5/5 (13)	3/5 (7)

[a]Control animals injected with buffered saline and complete Freund's adjuvant gave negative reactions with all test antigens.

[b]Immediate skin reactions observed after 2 hr; delayed skin reactions observed after 24 hr.

[c]Molecular weight 6300.

[d]Ratio of responders to total number of animals.

[e]Average reaction diameter in mm. Skin reactions (immediate or delayed) with an average diameter of 5 mm or less were considered negative.

[f]Reduced and carboxymethylated (RCM) Ascaris cuticle collagen.

poly(Pro-Gly-Pro) of low molecular weight (MW, 915). When preparations of higher-molecular-weight poly(Pro-Gly-Pro) were used for immunization, a strong cross-reactivity of both immediate and delayed types was observed with collagens. These results corroborate the finding that the collagenlike conformation of poly(Pro-Gly-Pro) depends on the molecular weight of the polymer (Figs. 1 and 2).

Although cross-reactivity between poly(Pro-Gly-Pro) and collagens was demonstrated by skin reactions in guinea pigs, it was quite difficult to detect such cross-reactions in several assays for humoral antibodies. Rabbit antipoly-(Pro-Gly-Pro)-ovalbumin serum was reacted with rat-tail collagen, guinea-pig-skin collagen, earthworm-cuticle collagen and reduced and carboxymethylated (RCM) *Ascaris*-cuticle collagen (25). Of the collagens tested, only the collagen from *Ascaris* cross-precipitated with antipoly(Poly-Gly-Pro) antibodies (24). In addition, collagen from *Ascaris* was capable of inhibiting the inactivation of poly(Pro-Gly-Pro)-T4 preparations by antipoly(Pro-Gly-Pro)-ovalbumin serum.

Of several anticollagen sera that were tested for their capacity to inactivate poly(Pro-Gly-Pro)-coated bacteriophage T4, only anti-*Ascaris* collagen inactivated the modified bacteriophage. Antibodies from anti-*Ascaris* collagen serum could be precipitated also by poly(Pro-Gly-Pro)-conjugated proteins. It should be pointed out that *Ascaris* collagen, which has been shown to be highly immunogenic in rabbits (26), has several unique structural features. This collagen has a triple-helical structure formed by a single polypeptide chain that folds back upon itself to form the collagenlike helix. Another prominent feature of *Ascaris*-cuticle collagen is that it contains more proline than any other known collagen. The distribution of pyrrolidine residues in triplets of various collagens of known composition was estimated statistically by Josse and Harrington (27), who have calculated that the highest incidence of dipyrrolidines appeared in *Ascaris*-cuticle collagen. This unique feature as well as the great phylogenetical distance of *Ascaris*-cuticle collagen from the other collagens tested, seems to contribute much to the high immunogenicity (26) and to the cross-reactivity of this collagen with poly(Pro-Gly-Pro) at both humoral and cellular levels.

Cytotoxic activity of antibodies to poly(Pro-Gly-Pro) on cells in tissue culture. In view of the immunological cross-reactivity between the synthetic collagen model poly(Pro-Gly-Pro) and natural collagens it was of interest to find out whether it is possible to apply antibodies to the synthetic model for biological studies of collagen. We have studied the effect of antibodies specific to the collagenlike polypeptide on fibroblasts (28) because fibroblasts are known to secrete collagen and collagen precursors (29–31). Antiserum to chick-tendon collagen, in the presence of complement, was shown (32, 33) to have a cytotoxic effect on chick-embryo fibroblasts.

Using the *in vitro* system of primary cultures of rat-muscle cells developed by Yaffe (34–36), we tested the effect of antibodies to poly(Pro-Gly-Pro) and antiserum to acid-soluble rat-tail-tendon collagen on muscle cells in tissue cultures

(28). In the presence of complement both the antibodies to the ordered polymer and the anticollagen serum were toxic to the cells in culture. The addition of complement (1:10, final dilution) caused the sequential appearance of irregularities in shape and granulation within the cell. Later, general cell damage, lysis, and complete destruction of the cultures were observed. It seems that the mononucleated cells were somewhat more susceptible to the cytotoxic activity than the fibers, although the contraction of the fibers stopped upon addition of complement in the presence of antisera to collagen or antibodies to poly(Pro-Gly-Pro). No cytotoxic effect on the culture was observed when antiserum to a non-related multichain synthetic polypeptide, poly(Tyr, Glu)-poly(DLAla)--poly(Lys), was used. Progressive cytotoxic effect on the cells in culture could be shown with time as well as with increasing concentrations of the specific antibodies. The cytotoxic activity of the specific antibodies was demonstrated also by counting the number of nuclei within cells in the cultures and by the ^{51}Cr release method (28).

The interaction of antibodies to the synthetic periodic polypeptide poly(Pro-Gly-Pro) with some cell membranes suggests localization of similar structures in fibroblast membranes. There may be several advantages in using antibodies to the synthetic polymer rather than antibodies to collagen in such a system. With antibodies to the synthetic collagenlike polymer, the antibodies are certain to be specific just to collagen and not to other protein contaminants that may accompany preparations of natural collagens. In addition, the barrier of species specificity is overcome and the same reagent, namely, antibodies to the synthetic model, may be used for fibroblasts from any source.

Recently, in collaboration with Duksin and Maoz, we have demonstrated that antiserum to rat-tail collagen or to poly(Pro-Gly-Pro), in the presence of complement, has a differential cytotoxic effect on rat osteoblasts (37) and rat-skin fibroblasts grown in cell culture. The extent of the cytotoxic effect was estimated directly by counting the viable cells in the presence of Trypan-Blue and by the ^{51}Cr release method. Under the test conditions, up to 40% and 53% of skin fibroblasts were killed in the presence of anticollagen serum and antipoly(Pro-Gly-Pro) serum, respectively, whereas only 7% and 0% of the osteoblasts were similarly affected by these antisera. This cytotoxicity could be detected only for cells pretreated with trypsin. Normal rabbit serum and rabbit antiserum against a nonrelated antigen were nontoxic toward both types of cells. The differential toxicity towards osteoblasts and skin fibroblasts is probably due to either antigenic differences between bone and skin collagen or a different susceptibility of the cell membrane of osteoblasts. The above system is an appropriate tool for obtaining osteoblast cultures free of fibroblasts and, therefore, it is important for the study of isolated bone-cell functions in culture.

Fractionation of mouse lymphocyte populations on collagen fibers and gels.
The immunological cross-reactivity between poly(Pro-Gly-Pro) and collagen and the availability of collagen in different insoluble forms, such as fibers and gels,

suggested an attractive technique for fractionation of spleen cells from mice immunized with poly(Pro-Gly-Pro) and with collagen. In collaboration with Maoz, we have used collagen fibers and gels for fractionation of mouse lymphocyte populations (38). Cell fractionation on collagen fibers was performed in a procedure similar to that described for cell fractionation on nylon fibers (39, 40). Spleen cells from mice immunized with collagenlike poly(Pro-Gly-Pro) were bound specifically to collagen fibers. The number of cells was determined by counting those cells seen on the edges of a 15 mm segment of the fiber. The number of fiber-binding cells from immunized mice was five or six times higher than that observed in nonimmunized animals (Table 2). The specifically bound cells were recovered by mechanical plucking of the fibers. Binding to the fibers was specifically inhibited (90%) by incubation of the cells together with either the immunizing antigen, poly(Pro-Gly-Pro), or with the respective antiserum.

Spleen cells from mice immunized with poly(Pro-Gly-Pro) were bound also to collagen gels. The cells were recovered from the gels after collagenase digestion of the collagen gels. The amount of gel binding cells recovered from cell suspensions of immunized mice was three or four times greater than that obtained from unimmunized animals (38).

In an attempt to extend the applicability of collagen and gelatin gels for fractionation of lymphoid cells from mice immunized with other antigens, these solid supports were conjugated with other antigens. Thus, lymphoid cells from mice immunized with DNP-protein conjugates, as well as mouse tumor cells (MOPC 315), were bound to DNP-substituted collagen and gelatin gels. Binding of spleen cells from mice immunized with DNP-protein conjugates to the substituted gels was about threefold higher than that obtained with unimmunized animals. A higher degree of specific binding was observed with the tumor cells.

Table 2. Binding of spleen cells from mice immunized with poly(Pro-Gly-Pro) to collagen fibers

Number of Cells Applied per Plate	Number of Cells Bound[a]	
	Nonimmunized	Immunized[b]
0.6×10^7	0	0
1.2×10^7	0	89
1.5×10^7	0	172
1.8×10^7	57	311
2.1×10^7	97	451
2.1×10^{7c}	0	55

[a]Expressed as number of cells bound to the edge of 15 mm fiber segment. Values represent an average of five determinations within each dish.

[b]Balb/c mice were immunized with poly(Pro-Gly-Pro).

[c]Cells were incubated in the presence of poly(Pro-Gly-Pro), 50 μg/ml.

The advantage of using gelatin gels or derived gelatin gels is that they melt easily at 37°C and the bound cells are released.

The gel fractionation method suggests a convenient technique for studying functions of specifically fractionated lymphocytes. In view of the finding (41) that collagen and poly(Pro-Gly-Pro) are thymus-independent antigens and that gelatin and poly(Pro^{66}Gly^{34}) are thymus-dependent antigens (see below), it seems that this collagen-gelatin fractionation method provides an intriguing system for studying the properties of T cells and B cells and their interrelationships. Hence, it may lead to a better understanding of the role of these cells in the immune response.

Thymus independence of poly(Pro-Gly-Pro) and collagen and their effect on the requirement for cell cooperation in the immune response to T-dependent antigens. In view of the unique structural properties of poly(Pro-Gly-Pro) and of collagen, it was of interest to find out what types of cells participate in the immune response to these two antigens, or more specifically, what is the role of the thymus in this immune response. For most antigens, thymus cells and bone-marrow cells are required for an efficient immune response. However, there is a group of antigens for which it has been demonstrated that an immune response can be obtained without the cooperation of thymus cells. This group includes polymerized flagellin (42, 43), pneumococcal polysaccharide (44), polyvinyl-pyrrolidone (45) and slowly metabolized synthetic antigens composed mostly or partially of D amino acids (46). The common denominators for all these antigens are identical repeating antigenic determinants and slow metabolism (46). Thus, it seemed likely that the ordered polymer having both repeating sequence and unique conformation, and possibly collagen itself, may be suitable candidates for thymus-independent antigens.

We have studied the role of the thymus in the immune response to the ordered poly(Pro-Gly-Pro) and the random poly(Pro^{66}Gly^{34}) for a possible correlation with structural features of the immunogen (41). Heavily irradiated recipients were injected with thymocytes, marrow cells, or a mixture of both cell populations and were immunized with the polymers. An efficient immune response to poly(Pro-Gly-Pro) was elicited in the absence of transferred thymocytes, but an effective response to the random poly(Pro^{66}Gly^{34}) was dependent on cooperation between thymus and marrow cells (Table 3). Similar transfer experiments revealed that native triple-helical collagen (from rat-tail tendon or *Ascaris* cuticle) is a thymus-independent antigen, whereas in order to elicit a response to its denatured product (gelatin), thymus cells are required (Table 4).

To determine whether the thymus independence of poly(Pro-Gly-Pro) and collagen can also affect the pattern of the immune response to otherwise thymus-dependent antigenic stimuli, we have studied, in collaboration with Mozes, the effect of poly(Pro-Gly-Pro) and collagen on the need for cell cooperation in the immune response to thymus-dependent antigens. Irradiated recipient mice that had been transplanted with either bone-marrow cells or a mixture of bone-marrow

Table 3. Thymus-marrow cell cooperation in the immune response to poly(Pro-Gly-Pro) and poly(Pro^{66}Gly^{34})^a

Mouse Strain	Immunogen	Assaying Antigen	Assayed by	Cells Injected per Recipient[b]		
				10^8 Thymocytes	2×10^7 Marrow Cells	10^8 Thymocytes and 2×10^7 Marrow Cells
BALB/c	poly(Pro-Gly-Pro)	poly(Pro-Gly-Pro)-RNase	PFC[c]	0 (2)	73 (30)	71 (24)
			HA[d]	0 (2)	73 (30)	65 (23)
BALB/c (Tx)[e]	poly(Pro-Gly-Pro)	poly(Pro-Gly-Pro)-Rnase	PFC[c]	0 (2)	59 (22)	55 (22)
			HA[d]	0 (2)	59 (22)	48 (21)
SWR	poly(Pro-Gly-Pro)	poly(Pro-Gly-Pro)-Rnase	PFC[c]	15 (13)	53 (15)	64 (11)
			HA[d]	0 (13)	57 (14)	44 (11)
C3H.SW	poly(Pro-Gly-Pro)	poly(Pro-Gly-Pro)-RNase	PFC[c]		73 (11)	67 (9)
			HA[d]		55 (11)	44 (9)
BALB/c	poly(Pro^{66}Gly^{34})	poly(Pro^{66}Gly^{34})-RNase	PFC[c]		30 (27)	72 (21)
			HA[d]		42 (19)	79 (19)
C3H.SW	poly(Pro^{66}Gly^{34})	poly(Pro^{66}Gly^{34})-RNase	PFC[c]	11 (9)	32 (22)	67 (18)
			HA[d]	18 (11)	27 (22)	76 (17)

[a]Reprinted with permission from Fuchs *et al.*, *J. Exp. Med.*, **139**, 148 (1974). Copyright by Rockefeller University Press.

[b]Percentage of syngeneic irradiated recipients producing detectable antibody titers. Number of mice tested is given in parentheses.

[c]Hemolytic plaque-forming cell assay. Results higher than 100 plaques per spleen were considered positive.

[d]Passive microhemagglutination assay. Hemagglutination titer at dilutions greater than 1/4 was considered positive.

[e]Thymectomized.

Table 4. Thymus–bone-marrow cell cooperation in the immune response to collagen and gelatin[a]

Mouse Strain	Immunogen and Assaying Antigen	Assayed by	Cells Injected per Recipient[b]		
			10^8 Thymocytes	2×10^7 Marrow Cells	10^8 Thymocytes and 2×10^7 Marrow Cells
SWR	RCM *Ascaris* collagen	PFC[c] HA[d]	21 (28) 19 (31)	74 (38) 70 (40)	70 (23) 57 (23)
SWR	rat-tail collagen	PFC[c] HA[d]	8 (13) 23 (30)	70 (10) 77 (35)	62 (13) 83 (30)
SWR	rat-tail gelatin	PFC[c] HA[d]	17 (23) 11 (26)	12 (32) 22 (32)	70 (23) 83 (23)

[a]Reprinted with permission from Fuchs *et al., J. Exp. Med.,* **139,** 148 (1974). Copyright by Rockefeller University Press.

[b]Percentage of syngeneic irradiated recipients producing detectable antibody titers. Number of mice tested is given in parentheses.

[c]Hemolytic plaque-forming cell assay. Results higher than 100 plaques per spleen were considered positive.

[d]Passive microhemagglutination assay. Hemagglutination titer at dilutions greater than 1/4 was considered positive.

and thymus cells were immunized with poly(Pro-Gly-Pro) covalently conjugated to thymus-dependent ovalbumin [poly(Pro-Gly-Pro)-ovalbumin] or with a mixture of poly(Pro-Gly-Pro) and ovalbumin. In both cases, an effective response toward ovalbumin was observed in the absence of thymocytes, as was found for thymus-independent poly(Pro-Gly-Pro) (Table 5). The effect on ovalbumin was the same when a mixture of collagen and ovalbumin was used for immunization (Table 6). On the other hand, when irradiated reconstituted mice were immunized with a mixture of ovalbumin and the thymus-dependent gelatin, cell-to-cell cooperation was required for an immune response to both immunogens. Studies designed to elucidate the possible biological meaning of the effect of thymus-independent antigens on thymus-dependent antigens are now in progress.

ACKNOWLEDGMENT

Part of the work reported here was supported by Grant 1 RO1 AI11405-01 from the National Institutes of Health, U.S. Public Health Service.

Table 5. The effect of poly(Pro-Gly-Pro) on the immune response to ovalbumin[a]

Immunogen	Assaying Antigen	Assayed by	Cells Injected per Recipient[b]		
			10^8 Thymocytes	2×10^7 Marrow Cells	10^8 Thymocytes and 2×10^7 Marrow Cells
Poly(Pro-Gly-Pro)-ovalbumin	poly(Pro-Gly-Pro)-RNase	PFC[c]	21 (14)	55 (11)	43 (14)
		HA[d]	15 (13)	63 (19)	36 (25)
	ovalbumin	PFC[c]	21 (14)	55 (11)	43 (14)
		HA[d]	8 (13)	63 (19)	27 (26)
Poly(Pro-Gly-Pro)-ovalbumin	poly(Pro-Gly-Pro)-RNase	PFC[c]	0 (6)	58 (12)	67 (12)
		HA[d]	0 (6)	64 (11)	42 (12)
	ovalbumin	PFC[c]	0 (6)	67 (12)	67 (12)
		HA[d]	33 (6)	70 (10)	33 (12)
Ovalbumin	ovalbumin	PFC[c]	0 (3)	7 (14)	55 (11)
		HA[d]			

[a]Transfer experiments were performed in BALB/c mice.

[b]Percentage of syngeneic irradiated recipients producing detectable antibody titers. Number of mice tested is given in parentheses.

[c]Hemolytic plaque-forming cell assay. Results higher than 100 plaques per spleen were considered positive.

[d]Passive microhemagglutination assay. Hemagglutination titer at dilutions greater than 1/4 was considered positive.

Table 6. The effect of collagen and gelatin on the immune response to ovalbumin[a]

Immunogen	Assaying Antigen	Assayed by	Cells Injected per Recipient[b]		
			10[8] Thymocytes	2 × 10[7] Marrow Cells	10[8] Thymocytes and 2 × 10[7] Marrow Cells
Collagen-ovalbumin	collagen	PFC[c]		73 (11)	45 (11)
		HA[d]	12 (25)	90 (28)	74 (23)
	ovalbumin	PFC[c]		54 (11)	9 (11)
		HA[d]	20 (25)	74 (30)	52 (25)
Gelatin-ovalbumin	gelatin	PFC[c]		17 (12)	92 (13)
		HA[d]	27 (11)	27 (22)	81 (21)
	ovalbumin	PFC[c]		33 (12)	62 (13)
		HA[d]	6 (17)	14 (22)	62 (21)
Ovalbumin	ovalbumin	PFC[c]		12 (18)	50 (14)
		HA[d]	15 (13)	22 (18)	68 (22)

[a]Transfer experiments were performed in SWR mice.
[b]Percentage of syngeneic irradiated recipients producing detectable antibody titers.   Number of mice tested is given in parentheses.
[c]Hemolytic plaque-forming cell assay.   Results higher than 100 plaques per spleen were considered positive.
[d]Passive microhemagglutination assay.   Hemagglutination titer at dilutions greater than 1/4 was considered positive.

REFERENCES

1. M. Sela, E. Katchalski, and A.L. Olitzki, *Science*, **123**, 1129 (1956).
2. M. Sela and R. Arnon, *Biochem. J.*, **75**, 91 (1960).
3. R. Arnon and M. Sela, *Biochem. J.*, **75**, 103 (1960).
4. M. Sela and R. Arnon, *Biochem. J.*, **77**, 394 (1960).
5. S. Fuchs and M. Sela, *Biochem. J.*, **93**, 566 (1964).
6. M. Sela, *Science*, **166**, 1365 (1969).
7. H.E. Jasin and L.E. Glynn, *Immunology*, **8**, 95 (1965).
8. H.E. Jasin and L.E. Glynn, *Immunology*, **8**, 260 (1965).
9. P.C. Brown and L.E. Glynn, *Immunology*, **15**, 589 (1968).
10. C.B. Anfinsen, *Biochem. J.*, **128**, 737 (1972).
11. M. Sela, B. Schechter, I. Schechter and F. Borek, *Cold Spring Harbor Symposia on Quantitative Biology*, **32**, 537 (1967).
12. B. Schechter, I. Schechter, J. Ramachandran, A. Conway-Jacobs, M. Sela, E. Benjamini, and M. Shimizu, *Eur. J. Biochem.*, **20**, 309 (1971).
13. M. Sela, *Harvey Lectures* 1971–1972, Academic Press, New York, **67**, 213 (1973).
14. J. Engel, J. Kurtz, E. Katchalski, and A. Berger, *J. Mol. Biol.*, **17**, 255 (1966).
15. J.P. Carver and E.R. Blout, in *Treatise on Collagen*, Vol. 1, G.M. Ramachandran, ed., Academic Press, New York, 1967, p. 441.
16. K.A. Piez and W. Traub, *Adv. Prot. Chem.*, **25**, 243 (1971).
17. D.M. Segal, *J. Mol. Biol.*, **43**, 497 (1969).
18. W. Traub and A. Yonath, *Israel J. Chem.*, **3**, 43 (1965).
19. W. Traub and A. Yonath, *J. Mol. Biol.*, **16**, 404 (1966).
20. F. Borek, J. Kurtz, and M. Sela, *Biochim. Biophys. Acta*, **188**, 314 (1969).
21. A. Maoz, S. Fuchs, and M. Sela, *Biochemistry*, **12**, 4238 (1973).
22. J. Haimovich and M. Sela, *J. Immunol.*, **94**, 338 (1966).
23. O. Mäkelä, *Immunology*, **10**, 81 (1966).
24. A. Maoz, S. Fuchs, and M. Sela, *Biochemistry*, **12**, 4246 (1973).
25. O.W. McBride and W.F. Harrington, *Biochemistry*, **6**, 1484 (1967).
26. S. Fuchs and W.F. Harrington, *Biochim. Biophys. Acta*, **221**, 119 (1970).
27. J. Josse and W.F. Harrington, *J. Mol. Biol.*, **9**, 269 (1964).
28. A. Maoz, H. Dym, S. Fuchs, and M. Sela, *Eur. J. Immunol.*, **3**, 839 (1973).
29. R.L. Church, S.E. Pfeiffer, and M.L. Tanzer, *Proc. Nat. Acad. Sci. U.S.*, **68**, 2638 (1971).
30. P.B. Ramalay and J. Rosenbloom, *FEBS Lett.*, **15**, 59 (1971).
31. D.L. Layman, E.A. McGoodwin, and G.R. Martin, *Proc. Nat. Acad. Sci. U.S.*, **68**, 454 (1971).
32. L. Lustig, H. Constantini, and R.E. Mancini, *Proc. Soc. Exptl. Biol. Med.*, **130**, 283 (1969).
33. L. Lustig, *Proc. Soc. Exptl. Biol. Med.*, **133**, 207 (1970).
34. D. Yaffe, *Current Topics in Developmental Biol.*, **4**, 37 (1969).
35. D. Yaffe, *Expt. Cell Res.*, **66**, 33 (1971).
36. D. Yaffe and S. Fuchs, *Develop. Biol.*, **15**, 33 (1967).
37. I. Binderman, D. Duksin, A. Harell, E. Katchalski, and L. Sachs, *J. Cell. Biol.*, in press.

38. A. Maoz and S. Fuchs, to be submitted.
39. G.M. Edelman, U. Rutishauser, and C.F. Millette, *Proc. Nat. Acad. Sci. U.S.*, **68**, 2153 (1971).
40. U. Rutishauser, C.F. Millette, and G.M. Edelman, *Proc. Nat. Acad. Sci. U.S.*, **69**, 1596 (1972).
41. S. Fuchs, E. Mozes, A. Maoz, and M. Sela, *J. Exp. Med.*, **139**, 148 (1974).
42. W.D. Armstrong, E. Diener, and G.R. Shellam, *J. Exp. Med.*, **129**, 393 (1969).
43. M. Feldmann and A. Basten, *J. Exp. Med.*, **134**, 103 (1971).
44. J.G. Howard, G.H. Christie, B.M. Courtenay, E. Leuchars, and J.S. Davies, *Cell. Immunol.*, **2**, 614 (1971).
45. B. Andersson and H. Blomgren, *Cell. Immunol.*, **2**, 411 (1971).
46. M. Sela, E. Mozes, and G.M. Shearer, *Proc. Natl. Acad. Sci. U.S.*, **69**, 2696 (1972).

THE USE OF SYNTHETIC POLYPEPTIDE IMMUNOGENS IN THE STUDY OF GENETIC CONTROL OF IMMUNE RESPONSE

EDNA MOZES, Department of Chemical Immunology, The Weizmann Institute of Science, Rehovot, Israel

SYNOPSIS: Synthetic polypeptide antigens were used to study the genetic control of immune responsiveness at the molecular and cellular levels. Genetic regulation of immune response potential of mice to branched-chain synthetic polypeptides built on multichain poly(DL-alanine) (A--L) and on multichain poly(L-proline) (Pro--L) is expressed by quantitative, autosomal dominant traits. The antibody responses to synthetic polypeptides built on A--L are specific to the short sequences of amino acids attached to A--L and, like the response to poly(Tyr, Glu)-poly(Pro)--poly(Lys), which is directed to multichain polyproline, they are unigenically controlled. On the other hand, two distinct gene loci regulate the ability to respond to poly(Phe, Glu)-poly(Pro)—poly(Lys), and at least three genes control the response potential to another polypeptide in this series, namely, poly(His, Glu)-poly(Pro)—poly(Lys). Thus, although synthetic polypeptides are fairly simple in structure, they can be designed to provide good models for natural antigens possessing multiple antigenic determinants that provoke an immune response controlled by several different genes.

Cellular analysis of antibody responses to synthetic polypeptides has demonstrated that the genetic defect in immune responsiveness is reflected at the thymus cell level in situations in which the poor response is due to a carrier defect, as is the case in SJL mice for determinants attached to multichain poly(DL-alanine), to which these mice are nonresponders. On the other hand, when the genetic defect is strictly at the determinant level, it is expressed in the marrow-derived antibody-forming cell precursors.

Using ordered tetrapeptides of tyrosine and glutamic acid, Tyr-Tyr-Glu-Glu was found to be a major determinant in the random polypeptide poly(Tyr, Glu)-poly(DLAla)--poly(Lys) or (T, G)-A--L, which is responsible for the phenotypic expression of antibody response to this immunogen. This finding provides new approaches for studies of the genetic defect in low-responder mice with regard to (T, G)-A--L.

INTRODUCTION

The ability of an individual to elicit an antibody response to a specific immunogen is subject to genetically determined factors. The phenomenon of immune responsiveness is a complex process involving many steps and at least two, and probably three, cell populations. Genetic control could be exerted at several different points in this process. Results of investigations reported during the last decade have indicated that the immune response potentials of several rodent species to many immunogenic systems are, indeed, genetically regulated (1, 2). Nevertheless, the most significant progress toward elucidation of the genetic and cellular mechanisms involved in the regulation of specific immune responses has been made using synthetic polypeptide antigens. These immunogens possess a restricted number of antigenic determinants, and their composition, size, shape, optical configuration, and so on, are more easily controlled than those of naturally occurring proteins. Therefore they have been used extensively for studying the molecular and cellular basis of immunological responsiveness.

GENETIC CONTROL OF IMMUNE RESPONSES TO SYNTHETIC POLYPEPTIDE ANTIGENS

The first example of unigenic control of an immune response is the PLL gene in guinea pigs. This gene controls responsiveness to poly(L-lysine), poly(L-arginine), copolymers composed of L-glutamic acid and L-lysine, and to hapten conjugates of these polypeptides (3-5). Inbred strain-2 guinea pigs were found to be responders, whereas strain 13 and some randomly bred Hartly animals were nonresponders. Expression of the PLL gene has been characterized by cellular immunity (6, 7), as well as by production of specific antibodies.

Regulation of antibody responses to polypeptides built on multichain poly(DL-alanine). Evidence for determinant-specific genetic control of antibody response in inbred strains of mice was shown with a series of synthetic polypeptides built on poly(DL-Ala)-poly(Lys) (denoted A--L). When short, random sequences of tyrosine and glutamic acid were added to A--L to produce poly(Tyr, Glu)-poly(DLAla)--poly(Lys), designated (T, G)-A--L (8), immunized C57 mice responded with tenfold higher antibody titers than CBA mice. When the tyrosine in (T, G)-A--L was replaced with histidine, the resulting poly(His, Glu)-poly(DLAla)--poly(Lys), or (H, G)-A--L (9), elicited a poor response in C57 mice, whereas CBA mice responded well. Both strains responded well to a third branched polymer, poly(Phe, Glu)-poly(DLAla)--poly(Lys), or (Phe, G)-A--L (9). The F_1 hybrids (C57 × CBA) responded well to all three immunogens. Backcross progeny segregated in response to (T, G)-A--L and (H, G)-A--L as a 1:1 mixture of the F_1 and the respective homozygous parent animals (10, 11). The number of control experiments precluded dose response differences (10) or different responses due to adjuvants, age, or sex. Thus, the results indicate that the ability of mice to respond to this series of synthetic polypeptides is a genetically controlled, quantitative dominant trait. Because substitution of histidine for tyrosine in the antigenic determinant causes a reversal of the high- and low-responder strains, this genetic trait, which has been designated Ir-1 (immune response-1), may be concerned with reactivity to the antigenic determinants. Further studies have shown that the ability to respond to these three related, branched synthetic polypeptide antigens is closely linked to the major histocompatibility (H-2) locus of the mice (12, 13).

Regulation of antibody responses to synthetic polypeptides built on multi-chain polyproline. Genetic control of immunological responsiveness has also been demonstrated for another series of branched-chain synthetic polypeptides built on multichain poly(L-proline) (Pro--L) (14). The immune response of inbred mice to poly(Tyr, Glu)-poly(Pro)--poly(Lys) [(T, G)-Pro--L] differs from the response of the same strains to (T, G)-A--L. Mice of the SJL strain, which respond poorly to the polypeptides built on A–L, are high responders to (T, G)-Pro--L, indicating that the prolyl peptides play an important role in the antigenic properties of this polymer (15). Further analysis of the antibodies produced upon immunization with (T, G)-Pro--L indicated that, in this case, the immune response is specific to the Pro--L region of the immunogen and not to the short peptides of tyrosine and glutamic acid as occurs after immunization with (T, G)-A--L. This is probably due to intramolecular competition between the Pro--L region and the (T, G) moiety of (T, G)-Pro--L. It is not likely that the (T, G) peptides are "hidden" within the (T, G)-Pro--L macromolecule. Therefore no antibodies specific for the (T, G) determinants are elicited upon immunization with this immunogen, because antibodies to (T, G)-A--L cross-react well with (T, G)-Pro--L.

Genetic analysis of the immune response to (T, G)-Pro--L carried out in high-responder SJL and DBA/1 low-responder mice as well as in the F_1 hybrids, F_2 generation and backcross mice is shown in Fig. 1. The response of the (SJL × DBA/1) F_1 mice was intermediate between the responses of the parental strains, but it is closer to that of the high-responder mice. About 25% of the [(SJL × DBA/1) F_1 × (SJL × DBA/1) F_1] F_2 mice were low responders to (T, G)-Pro--L whereas all the others were intermediate and high responders. Similarly, (SJL × DBA/1) F_1 × SJL and (SJL × DBA/1) F_1 × DBA/1 backcross mice were intermediate and high responders and intermediate and low responders, respectively. These results indicate that the response potential to (T, G)-Pro--L is a unigenic dominant trait (Ir-3), as was found for the three polypeptides derived from A--L. In contrast to responses controlled by the Ir-1 gene, however, no linkage was found between the ability to respond to (T, G)-Pro--L and the H-2 strain (16).

Definite genetic control of the specificity of the antibodies produced against the same polypeptide antigen in two genetically different strains of mice was observed in the immune response to poly(Phe, Glu)-poly(Pro)--poly(Lys) [abbreviated (Phe, G)-Pro--L] (16). This control is the result of the interaction of two gene loci, each of which affects the response to (Phe, G)-Pro--L. Table 1 shows that DBA/1 and SJL mice respond equally well to (Phe, G)-Pro--L, although DBA/1 mice produce antibodies to the (Phe, G) part of the polypeptide and SJL mice make antibodies to the Pro--L region. There is no linkage between the ability to respond to the Pro--L portion of (Phe, G)-Pro--L and H-2, whereas the response potential for the (Phe, G) part of the same immunogen is closely linked to H-2 (16). These results indicate that different loci control the immune response to antigenic determinants of different amino acid composition and are in agreement with the hypothesis that inbred animals can produce similar quantities

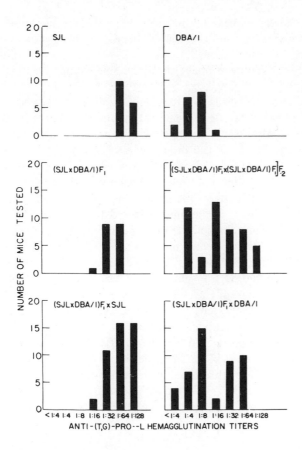

Figure 1. The antibody response of mice immunized with (T, G)-Pro--L and assayed with immunogen-coated SRBC according to the passive microhemagglutination technique. The horizontal axis is a $\log_2$ plot of hemagglutination titer, whereas the vertical axis gives the number of animals exhibiting a given antibody titer.

Table 1. Immune responses of SJL and DBA/1 mice to (Phe, G)-Pro--L

Mouse Strain	Secondary Responses[a] Assayed by		
	(Phe, G)-Pro--L	(T, G)-Pro--L	(Phe, G)-A--L
SJL	8.0	8.0	2.5
DBA/1	9.0	3.0	8.0

[a]Average $\log_2$ of hemagglutination titer. Antigen-coated SRBC used for the hemagglutination tests.

of antibodies against a natural complex immunogen, due to the fact that these antigens possess many determinants and that the various responding strains produce antibodies with different specificities.

Recently, we studied the immune response of mice to a new synthetic polypeptide, poly(His, Glu)-poly(Pro)--poly(Lys), designated (H, G)-Pro--L (17). This polymer is a potent immunogen that provokes an immune response controlled by several different genes and therefore provides a good model for natural immunopotent antigens. Most of the mouse strains studied responded well to (H, G)-Pro--L, but the specificities of the antibodies elicited in the various strains differed (Table 2). Strains C3H/HeJ and AKR/Cu, which are high responders to (H, G)-A--L, produced antibodies specific mainly to the (H, G) determinant, whereas SJL/J mice elicited antibodies specific to the Pro--L moiety exclusively. A third group of strains, including BALB/c, DBA/2, and DBA/1, responded to determinants characterized by a combination of the above specificities. Their high response could not be attributed to one of these specificities alone because anti(H, G)-Pro--L antibodies elicited in these strains did not cross-react well with either (H, G)-A--L or (T, G)-Pro--L (Table 2). The response potential to the (H, G) part of the immunogen was found to be linked to the major histocompatibility (H-2) locus of the mouse, as was reported for (H, G)-A--L, whereas the capacity to produce antibodies to the polyproline region of (H, G)-Pro--L or to determinants in which the (H, G) and Pro--L moieties participate was found not be linked to H-2. Thus, the ability to respond to (H, G)-Pro--L is controlled by at least three immune response genes (18).

Table 2. Immune responses of inbred mouse strains to (H, G)-Pro--L

Mouse Strain	H-2 Type	Secondary Responses[a] Assayed by		
		(H, G)-Pro--L	(H, G)-A--L	(T, G)-Pro--L
C57BL/6	b	4.75	1.1	3.0
C3H.SW	b	6.0	2.6	3.1
BALB/c	d	6.3	1.4	2.2
DBA/2	d	6.2	2.6	2.8
AKR/Cu	k	9.0	6.5	4.6
C3H/HeJ	k	8.5	6.0	4.0
DBA/1	q	6.9	2.0	2.2
SJL	s	5.3	2.3	5.5

[a] Average $\log_2$ of hemagglutination titer. Antigen-coated SRBC used for the hemagglutination tests.

Immune response potential to a polypeptide composed of D-amino acids.
The contribution to antigenicity of the optical configuration of the component
amino acids in a macromolecule has been studied in the last few years in rabbits
(14, 19–21) and mice (22–24). Immunogenicity differences between polypeptides composed of L-amino acids and those composed of D-amino acids and in
the need for cell-to-cell cooperation to provoke an efficient immune response to
these immunogens (25) were attributed to the metabolic fate of the polypeptides
of D and L configurations (26, 27).

To find out the extent of the role played by the optical configuration of the
amino acids comprising an antigen in the genetic control of immune responsiveness to the immunogens under investigation, we tested the immune response
potential of different inbred mouse strains to poly(DTyr, DGlu)-poly(DPro)--
poly(DLys). Table 3 gives the antibody responses of mice to this polypeptide,
as well as to the responses of the same mouse strains to the L-amino acid form,
poly(Tyr, Glu)-poly(Pro)--poly(Lys). The results presented in Table 4 show differences in the response patterns of inbred strains to the two immunogens.
Thus, DBA/1 and SWR mice, which are low responders to the immunogen composed of L-amino acids are high responders to poly(DTyr, DGlu)-poly(DPro)--
poly(DLys). On the other hand, C57Bl/6 mice are the lowest responders to
poly(DTyr, DGlu)-poly(DPro)--poly(DLys), whereas they respond well to
poly(Tyr, Glu)-poly(Pro)--poly(Lys). These results indicate not only that the
amino acid sequences in an immunogen play an important role in the genetic
control of the immune response, but they also illustrate that the optical configuration of the amino acids composing the immunogenic macromolecule is
important in determining the pattern of the immune responses of inbred strains
of mice (24, 28).

Table 3. Immune response of inbred mouse strains to multichain poly(Tyr, Glu)-
polyprolines composed of L or D amino acids

Mouse Strain	H-2	Secondary Antibody Response[a] to	
		poly(Tyr, Glu)- poly(Pro)--poly(Lys)	poly(DTyr, DGlu)- poly(DPro)--poly(DLys)
C57BL/6	b	5.0	2.2
C3H.SW	b	5.5	5.6
DBA/2	d	2.0	2.8
AKR/Cu	k	_[b]	7.7
C3H/HeJ	k	5.2	6.1
DBA/1	q	2.0	4.9
SWR	q	1.4	5.4
SJL	s	7.5	3.8

[a] Average $\log_2$ of hemagglutination titer. Antigen-coated SRBC used for the
hemagglutination tests.
[b] Not done.

Table 4. Contribution of thymocytes and bone-marrow cells to the antibody responses to (T, G)-A–L as a function of mouse strain

Number of Cells Transplanted (10^6)		Percentage of Responses Detected in Syngeneic Recipients[a]		
Thymus	Marrow	C3H.SW (high responder)	C3H/HeJ (low responder)	SJL/J (low responder)
100	0.5	40 (10)[b]	15 (30)	14 (14)
100	2.0	45 (11)	21 (14)	21 (44)
100	4.0	58 (34)	41 (17)	27 (11)
100	8.0	68 (35)	43 (21)	36 (25)
100	20.0	89 (9)	67 (12)	36 (14)
100	30.0		88 (8)	38 (16)
0.25	20.0	31 (16)	29 (7)	–
1.0	20.0	50 (10)	50 (12)	20 (15)
2.5	20.0	53 (15)	56 (16)	29 (14)
5.0	20.0	54 (13)	67 (12)	0 (12)
10.0	20.0	64 (11)	65 (17)	16 (19)

[a] Donor-derived responses showing $\log_2$ of hemagglutination titers higher than 2 in recipient sera.
[b] Number of recipients tested shown in parentheses.

CELLULAR ANALYSIS OF THE GENETIC CONTROL OF ANTIBODY RESPONSES

In order to understand the mechanisms responsible for control of immunity, it is necessary to establish whether the genetic factors involved have a cellular basis and, if so, to determine its nature. Our knowledge of the cellular aspects of immune phenomena has been increased in recent years by the realization that cooperation between functionally distinct populations of immunocompetent cells appears to be a prerequisite for specific antibody production (29-33). Using the limiting-dilution cell-titration technique, we have established that low responsiveness to the different immunopotent regions in synthetic polypeptides can be correlated with a reduced number of stimulated splenic antigen-sensitive units (34-36). The limiting-dilution approach is based on the fact that a relatively low number of cells injected into a group of heavily irradiated syngeneic recipients will generate donor-derived responses that are detected in only a portion of the recipients and that there is an increase in the fraction of responding hosts proportional to the increase in the number of cells injected. Plotting the percentage of positive responses in recipients as a function of the number of cells transferred gives a curve that conforms to the predictions of the Poisson model. Therefore this statistical approach has been used to estimate the frequencies of immunocompetent precursors relevant for the immunogens studied in spleen (34-36), bone marrow, and thymus (36-39).

The spleens of mice are considered to contain immunocompetent cells of both thymus and marrow origin, and because a direct correlation was found between immune response potential and the relative numbers of splenic antigen-sensitive units, it was important to determine whether the genetic defects responsible for low responsiveness are associated with a thymus- or marrow-derived cell population. Irradiated SJL and DBA/1 mice were injected with graded inocula of syngeneic marrow cells (5×10^4-2×10^7) to which an excess of 10^8 syngeneic thymocytes had been added. The recipients were immunized with (Phe, G)-Pro--L. Because this polypeptide possesses two immunopotent regions (15), the (Phe, G) portion to which the DBA/1 strain is a high responder and SJL mice are low responders and the Pro--L region to which the SJL strain responds well and DBA/1 mice respond poorly, precursor cell frequency for both specificities was compared within the same strain. The results, which are graphically illustrated in the upper portions of Figs. 2 and 3, indicate that the low response of the SJL strain to the (Phe, G) specificity and of DBA/1 mice to the Pro--L region can be attributed to a fivefold reduced number of monospecific precursors in the bone marrow of these two strains. On the other hand, injecting irradiated mice

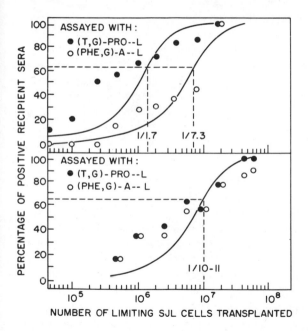

Figure 2. Positive sera in SJL recipients assayed with (Phe, G)-A–L (O) or (T, G)-Pro–L (●) after irradiation and injection with (Phe, G)-Pro–L and either 10^8 thymocytes and graded numbers of bone-marrow cells (upper) or 2×10^7 bone-marrow cells and graded numbers of thymocytes (lower) from syngeneic unimmunized donors. The ratios indicate precursor cell frequency ($\times 10^{-6}$). [Reprinted with permission from Mozes and Shearer, *Current Topics in Microbiology and Immunology*, **59**, 167 (1972). Copyright by Springer-Verlag.]

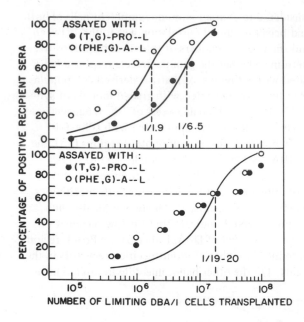

Figure 3. Positive sera in DBA/1 recipients assayed with (Phe, G)-A–L (O) or (T, G)-Pro–L (●) after irradiation and injection with (Phe, G)-Pro–L and either 10^8 thymocytes and graded numbers of bone-marrow cells (upper) or 2×10^7 bone-marrow cells and graded numbers of thymocytes (lower) from syngeneic unimmunized donors. The ratios indicate precursor cell frequency ($\times 10^{-6}$). [Reprinted with permission from Mozes and Shearer, *Current Topics in Microbiology and Immunology*, **59**, 167 (1972). Copyright by Springer-Verlag.]

with a mixture of graded inocula of thymocytes (0.5-100 $\times 10^6$) and 2×10^7 syngeneic bone-marrow cells together with (Phe, G)-Pro--L did not give rise to detectable differences in the frequency of relevant thymocytes for either specificity in these two mouse strains (Figs. 2 and 3, lower portions). Thus, thymus cells are not involved in generating the low responses observed in poor-responder mouse strains for these two specificities on (Phe, G)-Pro--L (37).

The response potential to (Phe, G), which is H-2 linked, can be tested when the specificity is attached to a different carrier, namely, multichain poly(DL-alanine). Limiting-dilution experiments were performed using (Phe,G)-A--L as the immunogen. In this case, the DBA/1 strain was the high responder and SJL mice were the low responders (Table 1). The results illustrated in Fig. 4 show that, in addition to the fivefold differences detected in the marrow dilutions (upper portion of Fig. 4), striking differences were detected for the thymocyte dilutions in high and low responders to (Phe, G)-A--L (lower portion of Fig. 4) (36). The observation that the genetic defect to (Phe, G) in SJL mice was reflected only in the number of marrow precursors when these peptides were attached to poly(L-proline), whereas the defect was reflected in the number of

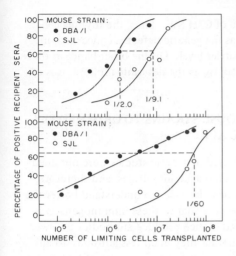

Figure 4. Positive sera in SJL (○) or DBA/1 (●) recipients assayed with (Phe, G)-A–L after irradiation and injection with (Phe, G)-A–L and either 10^8 thymo-cytes and graded numbers of bone-marrow cells (upper) or 2×10^7 bone-marrow cells and graded numbers of thymocytes (lower) from syngeneic unimmunized donors. The ratios indicate precursor cell frequency ($\times 10^{-6}$). [Reprinted with permission from Mozes and Shearer, *Current Topics in Microbiology and Immunology*, **59**, 167 (1972). Copyright by Springer-Verlag.]

both marrow and thymus immunocompetent cells when (Phe, G) was attached to poly(DL-alanine), stressed the importance of the chemical structure of the entire immunogenic macromolecule in the expression of these genetic defects for a given specificity at the cell level.

Strain-SJL mice do not respond to any of the multichain synthetic polypep-tides derived from multichain poly(DL-alanine) nor to natural determinants attached to A--L (2). It was of interest, therefore, to investigate, at the cell level, the immune response to another polypeptide built on A--L, (T, G)-A--L, in SJL mice compared to other low-responding strains that respond normally to anti-genic determinants attached to the A--L carrier. Results of limiting-dilution experiments using C3H.SW mice as high responders and C3H/He and SJL mice as low responders are given in Table 4. When the limiting cell type was of mar-row origin, the response frequencies to (T, G)-A--L observed in C3H/He low responders were considerably lower than those obtained for C3H.SW. The per-centage of responses in SJL recipients was lower than that obtained in C3H/He mice and did not exceed 36%. These results indicate that the genetic defect in the immune response potential of C3H/He and SJL mice to (T, G)-A--L is re-flected in the marrow-derived precursors. Limiting dilutions of thymocytes (Table 4) demonstrated a helper-cell defect in the low-responder SJL strain only and not in C3H/He poor responders, because the response frequencies for

C3H/He thymocytes were similar to those of C3H.SW. Thus, the thymus cell seems to be of crucial importance, as far as the genetic defect is concerned, in those cases in which the defect is at the carrier level. On the other hand, in those situations in which the genetic defect is strictly at the determinant level, it is reflected mainly in bone-marrow cells.

ANTIBODY RESPONSE OF INBRED MOUSE STRAINS TO ORDERED PEPTIDES OF TYROSINE AND GLUTAMIC ACID

The synthetic polypeptides commonly used in genetic studies were obtained by random polymerization of N-carboxyamino acid anhydrides, and therefore, they possess different sequence combinations. In order to understand better the genetic regulation of the immune response potential to these immunogens, it is important to establish whether low-responder mice produce antibodies to fewer determinants than the high responders, whether they elicit less antibody to all the determinants, or whether the antibodies produced by them are of lower affinity. It was desirable, therefore, to prepare a series of ordered peptides that would represent most of the possible combinations in the random polypeptides. Three-ordered tetrapeptides of tyrosine and glutamic acids were prepared and attached to multichain poly(DL-alanine) and multichain poly-L-proline (39). Tyrosine and glutamic acid were chosen for the ordered peptides because the random polypeptide (T, G)-A--L has been the most extensively used in studies of the genetic control of immune responsiveness. The immune response potential to these ordered peptides was followed in five inbred mouse strains that include low and high responders to (T, G)-A--L. The results, shown in Table 5, indicate that only one of the three antigenic determinants, namely, T-T-G-G, resembles the random peptide (T, G) in the pattern of immune responses that it elicited. Thus, C3H.SW and C57BL/6 mice, which are high responders to (T, G)-A--L, responded well to T-T-G-G on A--L, whereas C3H/HeJ, AKR/Cu and SJL mice, the low responders to (T, G)-A--L, responded poorly to the ordered (T-T-G-G)-A--L as well. Furthermore, the antibodies produced against (T, G)-A--L cross-reacted well with (T, G)-A--L and (T, G)-Pro--L, suggesting that T-T-G-G is an important determinant of (T, G)-A--L. That no significant antibodies specific to T-T-G-G were detected upon immunization with (T-T-G-G)-Pro--L (Table 5) confirms the similarity of this ordered peptide to the random (T, G), because immunization with the random (T, G)-Pro--L provokes antipolyproline antibodies exclusively.

In inhibition experiments of hemolytic plaque-forming cells to (T, G)-A--L, 50% inhibition was obtained with (T, G)-A--L at a molar concentration of 10^{-6}, whereas the same percentage of inhibition was observed with a concentration of 10^{-7} (T-T-G-G)-Pro--L (40). The fact that (T-T-G-G)-Pro--L is a more efficient inhibitor of the specific reaction with (T, G)-A--L than the homologous antigen in which the T-T-G-G determinant is less concentrated is compatible with the idea that T-T-G-G is a major determinant of (T, G)-A--L.

Table 5. Immune responses of inbred mouse strains to synthetic peptides of known sequence [from Mozes et al. (39)]

Immunogen	Assaying Antigen	Secondary Antibody Responses[a] of				
		C3H.SW	C57BL/6	AKR/Cu	C3H/HeJ	SJL/J
(T-T-G-G)-A--L	(T-T-G-G)-A--L	9.0	_b	2.0	2.0	2.0
(T-T-G-G)-A--L	(T-T-G-G)-Pro--L	8.0	9.0	2.0	2.0	1.0
(T-T-G-G)-A--L	(T,G)-A--L	5.6	5.7	2.0	2.0	_b
(T-T-G-G)-A--L	(T,G)-Pro-L	5.0	4.5	_b	_b	_b
(T-T-G-G)-Pro--L	(T-T-G-G)-Pro--L	3.0	3.0	4.0	4.0	6.0
(T-T-G-G)-Pro--L	(T-T-G-G)-A--L	2.0	2.0	2.0	2.0	2.0
(T-T-G-G)-Pro--L	(T,G)-Pro--L	3.0	3.0	4.0	4.0	5.0
(T-T-G-G)-Pro--L	(T,G)-A--L	<2	<2	<2	<2	<2
(T-G-T-G)-A--L	(T-G-T-G)-A--L	6.2	5.0	7.0	2.0	2.0
(T-G-T-G)-A--L	(T-G-T-G)-Pro--L	5.7	3.6	4.7	3.0	2.0
(T-G-T-G)-A--L	(T,G)-A--L	<2	<2	<2	<2	<2
(T-G-T-G)-Pro--L	(T-G-T-G)-Pro--L	5.5	3.4	4.5	4.0	4.5
(T-G-T-G)-Pro--L	(T-G-T-G)-A--L	3.0	2.0	4.0	2.3	2.0
(G-T-T-G)-A--L	(G-T-T-G)-A--L	3.2	3.2	3.6	2.4	2.0
(G-T-T-G)-A--L	(G-T-T-G)-Pro--L	2.0	3.6	3.6	2.5	3.0
(G-T-T-G)-Pro--L	(G-T-T-G)-Pro--L	3.0	4.0	5.0	3.0	4.6
(G-T-T-G)-Pro--L	(G-T-T-G)-A--L	3.0	2.2	4.0	2.6	4.0
(G-T-T-G)-Pro--L	(T, G)-A--L	<2	<2	<2	<2	<2

[a] Average $\log_2$ of hemagglutination titer. Antigen-coated SRBC used for the hemagglutination tests.
[b] Not done.

The immune response potentials to T-G-T-G and G-T-T-G attached to the multichain polypeptides were different from those observed after immunization with the random (T,G)-A--L and (T,G)-Pro--L. As can be seen in Table 5, AKR/ Cu mice, which are low responders to (T,G)-A--L, responded well to (T-G-T-G)- A--L and (T-G-T-G)-Pro--L. Moreover, SJL/J mice, which responded poorly to all the previously tested determinants that had been conjugated either with A--L or Pro--L (1, 2), responded to G-T-T-G when immunized with (G-T-T-G)-Pro--L. No cross reaction was detected between the antibodies provoked against either T-G-T-G or G-T-T-G and (T,G)-A--L (Table 5), so it can be concluded that these determinants are of minor importance or do not exist at all in the random polypeptide.

It is not yet known whether the low hemagglutination titers found in C3H/ HeJ and AKR/Cu mice to (T-T-G-G)-A--L are due to low amounts of antibodies or to antibodies with low affinity. Although no affinity measurements could be done on antibodies elicited to the random (T,G)-A--L due to the possible

heterogeneity of the antigenic determinants before it was established that
T-T-G-G is an important determinant of (T,G)-A--L, these studies now can be
performed.

The studies summarized here illustrate the usefulness of synthetic immunogens that possess a limited number of antigenic determinants in the detection
and interpretation of genetic variations in immune responses, which are not
easily observable with complex natural antigens. Studies with ordered peptides
of known sequences are especially important in elucidating the nature of the
genetic defect at the molecular level. A detailed analysis of the genetic regulation of immune responsiveness at the cellular level permitted the assignment of
a role to the various cell types in the expression of the genetic control.

ACKNOWLEDGMENTS

Part of the work reported here was supported by Agreement 06-035 and
Grant No. 1 RO1 AI11405-01 from the National Institutes of Health, U. S.
Public Health Service.

REFERENCES

1.   H.O. McDevitt and B. Benacerraf, *Advanc. Immunol.,* **11**, 31 (1969).
2.   E. Mozes and G.M. Shearer, *Current Topics in Microbiology and Immu-
     nology,* **59**, 167 (1972).
3.   B.B. Levine, A. Ojeda, and B. Benacerraf, *J. Exp. Med.,* **118**, 953 (1963).
4.   S. Ben-Efraim and P.H. Maurer, *J. Immunol.,* **97**, 577 (1966).
5.   B. Benacerraf, H.G. Bluestein, I. Green, and L. Ellman, *Progr. Immunol.,*
     1, 485 (1971).
6.   I. Green, W.E. Paul, and B. Benacerraf, *J. Exp. Med.,* **123**, 859 (1966).
7.   I. Green, W.E. Paul, and B. Benacerraf, *J. Exp. Med.,* **127**, 43 (1968).
8.   M. Sela, S. Fuchs, and R. Arnon, *Biochem. J.,* **85**, 223 (1962).
9.   S. Fuchs and M. Sela, *Biochem. J.,* **93**, 566 (1964).
10. H.O. McDevitt and M. Sela, *J. Exp. Med.,* **122**, 517 (1965).
11. H.O. McDevitt and M. Sela, *J. Exp. Med.,* **126**, 969 (1967).
12. H.O. McDevitt and M.L. Tyan, *J. Exp. Med.,* **128**, 1 (1968).
13. H.O. McDevitt and A. Chinitz, *Science,* **163**, 1207 (1969).
14. J.-C. Jaton and M. Sela, *J. Biol. Chem.,* **243**, 5616 (1968).
15. E. Mozes, H.O. McDevitt, J.-C. Jaton, and M. Sela, *J. Exp. Med.,* **130**, 493
     (1969).
16. E. Mozes, H.O. McDevitt, J.-C. Jaton, and M. Sela, *J. Exp. Med.,* **130**, 1263
     (1969).
17. S. Shaltiel, E. Mozes, and M. Sela, *Israel J. Chem.,* **10**, 627 (1972).
18. E. Mozes, S. Shaltiel, and M. Sela, *Eur. J. Immunol.,* in press.
19. T.J. Gill III, H.W. Kunz, J.J. Gould, and P. Doty, *J. Biol. Chem.,* **239**,
     1107 (1964).
20. T.J. Gill III, H.W. Kunz, and D.S. Papermaster, *J. Biol. Chem.,* **242**, 3308
     (1967).

21. Y. Stupp and M. Sela, *Biochem. Biophys. Acta,* **140**, 349 (1967).
22. C.A. Janeway, Jr., and M. Sela, *Immunology,* **13**, 29 (1967).
23. J. Medlin, J.H. Humphrey, and M. Sela, *Folia Biologica,* **16**, 145 (1970).
24. E. Mozes, M. Sela, and H.O. McDevitt, *Eur. J. Immunol.,* **3**, 1 (1973).
25. M. Sela, E. Mozes, and G.M. Shearer, *Proc. Nat. Acad. Sci. U.S.,* **69**, 2696 (1972).
26. J. Medlin, J.H. Humphrey, and M. Sela, *Folia Biologica,* **16**, 156 (1970).
27. T.J. Gill III, *Current Topics in Microbiology and Immunology,* **54**, 19 (1971).
28. A.-M. Schmitt, E. Mozes, and M. Sela, unpublished results.
29. H.N. Claman, E.A. Chaperon, and R.F. Triplett, *J. Immunol.,* **97**, 828 (1966).
30. G.F. Mitchell and J.F.A.P. Miller, *J. Exp. Med.,* **128**, 821 (1968).
31. R.B. Taylor, *Transplant. Rev.,* **1**, 114 (1969).
32. R.M. Gorczynski, R.G. Miller, and R.A. Phillips, *J. Exp. Med.,* **134**, 1201 (1971).
33. M. Feldmann, *J. Exp. Med.,* **135**, 1049 (1972).
34. E. Mozes, G.M. Shearer, and M. Sela, *J. Exp. Med.,* **132**, 613 (1970).
35. G.M. Shearer, E. Mozes, and M. Sela, *J. Exp. Med.,* **133**, 216 (1971).
36. G.M. Shearer, E. Mozes, and M. Sela, *J. Exp. Med.,* **135**, 1009 (1972).
37. E. Mozes and G.M. Shearer, *J. Exp. Med.,* **134**, 141 (1971).
38. L. Lichtenberg, E. Mozes, G.M. Shearer, and M. Sela, *Eur. J. Immunol.,* in press.
39. E. Mozes and M. Sela, *Proc. Nat. Acad. Sci. U.S.,* **71**, 1574 (1974).
40. E. Mozes, M. Schwartz, and M. Sela, *J. Exp. Med.,* in press.
41. M. Schwartz, E. Mozes, and M. Sela, unpublished results.

GENETIC CONTROL OF IMMUNE RESPONSE TO BRANCHED POLYPEPTIDES WITH DEFINED OLIGOPEPTIDE SIDE CHAINS IN RATS AND MICE

E. LIEHL, E. RÜDE,[*] M. MEYER-DELIUS, and E. GÜNTHER, Max-Planck-Institut für Immunbiologie, 78 Freiburg-Zähringen, Germany

SYNOPSIS: The immune response of mice and rats to the branched synthetic polypeptide poly(Tyr,Glu)-poly(Ala)–poly(Lys), denoted (T,G)-A–L, is specifically controlled by a dominant, autosomal immune response gene(s). In an attempt to identify the oligopeptides responsible for the specificity of this control, various analogs of (T,G)-A–L were synthesized in which the (T,G) copolymers at the end of the poly(DL-alanine) side chains of A–L were replaced by defined oligotyrosine peptides Tyr_1 to Tyr_6. In rats the immune response to some of the resulting polypeptides also was genetically controlled, but the difference between the high- and low-responder strains was smaller than in the case of (T,G)-A–L. Genetic control of the antibody response in mice to the Tyr_2- and Tyr_4-carrying polypeptides was as effective as that to (T,G)-A–L. Therefore it appears that oligotyrosine peptides, for instance, dityrosine, possibly in combination with alanine residues of the backbone, are typical of structures that play a role in the recognition of these antigens in high-responder strains.

INTRODUCTION

Linear and branched synthetic polypeptides have been used extensively as model antigens for studies on the structural requirements for immunogenicity. Of particular interest in this respect was the finding that the immune response of guinea pigs, mice, and rats to a variety of such antigens is specifically controlled by autosomal, dominant genes termed immune response genes (Ir genes). A characteristic feature of several of these Ir genes is their linkage with the major histocompatibility gene complex of the respective species (1). This feature was first demonstrated for the so-called Ir-1 genes of mice, which determine the ability to produce relatively high concentrations of antibody upon immunization with one of three related synthetic polypeptides: poly(Tyr,Glu)-poly(Ala)-- poly(Lys), abbreviated (T,G)-A--L; poly(His,Glu)-poly(Ala)--poly(Lys), or (H,G)-A--L; and poly(Phe,Glu)-poly(Ala)--poly(Lys), or (Phe,G)-A--L (2). These polypeptides, developed by Sela and coworkers (3, 4), are derived from a common branched polymer denoted A--L, in which poly(DL-alanine) side chains were attached to the ε-amino groups of a poly(L-lysine) backbone.

---

[*]Address correspondence to E. Rüde, Max-Planck-Institut für Immunbiologie, 78 Freiburg-Zähringen, Stübeweg 51, Germany

Then the side chains of A--L were elongated with short copolymers containing L-glutamic acid combined with L-tyrosine, L-histidine, or L-phenylalanine (Fig. 1).

In rats the immune response to the same antigens was found to be controlled also by dominant Ir genes linked to the major histocompatibility system (in rats termed the H-1 system) (5). All available data indicate that a close genetical and functional homology exists between these genes and the Ir-1 genes in mice (6). The linkage of Ir-genes to the major H system of various species has been described not only for synthetic antigens, but also for an increasing number of natural protein antigens (1).

Another characteristic of Ir genes is their specificity with respect to a particular antigen. This specificity is illustrated in Table 1 by the response pattern of various rat strains to the branched polypeptide antigens. Thus, rats carrying the H-1^a allele are high responders to (T,G)-A--L, but low responders to (H,G)-A--L, although the opposite is true for rats of the H-1^W type. It is reasonable to assume, therefore, that Ir genes may affect the specific recognition of certain determinants of these antigens by immunocompetent cells. Despite extensive studies, however, the cell type(s) in which Ir genes are expressed have not yet been unequivocally identified. Several experiments indicate that low responsiveness due to these genes cannot be explained by the lack of the respective antibody-producing cells. Instead, most of the data fit the hypothesis that Ir genes linked to the major H system control functions involved in the stimulation of T lymphocytes by these antigens (7, 8). On the other hand, cell-transfer experiments carried out by Mozes and Shearer (9) by the limiting-dilution method provide arguments for the expression of Ir-1 genes in bone-marrow-derived cells, but not in thymus cells that contain the precursors of T lymphocytes.

Little is known about the type and the specificity of the determinants that effectively stimulate T lymphocytes. In view of the possible involvement of T lymphocytes in the control of immune responsiveness by Ir-1 genes, it is of great interest to characterize the structures or determinants responsible for the recognition of these polypeptide antigens in high-responder strains more precisely and to compare them with the determinants to which antibodies are produced.

In order to carry out such studies it was necessary to synthesize analogs of polypeptides such as (T,G)-A--L with more defined side chains and to test whether they would still be subject to the same type of genetic control of immune responsiveness as (T,G)-A--L itself. The overall structure of (T,G)-A--L is relatively simple. Due to the method of synthesis by copolymerization, however, the amino acid sequence of its side chains is rather complex and a variety of different oligopeptide determinants is certainly present. Therefore, the possibility exists that in addition to one oligopeptide determinant, several related ones may act synergistically and may be responsible for genetic control.

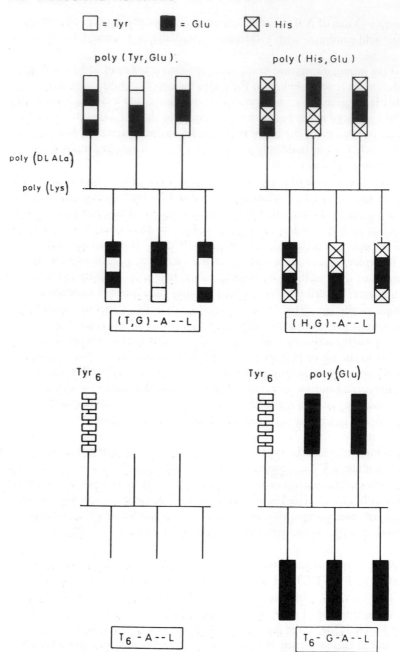

Figure 1. Schematic representation of branched synthetic polypeptide antigens. [Reprinted with permission from Rüde *et al.*, *Behring Institute Mitteilungen*, **53**, 53 (1973). Copyright by Behringwerke Aktiengesellschaft.]

Table 1. Response pattern of various inbred rat strains to three synthetic polypeptide antigens

Strain	H-1 type	(T,G)-A–L	(H,G)-A–L	(Phe,G)-A–L
L .AVN	a	high	low	high
L .BDV	d	low	low	high
LEWIS	l	low	medium	high
L .WP	w	low	high	–

METHODS

Antigens. Oligopeptides were prepared by classical stepwise methods using the *N*-hydroxysuccinimide ester for activation of the carboxyl group (10, 11). The γ-carboxyl group of glutamic acid and the phenolic hydroxyl group of tyrosine were always protected as tert-butyl esters or ethers, respectively. For chain elongation the carbobenzoxy group was used as N-terminal blocking group whereas the N terminus of the final oligopeptide that was to be coupled to the A--L backbone was protected by the tert-butyloxycarbonyl residue. After these fully protected oligopeptides were coupled to some of the terminal amino groups of A--L, the tert-butyl groups were removed by treatment with trifluoroacetic acid (12). In some cases L-glutamic acid oligopeptides were attached to the remaining free poly(DL-alanine) side chains (Fig. 1) by polymerization of the N-carboxyanhydride of L-glutamic acid (13) before deprotection. The amino acid composition of the various polypeptides is presented in Table 2. The same batch of the branched backbone A--L was used as starting material for the attachment of oligopeptides. Its molecular weight was 96,500; on the average, each molecule contained 75 poly(DL-alanine) side chains.

Table 2. Analytical properties of some polypeptides with defined oligopeptide side chains

Compound	Molar Ratio of Amino Acid Residues					Number of Oligopeptides per 10 Side Chains of A–L
	Lys	Ala	Tyr	Glu	Gly	
(T,G)-A--L/242	1	15,1	1,60	2,66		
Tyr-Glu-Ala-Gly-A–L	1	15,1	0,77	0,81	0,81	8,1
T_6-A–L/1-17	1	15,0	0,98			1,6
T_6-G-A--L/1-17	1	15,1	0,96	4,69		1,6 (Tyr_6) 8,4 ($Glu_{5,5}$)
T_4-A--L/101	1	16,3	1,39			3,5
T_4-A--L/1-17	1	16,4	0,21			0,5
T_2-A--L/2-17	1	14,6	0,86			4,3
T_1-A--L/1-17	1	14,5	0,54			5,4

Immunization. Rats of the two H-1 congenic inbred strains L .AVN(H-1[a]) and L .BDV(H-1[d]) received a primary injection with 200 μg of the polypeptide in complete Freund's adjuvant and a secondary injection of 200 μg in saline on day 21. Serum samples taken 10 days after secondary immunization were assayed for antibodies. Mice of the H-2 congenic strains C3HDiSn(H-2[k]) and C3H .SWSn(H-2[b]) (Jackson, Bar Harbor) were immunized in the same way, except that doses of 10 μg were used for priming and for secondary challenge.

Antibody assay. To determine antibody titers, the polypeptide antigens were labeled with ^{125}I (14) and were used in a coprecipitation radioimmunoassay at a final concentration of 4×10^{-8} with respect to the average polypeptide side chains (15). From the titration curves of individual sera, the serum titers at which 33% of the antigen was bound (titers$_{33}$) were obtained graphically. The results are expressed as geometric means of titers$_{33}$ within one group of animals.

DISCUSSION OF RESULTS

Synthesis of antigens. For initial studies with more defined analogs of (T,G)-A--L, various oligopeptides were attached to the side chains of the A--L backbone in order to replace the (T,G) copolymers that are composed, on the average, of about 1 to 6 tyrosine and 3 glutamic acid residues per side chain.

From the specificity of the response pattern to the branched polypeptide antigens in different strains of rats and mice (Table 1), it is obvious that the aromatic amino acid residues, [tyrosine in the case of (T,G)-A--L] play an essential role in genetically controlled recognition. This conclusion is also supported by the observation that the coupling of only a few *p*-azobenzenearsonate groups to the tyrosine residues of (T,G)-A--L (1.7 per 10 side chains) leads to a considerable decrease in immunogenicity in high-responder mice (16). On the other hand, the attachment of the same number of 2,4-dinitrophenyl groups to the N termini of (T,G)-A--L does not influence the large difference in the antibody response between high- and low-responder mice (17).

Therefore all oligopeptides tested in the present study contained at least one tyrosine residue. In addition to the tetrapeptide Tyr-Glu-Ala-Gly, which proved to be completely ineffective, these included a series of oligotyrosine peptides up to the hexamer. In the case of Tyr-Glu-Ala-Gly, the available side chains of A--L were almost fully substituted, whereas the oligotyrosines, especially the higher ones, were bound to only about 15 to 30% of the A--L side chains in order to retain solubility in water (Table 2).

The attachment of prefabricated, purified oligopeptides to the A--L backbone, compared to the possibility of growing these peptides directly on A--L as a macromolecular support, has the advantage that there will be no ambiguity with regard to the amino acid sequence of the final product. In addition, different peptides may be bound to the same backbone, which may be important in view of the possibility that several determinants may contribute to the specificity of genetic control. For instance, peptides composed, on the average, of 5.5 glutamic acid residues were bound by polymerization of the N-carboxyanhydride of glutamic acid to the remaining free poly(DL-alanine) side chains of T_6-A--L. The resulting polypeptide (T_6-G-A--L; see Fig. 1) had a similar amino acid composition to (T,G)-A--L, but tyrosine and glutamic acid peptides were arranged on different side chains.

Immunogenicity of (T,G)-A--L analogs in rats. The results of immunization experiments carried out in rats with some of the (T,G)-A--L analogs are presented in Fig. 2. Significantly higher responses were observed in L .AVN rats than in L .BDV rats to polypeptides containing hexa- and tetratyrosine. However, the titer differences between the two strains (7, 5-fold for T_6-A--L/1-17 and 15-fold for T_4-A--L/101) were considerably smaller than for (T,G)-A--L (87-fold).

The finding that T_4-A--L/1-17 carrying only 0.5 Tyr_4 peptides per 10 side chains is a much weaker immunogen mainly in L .AVN rats than T_4-A--L/101 with 3.5 Tyr peptides per 10 side chains suggests that the density of the oligotyrosines plays an important role for the recognition of these antigens in high-responder animals. Studies are under way to test this possibility.

The attachment of glutamic acid peptides to the free poly(DL-alanine) side chains of T_6-A--L/1-17 was found to have no effect on immunogenicity. Immunization of several other rat strains with T_6-G-A--L/1-17 gave a response pattern identical to that of (T,G)-A--L [i.e., only H-1[a] type strains that are high responders to (T,G)-A--L also gave a significantly higher response to T_6-G-A--L]. Preliminary experiments on the specificity of the antibodies produced against T_6-G-A--L indicate that a considerable fraction of the antibodies is directed against the glutamic acid peptides. Thus, glutamic acid influences the serological specificity of this polypeptide, but does not change its immunogenicity.

After immunization with T_2-A--L/2-17, both L .AVN and L .BDV rats showed only traces of antibodies. Similarly, Tyr-Glu-Ala-Gly-A--L was only weakly immunogenic in ten different rat strains including L .AVN. In the case of Tyr-Glu-Ala-Gly-A--L, this lack of immunogenicity could eventually be due to the presence of a glycine residue that links Tyr-Glu-Ala to the A--L backbone and that was originally used to avoid any racemization during the coupling reaction.

Because L .AVN and L .BDV rats, which were used for the experiments presented in Fig. 2, are congenic (i.e., they have the same genetic background, but differ with respect to the H-1 gene complex), it can be concluded that the response to T_6-A--L/1-17, T_6-G-A--L/1-17 and T_4-A--L/101 is controlled by a genetic factor linked to the H-1 system. No sex-linked differences of the antibody response to these antigens have been observed.

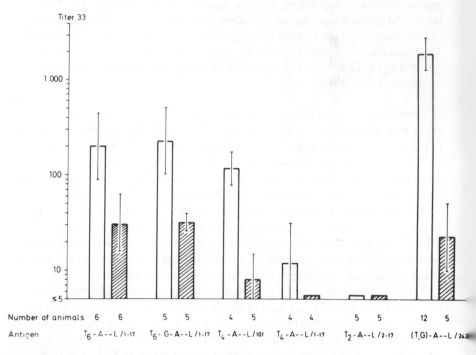

Figure 2. Secondary antibody response of the H-1 congenic rat strains
L .AVN (H-1[a]) and L .BDV (H-1[d]) to (T,G)-A–L and to branched polypeptides with oligotyrosine side chains. Open columns, L .AVN; hatched
columns, L .BDV; geometric means x:standard deviations of titers of
individual animals are given.

Immunogenicity of (T,G)-A--L analogs in mice. Like the rats in the
experiments described above, two strains of mice congenic with respect to the
major H system (the H-2 system) were immunized with three of the polypeptides carrying oligotyrosine peptides (Fig. 3). Differences in the antibody
response of the two strains, therefore, must be due to a gene or genes linked
to the H-2 gene complex.

In the case of T_4-A--L/101, it is remarkable that C3H low-responder mice
showed quite high titers to this antigen, but an additional 68-fold titer increase
due to a genetic factor associated with the H-2[b] allele was observed in C3H.SW
high-responder mice. Although the antibody titers to T_2-A--L/2-17 were generally much lower, the relative difference between C3H (H-2[k]) and C3H.SW(H-2[b])
was even larger (89-fold). A smaller but still significant difference (8-fold) was
seen in mice of the two strains immunized with T_1-A--L/1-17.

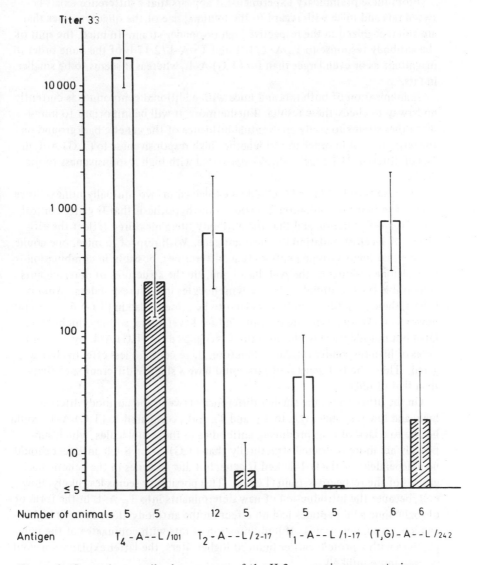

Figure 3. Secondary antibody response of the H-2 congenic mouse strains C3H .SW(H-2^b) and C3H(H-2^k) to (T,G)-A–L and to branched polypeptides with oligotyrosine side chains. Open columns, C3H .SW; hatched columns, C3H; geometric means x:standard deviations of titers of individual animals are given.

From these preliminary experiments it appears that a difference exists between rats and mice with regard to the minimal size of the oligotyrosines that are still recognized in the respective high-responder strain. In mice, the split of the antibody response to T_4-A--L/101 and T_2-A--L/2-17 is of the same order of magnitude as or even larger than for (T,G)-A--L, whereas it seems to be smaller in rats.

Immunization of both rats and mice with additional compounds is currently underway to check these results. Furthermore, it will be important to immunize other strains to study an eventual influence of the genetic background on the response and in order to test whether high responsiveness to (T,G)-A--L in rats of different H-1 type is always associated with high responsiveness to the analogs.

The results obtained so far could be explained in two mutually nonexclusive ways. The first interpretation is based on the hypothesis that Ir genes control specific T-cell function and that the antibody titers measured reflect the efficiency of T-cell stimulation by these antigens. With respect to mice, one could assume that oligotyrosine peptides (e.g., dityrosine), possibly in combination with alanine residues of the A--L backbone, are the structures or determinants responsible for the stimulation of T lymphocytes in high responders. Among others, these structures can be expected to be present also in (T,G)-A--L. In rats, however, these oligotyrosine peptides induce lower titers, so they might be related but not identical to the structures recognized on (T,G)-A--L by T lymphocytes of high-responder rats, and therefore, these cells are less effectively triggered. Thus, the Ir-1 gene(s) of rats could have a slightly different specificity than that of mice.

On the other hand, the smaller difference between the antibody titers of high- and low-responder rats to T_4- and T_6-A--L, compared to (T,G)-A--L, could be due to a lack of cells producing antibodies to these molecules, which supposedly are more restricted structurally than (T,G)-A--L. Such an effect should be independent of the H-1–linked Ir gene, but due to genes in the genetic background of the respective strain (Lewis). This possibility seems less likely, however, because the introduction of new determinants into T_6-A--L in the form of oligoglutamic acid peptides had no effect on the antibody titers in both high- and low-responder strains. If immunization of rats with conjugates of the polypeptides with a protein carrier induced higher titers, the latter explanation would be even more unlikely.

ACKNOWLEDGMENTS

This work was supported by the Deutsche Forschungsgemeinschaft. The excellent technical assistance of M. Müller and R. Thumb is gratefully acknowledged.

REFERENCES

1. B. Benacerraf and H.O. McDevitt, *Science,* **175,** 273 (1972).
2. H.O. McDevitt and A. Chinitz, *Science,* **163,** 1208 (1969).
3. M. Sela and R. Arnon, *Biochim. Biophys. Acta,* **40,** 382 (1960).
4. M. Sela, S. Fuchs, and R. Arnon, *Biochem. J.,* **85,** 223 (1962).
5. E. Günther, E. Rüde, M. Meyer-Delius, and O. Stark, *Transpl. Proc.,* **5,** 1467 (1973).
6. E. Rüde, E. Günther, and E. Liehl, *Behring Institute Mitteilungen,* **53,** 53 (1973).
7. B. Benacerraf, *Ann. d'Immunologie* (Inst. Pasteur), **125c,** 143 (1974).
8. H.O. McDevitt, K.B. Bechtol, J.H. Freed, G.J. Hämmerling, and P. Lonai, *Ann. d'Immunologie* (Inst. Pasteur), **125c,** 175 (1974).
9. E. Mozes and G.M. Shearer, *Current Topics Microbiol. Immunol.,* **59,** 167 (1972).
10. G.W. Anderson, J.E. Zimmermann, and F.M. Callahan, *J. Am. Chem. Soc.,* **86,** 1839 (1964).
11. F. Weygand, D. Hoffmann, and E. Wünsch, *Z. Naturforschg.,* **216,** 426 (1966).
12. H. Kappeler and R. Schwyzer, *Helv. Chim. Acta,* **44,** 1136 (1961).
13. R. Hirschmann, H. Schwam, R.G. Strachan, E.F. Schoenewaldt, H. Barkemeyer, S.M. Miller, J.B. Conn, V. Garsky, D.F. Veber, and R.G. Denkewalter, *J. Am. Chem. Soc.,* **93,** 2746 (1971).
14. E. Rüde, M. Meyer-Delius, and M.-L. Gundelach, *Eur. J. Immunol.,* **1,** 113 (1971).
15. E. Günther, E. Rüde, and O. Stark, *Eur. J. Immunol.,* **2,** 151 (1972).
16. A. Otto, E. Rüde, unpublished results.
17. J. Wrede, E. Rüde, R. Thumb, and M. Meyer-Delius, *Eur. J. Immunol.,* **3,** 798 (1973).

THE CHEMICAL BASIS AND POSSIBLE ROLE OF CARBAMINO HOMEOSTATIC MECHANISMS

JON S. MORROW, RUTH S. GURD, and FRANK R.N. GURD, Department of Chemistry and The Medical Sciences Program, Indiana University, Bloomington, Ind. 47401

SYNOPSIS: The development of ^{13}C Fourier transform nuclear magnetic resonance spectrometers has made possible the direct observation of carbamino derivative formation in a variety of compounds containing free amino groups. Studies of reactions with [8-homolysine]-vasopressin and [5-isoleucine]-angiotensin II have shown that both can be converted approximately 30% into the carbamino form under physiological conditions. Earlier reports that the action of angiotensin is potentiated by injection in alkaline solution fit the hypothesis that the carbamino derivative is more active than the free peptide. The properties of the carbamino reaction scheme suitable to a role in homeostatic mechanisms are described.

INTRODUCTION

CO_2 homeostasis. The physiological importance of CO_2 homeostasis was recognized many years ago (1, 2). Henriques recognized the importance of the formation of carbamino compounds derived by the combination of CO_2 with hemoglobin amino groups (3). Although the primary role of the erythrocyte in CO_2 transport was recognized to be the catalysis by its carbonic anhydrase (4) of the hydration of CO_2 to H_2CO_3 followed by rapid ionization to bicarbonate, the presence of carbamino derivatives was positively established (5). It was further recognized that the carbamino derivatives have a role in controlling the oxygen affinity of hemoglobin (5). Stadie and O'Brien, among others, studied the hemoglobin system in detail (6) and showed the generality of the chemistry with the simple glycine derivative as their model (7).

The CO_2-bicarbonate buffer system is dominant in extracellular fluid (1, 2, 8). The concentrations of its components are regulated by the combined actions of lungs, kidneys, and the elements of cellular respiration, linked by the highly adaptable and responsive cardiovascular system. Significantly, physiological regulation is more immediately responsive to changes in CO_2 pressure than to changes in O_2 pressure (2, 9). Not only are the lungs responsible for maintaining vascular CO_2 levels, they also modulate (10–12) other mechanisms controlling tissue perfusion by adjustment of cardiac output and peripheral resistance and, thus, ultimately the local levels of both O_2 and CO_2 (10–16).

Hastings and coworkers have shown that homeostasis with respect to CO_2 pressure is more finely maintained even than homeostasis with respect to pH (17). Hastings has drawn attention to a number of metabolic pathways that are sensitive specifically to CO_2 pressure or to bicarbonate ion (18). One is tempted to speculate on regulatory roles for each of these species at the level of key enzymes in central pathways in both animals and plants (19-23).

Carbamino compounds. Chemically, CO_2 can make its presence felt through formation of a carbamino derivative. The reaction involves the unprotonated form of the amine. The carbamic acid formed is strong enough to dissociate a proton completely in the physiological pH range (24-27). We suggest that the free amine and the carbamate forms could be related as agonist and antagonist with respect to a receptor site; furthermore, carbamino formation may directly alter the conformation of some molecules so that their physiological effects differ.

The requirement for the free base form of the amine as a reactant places a lower limit on the pH stability range of the carbamino compound. The fact that CO_2 rather than HCO_3^- is the reactive species places an upper limit on the pH range (6, 7); this limit may be breached temporarily before equilibrium is approached (7). For a given amine with proton dissociation constant K_z there will be a carbamino formation equilibrium described by a constant K_c, where

$$K_c = \frac{[\text{R-NHCOO}^-]\,[\text{H}^+]}{[\text{R-NH}_2]\,[CO_2]} \tag{1}$$

The mole fraction, Z, of the total amine in the carbamino form is given by

$$Z = \frac{K_c K_z\,[CO_2]_{\text{free}}}{K_c K_z\,[CO_2]_{\text{free}} + K_z\,[\text{H}^+] + [\text{H}^+]^2} \tag{2}$$

For a given value of K_c, the lower the pK_z of the amino group the lower the pH at which the derivative will form. For given values of K_c and K_z the formation of the carbamino derivative is a function of CO_2 concentration and pH. It is on this basis that carbamino formation is proposed as the monitoring system for CO_2 homeostasis.

Observation by ^{13}C NMR. Carbon-13 nuclear magnetic resonance (NMR) spectroscopy is well adapted to observing the CO_2-carbamino equilibria (28-32). Carbamino-derivative formation can be observed in any molecule containing a free amino group of the usual basicity (32) and also dissolved CO_2 and the bicarbonate-carbonate component.

The latter interconvert sufficiently rapidly that they are represented by a single resonance. Exchange of CO_2 with the amine is often too slow to affect the NMR observations (32). The interaction of the CO_2 adduct with the neighboring parts of the molecule can be analyzed (32). The use of CO_2 enriched with respect to ^{13}C has made possible the study of the carbamino derivatives of both chains of hemoglobin (29-31, 33).

Vasoactive peptides. The present work follows directly a general study of the CO_2 adducts of amino acids, peptides, and sperm-whale myoglobin (32). Both ^{13}C and ^1H spectrometers were used to evaluate chemical-shift parameters, establish methods of quantification, and explore conformational consequences. As expected, the effect of low values of the pK_z of the amino group was to facilitate carbamino formation at a relatively low pH. The pK_c for glycine agreed with the earlier estimate of Stadie and O'Brien (7). The lower pK_z characteristic of the terminal amino group of most proteins, for example, shifts the curve of carbamino formation into the physiological pH range (32). All such compounds with unblocked amino groups that we have examined have formed the carbamino derivatives near physiological pH. Most values of pK_c fall in the range 4.6 to 5.2, in agreement with the findings of Caplow (26). It follows [Eq. (2)] that a significant proportion of the unblocked α-amino groups of peptides and proteins will convert to the carbamino form, even at physiological pCO_2 values. For many combinations of pK_z with pK_c, the value of Z [Eq. (2)] is sensitive to small variations in pH and pCO_2.

There are many potential avenues for exploring the concept that control of CO_2 homeostasis through formation of carbamino compounds is central to general regulatory mechanisms. Some of these are now being investigated. A particularly attractive one is that of the reactivity of the vasoactive principles involved in cardiovascular and fluid adjustments. These principles are elaborated or activated in situations, both physiological and pathological, characterized by altered pCO_2. Compounds to consider are those involved in the renin-angiotensin system, kinins, vasopressins, histamine, serotonin, and other, less well-characterized compounds. Carbamino formation could lead to activation or inactivation of the compound itself, affect synthesis or degradation, or even convert it into an antagonist.

The present preliminary report deals with members of the vasopressin and angiotensin series. Although the antidiuretic action of the vasopressins in man appears functionally dominant, the development of synthetic analogs has shown that the α-amino group is associated relatively with the vasopressor activity and that its absence is compatible with full antidiuretic function (34). M. Bodanszky kindly supplied us with [8-homolysine]-vasopressin, an analog of the natural hormone that has indistinguishable activity in the usual assays (35). M. Manning generously made available several pairs of analogs, with and without the intact α-amino group. The pair [4-valine]-arginine vasopressin (M.M. 5-73) and [1-deamino-4-valine]-arginine vasopressin (M.M. 5-37) is included in this report (34). F.M. Bumpus very generously provided [5-isoleucine]-angiotensin II (36).

EXPERIMENTAL PROCEDURE

The ^{13}C Fourier transform NMR measurements were made as described previously (37). The peptides were dissolved in degassed water and sufficient sodium bicarbonate and carbonate, either labeled with ^{13}C (0.87 mole fraction ^{13}C) (Bio-Rad) or at natural abundance, was added to achieve the desired pH. The sealed samples were equilibrated at least one hour before the spectral measurements commenced. Total carbonates, degree of enrichment of the equilibrated CO_2 gas, and pH values were measured at the conclusion of the NMR experiments. Chemical shifts are expressed in parts per million (ppm) upfield of external CS_2, with dioxane (2%) included in the experimental solutions as an internal standard at 126.2 ppm. The temperature of measurement was $30 \pm 1°C$.

RESULTS

Vasopressins. The spectrum of [8-L-homolysine]-vasopressin equilibrated with $^{13}CO_2$ at pH 6.62, total carbonates 52 mM, showed the carbamino resonance at 30.0 ppm, the bicarbonate-carbonate resonance at 32.6 ppm, and the CO_2 resonance at 68.3 ppm. The chemical shifts of the various resonances are in line with past experience (30–32). The shift for the carbamino derivative is somewhat distinctive, in that precisely the same chemical shift has been observed for the carbamino adduct of proline and is near that of the α chain of hemoglobin (32, 33). The natural abundance signals from the peptide itself were comparable to the measurements of Lyerla and Freedman on lysine vasopressin (38). The chemical shifts generally fit with previous studies on small peptides (37, 39–41). Spectra were also obtained at pH 7.22 and pH 7.62. The chemical shift of the carbamino carbon was invariant with pH.

The chemical shift of the carbamino adduct again was found to be 30.0 ppm for [4-valine]-arginine vasopressin at pH 8.24. With [1-deamino-4-valine]-arginine vasopressin at pH 7.44 no carbamino adduct was observed. These results confirm the assignment of the resonance at 30.0 ppm in the present study to the carbamino adduct of the α-amino group and also confirm that the reaction with the homolysine vasopressin was confined to the α-amino group under these conditions.

The carbamino signals for the [8-L-homolysine]-vasopressin obtained at each pH value were integrated digitally and related to the integral values observed for the bicarbonate-carbonate and dissolved CO_2 resonances, and the known concentrations of total carbonates. Equation (2) was adapted to deal directly with pairs of experiments to yield an expression independent of total amine concentration leading to values of K_c. The value of pK_z for the experimental conditions was estimated to be 6.4 (42). A 40° pulse width was used for the measurements to minimize errors resulting from differences in relaxation time for the various resonances. Small corrections were applied assuming spin-lattice relaxation times of 1 sec for the carbamino resonance and 5 sec for the other resonances. The estimated value of pK_c was 5.0. We expect to refine the estimate after studying a larger sample.

Angiotensin II. The sample of [5-isoleucine]-angiotensin II was studied in a similar way. The experimental conditions involved a 90° pulse width with a recycle time of 2.5 seconds at 30° and pH 7.39, pH 7.84, and pH 8.05. The carbamino adduct resonance was identified at 29.5 ppm, upfield of that seen in the carbamino adduct of free aspartic acid, 28.8 ppm (32), and it corresponds to the NH_2-terminal amino acid in angiotensin. The carbamino signals at the three pH values were integrated as before. Separate measurements established that the relaxation time of bicarbonate was similar to that of the carbamino resonance in these solutions (30), hence no correction for differential saturation was necessary. The value of pK_c was approximately 4.7, assuming a pK_z of 7.6. This last value was estimated from the value for an asparagine analog (43), adjusted for the difference in pK_z between aspartic acid and asparagine (44).

DISCUSSION

Stability ranges of carbamino derivatives. The approximate value of pK_c found for [8-homolysine]-vasopressin, 5.0, taken with the pK_z of 6.4, was used to compute the mole fraction, Z, of the peptide in the carbamino form [Eq. (2)]. The results are shown in Fig. 1. The curves for various values of Z are identified on the right-hand ordinate. The left-hand ordinate shows bicarbonate concentration, the abscissa pH values. The curves that are concave upward represent a series of isobars for the pressure of CO_2.

The graph in Fig. 1 shows that for a normal arterial pCO_2 of 40 mm Hg, 24 mM bicarbonate, and pH 7.41, the mole fraction of the vasopressin in the form of the carbamino derivatives is near 0.3. For a normal venous pCO_2 of 47 mm Hg, 28 mM bicarbonate, and pH 7.38, the mole fraction of the carbamino derivative rises somewhat, but remains near 0.3. The small A-V difference in Z reflects primarily the low value of pK_z, which minimizes the pH-sensitivity. Compounds of this sort, therefore, are relatively more sensitive to variations in pCO_2 than to variations in pH within the physiologic range.

The upper left region of Fig. 1 corresponds to conditions of respiratory acidosis, and the lower right to respiratory alkalosis. The adjustments in both directions from the normal are accompanied by changes in Z that are similar to those that will accompany metabolic acidosis (lower left) and metabolic alkalosis (upper right), in each case a reflection of the accompanying pCO_2 alteration.

The corresponding figure for the angiotensin-II system is shown in Fig. 2. The values of pK_z and pK_c used here are 7.6 and 4.7, respectively. The main difference from Fig. 1 is the displacement of the curves for the carbamino mole fraction, Z, to higher pH values, a consequence of the higher pK_z that makes Z a sensitive function of both pH and pCO_2. The value of Z under the typical arterial and venous conditions is designated as before. With the widely studied analog, [1-asparagine]-angiotensin II, pK_z will be lower (43) and pK_c probably will be lower also. Given these conditions, the [1-asparagine]-angiotensin II would be expected to form the carbamino derivative more fully over a range of conditions than does the compound studied here.

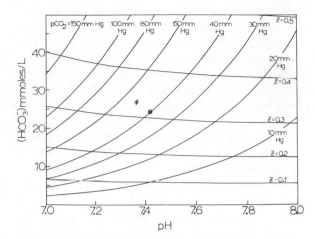

Figure 1. Relationship between pH, pCO_2, bicarbonate concentration, and mole fraction (Z) of [8-homolysine] vasopressin in the carbamino form, under conditions approximating the physiological range. Approximate normal arterial conditions, ⊕; approximate mixed venous conditions ϕ.

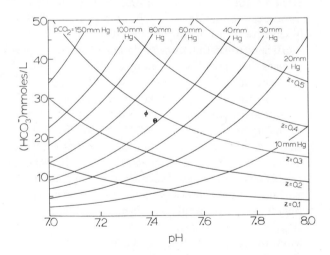

Figure 2. Relationship between pH, CO_2, bicarbonate concentration, and mole fraction (Z) of [5-Ile] angiotensin II in the carbamino form. Normal arterial conditions, ⊕; venous conditions ϕ. The enhanced steepness of the carbamino isobars (Z) is due primarily to the increased value of pK_Z for angiotensin relative to that of vasopressin.

By comparison with the vasopressin (Fig. 1), the higher pK_z of the natural angiotensin compound together with the lowered pK_c, yields an equilibrium that is highly sensitive to the conditions of metabolic acidosis or alkalosis, but much less sensitive to the corresponding respiratory derangements. These two vasoactive compounds demonstrate in their interactions with CO_2 and hydrogen ions molecular mechanisms that respond with different emphasis within a common pattern, potentially a basis for separate but concurrent homeostatic mechanisms.

The observation discussed earlier that raising the pH abruptly favors carbamino formation by the preferential reaction of CO_2 with the basic form of the amino group may provide an explanation for the potentiation of the pressor effects of angiotensin in alkaline solution. Paladini and coworkers showed that increases of $66 \pm 11\%$ in pressor activity resulted from adjustment of the peptide sample to pH 11.7 before injection (45–47). These observations suggest that the carbamino form of the angiotensin is more active than the simple amine. Angiotensin II may be reaching receptors before it is relatively slowly converted back to the descarbamino form. The time ranges of both the pressor response and the partial decarboxylation are probably measured in seconds (13, 26). The design of the experiment appears to rule out effects of such factors as sodium loading (48). Compounds characterized by still higher values of pK_z show less tendency in the physiological range to form carbamino derivative, but the derivative formation becomes increasingly sensitive to pH in contrast to pCO_2. The level of the carbamino derivative, if sufficient to be functionally significant, would represent the concentration of a species that is readily altered by small pH changes, as well as by pCO_2 changes, yet persists for long periods on the time scale of molecular motions such as those involved in the processes of binding to a receptor site or producing a conformational change. In principle, it is not essential that a large proportion of the amine in question should be converted to the carbamino derivative, only that it should be functionally distinct and effective in this form. Histamine, with its higher pK_z, is illustrative of a molecule that forms a carbamino derivative (32) with a sharp pH dependence of the (small) value of Z in the physiological range of pH and pCO_2. The same argument applies to some neurotransmitters that we have found will form carbamino compounds (32).

Attention should be drawn to the ability of phenylhydrazine, with a pK_z below the range for alkylamines (26), to form a carbamino derivative even below pH 6. Hence, Z for a given pCO_2 should show reduced sensitivity to pH. It is interesting that substituted hydrazines are among the therapeutic inhibitors of monamine oxidase (49).

Some further considerations. Addition of CO_2 to the amino group with dissociation of a proton has obvious consequences. The change in charge at the NH_2 terminus of a peptide can have strong effects (37–41). The release of the proton and formation of the anion result from the arrival of CO_2, a fat-soluble compound that moves freely through membranous barriers. This characteristic provides a mechanism for a redistribution of molecular species in a relatively hydrophobic environment.

The suggestion has already been made that the carbamino group should have some binding capacity toward alkaline-earth metal ions, such as Mg^{2+} and Ca^{2+}, that bind poorly to the amino group proper (31). It is interesting that in studying the carbonic anhydrase model behavior of metal-peptide complexes (50), Breslow (51) observed that the Zn^{2+} complex gave way to carbamino derivative formation by the peptide, whereas the more stable Cu^{2+}-peptide complex remained unchanged and maintained its catalytic role. It may be pertinent in this connection to point out that storage of some neurotransmitters, namely, epinephrine and serotonin, is effected as complexes with Mg^{2+} or Ca^{2+} and ATP (52). These amines could easily be in the carbamino form in such mixed complexes.

Interaction involving steric and electrostatic interactions between the $-CO_2^-$ moiety and the side chain components has been observed for carbamino proline and for carbamino peptides with valine in the NH_2-terminal position (32) with vasopressin and possibly with angiotensin II. The NH_2-terminal residue may not be rigidly fixed in relation to the rest of the molecule (53). This situation may be altered by carbamino formation, as suggested by the anomalous chemical shift (32). Considerable attention has been given to internal hydrogen bonding and other possible stabilization of particular conformations (43, 53, 54), and it is our intention to explore rotational motions in these compounds by ^{13}C NMR relaxation techniques.

The parallel between the CO_2 and the cyanate reactions should not be overlooked, particularly because the cyanate derivative of oxytocin has shown reduced or even inhibitory activity (55).

The emphasis on small molecular hormones and mediators should not divert attention from the possibilities for regulation at the enzyme and receptor level. Carbamino formation appears to be as widespread among proteins as among peptides. These include serum albumin (6), several hemoglobins (6, 30, 31, 33, 56), sperm-whale myoglobin (30, 32), and β-glutamate-aspartate transaminase (57). The converse cases in which α-amino groups in peptides or proteins are blocked may have special significance (58). The various hypothalamic regulatory factors should be examined with these contrasts in mind (59).

Close consideration should be given to mechanisms in which the slow equilibration of increased levels of CO_2 with bicarbonate may allow transient processes to exaggerate changes in carbamino levels intracellularly, in interstitial fluids or in the vascular system. Clues to such mechanisms may lie in the strikingly restricted distribution of carbonic anhydrase (60). Care must be taken to recognize direct bicarbonate-ion effects (18) whose role may easily be greater than has been recognized.

Overview and conclusions. The analytical discrimination and sensitivity of the ^{13}C NMR method makes it possible to show the widespread formation of carbamino derivatives, particularly with enriched $^{13}CO_2$. At the pCO_2 values in the body, the free α-amino groups of proteins, peptides, and amino acids containing them will be partly converted to the carbamino form.

A consequence of the interplay of pCO_2, pH, pK_c, and pK_z is that, in the body, many compounds with free α-amino groups exist in more than one form, with possible further consequences for interactions of many kinds including binding, assembly and membrane permeability. As shown for hemoglobin, carbamino formation can have consequences for allosteric regulation of function (5, 25, 27, 33, 61).

Some competitive relationship between the two forms of a peptide is also easy to imagine as a possible means of amplification. Amplification through several cascading steps must be considered in addition. Finally, the patterns illustrated in Figs. 1 and 2 show the variability of response to both pCO_2 and pH that reflects differences in pK_c and, especially, pK_z between one amine compound and another. Concurrent response through two or more systems based on the same fundamental carbamino mechanism could permit very fine physiological regulation.

ACKNOWLEDGMENTS

The generosity and helpfulness of M. Bodanszky, F.M. Bumpus and M. Manning are gratefully acknowledged. We are most grateful for the encouragement and interest of A.B. Hastings and J.T. Edsall.

This is the 62nd paper in a series dealing with coordination complexes and catalytic activity of proteins and related substances. This work was supported by National Institutes of Health Research Grant HL-05556 and Training Grant GM 1046 (J.S.M.).

REFERENCES

1. L.J. Henderson, *Blood: A Study in General Physiology*, Yale University Press, New Haven, Conn., 1928.
2. C. Lovatt Evans, *Starling's Principles of Human Physiology*, Twelfth Edition, Lea and Febiger, Philadelphia, 1956, p. 729.
3. O.M. Henriques, *Biochem. Z.*, **200**, 1 (1928).
4. N.V. Meldrum and F.J.W. Roughton, *J. Physiol.*, **80**, 113 (1933).
5. F.J.W. Roughton, in *Handbook of Physiology*, Vol. 1, Section 3, W.O. Fenn and N. Rahn, eds., American Physiological Society, Washington, D.C., 1964, p. 767.
6. W.C. Stadie and H. O'Brien, *J. Biol. Chem.*, **117**, 439 (1937).
7. W.C. Stadie and H. O'Brien, *J. Biol. Chem.*, **112**, 723 (1936).
8. J.L. Gamble, *Chemical Anatomy, Physiology and Pathology of Extracellular Fluid*, Harvard University Press, Cambridge, Mass., 1950.

9. H. Wollman and R.D. Dripps, in *The Pharmacological Basis of Therapeutics*, Fourth Edition, L.S. Goodman and A. Gilman, eds., Macmillan, New York, 1970, p. 908.
10. H.O. Heinemann, *Fed. Proc.*, **32**, 1955 (1973).
11. S.I. Said, *Fed. Proc.*, **32**, 1972 (1973).
12. S.I. Said, in *Pathophysiology: Altered Regulatory Mechanisms in Disease*, E.D. Frohlich, ed., Lippincott, Philadelphia, 1972, p. 167.
13. W.W. Douglas, in *The Pharmacological Basis of Therapeutics*, Fourth Edition, L.S. Goodman and A. Gilman, eds., Macmillan, New York, 1970, p. 620.
14. G.E. Sander and C.G. Huggins, *Ann. Rev. Pharmacol.*, **12**, 227 (1972).
15. D.L. Wilhelm, *Ann. Rev. Med.*, **22**, 63 (1971).
16. A.M. Lefer, *Fed. Proc.*, **29**, 1836 (1970).
17. N.W. Shock and A.B. Hastings, *J. Biol. Chem.*, **112**, 239 (1935).
18. A.B. Hastings, *Ann. Rev. Biochem.*, **39**, 1 (1970).
19. A.B. Hastings and E.B. Dowdle, *Trans. Assoc. Amer. Physicians*, **73**, 240 (1960).
20. W.J. Longmore, A.B. Hastings, and E.S. Harrison, *Proc. Nat. Acad. Sci. U.S.*, **52**, 1040 (1964).
21. Y. Kaziro, L.F. Hass, P.D. Boyer, and S. Ochoa, *J. Biol. Chem.*, **237**, 1460 (1962).
22. T.G. Cooper, T.T. Tchen, H.G. Wood, and C.R. Benedict, *J. Biol. Chem.*, **243**, 3857 (1968).
23. M.D. Lane and D.R. Halenz, *Biochem. Biophys. Res. Comm.*, **2**, 436 (1960).
24. J.K.W. Ferguson, *J. Physiol.*, **88**, 40 (1936).
25. F.J.W. Roughton and L. Rossi-Bernardi, *Proc. Roy. Soc.* (London), **B164**, 381 (1966).
26. M. Caplow, *J. Am. Chem. Soc.*, **90**, 6795 (1968).
27. L. Rossi-Bernardi, M. Pace, F.J.W. Roughton, and L. Van Kempen, in *Carbon Dioxide: Chemical, Biochemical and Physiological Aspects*, R.E. Forster, J.T. Edsall, A.G. Otis, and F.J.W. Roughton, eds., NASA SP-188, Washington, D.C., 1969, p. 65.
28. A. Patterson, Jr., and R. Ettinger, *Z. Electrochem.*, **64**, 98 (1960).
29. N.A. Matwiyoff and T.E. Needham, *Biochem. Biophys. Res. Commun.*, **49**, 1158 (1972).
30. J.S. Morrow, P. Keim, R.B. Visscher, R.C. Marshall, and F.R.N. Gurd, *Proc. Nat. Acad. Sci. U.S.*, **70**, 1414 (1973).
31. F.R.N. Gurd, J.S. Morrow, P. Keim, R.B. Visscher, and R.C. Marshall, in *Protein-Metal Interactions*, M. Friedman, ed., Plenum Press, New York, in press.
32. J.S. Morrow, P. Keim, and F.R.N. Gurd, *J. Biol. Chem.*, in press.
33. J.S. Morrow and F.R.N. Gurd, in preparation.
34. M. Manning, L. Balaspiri, M. Acosta, and W.H. Sawyer, *J. Med. Chem.*, **16**, 975 (1973).
35. M. Bodanszky and G. Lindeberg, *J. Med. Chem.*, **14**, 1197 (1971).
36. M.C. Khosla, M.M. Hall, R.R. Smeby, and F.M. Bumpus, *J. Med. Chem.*, **16**, 829 (1973).

37. P. Keim, R.A. Vigna, J.S. Morrow, R.C. Marshall, and F.R.N. Gurd, *J. Biol. Chem.*, **248**, 7811 (1973).
38. J.R. Lyerla, Jr., and M.H. Freedman, *J. Biol. Chem.*, **247**, 8183 (1972).
39. F.R.N. Gurd, P. Keim, V. Glushko, P.J. Lawson, R.C. Marshall, A.M. Nigen, and R.A. Vigna, in *Chemistry and Biology of Peptides*, J. Meienhofer, ed., Ann Arbor Science Publishers, Ann Arbor, Mich., 1972, p. 45.
40. P. Keim, R.A. Vigna, R.C. Marshall, and F.R.N. Gurd, *J. Biol. Chem.*, **248**, 6104 (1973).
41. P. Keim, R.A. Vigna, A.M. Nigen, J.S. Morrow, and F.R.N. Gurd, *J. Biol. Chem.*, in press.
42. B.J. Campbell, F.S. Chu, and S. Hubbard, *Biochemistry*, **2**, 764 (1963).
43. T.B. Paiva, A.C.M. Paiva, and H.A. Scheraga, *Biochemistry*, **2**, 1327 (1963).
44. E.J. Cohn and J.T. Edsall, *Proteins, Amino Acids and Peptides as Ions and Dipolar Ions*, Reinhold, New York, 1943, p. 84.
45. A.C. Paladini, A.E. Delius, and M.T. Franze de Fernandez, *Biochem. Biophys. Acta*, **74**, 168 (1963).
46. A.L. Methot, P. Meyer, P. Biron, M.F. Lorain, G. Lagrue, and P. Milliez, *Nature* (London), **203**, 531 (1964).
47. L.C. Craig, E.J. Harfenist, and A.C. Paldini, *Biochemistry*, **3**, 764 (1964).
48. J.R. Blair-West and J.S. McKenzie, *Experientia*, **22**, 291 (1966).
49. C.R. Creveling and J.W. Daly, in *Biogenic Amines and Physiological Membranes in Drug Therapy*, J.H. Biel and L.G. Abood, eds., Marcel Dekker, New York, 1971, Part B, p. 355.
50. E. Breslow, in *The Biochemistry of Copper*, J. Peisach, P. Aisen, and W.E. Blumberg, eds., Academic Press, New York, 1966, p. 149.
51. *Ibid.*, see appended discussion to reference 50, p. 157.
52. K.H. Berneis, A. Pletscher, and M. DaPrada, *Brit. J. Pharmacol.*, **39**, 382 (1970).
53. M.P. Printz, H.P. Williams, and L.C. Craig, *Proc. Nat. Acad. Sci. U.S.*, **69**, 378 (1972).
54. J. Zimmer, W. Haar, W. Maure, H. Ruterjans, S. Fermandjian, and P. Fromageot, *Eur. J. Biochem.*, **29**, 80 (1972).
55. G.W. Bisset, B.J. Clark, I. Krejci, I. Polacek, J. Rudinger, *Brit. J. Pharmacol.*, **40**, 342 (1970).
56. J.V. Kilmartin and L. Rossi-Bernardi, *Biochem. J.*, **124**, 31 (1971).
57. J.S. Morrow, R. Harruff, W.T. Jenkins, and F.R.N. Gurd, unpublished results.
58. K. Narita, in *Protein Sequence Determination*, S.B. Needleman, ed., Springer-Verlag, New York, 1970, p. 83.
59. A.V. Schally, A. Arimura, and A.J. Kastin, *Science*, **179**, 341 (1973).
60. T.H. Maren, in *Oxygen Affinity of Hemoglobin and Red Cell Acid Base Status*, Alfred Benzon Symposium IV, P. Astrup and M. Rorth, eds., Munksgaard, Copenhagen, 1972, p. 418.
61. J.V. Kilmartin, J. Fogg, M. Luzzana, and L. Rossi-Bernardi, *J. Biol. Chem.*, **248**, 7039 (1973).

POLY(AMINO ACIDS) AS SUBSTRATES OF PROTEOLYTIC ENZYMES

A. YARON, Department of Biophysics, The Weizmann Institute of Science, Rehovot, Israel

SYNOPSIS: Poly(amino acids) were used as substrates for the detection and purification of new exopeptidases from bacteria. The enzymes, which were obtained in pure form and were characterized, include aminopeptidase P, clostridial aminopeptidase, dipeptidocarboxypeptidase, and a nonspecific dipeptidase. The possible application of the combined action of the two aminopeptidases to the sequential hydrolysis of polypeptides was demonstrated by accomplishing complete hydrolysis of bradykinin, a proline-rich nonapeptide. The concerted action of dipeptidocarboxypeptidase, which produces dipeptides from the carboxyl end, and of the dipeptidase, which consequently hydrolyzes the dipeptides formed, indicates an efficient mechanism by which polypeptides are hydrolyzed from the carboxyl end of the molecule. The unique nature of the peptide bond—in which proline is involved through its nitrogen—is pointed out, and the possible role of proline as a residue that protects terminal regions of polypeptides against the hydrolytic action of exopeptidases is discussed.

The highly specific mechanism by which proteolytic enzymes act upon protein substrates has been a subject of extensive investigation. The selective cleavage of susceptible peptide bonds in a protein permits not only a controlled catabolism, but also the formation of biologically active polypeptides, such as hormones and enzymes, from their inactive precursors (1).

The elucidation of the mechanism of the highly selective processes mentioned above requires model substrates and specific inhibitors, composed of polypeptide chains of varying amino acid composition. Low molecular weight peptides have been extensively used for specificity studies of proteolytic enzymes and the initial knowledge has grown as progress in synthetic methods allowed the preparation of larger peptides. It is evident that quite extensive regions of a polypeptide chain are recognized and specifically bound during proteolysis.

The observation that some peptide bonds, which are resistant to enzymic hydrolysis in a native protein, become susceptible upon denaturation, clearly demonstrates that in the interaction of a protein substrate with a proteolytic enzyme, macromolecular properties, such as conformation, play an important role. Model compounds of relatively high molecular weight became available with the development of synthetic poly(amino acids). When the first soluble poly(amino acid), namely, poly(L-lysine), was prepared, it was shown to be a good substrate for trypsin (2). Since then, poly(amino acids) have been extensively used for the characterization of the specificity of various proteolytic enzymes (3). The investigations were performed mainly with soluble polymers containing ionizable amino acid residues and with the nonionizable poly(Pro). Recently, soluble, sequence-ordered polymers made of amino acids that produce an insoluble homopolymer, such as $(Tyr\text{-}Ala\text{-}Glu)_n$ (4), $(Lys\text{-}Ala\text{-}Ala)_n$ (5), and $(Pro\text{-}Gly\text{-}Pro)_n$ (6), have also been used as substrates. A variety of macromolecular substrates thereby became available for specificity studies not only of proteolytic enzymes, but also of enzymes that modify certain amino acid residues in proteins after completion of the polypeptide biosynthesis (7).

Carboxy- and aminopeptidases are capable of attacking polypeptides specifically at the C- or N-terminal part of the polypeptide chain. The exact composition of the terminal region, however, further modulates the specificity requirements, and a variety of different exopeptidases designed to split off one or more terminal amino acid residues is obviously controlling the broad spectrum of very specific reactions. Poly(amino acids) are particularly useful for unraveling this "microspecificity."

THE USE OF POLY(AMINO ACIDS) AS A TOOL
FOR THE DETECTION OF NOVEL PROTEOLYTIC ENZYMES

The presence of proline residues in a polypeptide chain usually renders vicinal peptide bonds resistant to proteolysis. If the hydrolysis of such bonds is to occur, enzymes of the proper specificity must be involved in the process. A search for such enzymes requires proline-containing substrates of relatively high molecular weight, such as poly(Pro), poly(Pro-Gly-Pro), and their C- or N-terminal-substituted derivatives. Thus, incubation of poly(Pro) with a bacterial extract of *E. coli* B at pH 8.6 in the presence of Mn^{++} ions leads to the degradation of the polymer with the concomitant release of proline in the incubation solution. Because no oligopeptides were found at intermediate stages of the hydrolysis and because α-DNP-poly(Pro) was not degraded, the hydrolytic activity was attributed to an aminopeptidase that is capable of splitting Pro-Pro bonds, and we named it aminopeptidase P (8, 9).

Another enzymic activity was detected when poly(Pro-Gly-Pro) was incubated under similar conditions with the same bacterial extract, and proline and glycine were formed. Substitution at the N-terminal α-amine of the polymer had no effect on the rate of formation of the proline. On the other hand, esterification of its C-terminal carboxyl rendered the polymer resistant to hydrolysis, indicating degradation by an exopeptidase acting on the C-terminal region. Because only one equivalent of proline and glycine was formed, it seemed that a carboxypeptidase split the Gly-Pro and Pro-Gly links, but not the following Pro-Pro bond. Extensive purification through several steps resulted in a protein that appeared as a single band when analyzed by polyacrylamide gel electrophoresis. We subsequently found, however, that the protein consists of two enzymes: a dipeptidocarboxypeptidase (10), which releases the C-terminal glycylproline, and a dipeptidase (11), which, after additional purification, was shown to hydrolyze various dipeptides, including glycylproline. The very similar behavior of the two enzymes during ion-exchange chromatography, gel filtration, and even the highly discriminating polyacrylamide gel electrophoresis, indicates that the two enzymes are structurally very similar. Moreover, identical pH-activity curves have been found for the two purified enzymes, and it is reasonable to assume that in nature, too, the complementary action of these two enzymes represents an alternative pathway for the well-known hydrolysis of polypeptides by carboxypeptidases. It is pertinent to note that the combined action of the two enzymes did not cleave the Pro-Pro bond after removal of proline and glycine from poly(Pro-Gly-Pro). Such a bond is known to be resistant also to most carboxypeptidases.

The extracellular proteins produced by the anaerobic bacterium *Clostridium histolyticum* are a rich source of collagenase, which is known to act on peptide bonds in the proline-rich regions of collagen (12). We therefore tested this bacterium for enzymes capable of hydrolyzing proline-containing polypeptides and detected (13) an aminopeptidase that releases one equivalent of proline from poly(Pro-Gly-Pro) or from the corresponding methyl ester. The clostridial aminopeptidase was shown to release effectively various amino acids, including proline, from the N-terminal part of various polypeptides. The combined action of this clostridial aminopeptidase and the above-mentioned aminopeptidase P was found to be useful in the hydrolysis of proline-rich polypeptides, as exemplified below for bradykinin.

In the following sections, some properties of the exopeptidases detected through specific interaction with proline-containing poly(amino acids) are described.

Aminopeptidase P (8, 9). The enzyme was obtained in pure form in a procedure involving six purification steps. Then the amount of proline formed from poly(Pro) at pH 8.6 in the presence of Mn^{++} ions was determined.

In an attempt to use poly(Pro) to isolate the enzyme by affinity chromatography, a branched poly(amino acid) of the formula shown below was used (14). This polyvalent substrate, which carried an overall negative charge, was

$$\begin{bmatrix} -\text{Lys}- \\ | \\ (\text{Asp}_4) \\ | \\ (\text{Pro}_{10}) \end{bmatrix}_n$$

MP-Pro-Asp-P-Lys (where MP denotes "multipoly")

adsorbed at low ionic strength to a DEAE-cellulose column. A crude preparation of aminopeptidase P was applied and eluted with a linear gradient of salt concentration as described in Fig. 1. Comparison of the elution patterns obtained in the presence and absence of the polymer shows that although the enzyme was retarded by the polymeric substrate, the general protein elution pattern was the same. The electrostatic binding of the polymer to the adsorbent was not very strong, because it was found to be eluted with the enzyme. This demonstration

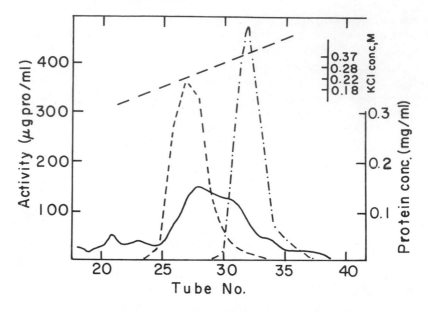

Figure 1. Elution pattern of a crude aminopeptidase P preparation, by ion-exchange chromatography on a column (0.8 X 28 cm): (– –) aminopeptidase P activity, as eluted from DEAE-cellulose; (–•–) aminopeptidase P activity, as eluted from DEAE-cellulose onto which the polyvalent substrate MPProAspPLys (20 mg) was adsorbed prior to chromatography; (———) total protein concentration (the same elution pattern was obtained in the presence and in the absence of the polyvalent substrate). In both experiments the salt gradient was linear from 0.1 M Tris buffer pH 8.6 (60 ml) to 0.7 M KCl in the same buffer (60 ml).

of tight binding of aminopeptidase P to the poly(Pro) side chains of the branched polymer is in accord with the results of specificity and inhibition studies that show that, for strong binding, a proline residue must be present in a polypeptide chain in position 2, following the N-terminal amino acid, which may also be a proline.

The specificity of aminopeptidase P was studied with a variety of peptides of different sizes, ranging from dipeptides such as Gly-Pro to large molecules such as reduced and carboxymethylated papain. The specificity was found to be strictly defined by two requirements: an N-terminal amino acid residue with a free α-amino group (which can be a secondary amine of proline) must be present and the second residue must be proline (not hydroxyproline). A typical example of the action of aminopeptidase P is the hydrolysis of bradykinin Arg-Pro-Pro-Gly-Phe-Ser-Pro-Phe-Arg (Fig. 2). One of the arginine residues is released rapidly, followed by a proline residue (one of the three proline residues present). No further hydrolysis occurs on prolonged incubation.

The strict specificity requirement for a proline residue led us to investigate some N-terminal proline-containing peptides as potential inhibitors of aminopeptidase P. Indeed these peptides acted as competitive inhibitors (see Table 1), the strongest one tested being Pro-Phe.

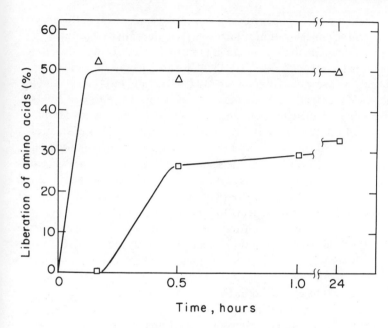

Figure 2. Time course of the action of aminopeptidase P (4.4 μg/ml) on brady-kinin (Arg-Pro-Pro-Gly-Phe-Ser-Pro-Phe-Arg, 0.65 mM) in 0.05 M veronal, pH 8.6 containing Mn-citrate (Mn^{++}, 2.5 × 10^{-3} M; citrate, 10^{-2} M) at 40°C; (△) ap-pearance of arginine; (□) appearance of proline.

Table 1. Kinetic parameters (aminopeptidase P)

Substrate or Competitive Inhibitor	K_m (M^{-1})	$K_i{}^a$ (M^{-1})	$k_3{}^b$ (sec^{-1})
Poly(Pro) (MW 6000)[c]	11,300	–	105
H-Pro-Pro-Ala-OH[d]	870	–	1210
H-Pro-Phe-Lys-OH[e]	–	8500	–
H-Pro-Phe-OH[e]	–	13,300	–
H-Pro-Ala-OH[e]	–	1000	–
H-Ala-Phe-Lys-OH[e]	–	<50[f]	–

[a] Reciprocal inhibition constant (1/K_i), obtained from Lineweaver–Burke plots. The composition of the reaction mixture was that of the standard assay. Substrate: Pro-Pro-Ala (10^{-3} to 10^{-4} M).
[b] Assumed MW 100,000.
[c] Enzyme concentration 0.3 μg/ml.
[d] Enzyme concentration 0.06 μg/ml.
[e] Enzyme concentration 0.057 μg/ml.
[f] No inhibition at 2.3 × 10^{-3} M peptide.

The specific activity of aminopeptidase P was found to depend on the enzyme concentration. As the concentration is increased, specific activity becomes lower, indicating a loss of activity through intermolecular interaction. In gel-filtration experiments, it was observed (9, 15) that the apparent molecular weight of the enzyme decreases from about 230,000 at 25 μg/ml (in the sample applied) to about 120,000 at 1.0 μg/ml, so it was assumed that only the 120,000 species is active. The calculated dependence of activity on enzyme concentration was compared to the experimental results and the dissociation constant of the system was calculated ($K = 10^{-9} M$). Further dissociation by denaturation was demonstrated (10) by polyacrylamide gel electrophoresis in the presence of sodium dodecylsulfate (SDS) and 2-mercaptoethanol. The result was a single component, the migration of which corresponds to a molecular weight of 60,000.

Although it is similar in many respects to other known aminopeptidases, aminopeptidase P is unique in its specificity and clears the way for the complete hydrolysis of polypeptides when used with other proteolytic enzymes. Stepwise degradation of polypeptides then becomes possible through use of aminopeptidase P in conjunction with another, nonspecific aminopeptidase, "clostridial aminopeptidase," as will be explained in the following section.

An aminopeptidase having a specificity similar to that of aminopeptidase P was partially purified (16) from the microsomal fraction of swine kidney. Preparations of prolidase (17), a dipeptidase specific for peptides of the type X-Pro and X-Hyp have been repeatedly reported (18–21) to be capable of splitting the X-Pro bond in larger peptides. It appears, therefore, that the aminopeptidase P activity is not restricted to the bacterial source, but occurs widely, providing the necessary step in polypeptide degradation needed for the catabolic, activating, and deactivating processes upon which various biologic processes depend.

Clostridial aminopeptidase (CAP) (13). The enzyme was purified 130 times in a two-step procedure (ion-exchange and gel-filtration chromatography) and shown to be homogeneous by polyacrylamide gel electrophoresis, immunodiffusion, and immunoelectrophoresis. The product thus obtained is devoid of endopeptidase activity, as demonstrated by its inability to digest the performic-acid-oxidized lysozyme. Furthermore, CAP does not hydrolyze N-blocked peptides such as Z-Ala$_3$, Z-Ala-Ala-Phe-Ala, and Ac-Gly-Phe-Ala.

Gel filtration of CAP on calibrated Sephadex G-200 indicates an apparent molecular weight of 340,000. On denaturation, a single component of molecular weight 63,000 is obtained, as indicated, by polyacrylamide gel electrophoresis in SDS and 2-mercaptoethanol.

The enzyme is inactive towards Pro-Gly-Pro-Pro when no metal ions are added or in the presence of Zn^{++}, Mg^{++}, Ba^{++}, Ca^{++}, Sr^{++}, or Cu^{++} at pH 8.6 in the 10^{-6}–$10^{-3} M$ range. Maximal activation is obtained in the presence of $5 \times 10^{-5} M$ Mn^{++}, and less activity is observed in the presence of Co^{++} (73%), Cd^{++} (15%), and Ni^{++} (12%) at their optimal concentrations. In the presence of $5 \times 10^{-5} M$ MnCl$_2$, complete inhibition is obtained with $10^{-5} M$ Zn^{++} or with $10^{-3} M$ Cu^{++}. A similar Zn^{++} effect is observed with aminopeptidase I

(22), but not with leucine aminopeptidase (LAP), which is a Zn^{++} enzyme. The pH optimum of CAP, in the presence of 5×10^{-5} M Mn^{++} and 0.05 M veronal is 8.6.

Specificity studies with 44 different peptides show that the enzyme acts as a typical aminopeptidase, splitting various amino acid residues from the amino end of peptides. The free α-amino group is essential because α-N-blocked peptides are resistant to CAP. Peptide bonds between an N-terminal amino acid residue and proline, which are known to be hydrolyzed by aminopeptidase P, are resistant to hydrolysis by CAP. N-terminal proline is released at a rate comparable to that of other amino acids bearing nonionizable side chains. N-terminal hydroxyproline is also cleaved efficiently ($Hyp \downarrow Ala-NH_2$), but N-terminal amino acid residues with ionized side chains are cleaved slowly. Naphthylamides of leucine and proline, as well as alanine amide and leucyl-p-nitroanilide, are hydrolyzed very slowly. Stereospecificity was demonstrated by the slow release of proline from Pro-Ala-DAla-Ala and by the resistance of the resulting tripeptide to further hydrolysis. CAP (50 μg/ml) was used to digest reduced and carboxymethylated insulin B chain (1 mM). An analysis of the incubation mixture indicates that 16 amino acid residues are released, including carboxymethylcysteine and histidine. The release of CMCys is the slowest and therefore rate-limiting.

Kinetic studies with five tripeptides of the general structure X-Gly-Gly, where X stands for Leu, Phe, Val, Ala, or Pro, shows the affinity of CAP for a hydrophobic side chain in the N-terminal amino acid residue of the substrate. This is indicated by the increase in K_m with increasing size of the hydrophobic side chain of the amino acid residue ($\overline{K}_m$ = 12.5, 54, and 155 M^{-1} for X = Ala, Val, and Leu, respectively). In fact, the leucyl peptide is the best of all substrates tested, and in this respect CAP resembles LAP. N-terminal proline is released quite efficiently, the tetrapeptide Pro-Gly-Pro-Pro ($\overline{K}_m$ = 225 M^{-1}, V_{max} = 6 μmol sec^{-1} mg^{-1}) being a better substrate than the tripeptide Pro-Gly-Pro ($\overline{K}_m$ = 50 M^{-1}, V_{max} = 10 μmol sec^{-1} mg^{-1}).

Due to the ability of CAP to release various N-terminal amino acid residues, including proline, this enzyme is of value for stepwise degradation of polypeptides and, when used with aminopeptidase P, complete hydrolysis may be expected. The combined action of the two enzymes is demonstrated by total enzymatic hydrolysis of the proline-rich nonapeptide bradykinin. No hydrolysis is observed when bradykinin is incubated either with CAP or with bovine lens aminopeptidase. If, however, the hormone is pre-incubated with aminopeptidase P, and CAP is added to the incubation mixture, complete hydrolysis is rapidly achieved. When LAP is added to the incubation mixture, a much slower degradation takes place because of the slow release of proline, as can be seen in Fig. 3. The initial release of Arg 1 and Pro 2 by aminopeptidase P is not shown in the figure. The residues Pro 3 to Pro 7 are released by the two enzymes rapidly and practically simultaneously. The dipeptide Phe^{8}Arg^9 is then hydrolyzed at a slower rate. The hydrolysis with CAP was complete within less than 30 min, but LAP with aminopeptidase P catalyzed the release of only 55% of the amino acids from the Pro-3 to Pro-7 portion and 38% from the remaining dipeptide

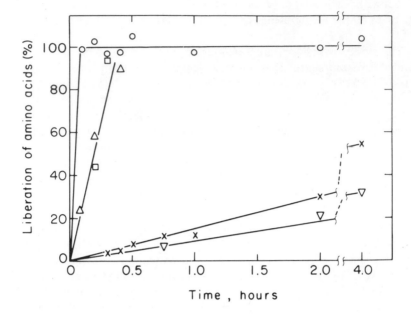

Figure 3. Enzymic hydrolysis of bradykinin $(0.65\ mM)$ under conditions given in legend to Fig. 2. Aminopeptidase P (6.5 μg/ml was added to the substrate solution and incubated for 90 min. At this time 100% of Arg 1 and Pro 2 were released by action of aminopeptidase P. Clostridial aminopeptidase (6 μg/ml) was then added and the incubation was continued. Amino acid analysis at different time intervals was performed. The releases of serine (o), phenylalanine 8 (△), and arginine 9 (□) are shown. An analogous experiment was performed in which bovine lens aminopeptidase was used instead of the clostridial enzyme, and the releases of serine (x) and arginine 9 (▽) are shown. For the release of Pro 3, Gly 4, Phe 5, and Pro 7 the graphs are identical to the one for serine, indicating that all these amino acid residues are released practically at the same rate.

Phe[8]Arg[9] after 4 hours. This reflects the efficient action of CAP on N-terminal proline residues.

Dipeptidocarboxypeptidase (DCP) (10). This enzyme was obtained in homogeneous form through six purification steps (including hydroxyapatite column chromatography). To determine the amount of the dipeptide glycylproline formed from α-DNP(Pro-Gly-Pro)$_n$ the colorimetric ninhydrin method was used. The preparation obtained in the last purification step was shown to be homogeneous by polyacrylamide gel electrophoresis, immunodiffusion, and immunoelectrophoresis. DCP acts optimally at pH 8.2 and does not require the addition of a metal for its catalytic action. Exhaustive dialysis against $10^{-4}\ M$ EDTA does not affect the specific activity further. Addition of Co^{++}, however, results in a five- to eightfold increase over the activity of the enzyme treated with EDTA (optimum concentration $5 \times 10^{-5}\,M$). Activity was only slightly enhanced by the addition of sodium chloride.

Substrate specificity was determined with 30 different peptides and was found to involve the hydrolysis of the penultimate peptide bond of tetra- and higher peptides, as well as of N-blocked tripeptides (e.g., Z-Ala-Ala-Ala). One peptide bond is hydrolyzed in tetra- and pentapeptides (unsubstituted tripeptides are resistant to DCP), two peptide bonds are hydrolyzed successively in the hexapeptide (Lys-Ala-Ala)$_2$, and three in the nonapeptide bradykinin. A free carboxyl is required because Ala$_4$-NH$_2$ is not hydrolyzed. Stereospecificity is demonstrated by the resistance of Z-Ala$_3$-DAla to DCP. Peptide bonds in which the nitrogen of a proline participates are not hydrolyzed, nor are peptide bonds between two glycine residues in Z-Gly$_4$ and Z-Gly$_4$-Phe.

Kinetic studies were performed with an N-substituted tripeptide and seven different tetrapeptides (Table 2). The effect of size on the kinetic parameters can be seen by comparing Ala$_3$, which is not a substrate, to Boc-Ala$_3$ and Ala$_4$, which show increasing values of $\bar{K}_m$ and k_{cat}. No further change is obtained with Z-Ala$_4$. Replacing the penultimate alanine in Ala$_4$ with phenylalanine reduces $\bar{K}_m$, with a 1.7-fold increase of k_{cat}. Although replacement of the first alanine with glycine does not have an appreciable effect, the same change in position 2 increases $\bar{K}_m$ about twice. Replacing both the first and the second position causes a drop in $\bar{K}_m$ and k_{cat} that is much larger than would be expected for independent contribution of the residues. Tetralysine binds effectively to the enzyme and displays substrate inhibition.

Table 2. Kinetic parameters for the hydrolysis of peptides by DCP[a]

Substrate	$\bar{K}_m$ (M^{-1})	k_{cat}[b] (sec^{-1})	$C = \bar{K}_m k_{cat}$ $(M^{-1}\,sec^{-1})$
Boc-Ala$_3$	1400	34	47,800
Ala	2280	139	318,000
Z-Ala$_4$	2430	131	318,000
Ala-Ala-Phe-Ala	785	225	177,000
Gly-Ala-Phe-Ala	720	194	140,000
Ala-Gly-Phe-Ala	1640	156	256,000
Gly-Gly-Phe-Ala	165	116	19,220
Lys$_4$	~10,000[c]	~30[c]	~300,000

[a]The composition of the reaction solution was: substrate $(1.0 \times 10^{-4}\text{--}7.5 \times 10^{-3}\,M)$, enzyme $(1.0\text{--}6 \times 10^{-9}\,M)$, 0.05 M borate buffer pH 8.15, (no metal added) at 40°C. The reaction rate was followed spectrophotometrically at 225 nm. The difference in molar extinction coefficients $\Delta\epsilon_{225}$ accompanying the hydrolysis was determined with authentic mixtures of the product peptides and the substrates.
[b]Assumed MW 100,000.
[c]Substrate inhibition.

Product inhibition, on the other hand, was observed during the hydrolysis of bradykinin by DCP. The effect of several dipeptides on the hydrolysis rate of Z-Ala$_4$ was therefore investigated (11) by the pH-stat method. Competitive inhibition was found in each case. The inhibition constants with Ala-Ala, Lys-Ala, Ala-Lys and Phe-Arg were $\bar{K}_i$ = 2,840, 38,200, 92,000, and 1.3 × 10^6 M^{-1}, respectively, indicating high affinity for basic amino acid residues. Although product inhibition by dipeptides is competitive, noncompetitive inhibition of the mixed type was observed with Pro-Phe-Lys and Boc-Phe-Arg, which can be regarded as products obtained from a penta- and a blocked tetrapeptide, respectively.

The occurrence of DCP does not seem to be restricted to the bacterial source. An enzymic activity compatible with the mode of action of DCP has been repeatedly reported to be present in mammalian tissue and was studied in connection with conversion of the decapeptide angiotensin I to the active hypertensive octapeptide angiotensin II with the liberation of the C-terminal dipeptide His-Leu. This "converting enzyme" (23) was partially purified from hog plasma (24), hog lung (25), calf lung (26), and bovine kidney cortex (27) and shown to split C-terminal dipeptides not only from angiotensin I, but also from other peptides, including bradykinin.

It is pertinent to note that the penultimate residue in angiotensin II is proline (. . . Pro-Phe). This feature prevents further action by the converting enzyme (27), which, like DCP, does not cleave N-prolyl peptide bonds.

Further studies of DCP are needed to assess its usefulness for sequence studies and to elucidate the relation of this enzyme to other carboxypeptidases, in particular to carboxypeptidase B, in view of the affinity of DCP for basic amino acid residues.

CONCLUSIONS

The use of poly(amino acids) in a search for new proteolytic enzymes is a fruitful approach that has resulted in the detection, purification, and characterization of several exopeptidases. Studies with proline-rich substrates have helped to clarify how such peptides are metabolized. A pair of aminopeptidases became available that, in principle, can hydrolyze any polypeptide. The uniqueness of the peptide bond involving the nitrogen of proline came to light by demonstrating that its hydrolysis requires a specific aminopeptidase.

An efficient hydrolytic combination of dipeptidocarboxypeptidase (DCP) and a dipeptidase (DP) that cleaves polypeptides from the C-terminal end was observed in $E.$ $coli.$ An interesting regulatory mechanism is indicated by the fact that dipeptides produced by DCP are competitive inhibitors of this enzyme and, in turn, serve as substrates to DP.

It is of interest to note that in a number of proteins and other polypeptides (e.g., hormones, antibiotics) a proline residue is found in the N- or C-terminal region (28–33). Its presence there may well serve to protect the molecule against most exopeptidases. Such a case is the already-mentioned activation of angiotensin by a dipeptidyl carboxypeptidase, in which the presence of proline prevents further action of the enzyme on the formed hormone. Another example is the formation of bradykinin from its precursor, in which enzymic hydrolysis stops short at an N-terminal Arg-Pro sequence, resistant to the action of aminopeptidases but essential for hormonal activity (28).

ACKNOWLEDGMENT

The financial support of the Helena Rubinstein Foundation is gratefully acknowledged.

REFERENCES

1. P.D. Boyer, ed., *The Enzymes* (Third Edition), Vol. 3, Academic Press, New York, 1971.
2. E. Katchalski, I. Grossfeld, and M. Frankel, *J. Am. Chem. Soc.*, **70**, 2094 (1948).
3. E. Katchalski, M. Sela, H.I. Silman, and A. Berger, in *The Proteins* (Second Edition), H. Neurath, ed., Academic Press, New York, 1964, p. 405. H.I. Silman and M. Sela, in *Biological Macromolecules, Poly-α-Amino Acids, Protein Models for Conformational Studies,* G.D. Fasman, ed., Vol. 1, M. Dekker, New York, 1967, p. 605. A. Yaron, in *Handbook of Biochemistry, Selected Data for Molecular Biology* (Second Edition), H.A. Sober, ed., The Chemical Rubber Co., Cleveland, Ohio, 1970, p. C-133.
4. G.K. Garg and T.K. Virupaksha, *Eur. J. Biochem.*, **17**, 13 (1970).
5. A. Yaron, N. Tal, and A. Berger, *Biopolymers,* **11**, 2461 (1972).
6. E. Harper, A. Berger, and E. Katchalski, *Biopolymers,* **11**, 1607 (1972).
7. J. Rosenbloom and D.J. Prockop, in *Repair and Regeneration; the Scientific Basis of Surgical Practice,* J.E. Dunphy and W. Van Winkle, Jr., eds., McGraw-Hill, New York, 1969.
8. A. Yaron and D. Mlynar, *Biochem. Biophys. Res. Commun.* **32**, 658 (1968).
9. A. Yaron and A. Berger, in *Methods in Enzymology,* Vol. 19, G.E. Perlman and L. Lorand, eds., Academic Press, New York, 1970, p. 521.
10. A. Yaron, D. Mlynar, and A. Berger, *Biochem. Biophys. Res. Commun.,* **47**, 897 (1972).
11. A Yaron and D. Mlynar, unpublished data.
12. S. Seifter and E. Harper in *The Enzymes* (Third Edition), Vol. 3, P.D. Boyer, ed., Academic Press, New York, 1971, p. 662.
13. E. Kessler, Ph.D. thesis, The Weizmann Institute of Science, Rehovot. Published in part in A. Yaron and D. Mlynar, *Biochem. Biophys. Res. Commun.,* **32**, 658 (1968).
14. A. Yaron and A. Berger, *Biochem. Biophys. Acta,* **107**, 307 (1965).
15. R. Granoth, Ph.D. thesis, The Weizmann Institute of Science, Rehovot, Israel, to be published.
16. P. Dehm and A. Nordwig, *Eur. J. Biochem.*, **17**, 364 (1970).
17. N.C. Davis and E.L. Smith, *J. Biol. Chem.*, **224**, 261 (1957).
18. R.L. Hill and W.R. Schmidt, *J. Biol. Chem.*, **237**, 389 (1962).
19. C. Nolan and E. Smith, *J. Biol. Chem.*, **237**, 453 (1962).
20. R. Frater, A. Light, and E.L. Smith, *J. Biol. Chem.*, **240**, 253 (1965).
21. A. Light and J. Greenberg, *J. Biol. Chem.*, **240**, 258 (1965).

22. V.M. Vogt, *J. Biol. Chem.*, **245**, 4760 (1970).
23. L.T. Skeggs, Jr., J.R. Kahn, and N.P. Shumaway, *J. Exp. Med.*, **103**, 295 (1956).
24. H.Y.T. Yang, E.G. Erdös, and Y. Levin, *J. Pharmacol. Exp. Therap.*, **177**, 291 (1971).
25. F.E. Dorer, J.R. Kahn, K.E. Lentz, M. Levine, and L.T. Skeggs, *Circulation Research*, **31**, 356 (1972).
26. R.L. Stevens, E.R. Micalizzi, D.C. Fessler, and D.T. Pals, *Biochemistry*, **11**, 2999 (1972).
27. Y.E. Elisseeva, V.N. Orekhovich, L.V. Pavlikhina, and I.P. Alexeenko, *Clin. Chim. Acta*, **31**, 413 (1971).
28. V.K. Hopsu-Havu, K.K. Makinen, and G.G. Glenner, *Nature*, **212**, 1271 (1966).
29. G. Weitzel, K. Eisele, H. Zollner, and U. Weber, *Hoppe-Seyler's Z. Physiol. Chem.*, **350**, 741 (1969).
30. M.O. Dayhoff, ed., *Atlas of Protein Sequence and Structure*, Vol. 5, National Biomedical Research Foundation, Washington, D.C., 1972.
31. A. Areus, E. Reienbusch, E. Irion, O. Wagner, K. Bauer, and W. Kaufmann, *Z. Physiol. Chem.*, **351**, 197 (1970).
32. H. Maeda and J. Meienhofer, *FEBS Lett.*, **9**, 301 (1970).
33. S.H. Ferreira, D.C. Bartelt, and L.J. Greene, *Biochemistry*, **9**, 2583 (1970).

POLY(AMINO ACIDS) AS ENHANCERS IN THE CELLULAR UPTAKE OF MACROMOLECULES

HUGUES J.-P. RYSER, Department of Pathology, Boston University School of Medicine, Boston, Mass. 02118

SYNOPSIS: The penetration of human serum albumin into cells of an established tumor line grown in culture is markedly enhanced by homopolymers of basic poly(amino acids) at concentrations of 0.01 μg/ml and above. This effect increases linearly with the molecular weight of the polymers in the range of 1000 to 230,000. DEAE-Dextran shows a similar effect and relationship, although of a smaller magnitude. Poly(D-lysine) is more effective than its equivalent L isomer. This difference does not appear to be due to different susceptibilities to proteolysis and may be related to the formation of right- versus left-handed helices upon contact with the cell surface. The unexpectedly strong effect of two lysine:tyrosine copolymers (1:1 and 1:3) is explained in part by the formation of aggregates in neutral aqueous solutions. Aggregates of a poorly water-soluble polynucleotide, poly(inosinic acid):polyvinylcytosine, appear to have an enhancing effect by virtue of their size alone. Polystyrene particles also enhance albumin uptake at the cell surface. The relationship between effect and size in the supramolecular range was tested with polystyrene beads of eight different sizes. The enhancement per bead increased linearly with the bead volume in the range of 4×10^{-4} to $2 \times 10^2 \ \mu^3$ (diameters of 0.091 to 5.7 μ). The enhancing effect of basic poly(amino acids) on the uptake of macromolecules at the cell surface has been confirmed by several groups using different macromolecules and different cell types, including plant-cell protoplasts.

The biological property of poly(amino acids) under consideration here is their effect on the plasma membrane of mammalian cells, and more precisely, their ability to enhance the uptake of other macromolecules at the cell surface.

By virtue of their interest in the membrane transport of solutes such as ions, amino acids, and sugars, biophysicists studying transport kinetics evaded for a long time the question whether macromolecules such as proteins, polysaccharides and nucleic acids can penetrate cells. This question was answered by electron microscopists who demonstrated by using, among others, tracer molecules such as ferritin, that a protein of MW 600,000 was taken up from the medium by the process of pinocytosis and could be seen intact within intracellular vesicles and vacuoles (1, 2). The potential meaning of this phenomenon was underscored by virologists who found that cells can be infected not only by intact viruses, but also by purified nucleic acids isolated from viruses. The proof of cellular uptake in this case is established by the biological expression of the foreign macromolecule within its host cell (3, 4). Such experiments, however, were hampered by the extraordinarily low yield of infection. We know now at least two reasons for such a low yield. On the one hand, the cellular uptake of pure nucleic acids is much less efficient than that of the whole virus, because it is no longer guided by specific interactions with "viral receptors" at the cell surface. On the other hand, ingested macromolecules undergo intracellular digestion in secondary lysosomes (digestive vacuoles) and only a few escape destruction to express their biological potential (5, 6, 7). These obstacles greatly limit the chances of achieving genetic transformations in mammalian cells comparable to those achieved in bacterial cells. They also limit the potential use of macromolecules such as enzymes, interferon, growth factors, and polynucleotides as biological agents (5). It will be shown in this presentation that certain poly(amino acids) can enhance the cellular uptake of a large variety of macromolecules and can increase their chance of survival and expression in host cells. Most of our data on this subject have been obtained by studying the uptake and intracellular degradation of a simple protein—iodinated human serum albumin—in tumor cells grown in cultures.

EFFECT OF MOLECULAR CHARGE

These investigations started with the observation that histones, when added to the growth medium of mouse sarcoma cells in vitro, markedly enhanced the cellular binding of ^{131}I albumin by increasing both its surface adsorption and net uptake (8). Arginine-rich histones had a stronger effect than the lysine-rich ones. It appeared logical to test whether homopolymers of lysine and arginine would have comparable effects, and we have found that both poly(amino acids) enhanced the uptake of albumin, poly(Arg) having a greater effect than poly(Lys) and all other types of histones. All other available homopolymers of basic amino acids [i.e., poly(Orn), poly(His), and poly(DLys)] had an effect comparable to that of poly(Arg), but poly(Glu) and copolymers devoid of basic amino acids, [e.g., poly(Glu) and poly(Tyr)] did not enhance albumin uptake. This new function of poly(amino acids) is therefore related to their cationic character at physiological pH. The importance of the positive charge was further demonstrated by the fact that nonpeptidic polycations such as DEAE-Dextran also enhanced albumin uptake, but unsubstituted Dextran did not (9).

ROLE OF MOLECULAR SIZE

Poly(DLys), poly(Lys), poly(Orn), and DEAE-Dextran samples of different molecular weights were tested for their relative enhancing effects. A double log plot of enhancement against molecular weight yielded a positive, linear correlation (Fig. 1), showing that the largest molecules within a homologous series were the most effective ones and that DEAE-Dextran is less effective than the poly-(amino acids) tested (9). By extrapolation of the data in Fig. 1, the minimal molecular sizes required for action could be estimated to be of the order of 10^3 for poly(amino acids) and 10^4 for DEAE-Dextran. Smaller polyamines, such as spermine and spermidine, were without effect. Our interpretation of this

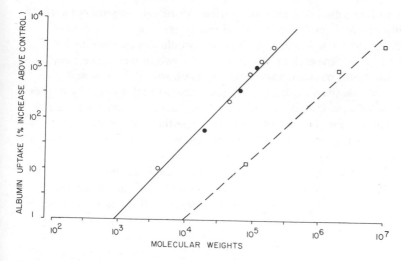

Figure 1. Activity versus molecular weight of polymers in homologous series of poly(Orn) (●), poly(DLys) (○), and DEAE-Dextran (□). Stimulation of albumin uptake is expressed in percent increases above the uptake of control cells. The experimental data, obtained with 3 μg/ml of each compound, were corrected for molarity to express the effect of a uniform concentration of 1.5^{-8} mole/l. (Reprinted with permission from Ryser, in *Proc. Fourth Int. Cong. Pharm.*, Vol. 3, 1970. Copyright by Schwabe.)

size-activity relationship is that the enhancing effect of polycations requires multiple, simultaneous interactions of their charged groups with the cell surface and that larger molecules make more interactions possible (9). The possibility must be entertained that upon interacting with a predominantly negatively charged cell surface, the basic homopolymers may undergo a coil-to-helix transition. Indeed, it has been shown that poly(Lys) changes its conformation from a random coil to an α helix when it interacts with phosphatidyl-L-serine in pH 7.0 solution (10).

POLY(LYS) VERSUS POLY(DLYS)

The parameters of charge and size, however, do not explain all the findings. For instance, poly(Lys) and poly(DLys) of comparable molecular weights (120,000 to 150,000) have effects of different magnitude. The L and D isomers enhance albumin uptake by factors of 1.5 and 15.8, respectively (7, 8). Immunologists were already aware that the two isomers differ in several antigenic properties. The explanation most commonly offered for such differences is that the two optical isomers have different susceptibilities to proteolytic enzymes and that the effect of poly(Lys) is decreased or modified by its rapid hydrolysis. Using ^{131}I albumin as substrate, we showed, however, that under the conditions of our experiments the medium is devoid of hydrolytic activity (11). Furthermore, polymers exert an effect only when present in the medium at the time of enhancement and their action is immediate (7). Although these observations suggest that polymers modify the outside surface of the cell, they do not exclude the possibility of a preferential hydrolysis of poly(Lys) at or in the cell membrane. Such a preferential degradation, however, would change the kinetics of enhancement and the time-effect relationship of the two isomers. The time curves of enhancement obtained with both isomers, however, have comparable shapes. Furthermore, the quantitative difference due to a differential hydrolysis would be expected to be concentration dependent, that is, less pronounced at high concentrations. Fig. 2 shows the opposite to be the case. In summary, there is no experimental evidence that the differences we observed are related to different rates of hydrolysis.

Coming back to the importance of the molecular size, it was conceivable (but not likely) that, in solution, poly(Lys) and poly(DLys) of similar molecular weights might form soluble supramolecular complexes of different sizes, hence

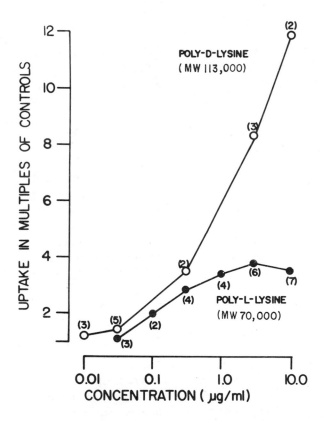

Figure 2. Semilogarithmic plotting of the dose-effect relationship for the two optical isomers of poly(Lys). Uptake of labeled albumin is expressed as a multiple of the control values obtained for each experiment. The points are averages of the number of experiments indicated in parentheses. Incubation was carried out for 2 hr at 37°C in Eagle's medium without serum. (Reprinted with permission from Ryser, in *Proc. Fourth Int. Cong. Pharm.*, Vol. 3, 1970. Copyright by Schwabe.)

different biological activities. This possibility was tested by passing solutions of both isomers through millipore filters of different pore sizes (0.22 and 1.2 μ). There was no loss of activity in either solution. Moreover, the two polylysines showed the same behavior when filtered in the presence of ^{131}I albumin. The difference in biological activity of the two isomers seems, therefore, to be due to specific interactions with optically active groups at the cell surface.

RELATIVE IMPORTANCE OF MOLECULAR CHARGE AND SIZE

In our experimental system, DEAE-Dextran MW 2×10^6 is less effective in enhancing albumin uptake than poly(Orn) MW 2×10^5. The lower charge density of the polysaccharide (due to a glucose:DEAE ratio of 2.7) is not compensated by a tenfold increase in size. It was puzzling, therefore, to find that a copolymer of Lys and Tyr (1:1), MW 41,500, had a greater enhancing effect than poly(Orn) MW 45,000 and 175,000 (Fig. 3), despite a 50% lowering of its charge density (7). Poly(Lys, Tyr) is less water-soluble than poly(Lys). It appeared conceivable, therefore, that in aqueous solutions the copolymer might form supramolecular aggregates that by virtue of their size, would have a greater enhancing effect. This possibility was tested by passing poly(Lys, Tyr) through filter paper and millipore membranes of pore sizes varying from 0.45 to 5 μ (7). Filtration through pores of 0.45 or 1.2 μ completely abolished the enhancing effect of the copolymer (Fig. 4).

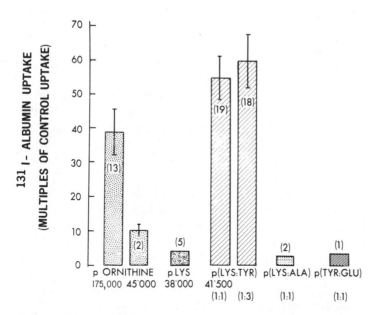

Figure 3. Enhancing effects of 3 μg/ml of various homo- and copolymers of L-amino acids. Figures above the column give the number of experiments. Figures below the columns give MW of respective compounds.

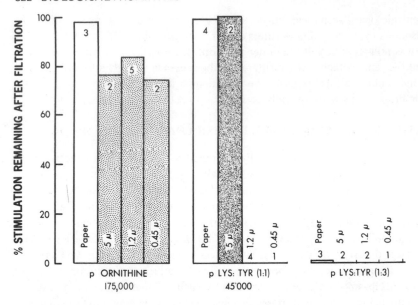

Figure 4. Increase in the effect of poly(Orn) and poly(Lys, Tyr) (1:1 and 1:3) following filtration through paper or millipore membranes of three different pore sizes (0.45, 1.2, and 5 μ). The full effect of 3 μg/ml of the various polymers on albumin uptake measured before filtration is set as 100%. The figures in the columns refer to the number of experiments or to the nature of the filter.

A copolymer of higher tyrosine content (Lys:Tyr = 1:3) and lesser solubility yielded a clear but viscous solution of even greater biological activity: It enhances the albumin uptake 60-fold, one of the largest effects observed with polymers of any kind. The activity was, however, totally prevented by filtration even through conventional filter paper or 5 μ millipore membranes (Fig. 4). It must be concluded that under our experimental conditions, the tyrosine copolymers formed soluble aggregates and that in this case, a charge density that is two- to fourfold lower than in poly(Lys) was more than compensated by the increase in size. Thus, the positive correlation between molecular weight and biological activity of polycations, first observed with homopoly(amino acids) and DEAE-Dextran in the range of MW 4×10^3 to 1×10^7, extends to molecular aggregates.

The data obtained with DEAE-Dextran indicate that the property of enhancing the cellular penetration of albumin is not restricted to poly(amino acids). The data obtained with poly(Lys, Tyr) stress the importance of molecular—or supramolecular—size and raise the following two questions: Is polycationic character a critical requirement for the enhancing activity of supramolecular polymers? Is this size effect restricted to aggregates of poly(amino acids) or is it shared by other polymers of biological interest, such as polynucleotides?

INTERFERON-INDUCING POLYNUCLEOTIDES

It was once believed that the ability of polyinosinic-polycytidylic acid (pI:pC) to induce interferon was limited by strict structural requirements, but it has been shown (12) that an analog of pI:pC, namely, a complex of polyinosinic acid and poly(1-vinylcytosine) (pI:pVC) can also induce strong interferon production. The water solubility of pVC is lower than that of pC, so the former is stored in 25% propylene glycol. When dialyzed against phosphate buffer or added to the culture medium, the polymer precipitates as an opalescent colloidal solution. Alerted by the striking inverse correlation between solubility and enhancing effects of poly(Lys, Tyr), we planned to determine whether pI:pVC owed part of its biological activity to a lowered solubility. We tested whether pI:pVC would enhance albumin uptake in the system previously used to measure the enhancing effect of poly(amino acids). Fig. 5 shows that 5 μg/ml pI:pVC caused a 40-fold increase of albumin uptake in the course of a 60-minute exposure. When opalescent aqueous pI:pVC solutions were filtered through millipore membranes

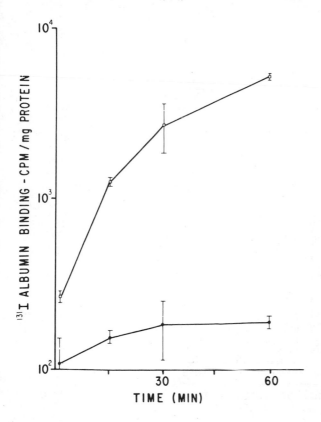

Figure 5. Time curve of albumin uptake in the absence (●) and presence (□) of 5 μg/ml pI:pVC. The points are means of four measurements. Vertical bars give standard deviations. The pI:pVC preparation was dialyzed against phosphate buffer prior to use.

(0.65 μ pore) prior to their use, they lost up to 85% of their enhancing activity (Fig. 6). Poly I:poly C, on the other hand, has no enhancing effect at any tested concentration. By analogy to our prior observations, this result suggests that the effect of pI:pVC on albumin uptake is related to the formation of a colloidal precipitate in aqueous solution and to the presence of insoluble supramolecular complexes. Thus, the answers to our questions regarding the generality of the effect of charge and size are, respectively: pI:pVC, although poorly ionized at neutral pH, is a weak polyanion and, therefore, supramolecular polymers do not need cationic groups to act as enhancers; supramolecular polymers other than poly(amino acids) can enhance the cellular uptake of foreign macromolecules.

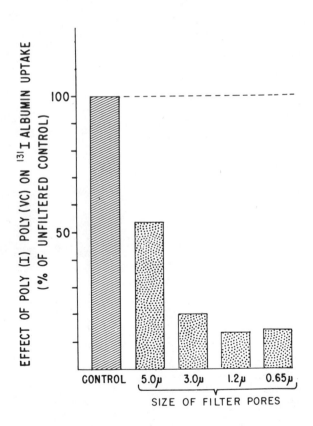

Figure 6. Decrease in the effect of pI:pVC following filtration through milli-pores of four different sizes. The full effect of the same pI:pVC solution (5 μg/ml) on albumin uptake measured before filtration is set as 100%.

POLYSTYRENE BEADS

Poorly water-soluble poly(amino acids) and polynucleotides have an obvious drawback as experimental tools in our system: They form unstable and heterogeneous solutions containing complexes of unknown size and size distribution. Isolating fractions of known size ranges is technically difficult and is made unreliable by possible reaggregation. It did not seem promising, therefore, to pursue with these compounds the analysis of a relationship between effect and supramolecular sizes. The only polymers available to us in the form of stable aggregates of homogeneous distribution were latex beads. Aqueous suspensions of polystyrene beads of eight different diameters ranging from 0.091 to 5.7 μ were tested for their ability to influence the cellular uptake of albumin. All sizes added to the culture medium in comparable weights per volume were found to cause a time-and-concentration-dependent enhancement (13). It is evident, when concentration is expressed as number of particles per ml, that the largest beads are the most effective enhancers. Identical effects can be achieved with 0.091 and 1.101 μ beads, but to do so requires a thousand times higher concentration (number of beads/ml) of the smaller beads.

The most meaningful mode of representation to test the size-effect relationship appeared to be a double log plot of the albumin enhancement per bead versus bead volume. The enhancement per bead is obtained by dividing the total increase in albumin uptake above control by the number of beads responsible for that increase. It is expressed in ng per mg cell protein per bead. The bead volume is expressed in μ^3. The points on Fig. 7 represent the effect of beads of eight different sizes. The points fall on a straight line, indicating a positive, linear correlation between the effect of the beads and their size (13). This relationship is strikingly similar to the one between effect and molecular size of poly(amino acids) and DEAE-Dextran (see Fig. 1). It must be kept in mind that the two correlations shown in Figs. 1 and 7 represent the effects of different compounds, that the polycations act by virtue of their charge as well as their size, and therefore, that the abscissas of the two plots cannot be expected to represent a continuum. The exact properties of the bead surface are not known because the polystyrene particles available commercially are coated to yield stable suspensions and it is not known whether the same suspension-stabilizing procedure is used for all sizes tested in these experiments. It is somewhat unlikely, however, that coating by itself could account for the striking relationship shown in Fig. 7. According to this graph, the enhancement caused by a bead of $10^{-4} \mu^3$ is of the order of 10^{-10} ng per mg cell protein per hour. It is likely that the correlation would hold for smaller volumes. One may wonder how it would hold for volumes larger than $10^2 \mu^3$, that is, for beads of a diameter greater than 5.7 μ. The results obtained with polystyrene beads confirm that size alone can be a parameter determining the biological effect of a polymer on the cell membrane.

In our experiments with poly(amino acids), it was shown that the polymers were taken up by cells and that their rate of uptake closely reflected their enhancing effect. In experiments with polystyrene beads, light-microscopic and

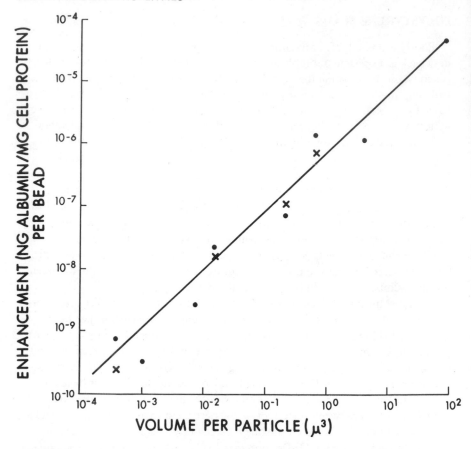

Figure 7. Enhancement of albumin uptake by latex beads of different sizes. The ordinate gives the enhancement per bead. The abscissa gives the volume of eight different beads, used in two experiments. The bead sizes were: polystyrene 0.091, 0.126, 0.234, 0.312, 0.760, and 1.101 μ; polyvinyltoluene 2.02 μ; and styrenedivinylbenzene 5.7 μ. Results of two different experiments are given by different symbols. The albumin uptake was measured for 60 minutes in the presence of 100 μg/ml polymers (12).

EM studies have shown that beads can be taken up by cells. This penetration, however, is a slow process that does not appear to be correlated with the instantaneous and strong enhancement of albumin uptake. By analogy, we may question whether the uptake of poly(amino acids) is coincidental to or critical for the enhancing effect. Recent data obtained with DEAE-Dextran and DNA suggest that uptake of the enhancer is not a critical requirement. In spite of its strong enhancement of the cellular uptake of DNA, labeled DEAE-Dextran was found to penetrate cells to a negligible extent (14). The mechanism by which poly(amino acids), substituted dextrans and polystyrene beads enhance the uptake of foreign macromolecules is not yet elucidated. Nor is it known whether all three polymers share the same mode of action. The enhancements caused by poly(amino acids) in molecular solutions differ from those caused by poly(amino acids) and polynucleotide aggregates in that the former are not influenced at all by metabolic inhibitors (7, 15). Enhancement thus appears to require less energy in the first instance.

Our data describe mostly the cellular uptake of albumin, but basic poly(amino acids) have been used to enhance the uptake of other proteins—including interferon and bacterial toxin—of polynucleotides, and of nucleic acids. It has been shown in several instances that basic poly(amino acids) enhance not only the cellular uptake, but the biological activity of a foreign macromolecule (5). Furthermore, such experiments have now been carried out in a large variety of cells. Perhaps the most unexpected extension of this work is in the direction of plant virology. Takebe and Otsuki (16) have shown that RNA isolated from tobacco mosaic virus can infect cell protoplasts made from tobacco leaves, but only when given together with enhancing concentrations of a basic poly(amino acid). This observation is of interest in two respects. First, it offers a new example in which the enhanced cellular penetration of a macromolecule is followed by an enhanced biological effect. Second, it stresses the generality of the phenomenon and the similarity of plant and animal cell membranes in their susceptibility and response to basic poly(amino acids).

ACKNOWLEDGMENTS

This work was supported by a U.S. Public Health Service grant (14551) from the National Cancer Institute. The experimental contributions of A. Bauer-Roberts, T.E. Termini, and S.W. Rothman are gratefully acknowledged. The lysine:tyrosine copolymers were a gift from Michael Sela.

REFERENCES

1. M.G. Farquhar and G.E. Palade, *J. Exp. Med.*, **114**, 699 (1961).
2. H. Ryser, J.B. Caulfield, and J.C. Aub, *J. Cell Biol.*, **14**, 255 (1962).
3. H.E. Alexander, G. Koch, M. Mountain, and O. van Damme, *J. Exp. Med.*, **108**, 493 (1958).
4. G.A. DiMayorca, B.E. Eddy, S.E. Stewart, W.S. Hunter, C. Friend, and A. Bendich, *Proc. Nat. Acad. Sci.*, U.S., **45**, 1805 (1959).
5. H.J.-P. Ryser, *Science*, **159**, 390 (1968).
6. M.-P. Gabathuler and H.J.-P. Ryser, *Proc. Roy. Soc.* (London), **B 173**, 95 (1969).
7. H.J.-P. Ryser, *Proc. Fourth Int. Cong. Pharm.*, **III**, 96, Schwabe, Basel, 1970.
8. H.J.-P. Ryser and R. Hancock, *Science*, **150**, 501 (1965).
9. H.J.-P. Ryser, *Nature*, **215**, 934 (1967).
10. G.G. Hammes and S.E. Schullery, *Biochem.*, **9**, 2555 (1970).
11. M.P. Petitpierre, Ph.D. thesis, University of Lausanne, 1974.
12. J. Pitha and P.M. Pitha, *Science*, **172**, 1146 (1971).
13. S.W. Rothman and H.J.-P. Ryser, *Fed. Proc.*, **33**, 556, Abstr. 1948, 1974.
14. E. Borenfreund, M. Steinglass, G.C. Korngold, and A. Bendich, *J. Nat. Cancer Inst.*, **51**, 1391 (1973).

15. H.J.-P. Ryser, M.-P. Gabathuler, and A.B. Roberts, "Uptake of Macromolecules at the Cell Surface," in *Biomembranes*, Vol. 2, L.A. Manson, ed., Plenum Press, New York, 1971, p. 197.
16. I. Takebe and Y. Otsuki, *Proc. Nat. Acad. Sci. U.S.*, **64**, 843 (1969).

LIST OF PARTICIPANTS

R. ARNON
Dept. of Chemical Immunology
Weizmann Institute
Rehovot, Israel

D. ATLAS
Dept. of Biophysics
Weizmann Institute
Rehovot, Israel

H. AUER
Dept. of Biochemistry
University of Rochester
Rochester, N.Y. 14642

E. BAER
Dept. of Macromolecular Science
Case Western Reserve University
Cleveland, Ohio 44106

M. BAYLEY
MRC, The Ridgeway
London NW2 1AA, England

E.R. BLOUT
Dept. of Biological Chemistry
Harvard Medical School
Boston, Mass. 02115

F.A. BOVEY
Bell Laboratories
Murray Hill, N.J. 07974

G. BLAUER
Dept. of Biological Chemistry
Hebrew University
Jerusalem, Israel

L. BRAND
Mergenthaler Lab. for Biology
Johns Hopkins University
Baltimore, Md. 21218

A.W. BURGESS
Dept. of Chemistry
Cornell University
Ithaca, N.Y. 14850

G. CARERI
Instituto di Fisica
"Guglielmo Marconi"
Rome, Italy

C.H. CHOTHIA
Dept. of Chemical Physics
Weizmann Institute
Rehovot, Israel

J.-L. DE COEN
Lab. de Chimie Biologique
Universite Libre de Bruxelles
1640, Rhode St. Genese, Belgium

P. COMBELAS
Institut Laue-Langevin
38042 Grenoble, France

A. COSANI
Institute of Organic Chemistry
Via Marzolo 1
Padova, Italy

C. CRANE-ROBINSON
Biophysics Laboratory
Portsmouth Polytechnic
Portsmouth PO1-2QG, England

T. CREIGHTON
MRC Lab. of Molecular Biology
Cambridge CB2 2QH, England

C.M. DEBER
Dept. of Biological Chemistry
Harvard Medical School
Boston, Mass. 02115

C. DEGANI
Isotope Dept.
Weizmann Institute
Rehovot, Israel

B.B. DOYLE
Dept. of Zoology
Lab. of Molecular Biophysics
Oxford OX1 3PS, England

H. EISENBERG
Polymer Dept.
Weizmann Institute
Rehovot, Israel

D. ELSON
Department of Biochemistry
Weizmann Institute
Rehovot, Israel

E. ELSON
Department of Chemistry
Cornell University
Ithaca, N.Y. 14850

P. ELSON
Department of Biochemistry
Weizmann Institute
Rehovot, Israel

J. ENGEL
Biozentrum der Universitat
Abt. Biophysikalische Chemie
CH-4056 Basel, Switzerland

A. ENGLERT
Universite Libre de Bruxelles
Service de Chimie Generale I
Brussels 5, Belgium

G.D. FASMAN
Graduate Dept. of Chemistry
Brandeis University
Waltham, Mass. 02154

S. FUCHS
Dept. of Chemical Immunology
Weizmann Institute
Rehovot, Israel

T.J. GILL III
Dept. of Pathology
Univ. of Pittsburgh School of
    Medicine
Pittsburgh, Pa. 15261

C. GILON
Dept. of Chemistry
University of California, San Diego
La Jolla, Calif. 92037

C. GLASER
Inst. of Medical Sciences
Pacific Medical Center
San Francisco, Calif. 94115

M. GOODMAN
Department of Chemistry
University of California, San Diego
La Jolla, Calif. 92037

F.R.N. GURD
Dept. of Chemistry and The Medical
    Sciences Program
Indiana University
Bloomington, Ind. 47401

A. HAGLER
Chemical Physics Dept.
Weizmann Institute
Rehovot, Israel

E. HASS
Chemical Physics Department
Weizmann Institute
Rehovot, Israel

H. VON HIPPEL
Institute of Molecular Biology
University of Oregon
Eugene, Ore. 97403

A.J. HOPFINGER
Dept. of Macromolecular Science
Case Western Reserve University
Cleveland, Ohio 44106

K. IMAHORI
Dept. of Agricultural Chemistry
The University of Tokyo
Tokyo, Japan

C. IRWING
Isotope Dept.
Weizmann Institute
Rehovot, Israel

G. JUNG
Institute of Chemistry
University of Tubingen
D-74 Tubingen
Germany

E. (KATCHALSKI) KATZIR
Department of Biophysics
Weizmann Institute
Rehovot, Israel

S. KOENIG
Isotope Dept.
Weizmann Institute
Rehovot, Israel

K.D. KOPPLE
SNAM Progetti
Laboratori Studi e Richerche di Base
000-15 Monterotondo (Rome), Italy

Y. KOYAMA
Faculty of Science
Kwansei Gakuin University
Nishinomiya, Japan

S. KRIMM
Harrison M. Randall Lab. of Physics
University of Michigan
Ann Arbor, Mich. 48104

J. KRUSEMAN
Nestle Products Technical Assistance
R & D Protein Laboratory
CH1814 La Tour de Peilz,
   Switzerland

A. LAPIDOT
Isotope Dept.
Weizmann Institute
Rehovot, Israel

M. LEVITT
Dept. of Chemical Physics
Weizmann Institute
Rchovot, Israel

S. LIFSON
Dept. of Chemical Physics
Weizmann Institute
Rehovot, Israel

G.P. LORENZI
Technisch Chemisches Laboratorium
ETH Zurich
CH-8000 Zurich, Switzerland

N. LOTAN
Dept. of Biophysics
Weizmann Institute
Rehovot, Israel

M.H. LOUCHEUX
Dept. of Chemistry
Lab. de Chimie Macromoleculaire
Universite des Sciences et Techniques
   de Lille
59 Villeneuve d'Ascq, France

Z. LUZ
Isotope Dept.
Weizmann Institute
Rehovot, Israel

V. MADISON
Dept. of Biological Chemistry
Harvard Medical School
Boston, Mass. 02115

B.R. MALCOLM
Dept. of Molecular Biology
University of Edinburgh
Edinburgh EH9 3JR, Scotland

P. MAURER
Department of Biochemistry
Jefferson Medical College
Philadelphia, Pa. 19107

W.G. MILLER
Department of Chemistry
University of Minnesota
Minneapolis, Minn. 55455

E. MOZES
Dept. of Chemical Immunology
Weizmann Institute
Rehovot, Israel

Y.P. MYER
Dept. of Chemistry
State University of New York at
   Albany
Albany, N.Y. 12203

F. NAIDER
Dept. of Pure and Applied Science
Richmond College, CUNY
Staten Island, N.Y. 10301

R.H. PAIN
The University of Newcastle
Dept. of Biochemistry
Newcastle Upon Tyne NE; 7RU,
  England

D.J. PATEL
Bell Laboratories
Murray Hill, N.J. 07974

I. PECHT
Dept. of Chemical Immunology
Weizmann Institute
Rehovot, Israel

E. PEGGION
Institute of Organic Chemistry
University of Padova
35100 Padova, Italy

G. PERLMANN
Rockefeller University
New York, N.Y. 10021

W.L. PETICOLAS
Institut Max von Laue-Langevin
38042 Grenoble Cedex, France

E. RALSTON
Lab. de Chimie Biologique
Universite Libre de Bruxelles
1640 Rhode St. Genese, Belgium

G.N. RAMACHANDRAN
Indian Institute of Science
Molecular Biophysics Unit
Bangalore-12 India

K. RASMUSSEN
Dept. of Chemical Physics
Weizmann Institute
Rehovot, Israel

J. REUBEN
Isotope Dept.
Weizmann Institute
Rehovot, Israel

M. RIGBI
Dept. of Biological Chemistry
Hebrew University
Jerusalem, Israel

W.B. RIPPON
Dept. of Macromolecular Science
Case Western Reserve University
Cleveland, Ohio 44108

V. RIZZO
Technisch-Chemisches Laboratorium
ETH Zurich
CH-8006 Zurich, Switzerland

R.S. ROCHE
Dept. of Chemistry
The University of Calgary
Calgary, Alberta, Canada

A. ROIG
Universidad Autonoma de Barcelona
Facultad de Ciencias de Baleares
Palma de Mallorca, Spain

L. ROMANIN-JACUR
Dept. of Structural Chemistry
Weizmann Institute
Rehovot, Israel

K. ROSENHECK
Polymer Dept.
Weizmann Institute
Rehovot, Israel

E. RÜDE
Max-Planck-Institut für
  Immunbiologie
78 Freiburg-Zahringen, Germany

J. RUDINGER
Institute of Molecular Biology and
  Biophysics
ETH Honggerberg
CH-8049 Zurich, Switzerland

H.J.-P. RYSER
Pathology Department
Boston University Medical School
Boston, Mass. 02118

A. SCATTURIN
Institute of Organic Chemistry
University of Padova
Padova, Italy

J.A. SCHELLMAN
Department of Chemistry
University of Oregon
Eugene, Ore. 97403

H.A. SCHERAGA
Dept. of Chemistry
Cornell University
Ithaca, N.Y. 14850

J. SCHLESSINGER
Chemical Physics Dept.
Weizmann Institute
Rehovot, Israel

M. SELA
National Institutes of Health
Bethesda, Md. 20014

H. SOBER
National Institutes of Health
Bethesda, Md. 20014

G. SPACH
Centre de Biophysique Moleculaire
  CNRS
La Source
L15-Orleans, France

I.Z. STEINBERG
Dept. of Chemical Physics
Weizmann Institute
Rehovot, Israel

M. SHEINBLATT
Dept. of Chemistry
Tel Aviv University
Tel Aviv, Israel

E. SUZUKI
Div. of Protein Chemistry CSIRO
Parkville, Victoria 3052
Australia

M. SWENSON
Dept. of Chemistry
Cornell University
Ithaca, N.Y. 14850

P. TEMUSSI
Laboratoria per la Chimica e Fisica di
  Molecole di Interesse Biol. del
  CNR
Naples, Italy

P. THAMM
Dept. of Peptide Chemistry
Max Planck Institute of Biochemistry
D-8000 Munich, Germany

C. TONIOLO
Institute of Organic Chemistry
University of Padova
Padova, Italy

D.A. TORCHIA
Polymer Crystal Physics Section
National Bureau of Standards
Washington, D.C. 20234

W. TRAUB
Dept. of Organic Chemistry
Weizmann Institute
Rehovot, Israel

D.W. URRY
Div. of Molecular Biophysics
Lab. of Molecular Biology
University of Alabama Medical Center
Birmingham, Ala. 35233

A.S. VERDINI
SNAM Progetti S.p.A.
Laboratori Richerche di Base
000-15 Monterotondo, Rome, Italy

A.G. WALTON
Department of Macromolecular
  Science
Case Western Reserve University
Cleveland, Ohio 44106

R.W. WOODY
Dept. of Chemistry
Arizona State University
Tempe, Arizona 85281

K. WÜTHRICH
Institute for Molecular Biology and
  Biophysics
ETH Honggerberg
CH-8049 Zurich, Switzerland

J.T. YANG
Cardiovascular Research Institute
University of California
San Francisco, Calif. 94122

A. YARON
Dept. of Biophysics
Weizmann Institute
Rehovot, Israel

A. YONATH
Dept. of Structural Chemistry
Weizmann Institute
Rehovot, Israel

R. ZANA
Centre de Recherches sur les
    Macromolecules
Strassbourg, France

A.R. ZEIGER
Thomas Jefferson University
Philadelphia, Pa. 19107